D1711174

PRODUCTION WINE ANALYSIS

PRODUCTION WINE ANALYSIS

BRUCE W. ZOECKLEIN
Department of Horticulture
Virginia Polytechnic Institute and State University

KENNETH C. FUGELSANG, M.A.

BARRY H. GUMP, Ph.D.

FRED S. NURY, Ph.D.
Department of Enology, Food Science and Nutrition
California State University, Fresno

An AVI Book
Published by Van Nostrand Reinhold
New York

An AVI Book
(AVI is an imprint of Van Nostrand Reinhold)

Library of Congress Catalog Card Number 89-31191
ISBN 0-442-23463-5

Printed in the United States of America

Van Nostrand Reinhold
115 Fifth Avenue
New York, New York 10003

Van Nostrand Reinhold International Company Limited
11 New Fetter Lane
London EC4P 4EE, England

Van Nostrand Reinhold
480 La Trobe Street
Melbourne, Victoria 3000, Australia

Nelson Canada
1120 Birchmount Road
Scarborough, Ontario M1K 5G4, Canada

16 15 14 13 12 11 10 9 8 7 6 5 4 3 2 1

Library of Congress Cataloging-in-Publication Data

Production wine analysis/Bruce Zoecklein . . . [et al.].
p. cm.
"An AVI book."
Bibliography: p.
Includes index.
ISBN 0-442-23463-5
1. Wine and wine making—Analysis. I. Zoecklein, Bruce.
TP548.5.A5P76 1989
663—dc20 89-31191
CIP

Contents

Preface

This text is designed to acquaint the reader with the commonly used procedures of juice and wine analysis as they are generally practiced in the industry, and as they are taught in the Department of Enology at California State University, Fresno. It is assumed that the reader has a basic preparation in the fields of chemistry and microbiology.

In developing material for this text, the authors have emphasized analyses as they would be carried out in a production laboratory. Realizing that different laboratories have different analytical capabilities, personnel as well as equipment, we have in many instances provided several different approaches to the same analysis. Throughout this book we have attempted to give special attention to practical considerations and the importance of these analyses in the total spectrum of winery operations. We hope the book's format will satisfy the interests of laboratory personnel as well as winemakers.

The process of making wine involves a series of concerns for the winemaker and staff of a winery. The first concerns are viticultural. Upon arrival of the fruit, its quality is assessed, grapes are processed and fermentation is begun. Almost immediately, and in many instances simultaneously, chemical and microbiological stability of the young and/or aging wine become important. Finally, problems do occur on occasion, and a number of what may be considered remedial techniques can be employed to produce an acceptable product.

We have used these general areas of concern as a loose framework in dividing the book into its major sections. There is, of course, much overlap between the sections; a given analytical test may be run for several reasons during the production of a wine. However, the major chapter divisions should guide the reader through the important areas of concern and provide some idea of when certain laboratory tests will be run. There is occasionally some difference in how terms are technically defined and how they are used in current practice in the industry, and we have attempted to make allowances for this. The reader will, therefore, find temperature values specified in degrees Celsius in some places and degrees Fahrenheit in others. Concentrations of low-level constituents are usually calculated in milligrams per liter (mg/L) although we are aware that much of the industry refers to these concentrations in parts-per-million (technically, mg/Kg) units. To make conversions between these temperature and concentration units, we have provided simple mathematical relationships in the Reagent Preparation section (Appendix II).

Each chapter is divided into introductory considerations of theory and application, intended to give the reader necessary background information, followed by a laboratory procedures section that outlines each analysis in detail. Each laboratory procedure concludes with a consideration of special precautions for the analysis in question. Appendixes covering chromatographic theory, reagent preparation, stains, and media preparation also are included.

List of Laboratory Procedures

Section I: Sampling, Fermentation, and Production Analyses

Section I

Sampling, Fermentation, and Production Analyses

This section begins with definitions of fruit maturity and quality starting in the vineyard, and describes measurement techniques for these factors. During initial processing and fermentation, analyses related to the production of ethanol, other alcohols, and structural considerations (including extract) are presented. This is also an appropriate place to deal with concepts of acidity, and corresponding measurements of pH, individual acid levels, and titratable and volatile acidity. Analyses for residual sugar levels found in fermented products, as well as the important phenolic compounds, are discussed in the context of their relationship to wine balance, body, astringency, aroma, flavor and color. Finally, levels of dissolved gases such as oxygen and carbon dioxide, are discussed, as well as the important concept of oxidation in juice and wine.

Chapter 1

Fruit Quality and Soluble Solids

MATURITY SAMPLING

Grape maturity assessment as well as fruit quality is critical to the successful production of palatable wines. Part of the grower's payment for the fruit often is based upon delivery of grapes at agreed-upon parameters; so care must be taken in the maturity monitoring process. Traditionally, degrees Brix (°B, weight percent sugar in the juice) and titratable acidity (g/L of acid in the juice) have been selected as the harvest parameters most important in monitoring maturity. Because of the importance of fruit maturity to ultimate wine palatability, field sampling of fruit must be performed in an objective and statistically acceptable manner.

Amerine and Roessler (1958 a, b) compared sample collection techniques and reported that berry sampling can provide an accurate and economical sampling technique. Theoretically, to be within plus or minus 1° B with a probability level of 0.05 (95 out of 100 samples), two lots of 100 berries each should be examined. In cases where one wishes to be within plus or minus 0.5° B 95% of the time, five lots of 100 berries should be collected.

Berry sampling is difficult because of the time required and variations in fruit chemistry. The chemical composition of grape berries differs with their position on the rachis, the location of the cluster on the vine, and the location of the vine in the vineyard. However, properly performed berry sampling can provide an accurate picture of the overall vineyard fruit chemistry. Jordan and Croser (1983) recommend collection of berries from the top, middle, and bottom of the cluster. Terminal berries on the rachis may be less mature than other berries. Berry sampling in various locations on the cluster may be significant in the case of larger clusters; but many persons, sample by selecting berries only from the middle of the rachis, a technique that may be acceptable in the case of varieties with small clusters. Additionally, in a berry sampling procedure the side of the cluster from which the berries are taken must be randomized. One should avoid selection of fruit from ends of rows or from isolated vines or those with obvious physiological or morphological differences. Vine (1981) suggested an alternating-row method of sample selection. Significant variations in soil type, shading, and so on, may play an important role in vine growth and therefore fruit maturity. Ideally, sampling should be designed to reflect differences in soil type, topography, and vine growth. For consistency, some recommend that selected vines within each block be targeted for sampling.

Samples should be collected at approximately the same time period each day. Jordan and Croser (1983) have made the following sampling recommendations: (1) edge rows and the first two vines in a row should not be sampled; (2) samples should be collected from both sides of the vine; (3) for each row, estimate the proportion of shaded bunches and sample according to that proportion. This proportion may vary with the side of the row sampled because of variation in leaf cover. Smart et al. (1977) demonstrated that white and red grapes exposed to direct sunlight may be as much as 8°C and 15°C, respectively, warmer than the ambient air temperature. Such exposure may have a dramatic effect on fruit composition. Kliewer (1985) summarizes these differences in Table 1-1. As the table shows, a sampling technique that does not consider the effect of exposure to the sun may not be accurate.

There is a natural tendency when one is picking individual berries to select those with the most eye appeal. These are often the more mature berries. Such bias in a vineyard sampling technique may result in Brix readings that are as much as 2° B higher than that measured by the winery after crushing the entire vineyard load (Kasimatis 1984).

Most growers begin sampling fruit several weeks prior to harvest. Initially, samples are taken once weekly, with a shift to more frequent intervals as harvest nears. Brix and titratable acidity are the traditional harvest parameters most important in monitoring maturity. Recently, the importance of pH, aroma, organic acid, etc., has been established.

Measurements of °B, titratable acidity, and pH, by themselves, may not be specific indicators of physiological maturity or potential wine character and pal-

Table 1-1 Comparison of shaded and control[1] Cabernet Sauvignon fruit.

Parameter measured	Control	Shade	Significance level
Harvest date	27 Sept.	19 Oct.	
°Brix	23.3	21.1	+
T.A. (g/L)	5.5	6.4	+
pH	3.44	3.74	++
K^+	2325	2510	+++
Malate (g/L)	1.65	2.84	++
Tartrate (g/L)	0.86	0.83	ns
Anthocyanins (g/g fruit)	0.98	0.56	+++

[1]Control treatment was a bilateral cordon 3-wire "T" trellis. Shaded treatment consisted of bunching the foliage around the fruit using bird netting.
SOURCE: Kliewer and Benz 1985.

atability. These parameters will vary considerably, depending upon the season, crop load, soil moisture, and so on. Maturity is clearly a multidimensional phenomenon, and should be viewed on a relative rather than an absolute basis.

There is considerable variation in the definition of optimum maturity for a particular wine grape variety. If maturity is defined as that point in time when the berry possesses the maximum varietal character consistent with the type and style of wine desired, then the assessment of varietal character is a logical adjunct and extension of conventional definitions of maturity. Further, the optimal maturity may vary somewhat with the style of wine to be produced. Long (1984a) lists the parameters she considers important in affecting fruit development and maturity assessment, which are summarized in Table 1-2. The relative importance of each parameter is predicated upon the type and style of wine to be produced as well as the ability of the vine to continue to mature the crop. Several of these parameters are described in more detail below.

Contribution of Juice Aroma

Aroma and flavor in wine arise from several sources, including the grape and an array of chemical changes that may occur during juice processing, fermentation, and aging of wine. As the fruit matures, the aroma develops from "green" underripe tones to the more complex characteristics of the particular cultivar.

The floral, citrus, honey, and melon aromas of Chardonnay as well as the black currant, blackberry, spinach, and bell pepper of Cabernet sauvignon are

Table 1-2. Factors to consider in maturity assessment and picking decisions.

Factors	Importance	Chapter Discussed in
°Brix	Final alcohol level	1
Sugar/berry ratio	Maturation/dehydration–rehydration	1
Aroma assessment	Varietal aroma/aroma intensity–maturity	1
Titratable acidity	Balance, style	4
pH	Balance, stability, etc.	4
Malic acid	Balance + malolactic involvements	4, 12
Vine condition	Potential for further ripening	4
Grape condition	Maturity, potential for further ripening	1
Color and phenol levels	Maturity, color, etc.	7

SOURCE: Long 1984.

properties that, although modified by processing and aging, have originated in the grape to a significant degree. Naturally the more neutral the grape, the more difficult it is to use aroma as a maturity adjunct. There is little doubt about the importance of grape aroma components to the palatability of the finished product. If grapes are harvested when varietal character is lacking or deficient, varietal character in the finished wine subsequently will be absent or reduced. Aroma evaluations are based upon subjective descriptors, such as those utilized in the aroma wheel developed by Noble et al. (1984, 1987), as well as estimations of aroma intensity.

Aroma evaluation of the juice can be used as a stylistic tool to note potential varietal character and intensity (Cootes, et al. 1981). If more aroma is required, assuming that the potential for more aroma development exists, harvest may be delayed. In California, Simi Winery developed an evaluation system using a 0 to 10 Intensity/Quality (I/Q) Scale (Table 1-3). A methodology for grape aroma evaluation is presented at the end of this chapter (see Procedure 1-1).

Procedures for rapid and easy chemical evaluations of certain aroma components (monoterpenes) have been developed (Dimitriadis and Bruer 1984; Dimitriadis and Williams 1984). Such methods may aid in reducing the subjectivity of aroma evaluation of certain cultivars such as Rieslings, Muscats, and Gewurztraminer.

Color and Phenols (see also Chapter 7)

According to Peynaud (1984), vinification is the art of fractional extraction of the grape. Because of the importance of phenolic compounds to color, this is of particular significance in red wines. There is a sound basis for using the level of total anthocyanins and total phenols in the fruit as an additional harvest criterion. However, anthocyanin levels appear to vary more widely than any other commonly measured compounds in grapes in response to environmental influences. Thus, it remains difficult to measure these levels on a routine basis, and to interpret the results in relation to winemaking.

Table 1-3. Rating scale for evaluation of grape maturity. Scores will be recorded as I/Q (Intensity /Quality), each on a scale of 1 to 10.

I:	No aroma at all	0 pts
	Neutral—aroma of neutral juice	1–2 pts
	Slightly varietal aroma	3–5 pts
	Distinctive varietal aroma	6–8 pts
	Very distinctive varietal aroma	9–10 pts
Q:	Spoiled or off quality	0 pts
	Broad, uninteresting, unpleasant aroma	1–2 pts
	Some pleasing varietal character	3–5 pts
	Distinctive varietal aroma	6–8 pts
	Very distinctive varietal characteristics	9–10 pts

SOURCE: Simi 1983.

Grower Input

Grower input as it relates maturity decisions refers to the intimate relationships between vineyard management, fruit chemistry, and ultimately, wine palatability. Such factors as vine population, training, trellis systems, hedging, crop load, spray schedule, fertilization, irrigation, and so on, will influence fruit ripening and maturity decisions. For example, overcropping delays maturity and can result in reductions in titratable acidity, malic acid, phenols, the levels of volatile components, and color, as well as causing an elevation in berry pH.

Sugar per Berry

Simple measurement of Brix, as described in Procedures 1-3 and 1-4, although a useful tool in wine production, has certain drawbacks when used by the grower and winery to evaluate crop maturation. First, Brix is defined as sucrose per 100 g juice. It is a measure of the grams of all soluble solids per 100 g juice, and thus includes pigments, acids, glycerol, and so on, as well as sugar. Because the fermentable sugar content of grape must accounts for around 95% of the total soluble solids, determination of °B provides a reasonable approximate measurement of sugar levels. However, this measurement involves a ratio (wt/wt) of sugar to water that may change because of the physiological condition of the fruit. Another problem one encounters when using standard methods of °B determination is that no consideration is given to the weight of fruit used in the sample preparation. Thus, it is entirely possible that over a period of days, °B may show no change, but in fact there may be major changes in the weight of fruit (either increases or decreases) comprising that sample. Obviously this information would be of value to both the grower and the vintner in selecting harvest dates to optimize fruit maturity.

Several wineries have evaluated alternatives to traditional interpretation of soluble solids measurements (°B). As compared with °B, which measures soluble solids (in g/100 g juice) of a collected sample, the concept of reporting sugar per berry utilizes the same initial Brix measurement but extends it such that the final value takes into account the weight of a berry sample.

For example, the data of Table 1-4 were collected from randomly selected 100-berry samples. The samples were taken from the same vineyard at five-day intervals. Upon sample preparation, the soluble solids content (°B) by refractometry of both samples was measured to be 22° B. The grower's conclusion was that there had been no change in the maturity of the fruit. However, sugar

Table 1-4. Sample harvest data for 1985 Barbara grapes.

	Sept. 5	Sept. 10
°B	22	22
Sample weight (grams)	112	118

per berry concentration calculations leads one to a different conclusion. Sugar per berry concentrations for the same samples are calculated to be:

$$\text{Sept. 5:}\quad \text{Soluble solids} = \frac{22}{100} = \frac{X}{112} \qquad \text{Sept. 10:}\quad \frac{22}{100} = \frac{X}{118}$$

$$X = 24.6 \text{ g}/100 \text{ berries} \qquad X = 26.0 \text{ g}/100 \text{ berries}$$

Dividing by 100 berries/sample:

$$\begin{array}{ll} \text{Soluble solids per berry (Sept. 5)} = 0.246 \text{ g}/\text{berry} & \text{Soluble solids per berry (Sept. 10)} = 0.260 \text{ g}/\text{berry} \end{array}$$

Results from typical data sets collected over several days are summarized in Table 1-5. It can be seen that increases in soluble solids per berry and increases in berry sample weight correspond to maturation. One can also see that the information obtained by consideration of sugar per berry is considerably more complete than that available by evaluation of Brix measurements alone.

Table 1-5. Evaluation of results from sugar/berry data.

Changes in Brix	Changes in Berry Weight		
	Decreases	No change	Increases
Increases	Maturation *and* dehydration	Maturation	(a) Major increase: maturation and dilution (b) Minor increase: maturation
No change	Dehydration	No change	Dilution
Decreases	Dehydration *and* sugar-export	Sugar export	Sugar export *and* dilution

SOURCE: Long (1984).

SAMPLE PROCESSING

Procedures used in berry sample preparation are critical and must be uniform from grower to vintner and from one sample period to the next. The chemical components of the grape berry are not uniformly distributed in the fruit, so sample preparation (i.e., degree of crushing or pressing) can have a profound effect on analytical results. One example of this relationship between degree of

Table 1-6. Changes In grape berry chemistry with the extent of pressing.

Press fraction	Titratable acidity, g/100 mL	pH	°Brix
#1 First juice extracted	1.3	3.0	19.5
#2 Juice	1.1	3.2	19.5
#3 Last juice extracted	0.95	3.4	19.5

SOURCE: Zoecklein 1986

pressing whole berries and sample analysis is presented in Table 1-6, where the degree of pressing is seen to have a significant effect on pH and titratable acidity. It follows that if sample preparation methods are not standardized between growers and vintners, significant variations in fruit analysis will occur and may have an impact on ultimate wine quality and palatability.

A variety of methods can be employed for small-lot processing. The simplest of these is the use of a cheesecloth bag, in which the berry samples are placed and then macerated by pounding. This is adequate only if the fruit has not been raisined, and if there is not a high degree of uneven ripening. Further, seed breakage may affect the results. The analysis results of juice derived from such a procedure correspond closely to those of grapes processed with limited skin contact time (i.e., most white wines).

Many individuals use hand presses. Such systems can be adequate if operated without seed breakage. Seed breakage, as would occur with the use of Waring-type blenders, may cause an elevation of 0.2 to 0.3 pH unit in the sample. This result corresponds to pH values expected of red varieties when red grapes with skin and seeds are fermented together for several days. The high sugar content of raisined berries may not be adequately represented by either sample processing procedure.

FRUIT EVALUATION

Upon arrival of grapes at the winery, representative samples are collected from lugs, gondolas, and so on, and examined for material-other-than-grape (MOG), which may include leaves, cane fragments, and so forth, as well as for defects that might be present in the form of mold and rot. Table 1-7 summarizes the results of the California Wine Grape Inspection Program for the 1986 season. As the results indicate, the majority of fruit inspected fell within the range of zero to 1.0% MOG.

Ongoing research is focusing on the development of new techniques that assesses grape quality. Quantification of mold, yeast, and bacterial metabolites in collected juice samples is evaluated using High Performance Liquid Chromatography (HPLC) (see Procedure 1-5). No doubt, the nature and the concen-

Table 1-7. Distribution of material-other-than-grape (MOG) in grapes delivered during the 1986 season.[1]

Levels of MOG (%)	% of total tonage inspected
0.0–1.0	89.38
1.1–2.0	10.12
2.1–3.0	0.46
> 3.1	0.04

[1]Excerpted from Wine Grape Inspection Report—1986, California Department of Food and Agriculture.

tration of microbial metabolites differ as a function of a myriad of factors, both biological and abiotic. Key indicators such as ethanol, glycerol gluconic acid, laccase and acetic acid, have been documented.

Glycerol is a by-product of mold metabolism that results from the breakdown of glucose and fructose sugars from the grape. Various molds, such as *Botrytis*, *Aspergillus*, *Penicillium*, *Rhizopus*, *Alternaria*, and so on, are commonly found on grapes and cause mold defects. Their metabolism may result in various levels of glycerol formation.

Molds alone are not the only cause of grape defects. Sour rot and sour bunch rot are primarily the result of yeasts and acetic acid bacterial growth. Acetic acid measurements thus are a good indicator of the extent of yeast action and bacterial infection. Measurement of ethanol in grape juice is an indicator of wild yeast fermentation of the fruit on the vine. However, the volatility of these components may effect analytical measurement at the winery.

The ability to measure key indicators and quickly determine grape berry degradation is a major step forward in monitoring grape quality. The HPLC method of analysis of glycerol, ethanol, and acetic acid, as developed by Kupina (1984), is presented at the end of this chapter. Further work in detection and quantification of mold and rot is continuing (including use of immunoassay techniques).

Laccase is an enzyme found in grapes degraded by the mold *Botrytis cinaeria*. Historically it has been included in the group of enzymes known as polyphenoloxidases. Today, the term polyphenoloxidase refers to the enzyme *o*-diphenol oxidase, which acts on the following substrates: ortho and vicinal (3,4,5-trihydroxy) OH groups. It can also act on monophenols by first converting them to corresponding *o*-diphenols. The laccase enzyme, in contrast, oxidizes *o*- and *p*-dihydroxy phenols, but does not hydroxylate monophenols (see Chapter 8). The measurement of laccase concentration has also been used as an indicator of deterioration due to mold growth. Procedure 1-2 outlines a method for laccase measurement.

Gluconic acid in juice and wine indicates mold contamination of the fruit. Gluconic acid is the aldonic acid of glucose. The aldehyde group of glucose has been oxidized to a carboxyl group. The transformation of glucose to gluconic acid is known to occur due to the action of the enzyme glucose oxidase.

Ribereau-Gayon 1988 suggested that the ratio of glycerol to gluconic acid be used for judging the quality of rot. In the case of extentive growth of *Botrytis cinerea* this ratio is the highest. Wines from 'clean' fruit should have gluconic acid levels to 0.5 g/L, from 'noble rot' 1-5 g/L and from sour rot and other degradative rots 5 g/L. See procedure 4-3 for the analysis of gluconic acid.

APPLICATION OF SOLUBLE SOLIDS DATA IN WINEMAKING

As the sugar content of grape musts accounts for 90 to 95% of the total soluble solids, soluble solids measurements provide a good approximation of the available fermentable sugar (Amerine 1965). The balance of nonfermentable species includes pectins, tannins, pigments, acids, and their salts. On a practical basis, soluble solids determinations provide the winemaker with an indication of fruit maturity, approximate harvest date, and potential alcohol yield, as well as utilization of the grapes in the total production program. Furthermore, °B measurements are routinely used to monitor fermentation rates, as a relative indication of sugar content in finished blends, and as a legal standard for certain wine types. Winemakers in certain regions of the United States rely on soluble solids measurements to dictate amelioration considerations.

Theoretically, a given weight of fermentable sugar should yield 51.1% (by weight) ethanol, according to the Gay-Lussac relationship:

$$C_6H_{12}O_6 \longrightarrow 2\,C_2H_5OH + 2\,CO_2 \qquad (1\text{-}1)$$

Thus, an initial 180 g glucose should produce 92 g ethanol and 88 g carbon dioxide upon complete fermentation. Practically speaking, the actual alcohol yield is generally less than the theoretical 51% because of losses during fermentation as well as diversion of available sugars to products other than ethanol. Despite this, many winemakers use 55% (0.55) as an estimate of potential alcohol. However, the 0.55 factor may be valid only in warm climates.

Many wineries in cool climate regions use higher conversion factors. Peterson (1979) notes that, in general, the cooler the grape-growing region the higher the conversion factor; in upper Monterey County (California), for example, winemakers may use figures as high as 0.62. Jones and Ough (1985) found that alcohol conversions ranged from 0.54 to 0.61. Differences were noted according to region and growing season, as well as slight variations between varieties.

In general, it has been found that grapes grown in cooler regions yield higher alcohol levels per °B than do grapes grown in warmer climates. Another factor that influences the alcohol/°B ratio is fermentation of partially raisined fruit. In this case, initial sugar extraction is incomplete, yielding erroneously low initial °B readings. With more complete extraction during pressing and fermentation, additional fermentable sugar is liberated, yielding higher-than-expected final alcohols. Values can be lower, primarily because of entrainment losses as well as formation of products other than alcohol during fermentation. More complete discussions of the nature and concentrations of these by-products of fermentation are presented, when appropriate, in subsequent chapters.

In certain winemaking regions of the United States and in most European wine-growing regions, amelioration, or chaptalization, of must or wine is permissible. Amelioration is defined as the addition of sugar and/or water to juice or wine, whereas chaptalization refers only to the addition of sugar. Sugar additions may be carried out either before, during, or after fermentation. Thus low-sugar grapes can be harvested and sufficient sugar added to yield the desired alcohol level. If the winemaker desires an alcohol level of at least 11.5% (vol/vol), and is working with a conversion ratio of 0.55, for example, he must have an initial Brix of 21.0, as seen in the following relationship:

$$\frac{11.5\% \text{ final alcohol}}{0.55} = 21.0 \text{ g}/100 \text{ g} \qquad (1\text{-}2)$$

The quantity of dry or liquid sugar needed to raise the Brix of a must may be determined from tables or formulae (see Table 1-8). When dextrose is used instead of sucrose, the 8% water of crystallization in the dextrose must be taken into account. (With anhydrous dextrose, this, of course, does not apply.)

In wines carrying California appellation, amelioration is normally permissible only in production of fruit wines, special naturals, and sparkling wines. In other cases, deficiencies in must sugar may be made up by the addition of grape juice concentrate. In cases where fruit maturity exceeds 22° B at crush, the winemaker is permitted to add sufficient water to bring the soluble solids (°B) level back to that level (CFR 240.361). For further discussion of amelioration as it relates to regulatory considerations, refer to Section 5.6, in the Code of Federal Regulations (Part 240, Title 27). California winemakers should also refer to the California Administrative Code (Title 17, 17,000-17, 135).

Soluble solids (°B) measurements are routinely used in winemaking to monitor the progress of fermentation. As the fermentation progresses, grape sugar content decreases with corresponding increases in alcohol. As a result of alcoholic fermentation, the fermenting juice has a lower density than that of the original. Pure water at 20°C has a °B reading of zero. A 20% sucrose solution at the same temperature has a reading of 20° B. The °B of a 95% solution of ethanol is approximately −7° B. From this, it can be seen that it is possible to have a finished dry wine with an apparently negative °B value. Therefore, one cannot use the hydrometer for an accurate measurement of sugar content because of the alcohol present in fermenting must and finished wine. Despite this, °B levels are used as industry standards for certain wine types. Based on *required alcohol content*, Table 1-9 summarizes °B values for some California wine types.

Table 1-8. Weight of sucrose solutions at 20°C (68°F).[1]

Sucrose % by weight	Specific gravity at 20°/20°C	Total pounds per gallon	Pounds solids per gallon	Pounds water per gallon
1.0	1.00387	8.35379	0.08354	8.27026
2.0	1.00777	8.38626	0.16773	8.21854
3.0	1.01170	8.41898	0.25257	8.16641
4.0	1.01565	8.45186	0.33807	8.11379
5.0	1.01964	8.48508	0.42425	8.06083
6.0	1.02366	8.51855	0.51111	8.00744
7.0	1.02771	8.55218	0.59865	7.95353
8.0	1.03179	8.58615	0.68689	7.89926
9.0	1.03589	8.62029	0.77583	7.84446
10.0	1.04002	8.65468	0.86547	7.78921
11.0	1.04418	8.68923	0.95582	7.73342
12.0	1.04836	8.72404	1.04688	7.67715
13.0	1.05257	8.75909	1.13868	7.62041
14.0	1.05681	8.79440	1.23122	7.56318
15.0	1.06110	8.83003	1.32451	7.50553
16.0	1.06541	8.86592	1.41855	7.44738
17.0	1.06975	8.90206	1.51335	7.38871
18.0	1.07412	8.93845	1.60892	7.32953
19.0	1.07853	8.97509	1.70527	7.26983
20.0	1.08297	9.01207	1.80241	7.20966
21.0	1.08743	9.04921	1.90033	7.14888
22.0	1.09193	9.08660	1.99905	7.08755
23.0	1.09645	9.12424	2.09858	7.02567
24.0	1.10101	9.16222	2.19893	6.96329
25.0	1.10562	9.20053	2.30013	6.90040
26.0	1.11025	9.23909	2.40216	6.83693
27.0	1.11492	9.27790	2.50503	6.77287
28.0	1.11962	9.31705	2.60877	6.70827
29.0	1.12435	9.35644	2.71337	6.64307
30.0	1.12913	9.39617	2.81885	6.57732
31.0	1.13392	9.43607	2.92518	6.51089
32.0	1.13876	9.47630	3.03241	6.44388
33.0	1.14363	9.51686	3.14056	6.37630
34.0	1.14854	9.55767	3.24961	6.30807
35.0	1.15348	9.59882	3.35959	6.23923
36.0	1.15845	9.64022	3.47048	6.16974
37.0	1.16348	9.68204	3.58235	6.09968
38.0	1.16852	9.72402	3.69513	6.02889
39.0	1.17362	9.76642	3.80890	5.95752
40.0	1.17875	9.80907	3.92363	5.88544
41.0	1.18390	9.85197	4.03931	5.81266
42.0	1.18910	9.89520	4.15599	5.73922

Table 1-8. (*Continued*)

Sucrose % by weight	Specific gravity at 20°/20°C	Total pounds per gallon	Pounds solids per gallon	Pounds water per gallon
43.0	1.19433	9.93877	4.27367	5.66510
44.0	1.19961	9.98267	4.39238	5.59030
45.0	1.20491	10.02683	4.51207	5.51476
46.0	1.21026	10.07131	4.63280	5.43851
47.0	1.21564	10.11613	4.75458	5.36155
48.0	1.22106	10.16121	4.87738	5.28383
49.0	1.22653	10.20669	5.00128	5.20541
50.0	1.23202	10.25243	5.12622	5.12622
51.0	1.23756	10.29850	5.25224	5.04627
52.0	1.24313	10.34483	5.37931	4.96552
53.0	1.24874	10.39157	5.50753	4.88404
54.0	1.25439	10.43856	5.63682	4.80174
55.0	1.26008	10.48588	5.76723	4.71865
56.0	1.26575	10.53312	5.89855	4.63457
57.0	1.27147	10.58070	6.03100	4.54970
58.0	1.27729	10.62910	6.16488	4.46422
59.0	1.28321	10.67835	6.30023	4.37812
60.0	1.28908	10.72726	6.43635	4.29090
61.0	1.29500	10.77650	6.57367	4.20284
62.0	1.30096	10.82608	6.71217	4.11391
63.0	1.30695	10.87591	6.85182	4.02409
64.0	1.31298	10.92615	6.99274	3.93341
65.0	1.31905	10.97665	7.13482	3.84183
66.0	1.32516	11.02748	7.27814	3.74934
67.0	1.33131	11.07864	7.42269	3.65595
68.0	1.33749	11.13005	7.56844	3.56162
69.0	1.34371	11.18189	7.71550	3.46638
70.0	1.34997	11.23397	7.86378	3.37019

[1]From Hoynak and Bollenback (1966).

Table 1-9. Apparent °B values for California wine as determined by hydrometer.[1]

Wine type	Apparent °B (g/100 g)
Angelica, white port, muscatel, port	>5.5
Tokay (dessert wine)	>3.5

[1]From California Administrative Code, Title 17, Section 17,010.

LABORATORY MEASUREMENTS OF SOLUBLE SOLIDS

Densimetric Procedures

Hydrometry

The absolute density of any substance, expressed in units of grams per cubic centimeter (g/cm^3) or grams per milliliter (g/mL), is defined in Equation 1-3:

$$\text{Density} = \frac{\text{Weight of substance}}{\text{Volume of substance}} \tag{1-3}$$

Direct measurement of volumes, as required by Equation 1-3, may present a problem, especially where gases are concerned. As a result, it becomes convenient to speak in terms of relative density or the ratio of the density of the substance to that of a recognized reference such as water. This relationship, known as the specific gravity, is expressed in Equation 1.4:

$$\text{Specific gravity} = \frac{\text{Weight of } x \text{ mL of substance}}{\text{Weight of } x \text{ mL of water}} \tag{1-4}$$

The density of water at 4°C is, for all practical purposes, 1.000 g/cm^3. Because the weight of any substance will change as a function of temperature, any complete definition of specific gravity must include the temperature at which the determination was made, as well as the reference temperature for water. Custom dictates that the temperature of the measured sample be noted above that of the reference. For example, the notation 15°/4°C indicates that the specific gravity of the solution in question was made at 15°C relative to water at 4°C.

Of what value is specific gravity to the laboratory analyst? The concentration of dissolved substances in solution is related to the specific gravity; so a determination of specific gravity can be used as a measure of concentration in terms of these dissolved species. One should not, however, assume a simple and direct correlation between observed specific gravity and concentration in all cases because molal volumes of substances in solution may vary in a complex and unpredictable manner. Tables are available that relate concentration of dissolved substances to apparent specific gravity; those most commonly encountered in analyses of wine are for Brix (Table 1-12) and alcohol (Table 2-5).

Printed tables usually are prepared with reference to only one or two standard temperatures; so one must either measure the specific gravity at the defined temperature or, alternatively, employ a temperature correction factor. Correction factors are calculated from the coefficient of cubical expansion, a function that varies with both temperature and concentration; so appreciable error may result. Therefore, such correction factors are best employed with less accurate measurements such as hydrometric determinations. For the most accurate work, it is recommended that the solution be brought to a defined temperature prior to measurement.

Hydrometric determinations are based on the principle that an object will displace an equivalent weight in any liquid in which it is placed:

$$\text{Mass of hydrometer} = \text{Mass displaced} \tag{1-5a}$$

Density is defined as units of mass per units of volume:

$$\text{Density} = \frac{\text{Mass}}{\text{Volume}} \tag{1-5b}$$

so it follows that mass displaced is equal to the product of the volume of liquid displaced times the density of that liquid:

$$\text{Mass} = \text{Density} \times \text{Volume} \tag{1-5c}$$

Substituting into Equation 1-5a, one obtains a relationship between the volume and density of the object (hydrometer) *and* the volume and density of the liquid in which the hydrometer floats:

$$V_1D_1 \text{ (hydrometer)} = V_2D_2 \text{ (liquid)} \tag{1-5d}$$

Rearranging Equation 1-5d:

$$\frac{D_1}{D_2} = \frac{V_2 \text{ displaced}}{V_1 \text{ displaced}} \tag{1-5e}$$

In the above relationship, V_1 and V_2 can also represent volumetric displacements by the same partially submerged object in two liquids of different densities, D_1 and D_2. Upon examination of Equation 1-5e, it can be seen that the volume displaced by the object in question is inversely proportional to its density. Hence a solution of high density will show less displacement than one of lower density. This can be extended to include the depths to which the object will sink if allowed to reach equilibrium in an upright cylinder:

$$\frac{D_1}{D_2} = \frac{H_2}{H_1} \tag{1-5f}$$

This relationship defines the basic principle of hydrometry.

The hydrometer (Fig. 1-1) consists of a calibrated scale within a glass tube

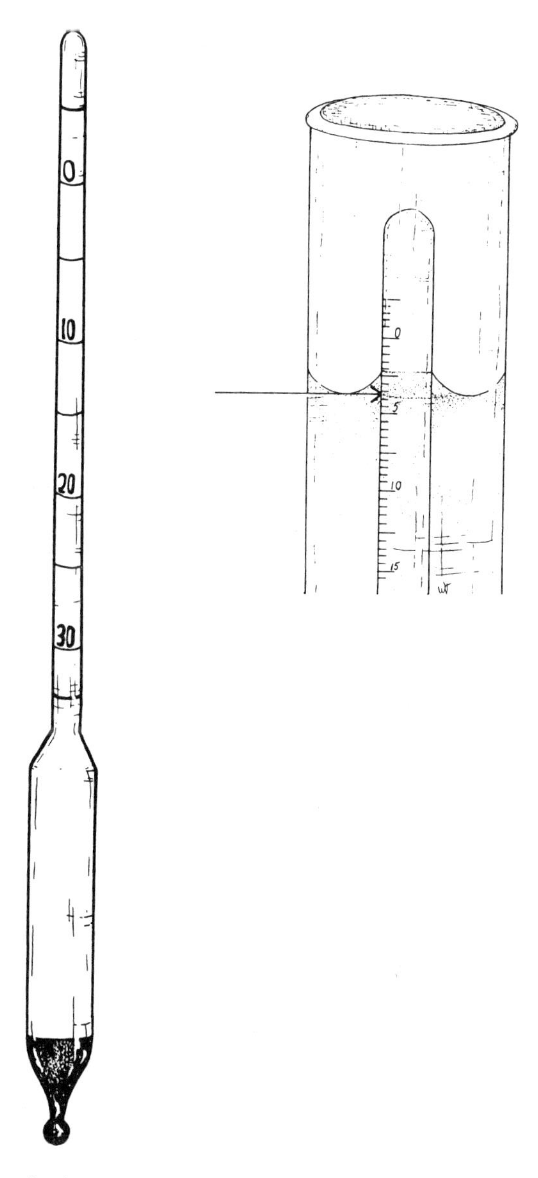

Fig. 1-1. Glass hydrometer designed to float upright in liquid. Scale can be calibrated to read specific gravity or concentration of some specific solute at the defined temperature of calibration. Hydrometers are read at the bottom of the miniscus.

that is usually constructed with a mercury or shot-filled terminal bulb to maintain it in an upright position. Hydrometers are available to read either specific gravity or the concentration of some component in solution. Examples of the latter include the familiar saccharometer and the salinometer. Saccharometers using either the Brix or the Balling scale are calibrated to read the concentration of sucrose (g/100 g per definition of °B) in aqueous solution; and, for the purposes of measurement, Brix and Balling scale hydrometers are identical. Vine (1981) offers a view of the relationship and differences between Brix and Balling. He reserves the term Brix for samples that contain no alcohol, and Balling for samples with alcohol present. The *American Journal of Enology and Viticulture* prefers the use of the term Brix (as °B) alone.

Refractometry

Refractometric determinations frequently are used in lieu of, or in addition to, laboratory hydrometric determinations of soluble solids. In principle, the passage of a ray of light from one medium to another of different optical density (different number of molecules interacting with the light) causes the incident ray to undergo a change in direction, or refraction. (See Fig. 1-2.) The angle formed by the light beam with the perpendicular at the plane of refraction is called the "angle of incidence." (This angle is identified as x in Equation 1-6.) The angle formed by the refracted ray and the perpendicular (angle y) is termed the "angle of refraction."

In Equation 1-6, the "index of refraction" (n) is defined as the ratio of the sine of the angle of incidence (x) to the sine of the angle of refraction (y):

$$n = \frac{\text{sine } \angle\, x \text{ (incidence)}}{\text{sine } \angle\, y \text{ (refraction)}} \tag{1-6}$$

Refraction of white light yields a host of prismatic colors of different wave-

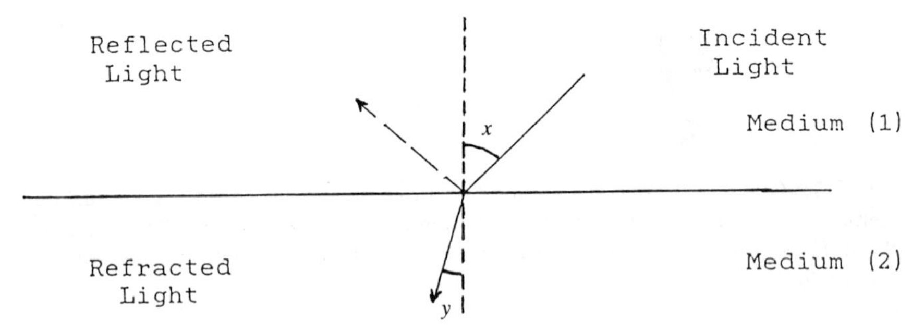

Fig. 1-2. Refraction of a beam of light resulting from passage from one medium into a second medium of greater density.

lengths (a phenomenon called "dispersion"). For a solution of defined composition, the refractive index would vary as a function of wavelength, increasing with decreasing wavelength. Therefore, a reference wavelength has been selected for refractive index measurement: monochromatic sodium light at 589 nm and 20°C. Thus, for a solution of constant composition at a defined wavelength of light (usually the sodium emission lines) and temperature, the index of refraction is a constant. As with hydrometric determinations, shorthand notation frequently is used. The designation n_D^{20} refers to a reading taken at 20°C using incident light of the D-line of sodium. Using this convention, the refractive index of water is defined as $n_D^{20} = 1.330$. Aqueous solutions of sugar have higher indexes of refraction. Other aqueous solutions, with the exception of methanol, also have higher refractive indexes that vary in a regular manner with concentration. (Note that the data in Tables 1-2 and 1-14, cited below, are based on values for solutions of pure sucrose. Where solutions contain significant amounts of dissolved solids other than sucrose, results should be recorded as percentage soluble solids.)

Conversions between Fahrenheit and Centigrade can be accomplished by using the following relationships:

$$°F = [(°C + 40)\tfrac{9}{5}] - 40$$
$$°C = [(°F + 40)\tfrac{5}{9}] - 40$$

Procedure 1-1. Juice Preparation for Aroma Evaluation

Juice separation from a grape sample for aroma evaluation must be done with care. Rapid enzymatic oxidation can occur if berries are warm, or if juicing has occurred. In addition to phenolic oxides, oxidation products include aldehydes, as a result of enzymatic oxidative cleavage of linoleic and linolenic acids to hexenal (Drawert, 1973). These aldehydes can produce grassy aromas that mask fruit characteristics and make aroma assessment very difficult. The following procedure is that of Jordan and Croser (1983).

I. Equipment
   - Cone-in-cone juicer, or potato ricer, or hand press
   - Sieve

II. Reagents
   - Pectolytic enzymes
   - Ascorbic acid
   - Sulfur dioxide (potassium metabisulfite or equivalent)
   - N_2 and CO_2 gases

III. Procedure
   1. Chill grape sample to <2°C.
   2. Press the chilled grape sample.

3. Estimate the quantity of juice that the sample will yield. Mix and then add 50 mg/L ascorbic acid and 30 mg/L sulfur dioxide. Add pectolytic enzyme at the supplier's recommended level.
4. Lightly sparge the juice sample with nitrogen, and pour it through a sieve into CO_2-filled sample bottles. Seal bottles and cold-settle at $<2°C$.
5. Decant the cold-clarified juice into CO_2-filled bottles, and carry out aroma/flavor evaluation and any chemical analyses desired.

IV. Supplemental Notes

1. Ascorbic acid (vitamin C) is an antioxidizing agent that along with sulfur dioxide will help minimize degradation of aroma components.
2. The sample preparation method (i.e., pressing, degree of pressing vs. crushing) affects the titratable acidity and pH.
3. Under optimum conditions, juices prepared using this methodology and stored at 0°C will remain viable for aroma/flavor assessment for up to several days.
4. Aroma and intensity are evaluated using the scale presented in Table 1-3.
5. As fruit matures, the aroma develops from "green" underripe tones to the floral complexes characteristic of the particular cultivar. Naturally, the more neutral the grape variety, the more difficult it is to use aroma as a maturity indicator. Several of the commonly used descriptors noted during maturation of Cabernet sauvignon, Pinot noir, Sauvignon blanc, and Chardonnay are listed in Table 1-10.

Table 1-10. Commonly used descriptors for maturation in four premium grape varieties.

Cabernet sauvignon	*Chardonnay*
Green–unripe–spinach	Green–unripe–citrus
Slightly herbaceous	Cucumber
Herbaceous	Melon, ripe figs
Minty, blackcurrant, blackberry	Honey
Pinot noir	*Sauvignon blanc*
Green–unripe	Green–unripe–grassy
Lightly herbaceous	Lightly herbaceous
Slightly brambly	Stalky/leafy
Spicy, rose-like, violets	Tobacco

Procedure 1-2. Mold (Botrytis) *Estimation via Laccase Assay*

Botrytis cinerea produces the oxidative enzyme laccase, which is partially responsible for oxidative deterioration of grapes. Dubourdieu et al. (1984) developed an assay for this enzyme that can be used to help establish fruit quality and as a means for grower compensation. Laccase values in grapes range from 0.1 to 7.8 nanomoles/mL (nmol/mL). Values above 4.6 may suggest oxidative problems with the must.

The procedure involves the reaction of syringaldazine in the presence of laccase to produce a colored quinone (Fig. 1-3). Laccase activity is measured at 20°C in a 0.1 *M* sodium acetate buffer solution at pH 5. The oxidized quinone formed has an absorption maximum at 530 nm.

I. Equipment
   Visible range spectrophotometer
   1.2-μm membrane filter or a syringe fitted with a glass wool plug
II. Reagents (See Appendix II)
   0.1% Syringaldazine prepared in absolute ethanol (sonicate to dissolve)
   Activated PVPP
   0.1 *M* sodium acetate
III. Procedure
   1. Transfer 5 mL of juice to an Erlenmeyer flask.
   2. Add 0.8 g of activated dry PVPP.
   3. After 10 minutes of contact, filter the sample through a 1.2-μm membrane filter or a syringe packed with glass wool.
      (For heavily colored wines, it may be necessary to repeat this treatment.)
   4. In a spectrophotometer cell, mix the following:
      a. 0.6 mL syringaldazine–ethanol solution
      b. 1.0 mL juice sample (diluted if necessary)
      c. 1.4 mL buffer

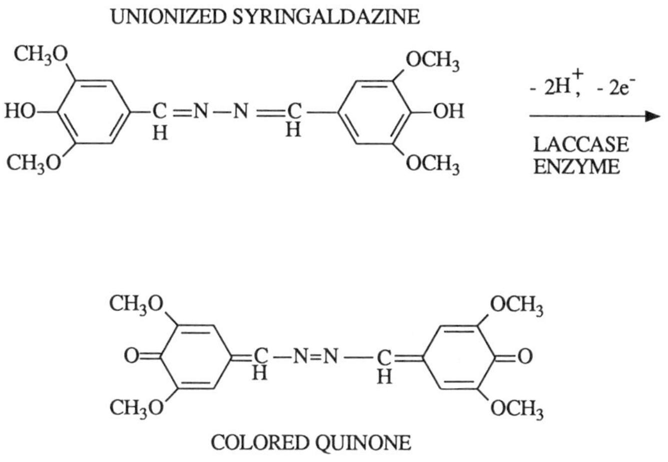

UNIONIZED SYRINGALDAZINE

$\xrightarrow[\text{LACCASE ENZYME}]{-2H^+,\ -2e^-}$

COLORED QUINONE

Fig. 1-3. Reaction of syringaldazine with laccase.

5. Read the change in absorbance at 530 nm over several minutes. The response should be linear for at least 10 minutes.
6. Determine ΔA_m, corresponding to the change in absorbance per minute.
7. Calculation:
   a. One laccase unit (U) is equal to the quantity of enzyme catalyzing the oxidation of one nanomole (1 nm) of syringaldazine per minute.
   b. ΔA_m is the recorded change in absorbance per minute at 530 nm.
   c. Laccase activity in laccase units (U) per mL (M/mL = nmol/mL) is defined as:

$$U\ (\text{nmol/mL}) = \frac{\Delta A_m \times 300}{6.5} \frac{\text{nmol}}{\text{mL}} \qquad (1\text{-}7a)$$

Rearranging the above equation:

$$U = \frac{\Delta A_m \times 3}{65{,}000} \frac{\text{mmol}}{\text{mL}} \qquad (1\text{-}7b)$$

IV. Supplemental Notes
1. Polyphenols interfere with the final reading of laccase activity. Best results are obtained by eliminating or reducing the levels of polyphenols present in juice by prior treatment with activated PVPP.
2. The above analysis has been developed in "kit" form for field assay of laccase.
3. It has been reported that laccase activity can also be measured in a closed system using an oxygen probe (J. Pineau, personal communication).

Procedure 1-3. Total Soluble Solids Determination by Hydrometry

Hydrometry essentially measures the specific gravity of the test solution at defined temperature. Brix hydrometers are calibrated against known wt/wt % sucrose solutions so that they can be read directly in °B rather than in typical density units. Glassware that is clean and dry (or rinsed with the sample) is required to avoid dilution and/or contamination with water or previous samples. The glass cylinder containing the sample should be about twice the diameter of the hydrometer in order to avoid frictional effects between the hydrometer and walls of the cylinder. Entrained air or carbon dioxide should be removed from the sample before hydrometer measurements are made.

I. Equipment
Brix or Balling hydrometer of varying scales
Hydrometer cylinder of approximately 250 mL capacity
Thermometer
Laboratory tissues

III. Procedure

1. Collect a homogeneous representative sample, removing gross particulates. Carefully transfer it to the hydrometer cylinder, taking care to create as few air bubbles as possible.
2. Record the temperature of the sample.
3. Select a hydrometer of appropriate scale range and immerse it into the solution with a gentle spinning motion.
4. When a constant flotation level is reached, take a reading at the bottom of the meniscus, as shown in Fig. 1-1.
5. If necessary, make temperature compensations using Table 1-11. If such a table is not available, an approximate correction factor of ± 0.06 for each degree above or below 20°C may be utilized. Alternatively, the sample may be allowed to reach 20°C prior to the measurement.

IV. Supplemental Notes

1. In unfermented juice (or fermenting wine), the buoying effects of entrapped air or carbon dioxide may act to produce erroneously high readings. Air and/or carbon dioxide can be reduced significantly by vacuum filtration of the sample. Readings should be taken as soon as the hydrometer reaches a constant level. The soluble solids content of a juice should not be taken to include excessive particulate matter. With highly turbid samples, accuracy can be improved by settling and decanting the sample prior to measurement.
2. Accuracy can also be improved by utilizing, whenever possible, the hydrometer with the shortest scale range. Commonly available scale ranges include:

−5 to +5 °B	12 to 18 °B
0 to +6 °B	18 to 24 °B
0 to 30 °B	30 to 70 °B

3. Because of the ever-present problem of faulty data manipulation, the authors recommend, whenever possible, adjusting the sample to the temperature at which the hydrometer is calibrated. When a water bath is used for this purpose, temperature stratification may be encountered. Therefore, samples should be thoroughly mixed prior to analysis.
4. Even though hydrometers are calibrated to 0.1° B, interpolation between markings is not recommended, and readings are made only to the nearest 0.1° B. Practically speaking, then, there is no value in reporting results beyond the first decimal place, even though temperature correction tables may indicate that more accurate results are possible.
5. A good practice is for laboratory personnel to calibrate new hydrometers against known sugar solutions to ensure accuracy.
6. Should °B not be available, a specific gravity hydrometer calibrated "for liquids heavier than water" may be used. For conversion to °B, see Table 1-12.

Table 1-11. Correction factors for hydrometer Brix measurements taken at temperatures other than 68°F.*

	Temperature	Degree Brix																				
		1	2	3	4	5	6	7	8	9	10	11	12	13	14	15	16	17	18	19	20	21
	°F																					
	50	0.17	0.17	0.18	0.19	0.20	0.20	0.21	0.21	0.22	0.22	0.22	0.23	0.23	0.24	0.24	0.24	0.25	0.25	0.25	0.26	0.26
	51	.16	.16	.17	.18	.19	.19	.20	.20	.21	.21	.21	.22	.22	.23	.23	.23	.24	.24	.24	.24	.25
	52	.15	.15	.16	.17	.18	.18	.18	.19	.19	.19	.19	.19	.20	.20	.21	.21	.22	.22	.22	.22	.23
	53	.14	.14	.14	.15	.16	.16	.16	.17	.17	.18	.18	.18	.18	.19	.19	.19	.19	.20	.20	.20	.20
To be subtracted from the indicated degree	54	.12	.12	.13	.14	.15	.15	.15	.16	.16	.16	.16	.17	.17	.17	.17	.17	.18	.18	.18	.18	.18
	55	.09	.09	.09	.10	.11	.11	.11	.12	.12	.12	.12	.13	.13	.13	.14	.14	.14	.14	.14	.14	.15
	56	.07	.07	.07	.08	.09	.09	.09	.09	.10	.10	.10	.10	.11	.11	.11	.11	.11	.11	.11	.11	.11
	57	.06	.06	.06	.07	.07	.07	.07	.07	.08	.08	.08	.08	.08	.09	.09	.09	.09	.10	.10	.10	.10
	58	.04	.04	.04	.05	.05	.05	.05	.05	.06	.06	.06	.06	.06	.07	.07	.07	.07	.07	.07	.07	.07
	59	.02	.02	.02	.02	.02	.02	.02	.02	.03	.03	.03	.03	.03	.03	.03	.03	.03	.03	.03	.03	.03
	60	----	----	----	----	----	----	----	----	----	----	----	----	----	----	----	----	----	----	----	----	----
	61	.02	.02	.02	.03	.03	.03	.03	.03	.03	.03	.03	.03	.03	.03	.03	.03	.03	.03	.03	.03	.03
	62	.06	.06	.07	.07	.08	.08	.08	.08	.08	.08	.09	.09	.09	.09	.09	.09	.09	.09	.09	.09	.09
	63	.09	.10	.10	.11	.11	.11	.11	.11	.11	.11	.12	.12	.12	.12	.12	.12	.12	.12	.12	.12	.12
	64	.12	.12	.13	.13	.14	.14	.14	.14	.15	.15	.15	.15	.16	.16	.17	.17	.17	.17	.17	.17	.17
	65	.14	.15	.15	.16	.17	.17	.17	.17	.18	.18	.18	.19	.19	.19	.20	.20	.20	.20	.20	.20	.20
	66	.17	.17	.18	.19	.20	.20	.21	.21	.21	.22	.22	.23	.23	.23	.24	.24	.24	.24	.24	.24	.24
	67	.20	.21	.21	.22	.23	.23	.23	.24	.24	.25	.25	.25	.26	.26	.27	.27	.27	.27	.27	.27	.27
	68	.22	.23	.24	.25	.26	.26	.27	.28	.28	.29	.29	.30	.30	.31	.31	.31	.31	.31	.31	.31	.31
	69	.25	.26	.27	.28	.29	.29	.30	.31	.31	.32	.32	.33	.33	.34	.34	.34	.34	.34	.34	.34	.34
	70	.28	.29	.30	.31	.32	.32	.33	.34	.34	.35	.35	.36	.36	.37	.37	.37	.37	.38	.38	.38	.38
	71	.31	.32	.33	.34	.35	.35	.36	.37	.37	.38	.38	.39	.39	.40	.40	.40	.40	.40	.40	.40	.41
	72	.34	.35	.36	.37	.38	.39	.40	.41	.42	.42	.42	.43	.43	.43	.44	.44	.44	.44	.44	.44	.44
	73	.37	.38	.40	.41	.42	.43	.44	.44	.45	.46	.46	.46	.46	.47	.47	.47	.47	.47	.47	.47	.47
	74	.40	.42	.43	.45	.46	.47	.48	.48	.49	.50	.50	.50	.51	.51	.51	.51	.51	.52	.52	.52	.52
	75	.44	.46	.47	.49	.50	.51	.52	.52	.53	.54	.54	.54	.55	.55	.56	.56	.56	.57	.57	.58	.58
	76	.47	.48	.50	.52	.53	.54	.55	.55	.55	.57	.57	.58	.58	.59	.60	.60	.61	.61	.61	.62	.62
	77	.50	.52	.54	.56	.57	.58	.59	.60	.60	.61	.61	.62	.62	.63	.63	.63	.64	.64	.64	.65	.65
	78	.53	.54	.56	.58	.60	.61	.62	.62	.63	.64	.64	.65	.65	.66	.66	.66	.67	.67	.68	.68	.68
	79	.57	.58	.60	.62	.64	.65	.66	.66	.67	.68	.68	.69	.69	.70	.70	.70	.71	.71	.71	.72	.72
To be added to the indicated degree	80	.60	.62	.64	.66	.68	.69	.70	.70	.71	.72	.72	.72	.73	.73	.74	.74	.75	.75	.75	.76	.76
	81	.63	.65	.67	.69	.71	.72	.73	.73	.74	.75	.75	.76	.76	.77	.78	.78	.79	.79	.79	.79	.79
	82	.67	.69	.71	.73	.76	.76	.77	.77	.78	.78	.79	.80	.80	.81	.82	.82	.83	.83	.84	.84	.84
	83	.72	.74	.76	.78	.81	.81	.81	.82	.82	.82	.83	.84	.85	.86	.87	.87	.87	.87	.87	.87	.88
	84	.77	.79	.81	.83	.86	.86	.86	.86	.87	.87	.88	.89	.90	.91	.92	.92	.93	.93	.93	.93	.94
	85	.83	.85	.87	.89	.91	.92	.92	.92	.93	.93	.94	.95	.96	.97	.98	.98	.98	.99	.99	.99	.99
	86	.88	.90	.92	.94	.96	.96	.97	.98	.98	.99	1.00	1.01	1.02	1.02	1.03	1.03	1.04	1.04	1.04	1.05	1.05
	87	.94	.96	.98	.99	1.01	1.02	1.03	1.03	1.04	1.05	1.05	1.06	1.07	1.07	1.08	1.08	1.09	1.10	1.10	1.11	1.11
	88	.99	1.01	1.03	1.04	1.06	1.07	1.08	1.09	1.10	1.11	1.11	1.12	1.12	1.13	1.13	1.14	1.15	1.15	1.16	1.17	1.17
	89	1.05	1.07	1.09	1.10	1.11	1.12	1.13	1.14	1.15	1.16	1.16	1.17	1.17	1.18	1.18	1.19	1.21	1.21	1.22	1.23	1.23
	90	1.11	1.13	1.15	1.16	1.17	1.18	1.19	1.20	1.21	1.22	1.22	1.23	1.23	1.24	1.24	1.25	1.27	1.28	1.29	1.30	1.30
	91	1.15	1.17	1.19	1.20	1.21	1.22	1.23	1.24	1.26	1.27	1.27	1.27	1.28	1.28	1.29	1.30	1.32	1.33	1.34	1.35	1.36
	92	1.20	1.22	1.24	1.25	1.26	1.27	1.28	1.30	1.31	1.32	1.32	1.33	1.33	1.34	1.34	1.35	1.37	1.38	1.39	1.40	1.41
	93	1.24	1.26	1.28	1.29	1.31	1.32	1.33	1.34	1.36	1.37	1.37	1.38	1.39	1.39	1.40	1.40	1.42	1.43	1.44	1.44	1.45
	94	1.29	1.31	1.33	1.35	1.36	1.37	1.39	1.40	1.41	1.42	1.42	1.43	1.44	1.44	1.45	1.45	1.47	1.47	1.48	1.49	1.50
	95	1.33	1.35	1.37	1.39	1.41	1.42	1.44	1.45	1.46	1.47	1.47	1.48	1.49	1.49	1.50	1.50	1.51	1.52	1.53	1.54	1.54
	96	1.38	1.40	1.42	1.44	1.46	1.47	1.49	1.50	1.51	1.52	1.52	1.53	1.54	1.54	1.55	1.56	1.56	1.57	1.57	1.58	1.58
	97	1.42	1.44	1.46	1.48	1.51	1.52	1.54	1.55	1.56	1.57	1.57	1.58	1.59	1.59	1.60	1.60	1.62	1.62	1.63	1.63	1.64
	98	1.47	1.49	1.51	1.54	1.56	1.57	1.58	1.59	1.60	1.62	1.63	1.63	1.64	1.64	1.65	1.65	1.66	1.67	1.67	1.68	1.68
	99	1.51	1.53	1.55	1.58	1.61	1.62	1.63	1.64	1.66	1.67	1.68	1.68	1.69	1.70	1.71	1.71	1.72	1.73	1.73	1.73	1.74
	100	1.55	1.57	1.59	1.62	1.66	1.67	1.68	1.69	1.72	1.72	1.73	1.73	1.74	1.76	1.77	1.77	1.78	1.79	1.79	1.79	1.80

N. B. The correction for indications between 21° and 30° Brix and 50° and 70°F. temperature may be made by adding 0.005 to each correction given in the above table, column 21°, for each increase in degree Brix. For instance, 25° Brix 79°F. correction to be added to column 21° and 70°, 0.02 + 0.38 = 0.40.

*(From Bureau of Alcohol, Tobacco, and Firearms, "Wine" (Part 240, Title 27)

Table 1-12. Degrees Brix and corresponding specific gravity readings of sugar solutions.[1]

°Brix or % by weight of sucrose	20°/20°	°Brix or % by weight of sucrose	20°/20°
0.0	1.00000	8.0	1.03176
0.2	1.00078	8.2	1.03258
0.4	1.00155	8.4	1.03340
0.6	1.00233	8.6	1.03422
0.8	1.00311	8.8	1.03504
1.0	1.00389	9.0	1.03586
1.2	1.00468	9.2	1.03668
1.4	1.00545	9.4	1.03750
1.6	1.00623	9.6	1.03833
1.8	1.00701	9.8	1.03915
2.0	1.00779	10.0	1.03998
2.2	1.00858	10.2	1.04081
2.4	1.00936	10.4	1.04164
2.6	1.01015	10.6	1.04247
2.8	1.01093	10.8	1.04330
3.0	1.01172	11.0	1.04413
3.2	1.01251	11.2	1.04497
3.4	1.01330	11.4	1.04580
3.6	1.01409	11.6	1.04664
3.8	1.01488	11.8	1.04747
4.0	1.01567	12.0	1.04831
4.2	1.01647	12.2	1.04915
4.4	1.01726	12.4	1.04999
4.6	1.01806	12.6	1.05084
4.8	1.01886	12.8	1.05168
5.0	1.01986	13.0	1.05252
5.2	1.02045	13.2	1.05337
5.4	1.02125	13.4	1.05422
5.6	1.02206	13.6	1.05506
5.8	1.02286	13.8	1.05591
6.0	1.02366	14.0	1.05677
6.2	1.02447	14.2	1.05762
6.4	1.02527	14.4	1.05847
6.6	1.02608	14.6	1.05933
6.8	1.02689	14.8	1.06018
7.0	1.02770	15.0	1.06104
7.2	1.02851	15.2	1.06190
7.4	1.02932	15.4	1.06276
7.6	1.03013	15.6	1.06362
7.8	1.03095	15.8	1.06448

Table 1-12. (*Continued*)

°Brix or % by weight of sucrose	20°/20°	°Brix or % by weight of sucrose	20°/20°
16.0	1.06534	21.0	1.08733
16.2	1.06621	21.2	1.08823
16.4	1.06707	21.4	1.08913
16.6	1.06794	21.6	1.09003
16.8	1.06881	21.8	1.09093
17.0	1.06968	22.0	1.09183
17.2	1.07055	22.2	1.09273
17.4	1.07142	22.4	1.09364
17.6	1.07229	22.6	1.09454
17.8	1.07317	22.8	1.09545
18.0	1.07404	23.0	1.09636
18.2	1.07492	23.2	1.09727
18.4	1.07580	23.4	1.09818
18.6	1.07668	23.6	1.09909
18.8	1.07756	23.8	1.10000
19.0	1.07844	24.0	1.10092
19.2	1.07932	24.2	1.10183
19.4	1.08021	24.4	1.10275
19.6	1.08110	24.6	1.10367
19.8	1.08198	24.8	1.10459
20.0	1.08287	25.0	1.10551
20.2	1.08376	25.2	1.10643
20.4	1.08465	25.4	1.10736
20.6	1.08554	25.6	1.10828
20.8	1.08644	25.8	1.10921

[1] *Official Methods of Analysis of the Association of Official Analytical Chemists*, 12 Edition (1975).

Procedure 1-4. Total Soluble Solids Determination by Refractometry

As the concentration of soluble solids increases, so too does the refractive index of the sample. The principle here is that the more molecules there are in a sample, the more the light passing through the sample is bent (refracted). Most refractometers have a °B scale (calibrated against sucrose solutions) along with the refractive index scale; so one may read °B directly in juice samples. Since all soluble solids (especially ethanol) in a juice sample affect the measured refractive index, fermenting samples give °B measurements that are erroneously high.

I. Equipment
   Hand-held refractometer, Abbe Refractometer, or equivalent digital refractometer (range: 1.300–1.700)
   Circulating water source if required

Light source
Wash bottle with distilled water
Cotton swabs
Laboratory tissue and lens paper

II. Procedure

1. On instruments utilizing water-cooled prisms, connect the laboratory water line to the intake nipple on the instrument and return water to the drain via the exit port.
2. Circulate water to constant temperature through the illuminating prism.
3. Follow the manufacturer's instructions for setup and operation of the instrument used.
4. Field-type refractometers must have their reading temperatures corrected to 68°F. In the absence of temperature-corrected refractometers, refer to Table 1-13.
5. All units must be standardized with distilled water such that the refractive index reads 1.330 with a corresponding sugar concentration of 0 °B.

Table 1-13. Correction factors for refractive index measurements taken at temperatures other than 20°C (68°F).

International Temperature Correction Table (1936) for the Normal Model of Refractometer Above and Below 20° C.

Temp. °C	Per cent Sucrose														
	0	5	10	15	20	25	30	35	40	45	50	55	60	65	70
	Subtract from the per cent Sucrose														
10	0.50	0.54	0.58	0.61	0.64	0.66	0.68	0.70	0.72	0.73	0.74	0.75	0.76	0.78	0.79
11	0.46	0.49	0.53	0.55	0.58	0.60	0.62	0.64	0.65	0.66	0.67	0.68	0.69	0.70	0.71
12	0.42	0.45	0.48	0.50	0.52	0.54	0.56	0.57	0.58	0.59	0.60	0.61	0.61	0.63	0.63
13	0.37	0.40	0.42	0.44	0.46	0.48	0.49	0.50	0.51	0.52	0.53	0.54	0.54	0.55	0.55
14	0.33	0.35	0.37	0.39	0.40	0.41	0.42	0.43	0.44	0.45	0.45	0.46	0.46	0.47	0.48
15	0.27	0.29	0.31	0.33	0.34	0.34	0.35	0.36	0.37	0.37	0.38	0.39	0.39	0.40	0.40
16	0.22	0.24	0.25	0.26	0.27	0.28	0.28	0.29	0.30	0.30	0.30	0.31	0.31	0.32	0.32
17	0.17	0.18	0.19	0.20	0.21	0.21	0.21	0.22	0.22	0.23	0.23	0.23	0.23	0.24	0.24
18	0.12	0.13	0.13	0.14	0.14	0.14	0.14	0.15	0.15	0.15	0.15	0.16	0.16	0.16	0.16
19	0.06	0.06	0.06	0.07	0.07	0.07	0.07	0.08	0.08	0.08	0.08	0.08	0.08	0.08	0.08
	Add to the per cent Sucrose														
21	0.06	0.07	0.07	0.07	0.07	0.08	0.08	0.08	0.08	0.08	0.08	0.08	0.08	0.08	0.08
22	0.13	0.13	0.14	0.14	0.15	0.15	0.15	0.15	0.15	0.16	0.16	0.16	0.16	0.16	0.16
23	0.19	0.20	0.21	0.22	0.22	0.23	0.23	0.23	0.23	0.24	0.24	0.24	0.24	0.24	0.24
24	0.26	0.27	0.28	0.29	0.30	0.30	0.31	0.31	0.31	0.31	0.31	0.32	0.32	0.32	0.32
25	0.33	0.35	0.36	0.37	0.38	0.38	0.39	0.40	0.40	0.40	0.40	0.40	0.40	0.40	0.40
26	0.40	0.42	0.43	0.44	0.45	0.46	0.47	0.48	0.48	0.48	0.48	0.48	0.48	0.48	0.48
27	0.48	0.50	0.52	0.53	0.54	0.55	0.55	0.56	0.56	0.56	0.56	0.56	0.56	0.56	0.56
28	0.56	0.57	0.60	0.61	0.62	0.63	0.63	0.64	0.64	0.64	0.64	0.64	0.64	0.64	0.64
29	0.64	0.66	0.68	0.69	0.71	0.72	0.72	0.73	0.73	0.73	0.73	0.73	0.73	0.73	0.73
30	0.72	0.74	0.77	0.78	0.79	0.80	0.80	0.81	0.81	0.81	0.81	0.81	0.81	0.81	0.81

From Bausch and Lomb Optical Company.

6. With the blotter and lens paper provided, dry the prisms.
7. Using a cotton swab, apply several drops of juice or appropriately diluted concentrate to the lens, and repeat the operations. *Caution:* The user should be aware that particulate matter in the sample may scratch the prisms.
8. Should the refractometer not be calibrated in °B, refractive index readings can be converted to equivalent % Sucrose values (Table 1-14).

III. Supplemental Notes

1. The refractive index is critically dependent upon the temperature of the solution measured. Unless corrected, small differences from the reference temperature should be expected to result in significant error. For convenience, laboratory refractometers normally have a thermometer incorporated into the body of the instrument. Field units without temperature compensation capacity provide only a rough approximation of the soluble solids content.

Table 1-14. International scale of refractive indexes of sucrose solutions at 20°C (68°F).

International Scale (1936) of Refractive Indices of Sucrose Solutions at 20° C.

Index	Per cent	Index	Per cent	Index	Per cent	Index	Per cent
1.3330	0	1.3723	25	1.4200	50	1.4774	75
1.3344	1	1.3740	26	1.4221	51	1.4799	76
1.3359	2	1.3758	27	1.4242	52	1.4825	77
1.3373	3	1.3775	28	1.4264	53	1.4850	78
1.3388	4	1.3793	29	1.4285	54	1.4876	79
1.3403	5	1.3811	30	1.4307	55	1.4901	80
1.3418	6	1.3829	31	1.4329	56	1.4927	81
1.3433	7	1.3847	32	1.4351	57	1.4954	82
1.3448	8	1.3865	33	1.4373	58	1.4980	83
1.3463	9	1.3883	34	1.4396	59	1.5007	84
1.3478	10	1.3902	35	1.4418	60	1.5033	85
1.3494	11	1.3920	36	1.4441	61		
1.3509	12	1.3939	37	1.4464	62		
1.3525	13	1.3958	38	1.4486	63		
1.3541	14	1.3978	39	1.4509	64		
1.3557	15	1.3997	40	1.4532	65		
1.3573	16	1.4016	41	1.4555	66		
1.3589	17	1.4036	42	1.4579	67		
1.3605	18	1.4056	43	1.4603	68		
1.3622	19	1.4076	44	1.4627	69		
1.3638	20	1.4096	45	1.4651	70		
1.3655	21	1.4117	46	1.4676	71		
1.3672	22	1.4137	47	1.4700	72		
1.3689	23	1.4158	48	1.4725	73		
1.3706	24	1.4179	49	1.4749	74		

From Bausch and Lomb Optical Company (1952).

2. Samples for analysis must be representative, homogeneous, and free of particulate matter that could damage prisms or impair accuracy.
3. In addition to temperature, alcohol is a major interference in accurate measurements of the refractive index. Therefore, the refractometer should not be used in soluble solids determinations of fermenting must or wine.

Procedure 1-5. HPLC Analysis of Glycerol, Acetic Acid, and Ethanol in Grape Juice

High Performance Liquid Chromatography (HPLC) can be used to quantify certain mold, yeast, and bacterial metabolites present in a juice sample. This procedure was developed to measure the amounts of glycerol, ethanol, and acetic acid in fruit. Filtered juice samples are injected into a special HPLC column that separates the three components from each other and from any other matrix compounds. Quantitation is accomplished by comparing component peak areas to those from standard solutions chromatographed in the same way.

I. Equipment
   HPLC unit with column oven ($>40°C$ capability) and refractive index detector
   Bio-Rad fruit quality column (or equivalent)
   Bio-Rad Cation H guard column (or equivalent)
   0.45-μm membrane filter
   50-μL syringe
II. Reagents (See Appendix II)
   Glycerol, acetic acid, ethanol standard solution
   Sulfuric acid ($\sim 0.002\ N$) mobile phase
III. Procedure
   1. Install the column and an appropriate guard column in the HPLC mobile phase flow stream.
   2. Begin pumping the mobile phase. Adjust the flow rate to 1.2 mL/min.
   3. Set the column oven temperature at 55°C, and allow the system to stabilize.
   4. Turn on the refractive index detector, and allow it to warm up and stabilize.
   5. Fruit samples are crushed and the juice passed through cheesecloth for primary filtration. An analysis sample is prepared by filtration through a 0.45-μm disposable membrane filter.
   6. Inject a 10-μL sample of the standard mixture. The peaks should be eluted in the order: sugars, glycerol, acetic acid, and ethanol. There may be a water dip occurring between the sugar and glycerol peaks. Record the peak heights or areas.

7. Inject 10 μL of the sample juice. Identify peaks of interest by comparison of retention times to those of the standards. Again record the peak heights or areas. (See Fig. 1-4 for a sample chromatogram.)
8. Concentrations of the components of interest may be calculated directly from the electronic integrator readouts or by use of the following equation:

$$\text{Conc (Unk)} = \frac{\text{Peak area (Unk)} \times \text{Conc (Std)}}{\text{Peak area (Std)}} \tag{1-8}$$

IV. Supplemental Notes

1. Temperature variation of the analytical column does not appear to be a critical factor in this analysis. Ambient and higher temperatures can be used with only minimal changes in the retention times. For consistency of operation, it is suggested that the column be operated at a constant temperature of 40°C or higher.
2. Variable sugar levels affect the glycerol response. As sugar levels increase, the fructose peak expands into the glycerol peak, resulting in a lower peak response. Thus, one would prefer to prepare calibration standards with sugar levels similar to those of the samples being analyzed.
3. The purpose of the guard column is to trap any compounds in the filtered juice sample that might damage the analytical column. As the guard column becomes used, resolution of the sample peaks (and pos-

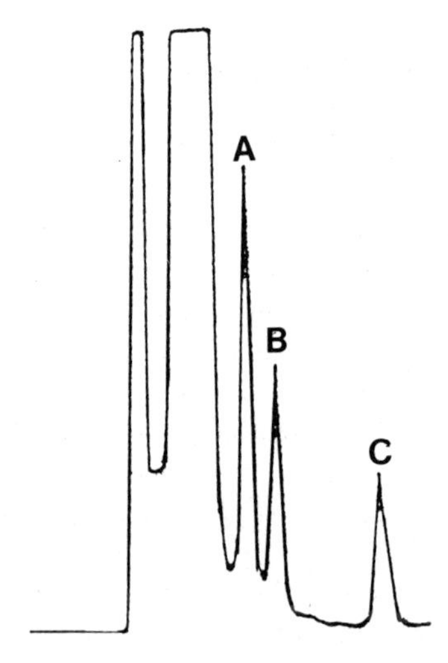

Fig. 1-4. Chromatogram of sample prepared from infected grapes on HPLC fruit quality column. Peaks for glycerol (*A*), acetic acid (*B*), and ethanol (*C*) are clearly present.

sibly the peak responses) decreases. Recalibration with standards is necessary, as is eventual replacement of the guard column.

Procedure 1-6. Monoterpene Analysis

A technique was developed by Dimitriadis and Williams (1984) to estimate the quantity of free volatile terpenes (FVT) and potential volatile monoterpenes (PVT) in grape juice. This in turn can provide an index of the flavor and flavor potential of the fruit to be established. Steam distillation of pH-neutral juice yields free aroma compounds, whereas steam distillation of low-pH juice yields monoterpenes derived from the polyol and glycosidically bound forms. Linalool is used as the reference monoterpene for constructing a calibration curve, and results are obtained in linalool equivalents. Essentially quantitative recovery of free volatile turpines is obtained with this method. However, recoveries of potential volatile monoturpenes are reported to be in the 55 to 88% range.

I. Equipment
   Cash still (or equivalent) for steam distillation of sample
   Waring blender (or equivalent)
   Water bath (60 ± 1°C)
   Visible spectrometer (608 nm)
   Pyrex test tubes (20 × 150 mm) with silicon septum lined screw caps
II. Reagents (See Appendix II)
   Linalool stock solution (1 mg/mL)
   Linalool standard (0.1 mg/mL)
   Vanillin: H_2SO_4 reagent (2% wt/wt)
   Sodium hydroxide (20% wt/vol)
   Phosphoric acid (50% vol/vol)
III. Procedure
   (a) Distillation of sample
      1. Homogenize ~500 g destemmed berries in a Waring blender for 10 to 20 sec. Filter the homogenate through cheesecloth.
      2. Immediately prior to distillation, adjust the pH of the homogenate to 6.6 to 6.8 with the 20% (wt/vol) sodium hydroxide solution.
      3. Transfer 100 mL of neutral juice to the inner chamber of a Cash still for steam distillation of the monoterpenes.
      4. Bring water in the still to boiling; collect the first 25 mL of condensate.
      5. Immediately add 10 mL of 50% (vol/vol) phosphoric acid reagent to the Cash still inner chamber to drop the sample pH (2.0–2.2).
      6. Continue steam distillation, and collect the next 40 mL of condensate.
      7. Mix each collected condensate sample to ensure homogeneity.

(b) Colorimetric development

1. Into a series of test tubes pipette 0.2, 0.5, 1.0, 1.5, and 2.0 mL of linalool standard. Bring the volume of each standard to 10 mL with distilled water. The amount of linalool present in each tube is listed in Table 1-15.
2. Into additional test tubes pipette 10-mL aliquots of each FVT and PVT distillate. Prepare a distilled water blank at the same time.
3. To each precooled test tube add 5 mL of the vanillin: sulfuric acid (2% wt/vol) reagent while agitating the tube in an ice bath.
4. Place test tubes for standards and FVT and PVT samples in a 60 ± 1°C water bath for 20 minutes.
5. Cool the tubes at 25°C for 5 minutes and read the absorbance on a spectrophotometer at 608 nm. *Note:* Absorbance readings should be taken within 20 minutes following the cooling operation.
6. Prepare a calibration curve of μg linalool vs. absorbance for the standards.
7. Read the amount of linalool equivalents contained in each distillate aliquot directly from the calibration curve.
8. Calculate amounts of linalool equivalents in original samples using the following equation:

$$\text{FVT or PVT } (\mu\text{g/mL}) = \mu\text{g linalool} \times \frac{V(\text{distillate})}{V(\text{aliquot})} \times \frac{1}{100 \text{ mL}} \qquad (1\text{-}9)$$

where:

μg linalool = amount read from calibration curve
V(distillate) = 25 or 40 mL for FVT and PVT, respectively
V(aliquot) = 10-mL aliquot taken from distillate
100 mL = original juice sample on which test is run

Table 1-15. Preparation of linalool calibration standards from 0.1 mg/mL standard solution.

Volume linalool standard (mL)	Final amount of linalool (μg) in calibration standard
0.2	20
0.5	50
1.0	100
1.5	150
2.0	200

IV. Supplemental Notes

1. Sulfur dioxide at normal use levels produces low values for PVT. To avoid this interference, add an equivalent amount of hydrogen peroxide to oxidize SO_2 *prior* to distillation.
2. Ethanol present in wine samples produces an enhanced color reaction. If wine samples are to be run by this procedure, standards should be prepared so that their ethanol content will match that of the distillates.
3. The potential volatile monoterpenes analyzed here include only volatile products derived from polyols and glycosides. Thus, values obtained are subject to the degree of dehydration, hydrolysis, etc., under the specific experimental conditions used.
4. It has been suggested that heat generated from the addition of sulfuric acid may cause terpene volatilization and, therefore inconsistent results. The use of an ice bath may resolve this potential difficulty.

Chapter 2

Alcoholometry

The alcohol content of a wine, influences its stability as well as its sensory properties. Wines are taxed, in large part, according to their alcohol levels. Careful monitoring of alcohol is important in stylistic wine production, as well as in carrying out accurate fortifications and in formulation of blends for bottling.

YEAST METABOLISM

Fermentation

Microorganisms have varying requirements for oxygen. At the two extremes are the microorganisms that require oxygen, the aerobes, and those that cannot survive in its presence, the anaerobes. Between these extremes are facultative microorganisms such as the yeasts, which are metabolically equipped to handle conditions where oxygen is either plentiful or limiting. Under oxidative (aerobic) conditions where concentrations of utilizable sugars are less than 3%, yeasts utilize the pathway outlined by the solid line in Fig. 2-1. Glucose is converted to pyruvate via the Embden-Meyerhoff-Parnas (EMP) pathway (glycolysis) and subsequently to carbon dioxide and water via the tricarboxylic acid (Krebs cycle) and cytochrome oxidase pathways. Alternately, where oxygen is limiting, and a fermentable carbohydrate source is present, fermentative metabolism ensues, with pyruvate being decarboxylated to acetaldehyde and subsequently reduced to ethanol. This fermentative pathway is outlined by dashed lines in Fig. 2-1.

Anaerobiosis is not the only condition that favors fermentatitive metabolism. Even under conditions of relatively high oxygen, the presence of glucose at levels of more than 5% inhibits the activity of respiratory enzyme systems. This example of catabolite repression is generally referred to as the "Crabtree" or "counter Pasteur effect." Under such conditions, fermentative rather than oxidative metabolism is observed. As glucose levels drop, TCA cycle enzymes are induced, and metabolism may shift from anaerobic (fermentative) to aerobic.

The energy differences between metabolic pathways are considerable. Anerobic metabolism yields only 56 kcal/mol of glucose, whereas aerobic metabolism (respiration) of this sugar produces 688 kcal/mol.

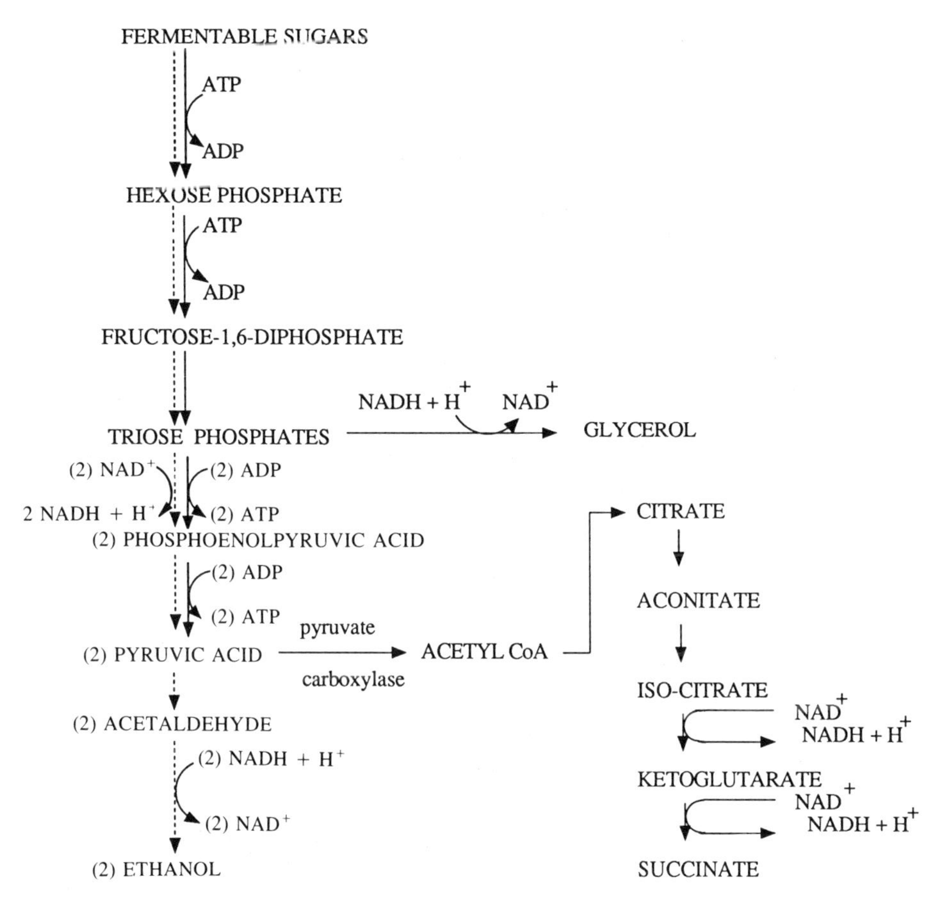

Fig. 2-1. Aerobic and fermentative metabolic pathways. Dashed lines represent fermentative metabolism; solid lines represent aerobic metabolism.

The importance of carbohydrate metabolism to the yeast is twofold. On the one hand, the cell is provided with a source of utilizable energy in the form of substrate-level phosphorylations involving adenosine triphosphate (ATP). Secondly, a variety of intermediate carbon compounds are formed in the process that are channeled into related biosynthetic pathways of the cell.

On an equivalent basis, glucose has a much greater energy potential than the energy generated by conversion of ATP to ADP. However, the importance of ATP lies not so much in its endogeneous energy content as in its direct involvement in energy-transfer reactions. The latter attribute resides in very reactive phosphate anhydride linkages.

The free energy of hydrolysis for ATP is reported as 7 kcal/mol, whereas for ADP and AMP the values are 6.4 and 3.0 kcal/mol, respectively (Machlis and Torrey 1956). Therefore, ATP serves as an "energy carrier" whereby the energy produced from biological oxidation is harnessed (conserved) in ATP synthesis rather than liberated immediately in the form of heat. As such, this stored energy is available to drive cellular reactions requiring the input of energy.

For brevity, only the salient points of the EMP pathway will be discussed further.

1. In addition to glucose, the monosaccharide fructose is fully fermentable. In this case, direct phosphorylation yields fructose-6-phosphate and subsequently fructose-1,6-diphosphate.

2. Formation of the trioses, dihydroxyacetone-phosphate and glyceraldehyde-3-phosphate, occupies a special position of importance to the winemaker—namely, in the formation of glycerol, which is discussed in greater detail later in this chapter.

3. The initial phosphorylation of glucose to form glucose-6-phosphate and fructose to yield fructose-1,6-diphosphate utilizes two molecules of ATP. Direct energy in the form of ATP is recovered at two locations in the EMP pathway. First, the oxidation of 1,3-diphosphoglyceric acid to 3-phosphoglyceric acid is accompanied by phosphorylation of ADP. A second molecule of ATP is recovered in the formation of pyruvate from phosphoenolpyruvate.

One molecule of glucose produces two molecules of the triose phosphate, so a total of four ATP's are generated per molecule of initial sugar. Subtracting the two ATP's utilized in initial phosphorylation reactions, the pathway yields a net gain of two ATP's. To emphasize the dramatic differences in energy yield between aerobic and anaerobic glycolysis, the major reaction sequences are summarized in Table 2-1.

The nucleotide NAD^+ is reduced to $NADH + H^+$ in formation of 1,3-diphosphoglyceric acid. Because intracellular concentrations of NAD^+ are low, the cell must regenerate this compound if metabolism is to continue. Under

Table 2-1. Adenosine triphosphate formation in aerobic and anaerobic metabolism.

Reaction	ATP Produced	
	Anaerobic phase	Aerobic phase
Glucose → Glucose-6-phosphate	−1	−1
Fructose-6-phosphate → Fructose-1,6-diphosphate	−1	−1
(2) 1,3-Diphosphoglyceric acid → 3-phospho-glyceric acid	+2	+2
(2) Phosphoenolpyruvate → 2 pyruvic acid	+2	+2
(2) $NADH + H^+ \rightarrow NAD^+$ (electron transport system)	0	+6
	2 ATP	8 ATP

aerobic conditions, reduced NAD is reoxidized in the cytochrome oxidase system of the cell, yielding an additional six ATP's. In the fermentative metabolic mode, however, NADH + H^+ is reoxidized in the reduction of acetaldehyde to ethanol. Fifty-six (56) kcal of energy (7 kcal/ATP) are potentially available in glycolysis, when coupled to oxidative pathways. However, during fermentation only 14 kcal (two ATP's) are produced. This amounts to a net efficiency with respect to ATP production of only 25% in anaerobic glycolysis.

Fermentations may be described as a series of redox reactions in which organic compounds (in this case sugars) are oxidized initially, and, at a later point, their products serve as terminal electron acceptors (Stanier, et al. 1976) (see also Chapter 8). Thus a fermentable compound must, ideally, be at some intermediary stage of oxidation; and carbohydrates fit this requirement. Furthermore, the end products and by-products of fermentation (ethanol and organic acids) are, themselves, at an intermediate stage of oxidation. As such, they may serve as reservoirs that may be available to the organisms growing under oxidative conditions (i.e., acetic acid bacteria).

In fermentation, acetaldehyde formed by decarboxylation of pyruvate may also combine with sulfite to form an addition product (Neuberg's second form of fermentation), as shown in Equation 2-1:

$$\text{GLUCOSE} + SO_3^{-2} \longrightarrow \text{GLYCEROL} + \text{ACETALDEHYDE} \bullet SO_3^{-} + CO_2 \quad (2\text{-}1)$$

Thus the addition of large quantities of bisulfite acts as a block, preventing acetaldehyde from operating as an hydrogen acceptor for NAD. Under these conditions, the triose dihydroxyacetone phosphate replaces acetaldehyde as a hydrogen acceptor, resulting in formation of glycerophosphate in amounts equivalent to the quantity of acetaldehyde bound. Hydrolysis of accumulated glycerophosphate by phosphate enzymes then leads to the formation of glycerol. It has been reported that NAD regeneration occurs in this manner in the early stages of fermentation before the concentration of acetaldehyde reaches the levels needed for alcohol dehydrogenase activity (Holzer et al. 1963). However, utilization of this pathway does not provide the cell with energy (Sols et al. 1971).

Additional alcohols of importance in winemaking include glycerol, methanol, and several 3- to 5-carbon alcohols collectively known as fusel oils. These may, on occasion, be important in sensory and regulatory considerations.

By-products of Fermentation

Glycerol

Glycerol is a normal by-product of alcoholic fermentation resulting from reduction of dihydroxyacetone phosphate (Fig. 2-2). Oura (1977) reports the major source of reduced pyridine dinucleotide in the above pathway to be from suc-

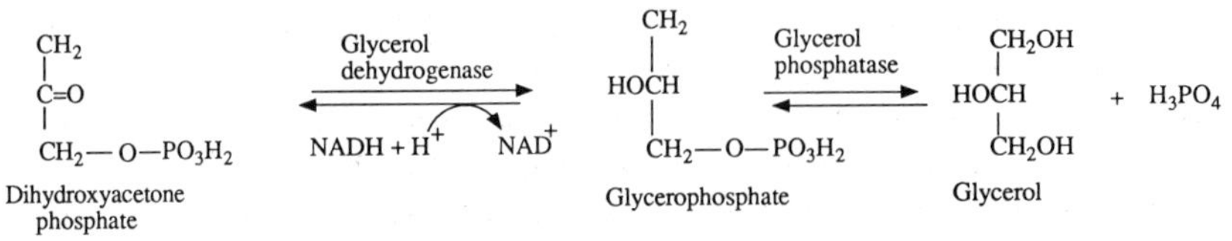

$$\underset{\text{Dihydroxyacetone phosphate}}{CH_2 - C{=}O - CH_2{-}O{-}PO_3H_2} \xrightleftharpoons[\text{NADH + H}^+ \rightarrow \text{NAD}^+]{\text{Glycerol dehydrogenase}} \underset{\text{Glycerophosphate}}{CH_2 - HOCH - CH_2{-}O{-}PO_3H_2} \xrightleftharpoons{\text{Glycerol phosphatase}} \underset{\text{Glycerol}}{CH_2OH - HOCH - CH_2OH} + H_3PO_4$$

Fig. 2-2. Glycerol production from reduction of dihydroxyacetone phosphate during alcoholic fermentation.

cinate production via the tricarboxylic acid cycle. In this case, ATP formation from fermentation apparently triggers pyruvate carboxylase, leading to formation of acetyl-CoA (see Fig. 2-1). Radler and Schutz (1982) suggest that glycerol formation results from competition between glycerol-3-phosphate dehydrogenase and alcohol dehydrogenase for available $NADH_2$. The activity of alcohol dehydrogenase appears to be fairly constant, but glycerol-3-phosphate dehydrogenase levels vary significantly between those strains producing larger amounts of glycerol and those that do not produce the end product.

Amerine and Ough (1980) reported glycerol concentrations in U.S. wines ranging from 1.9 to 14.7 g/L (average of 7.2 g/L). Because of its relatively high specific gravity (1.26), glycerol may contribute to the overall sensory perception of body in certain wine types. However, it is questionable if this contribution is significant at the alcohol concentrations of 10 to 12% normally found in table wine (Amerine and Roessler 1976).

The glycerol content of a wine produced from sound fruit may be affected by harvest and production parameters. Several of the more important considerations are:

1. Fermentation temperatures play an important role in glycerol formation. Workers have reported increased glycerol production with increased fermentation temperatures in the range of 15 to 25°C (Rankine 1955; Rankine and Bridson 1971).

2. Yeast strains are known to vary in their ability to produce glycerol. Glycerol formation was most rapid in the initial stages of fermentation. Radler and Schutz (1982) reported glycerol levels produced in several strains of *Sacch. cerevisiae* to range from 4.2 to 10.4 g/L.

3. Grape condition also affects the resultant glycerol content. Sweet wines, produced from botrytized grapes, are usually high in glycerol. Amerine and Roessler (1976) reported that wines produced from moldy grapes contained more than 1.5% (wt/wt) glycerol when compared with 0.5% (wt/wt) glycerol from sound fruit. The glycerol content of these wines varied from 10.2 to 12.5% of the total alcohol content as compared with 6.5 to 9.6% from sound grapes.

4. Sulfur dioxide affects glycerol production by binding acetaldehyde, thereby causing dihydroxyacetone phosphate to act as a substrate for the regeneration of oxidized nucleotide. However, the amount of sulfur dioxide needed to stim-

ulate significant increases in glycerol is far in excess of that used in routine winemaking practices.

Acetic acid bacteria, particularly *Gluconobacter oxydans*, are known to utilize glycerol oxidatively in formation of dihydroxyacetone. In the case of *G. oxydans*, this reaction is quantitative, and it is currently employed in commercial production of dihydroxyacetone. Although it has not been determined if this reaction can occur in wine, it is well established that formation occurs in infected grapes and must (Sponholz and Dittrich 1985). In their study, Sponholz and Dittrich reported levels of dihydroxyacetone reaching 260 mg / L in must infected with *G. oxydans*. Upon fermentation, this level was reduced to 133 mg / L in wine. In addition to potentially altering the sensory properties of wine, the accumulation of dihydroxyacetone provides another important substrate for binding free sulfur dioxide. For more details regarding acetic acid bacteria, the reader is referred to Chapters 5 and 12.

Methanol

Methyl alcohol is not a normal product of alcoholic fermentation (Amerine 1954). Rather, it results from the hydrolysis of pectin present in grapes by pectin methylesterase enzymes (PME). These enzymes hydrolyze the methyl ester group of galacturonic acid. Pectin is a polymer of galacturonic acid coupled via an alpha-1,4-glycosidic linkage (Fig. 2-3); approximately two-thirds of the carboxylic acid groups are esterified with methanol (Reed 1966).

Pectinases are used to enhance juice yield, color extraction and clarification. They may significantly increase the methanol content in the resultant wine. However, in the case of grape wines, increases generally are still below the federal government's limits of less than 1000 mg / L (0.1%).

Although the average methyl alcohol content reported is 100 mg / L (Amerine and Joslyn 1970), it is usually lower in white wines, and it may reach 150 mg / L in some reds. By comparison, Lee et al. (1979) report methanol formation in Niagara and Concord varieties to be as high as 200 mg / L to 250 mg / L, respectively.

Gnekow and Ough (1976) found that in white wine fermentations methanol content reached an early maximum level ranging from 18.9 mg / L to 40.5 mg / L

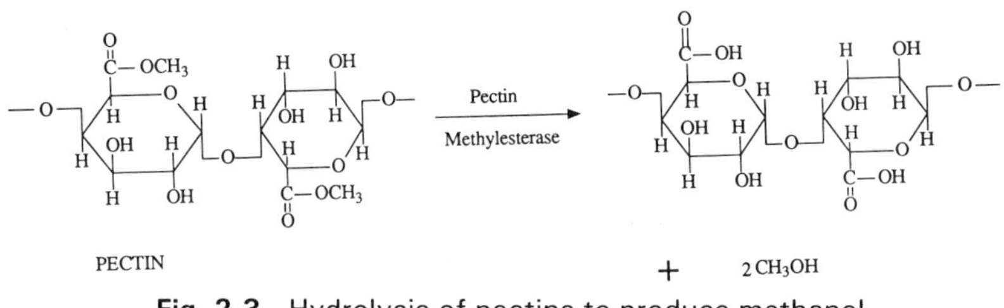

PECTIN → (Pectin Methylesterase) + $2\,CH_3OH$

Fig. 2-3. Hydrolysis of pectins to produce methanol.

for Emerald Riesling. In the case of red varieties, which are fermented on the skins, methanol formation continues throughout fermentation, yielding much higher levels (from 61.3 mg/L in Cabernet Sauvignon to 155.6 mg/L in Zinfandel). Wines made from fruit other than grapes may be especially high in methanol content, owing to their higher relative pectin levels. Distillates produced from plum and apricots have been reported to have methanol contents of 2000 to 5000 mg/L (Woidich and Pfannhauser 1974). Methanol analysis is not a routine wine industry practice, but it may be readily accomplished by gas chromatography. The sensory importance of methyl alcohol has not been completely evaluated, but fruity esters are known to form.

Higher Alcohols (Fusel Oils)

Alcohols of more than two carbons, commonly called fusel oils, are produced by yeasts during fermentation. The most frequently encountered fusel oils include isoamyl and "active amyl," isobutyl, and *n*-propyl alcohols. This group of alcohols may present problems in distillation, where they concentrate in the lower boiling "tails" fractions of distilled spirits. In cases of stills without fractionization capabilities or of malfunction in column still operation, significant concentrations of these alcohols may appear in the product. Depending upon the production objectives (e.g., brandy vs. neutral spirits), significant amounts may represent defects in the sensory interpretation of the distillate.

Quantitatively, fusel oils represent an important group of alcohols that may affect flavor. They may be present in wines at varying concentrations. Variables affecting the final concentration of fusel oils in wine include yeast strain, tem-

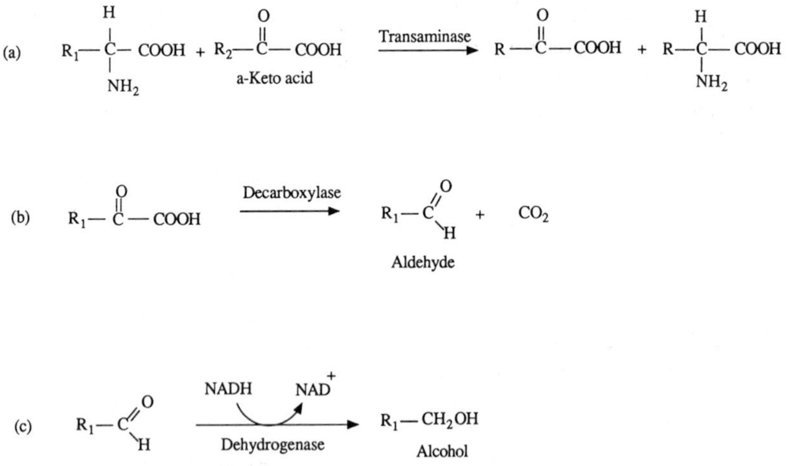

(a) $$R_1-\underset{NH_2}{\overset{H}{C}}-COOH + R_2-\overset{O}{\overset{\|}{C}}-COOH \xrightarrow{\text{Transaminase}} R-\overset{O}{\overset{\|}{C}}-COOH + R-\underset{NH_2}{\overset{H}{C}}-COOH$$

a-Keto acid

(b) $$R_1-\overset{O}{\overset{\|}{C}}-COOH \xrightarrow{\text{Decarboxylase}} R_1-C{\overset{O}{\underset{H}{\lessgtr}}} + CO_2$$

Aldehyde

(c) $$R_1-C{\overset{O}{\underset{H}{\lessgtr}}} \xrightarrow[\text{Dehydrogenase}]{NADH \rightarrow NAD^+} R_1-CH_2OH$$

Alcohol

Fig. 2-4. Generalized pathway for formation of higher alcohols ("fusel oils") from amino acid and alpha-keto acid precursors.

perature of fermentation, solids load during fermentation, and extent and timing of aeration. Klingshirn et al. (1987) report that total higher alcohol production increases with particle size. These workers speculate that oxygen entrapped within the particle matrix may stimulate production of fusel oils by yeast.

In table wines, the total fusel oil concentration is reported to range from 140 to 420 mg/L (Amerine and Ough 1980). In this range, fusel oils are believed to contribute to overall complexity, whereas at higher levels they may become objectionable. In dessert wines, total fusel oil concentration may range from near 100 mg/L to over 1000 mg/L. Values on the higher side reflect additions of brandy and/or lower-quality high proof.

Fusel oil formation was originally described by Ehrlich in 1907 as arising from transamination of the corresponding amino acid. As seen in Figs. 2-4 and 2-5, the Ehrlich mechanism involves initial transamination between an amino

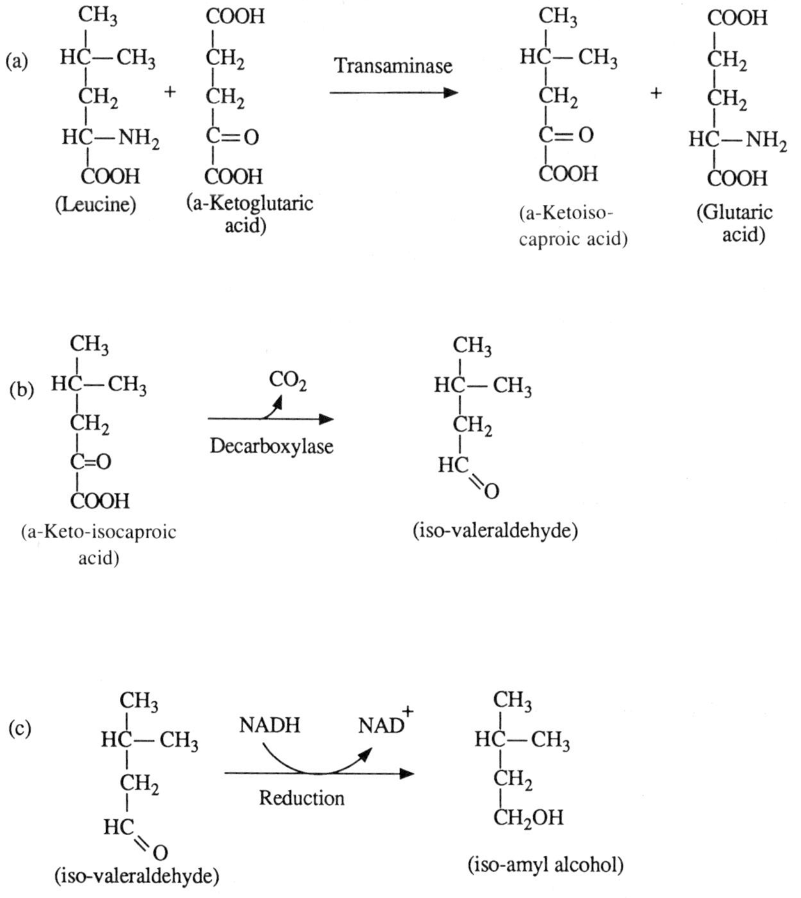

(a) $CH_3{-}CH(CH_3){-}CH_2{-}CH(NH_2){-}COOH$ (Leucine) + $HOOC{-}CH_2{-}CH_2{-}C(=O){-}COOH$ (a-Ketoglutaric acid) $\xrightarrow{\text{Transaminase}}$ $CH_3{-}CH(CH_3){-}CH_2{-}C(=O){-}COOH$ (a-Ketoiso-caproic acid) + $HOOC{-}CH_2{-}CH_2{-}CH(NH_2){-}COOH$ (Glutaric acid)

(b) $CH_3{-}CH(CH_3){-}CH_2{-}C(=O){-}COOH$ (a-Keto-isocaproic acid) $\xrightarrow[\text{Decarboxylase}]{CO_2}$ $CH_3{-}CH(CH_3){-}CH_2{-}CH{=}O$ (iso-valeraldehyde)

(c) $CH_3{-}CH(CH_3){-}CH_2{-}CH{=}O$ (iso-valeraldehyde) $\xrightarrow[\text{Reduction}]{NADH \rightarrow NAD^+}$ $CH_3{-}CH(CH_3){-}CH_2{-}CH_2OH$ (iso-amyl alcohol)

Fig. 2-5. Formation of isoamyl alcohol from leucine.

acid and an alpha-keto acid with subsequent decarboxylation and reduction. The final reduction step involves reoxidation of reduced nicotinamide adenine dinucleotide (Fig. 2-4). Figure 2-5 shows the specific case of the amino acid leucine. It is now known that fusel oils arise from both amino acid and carbohydrate sources. Using radioactive tracers, Reazin et al. (1970) found that 35% of higher alcohols arose from carbohydrates.

Ethanol Production

Yeast species and strains vary in their abilities to utilize carbohydrates in formation of alcohol and other by-products, as well as to grow in varying concentrations of alcohol. Most strains of *Saccharomyces cerevisiae* are inhibited at alcohol levels approaching 14 to 15% (vol/vol). However, several strains are more alcohol-tolerant. In addition, many strains of wine yeasts can be acclimatized to tolerate higher concentrations of alcohol through supplemented "syruped" fermentations.

The quantity of alcohol and carbon dioxide formed as well as the nature and concentration of by-products varies with yeast strain, temperature of fermentation, and extent of aeration. Fermentation temperature plays an important role in alcohol tolerance. In general, yeast fermentations at lower temperatures result in higher alcohol yields, due, in part, to reduced losses of alcohol resulting from evaporation and entrainment. Work by Warkentine and Nury (1963) suggests that alcohol loss by carbon dioxide entrainment is small (0.83% at 27°C). Generally, losses are correlated to fermentation temperature as well as surface-to-volume relationships in fermenters. Red wines often have lower alcohol levels than white wines with the same initial sugar content, primarily because of the practice of fermenting reds at higher temperatures in order to facilitate color extraction. Wines fermented at temperatures greater than 95°F sometimes have lower than expected alcohol levels due to slow or incomplete (stuck) fermentation as well as atmospheric evaporation. By comparison, fermenting a must under strongly reducing conditions (low oxygen content), in pressure fermentations, will yield slightly higher alcohol levels (Amerine et al. 1972). Under such conditions, the yeast cell channels more energy to fermentative metabolism than toward the cellular reproduction that normally occurs in the very early stages of fermentation.

DETERMINATION OF ALCOHOL CONTENT

The principal basis for wine taxation is alcohol content. Table 2-2 presents minimum alcohol contents for wine types sold in California. By comparison, federal regulations define wine as containing between 7 and 24% (vol/vol) alcohol. Within this range, table wines must contain between 7 and 14% alcohol and dessert wines between 14 and 24%. In the dessert wine category, sherry

Table 2-2. Alcohol minima for California wine types (sold in California).

Wine categories	Minimum alcohol content (vol/vol)
Red table (grape)	10.5
White and pink table (grape)	10.0
Sherry	17.0
All other dessert wines	18.0

From California Administrative Code 17.000-17.130.

must have a minimum alcohol content of 17%, whereas other dessert wine types (ports, muscatels, etc.) have 18% minima.

The above regulations have recently been amended to reflect current industry interest in the production of low-alcohol wines. Thus wineries may produce wines of alcohol content less than that indicated in Table 2-2, but such wines are considered experimental and must meet all other standards for California table wines. Further, these wines must not contain less than 7% alcohol per federal definition.

The sensory and physical properties of wine stability depend, in part, on alcohol content. Thus blending or additions of wine spirits (WSA) that result in changes in the final alcohol content may subsequently result in a change in wine stability (see Section III).

For the sake of discussion, procedures for the determination of alcohol in beverages may be divided into methods utilizing the physical characteristics of solutions (colligative properties) and those based on the chemical properties of alcohol.

Physical Methods

Ebulliometric Analysis (see Procedure 2-1)

Ebulliometry is the most commonly encountered procedure for determination of the alcohol content of aqueous solutions. A typical ebulliometer is schematically represented in Fig. 2-6. In principle, the analysis is based on the Raoult's law relationship of boiling point depression:

$$p = h \cdot x_1$$

where:

p = vapor pressure of solution
h = proportionality constant
x_1 = mole fraction of solvent

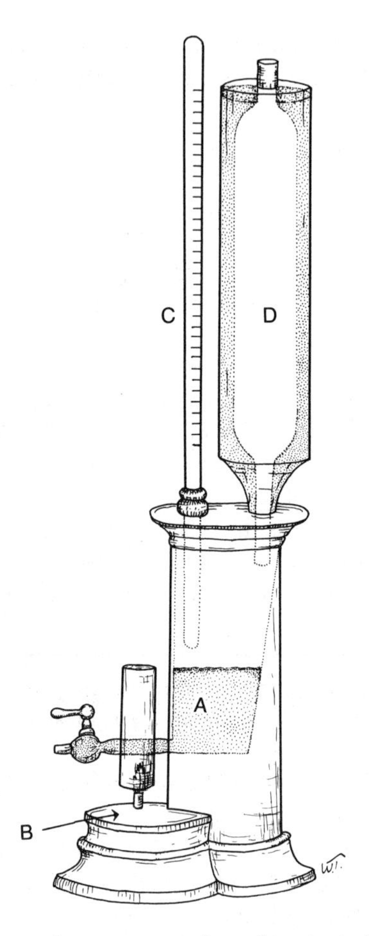

Fig. 2-6. Schematic representation of typical ebulliometer.

Thus, the vapor pressure of a solute (ethanol) in a solution will vary in a regular manner as a function of its concentration. As the concentration of the solute (alcohol) in the wine increases, the boiling point is reached at a temperature that is low relative to the boiling point of pure water.

Raoult's law effects can also be viewed from the perspective of a boiling point versus composition diagram for the water–ethanol system (Fig. 2-7). The lower curve on this diagram represents the boiling point of various water–ethanol mixtures at one atmosphere pressure. As the percent ethanol increases, the boiling point decreases. Because one does not always have one atmosphere pressure in the laboratory, the ebulliometric method utilizes a sliding scale that allows one to adjust (the left axis of Fig. 2-7) for the actual boiling point of pure water.

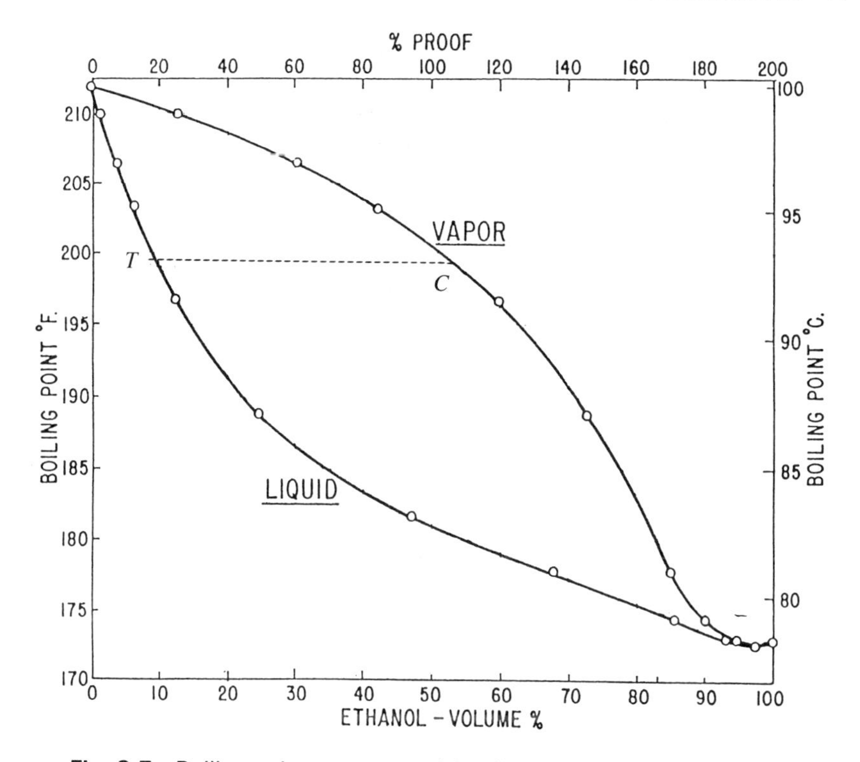

Fig. 2-7. Boiling point vs. composition for ethanol–water mixtures.

The upper curve in Fig. 2-7 represents the composition of the vapor phase in equilibrium with the boiling liquid phase. One can readily observe that a 10% (vol/vol) ethanol–water solution boiling at ~199°F (point *T* on Fig. 2-7) is in equilibrium with refluxing vapor of approximately 54% (vol/vol) ethanol (point *C* on Fig. 2-7). This demonstrates why the ebulliometer must be operated under conditions of total reflux. Any ethanol–water vapor that leaks out of the condenser takes with it a proportionately large amount of ethanol. This causes the composition of the liquid phase to change (decreasing ethanol concentration) and the boiling point of the sample gradually to creep toward that of pure water.

Although it is simple in theory, several interferences may be encountered in routine laboratory application of ebulliometry, the most important of these being the effect of sugars. According to the colligative properties of solutions, sugar molecules would be expected to cause a boiling point elevation (hence lower apparent alcohol levels). However, observation contradicts this, in that sweet wines usually boil at a temperature lower than expected, resulting in higher apparent alcohols. This behavior is due to the sugar–water matrix squeezing out the ethanol (increasing its vapor pressure). To reduce the errors attributed to sugar, slightly sweet wines may be diluted with water to a sugar level of less than 2%, yielding a boiling point of 96 to 100°C. Amerine (1954), however, points out that it is doubtful if dilution avoids the errors because the result must be multiplied by the dilution factor, which, in turn, multiplies the relative error.

In contrast to the "sugar effect," aldehydes, esters, and acids tend to raise the boiling point of a solution. Thus the combined effects of solution components producing boiling point depression and elevation result in a compromise in which the boiling point, in practice, approximates that of the actual alcohol content.

Hydrometry (see Procedure 2-2)

The alcohol content of a distillate from an accurately measured volume of sample may be determined hydrometrically. (For a general review of hydrometric procedures, see Chapter 1.) Reducing sugars are not distillable and hence not a problem in distillate analysis, but two common interferences, sulfur dioxide and acetic acid, may cause problems. At levels approaching 200 mg/L, sulfur dioxide is steam-distillable as sulfurous acid. As such, it may affect hydrometric determinations of alcohol by producing results lower than expected (0.2 to 0.5% vol/vol); so it is recommended that in cases where the sulfur dioxide content is known to be high, samples be neutralized prior to distillation.

Neutralization may be accomplished in two ways: (1) by addition of calcium hydroxide directly to the sample, or (2) by redistillation of a distillate that has been made alkaline. The latter is the procedure of choice. Abnormal levels of acetic acid (greater than 1.0 g/L) likewise present problems. Some workers suggest neutralization of such samples with 1 *N* NaOH prior to the distillation step. Finally, young wines and sweetened wines may foam excessively during distillation. This problem can be avoided by the addition of small amounts of an antifoaming agent prior to distillation. However, antifoaming agents may foul glassware with residues that are difficult to remove.

Gas Chromatography

For a general discussion on chromatography see Appendix I. Gas chromatography (GC) is a separation technique used to separate volatile components in the sample. For example, wine (or juice, distillate, etc.) is injected into a heated tube that is packed with a porous bed of specialized absorbent, through which an inert gas flows. Ethanol and other volatile components are vaporized and carried through the tube (referred to as the GC column) toward a detector that senses their presence. Because of differences in their interaction with the absorbent, different compounds migrate or travel through the column at different rates, and are separated by the time they reach the detector. To quantitate ethanol, one prepares standards of known concentration, injects them into the GC, and compares their detector responses to that of the unknown sample.

The same GC technique can be used for the analysis of fusel oils in distillate (Procedure 2-6). For this particular analysis, a column with a different specialized absorbent is utilized, as well as a different column temperature. Other uses of gas chromatography in the winery laboratory are being developed. It is a powerful analytical tool that will see more and more utility in the future.

High Performance Liquid Chromatography (HPLC)

HPLC is finding increasing use in the winery laboratory. The technique can be used to determine alcohol in harvested fruit as well as in finished wine. See Procedure 1-5.

Chemical Methods For Alcohol Determination

Dichromate Oxidation (see Procedures 2-3, 2-4)

Although several chemical methods for alcohol determination appear in the literature, the most important currently used procedure is based on the quantity of acid dichromate required to oxidize the alcohol to acetic acid:

$$2\,Cr_2O_7^{-2} + 3\,C_2H_5OH + 16H^{+} \longrightarrow 4\,Cr^{+3} + 3\,CH_3COOH + 11H_2O \tag{2-2}$$

The excess dichromate remaining upon completion of the reaction is titrated with ferrous ammonium sulfate (FAS):

$$Cr_2O_7^{-2} + 6\,Fe^{+2} + 14\,H^{+} \longrightarrow 2Cr^{+3} + 6\,Fe^{+3} + 7\,H_2O \tag{2-3}$$

The completeness of this redox reaction depends on the time and the concentration of reacting components, several workers having reported the optimum temperature for oxidation to be 60 to 65°C. The reaction is also critically dependent upon hydrogen ion concentration for the complete oxidation of alcohol to acetic acid rather than to mixtures of the acid and aldehyde intermediates.

Zimmerman (1963) reported that "the dichromate method is inherently more accurate than densimetric methods for wines." In support of this view, he pointed out that distillates normally contain volatile acids and sulfurous acid, which act to depress alcohol determinations based on densimetric considerations. These interferences, however, have no effect on redox titrations; so analytical results can be expected to run higher than those developed by densimetry by a factor of 0.24%.

Some laboratories use a modification of the micro-dichromate procedure. Instead of titrating the excess dichromate with FAS, the procedure developed by Caputi et al. (1968) measures absorbance spectrophotometrically. The absorbance of unknown alcohol solutions is then related to standards via a standard curve. Alternately, if the spectrophotometer has a concentration mode, the concentration (as % vol/vol) may be obtained directly by comparison of the unknown with an accurately prepared standard. The latter is usually prepared in the range of 4 to 20% (vol/vol).

Enzymatic Analysis

Ethanol can be oxidized in the presence of the enzyme alcohol dehydrogenase by nicotinamide-adenine dinucleotide (NAD) to produce NADH. This is a stoichiometric reaction under proper experimental conditions, and the NADH produced can be determined spectrophotometrically at 334 nm. Reagents for this determination may be obtained from a number of chemical supply houses (e.g., Sigma), and complete kits for the analysis are also available (e.g., Boehringer-Mannheim). One drawback in using this method is that it is necessary to quantitatively transfer very small volumes of reagent and sample.

Procedure 2-1. Alcohol Determination by Ebulliometry

Ebulliometry involves a measurement of the boiling point depression caused by the presence of alcohol in a wine sample. Mechanically, the boiling point of the sample is measured relative to the boiling point of pure water, and the difference is related to percent ethanol. Sugar (>2% r.s.) is the major interference in this determination.

I. Equipment
   Ebulliometer, mercury thermometer (°C), appropriate alcohol scale
   Distilled water and cold tap water source
   Microburner or alcohol lamp
II. Reagents (see Appendix II)
   Sodium hydroxide (1%) cleaning solution
III. Procedure (refer to Fig. 2-6)
   (a) Determination of the boiling point of water
      1. Add approximately 30 mL of distilled water to the boiling chamber. There is no need to add cold tap water to the condenser at this time.
      2. Insert the thermometer. Position the instrument over a flame. Prior to use, inspect the thermometer, making certain that the mercury column is not separated. If you elect to use a microburner rather than the alcohol lamp included with the kit, the ebulliometer should be positioned high enough above the flame to prevent excessive "bumping." The latter is a sign of overheating and impedes proper boiling point determination.
      3. When the thermometer reaches a stable point, allow 15 to 30 sec

for minor fluctuations to occur. At this time, take the boiling point reading and set the inner scale opposite 0.0% alcohol on the "Degre Alcoolique du Vin" outer scale.

4. Cool and drain the instrument.

(b) Determination of the boiling point of wine

5. Rinse the boiling chamber with a few milliliters of the wine to be analyzed, and drain it.
6. Place approximately 50 mL of wine in the boiling chamber.
7. Fill the condenser with cold tap water.
8. Insert the thermometer as shown in Fig. 2-6, and place the instrument over the heat source. Once again, check the mercury column to ensure continuity of the liquid.
9. When the thermometer reaches a stable level, allow 15 to 30 sec for changes, and take a reading.
10. Locate the boiling point of wine on the inner "Degres du Thermometre" scale and record the corresponding alcohol content (% vol/vol) on the outer scale.
11. Cool and thoroughly rinse the instrument.

IV. Supplemental Notes

1. Because the boiling point will vary with atmospheric pressure, it *must* be determined for the distilled water reference during each day's operation. It is recommended that boiling point determinations be carried out at least twice daily and more frequently during periods of unstable weather.
2. Reducing sugar levels exceeding 2% interfere with the procedure, yielding erroneously low boiling points (hence higher apparent alcohols). The effect of sugar may be overcome by carrying out an initial separation (distillation) and determining the boiling point of the collected distillate (see Procedure 2-2). Many laboratories simply dilute sweet wines to a level of less than 2% reducing sugar prior to analysis. Still others use a correction factor (reducing sugar $\times$ 0.05) and subtract it from their apparent alcohol content.
3. To prevent premature loss of alcohol, the cooling water in the condenser must be cold (ice water preferably). Boiling point readings should be taken per instruction. As the condenser water warms, alcohol is volatilized and lost, resulting in erroneously high boiling points and lower apparent alcohols.
4. In newly fermented wines, excessive foaming may be prevented by the addition of antifoaming agents.
5. During extended use, debris will coat the walls of the boiling chamber, reducing the efficiency of the unit. It is recommended that the boiling chamber be cleaned periodically with a boiling solution of 1% NaOH.

Procedure 2-2. Alcohol Determination by Distillation and Hydrometric Analysis

Distillation of the alcohol from a wine sample followed by measurement of the specific gravity of the ethanol–water distillate can provide an accurate determination of the original alcohol content. Volatile acidity and sulfur dioxide levels can interfere with this analysis. Control of the temperature of the initial wine sample and final distillate is critical to accurate measurements.

I. Equipment
   Distillation apparatus as seen in Fig. 2-8
   200-mL Kohlrausch receiving flask
   Alcohol hydrometer

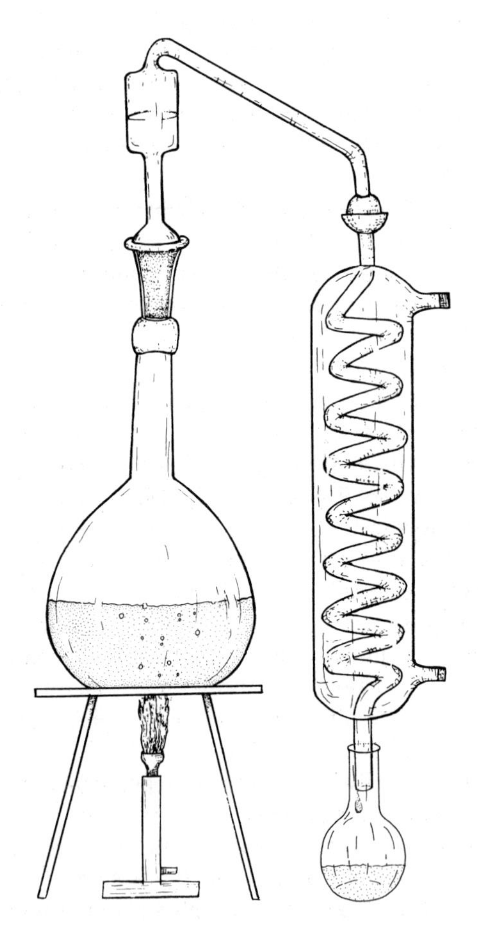

Fig. 2-8. Simple distillation apparatus for separation of alcohol.

Hydrometer cylinder
Distilled water

II. Reagents (see Appendix II)
Antifoaming agent (Tween 80 or equivalent)
2 *N* NaOH

III. Procedure

1. Using several milliliters of wine sample, rinse the 200-mL Kohlrausch receiving flask. Then fill the flask to within 1 cm of the volume mark with the wine sample.
2. Adjust the temperature to 20°C, and bring the sample to volume.
3. Transfer the contents to the Kjeldahl distillation flash, ensuring a complete transfer with several rinses of distilled water. To prevent charring, add 40 to 50 mL of distilled water to the distillation flask. In the case of young or sweet wines, a drop of antifoaming agent may be added to prevent excessive foam formation during the distillation.
4. Position and connect the flask as shown in Fig. 2-8, ensuring a tight seal between the expansion bulb, the flask, and the condenser.
5. Turn on the condenser water. Adjust the heat input to yield a moderate boiling.
6. Collect approximately 195 mL of distillate. If a subsequent extract analysis is desired, save the residue in the Kjeldahl flask. (Refer to Chapter 3 for details of extract determination.)
7. Adjust the temperature of the distillate to 20°C, and bring it to volume with distilled water. Mix the solution thoroughly.
8. Transfer the distillate to a hydrometer cylinder, and immerse a hydrometer of the appropriate range into the contents. *Caution*: Accurately calibrated alcohol hydrometers are expensive and easily broken.
9. After the hydrometer reaches equilibrium, take a reading. Record the temperature of the distillate, and make corrections as necessary using Tables 2-3 and 2-4.
10. Specific gravity hydrometers also can be used. Percent alcohol (vol/vol) is then determined by reference to Table 2-5.

IV. Supplemental Notes

1. Because two or more connections between component glassware are involved, they should be checked routinely for escaping vapors during distillation.
2. If the condenser capacity is insufficient, resulting in a warm or hot distillate, the receiving flask should be positioned in an ice bath.
3. As discussed, volatile acidities exceeding 0.10% and sulfur dioxide levels greater than 200 mg/L may interfere with this method. Therefore, wines with excessive amounts of either or both should be neutralized with 2 *N* NaOH prior to distillation. However, such neutralized wines *cannot* be used for subsequent extract analysis.

Table 2-3. Temperature corrections of alcohol hydrometers calibrated at 15.56°C (60°F) in volume % ethanol.

Observed alcohol content (vol. %)	To or from the observed: Add			Subtract														
	at 57°F 13.9°C	at 58°F 14.4°C	at 59°F 15.0°C	at 61°F 16.1°C	at 62°F 16.7°C	at 63°F 17.2°C	at 64°F 17.8°C	at 65°F 18.3°C	at 66°F 18.9°C	at 67°F 19.4°C	at 68°F 20.0°C	at 69°F 20.6°C	at 70°F 21.1°C	at 72°F 22.2°C	at 74°F 23.3°C	at 76°F 24.4°C	at 78°F 25.6°C	at 80°F 26.7°C
1	0.14	0.10	0.05	0.05	0.10	0.16	0.22	0.28	0.34	0.41	0.48	0.55	0.62	0.77	0.93			
2	0.14	0.10	0.05	0.05	0.11	0.17	0.23	0.29	0.35	0.42	0.48	0.56	0.63	0.78	0.94	1.10	1.28	1.46
3	0.14	0.10	0.05	0.06	0.12	.018	0.24	0.30	0.36	0.43	0.50	0.57	0.64	0.80	0.96	1.13	1.31	1.50
4	0.14	0.10	0.05	0.06	0.12	0.19	0.25	0.32	0.38	0.45	0.52	0.59	0.67	0.83	1.00	1.17	1.35	1.54
5	0.15	0.10	0.05	0.07	0.13	0.20	0.26	0.33	0.40	0.47	0.54	0.62	0.70	0.86	1.03	1.21	1.40	1.60
6	0.17	0.11	0.06	0.07	0.14	0.20	0.27	0.34	0.42	0.50	0.57	0.66	0.74	0.90	1.09	1.27	1.46	1.66
7	0.18	0.12	0.06	0.07	0.14	0.21	0.29	0.36	0.44	0.52	0.60	0.68	0.77	0.94	1.13	1.32	1.52	1.73
8	0.19	0.13	0.06	0.08	0.16	0.23	0.31	0.39	0.47	0.55	0.64	0.73	0.81	0.99	1.18	1.38	1.59	1.80
9	0.21	0.14	0.07	0.08	0.16	0.24	0.32	0.41	0.50	0.58	0.67	0.76	0.86	1.04	1.25	1.46	1.67	1.89
10	0.23	0.16	0.08	0.08	0.17	0.25	0.34	0.43	0.52	0.61	0.71	0.80	0.90	1.10	1.32	1.54	1.76	1.99
11	0.25	0.16	0.08	0.09	0.18	0.27	0.37	0.46	0.56	0.65	0.75	0.85	0.96	1.16	1.39	1.61	1.84	2.09
12	0.27	0.18	0.09	0.10	0.20	0.29	0.39	0.49	0.59	0.70	0.80	0.91	1.02	1.23	1.46	1.70	1.94	2.20
13	0.29	0.19	0.10	0.10	0.21	0.31	0.42	0.52	0.63	0.74	0.85	0.97	1.08	1.31	1.55	1.80	2.05	2.31
14	0.32	0.21	0.11	0.11	0.22	0.32	0.44	0.55	0.66	0.78	0.91	1.02	1.14	1.39	1.65	1.91	2.17	2.44
15	0.35	0.23	0.12	0.12	0.24	0.35	0.48	0.60	0.71	0.84	0.97	1.10	1.23	1.50	1.76	2.03	2.30	2.58
16	0.37	0.24	0.12	0.13	0.26	0.38	0.52	0.65	0.77	0.90	1.03	1.17	1.31	1.60	1.88	2.16	2.44	2.72
17	0.40	0.26	0.13	0.14	0.27	0.41	0.54	0.68	0.82	0.96	1.10	1.25	1.40	1.70	1.99	2.28	2.58	2.87
18	0.44	0.29	0.14	0.14	0.29	0.44	0.58	0.73	0.88	1.03	1.18	1.33	1.49	1.80	2.10	2.41	2.72	3.02
19	0.47	0.32	0.16	0.15	0.30	0.46	0.62	0.78	0.94	1.10	1.26	1.42	1.58	1.90	2.22	2.54	2.86	3.17
20	0.51	0.34	0.17	0.16	0.32	0.49	0.66	0.82	0.98	1.15	1.33	1.48	1.65	2.00	2.32	2.65	2.98	3.33
21	0.53	0.35	0.18	.017	0.34	0.51	0.68	0.85	1.02	1.20	1.38	1.54	1.72	2.06	2.41	2.76	3.10	3.45
22	0.56	0.38	0.19	0.17	0.36	0.53	0.71	0.90	1.07	1.25	1.44	1.61	1.78	2.13	2.48	2.84	3.20	3.56
23	0.58	0.40	0.20	0.18	0.37	0.55	0.74	0.92	1.11	1.30	1.49	1.66	1.84	2.20	2.56	2.93	3.30	3.67
24	0.60	0.40	0.20	0.18	0.38	0.56	0.77	0.96	1.16	1.35	1.54	1.72	1.91	2.27	2.65	3.03	3.40	3.78

U.S. Internal Revenue Service. "Regulations No. 7". U.S. Govt. Printing Office, Washington, D.C. 1945.

Table 2-4. Temperature corrections of alcohol hydrometers calibrated at 20°C (68°F) in volume % ethanol.

Temperature (°C) Add	Apparent degree of of alcoholic strength at t°C																	
	0	1	2	3	4	5	6	7	8	9	10	11	12	13	14	15	16	17
0	0.76	0.77	0.82	0.87	0.95	1.04	1.16	1.31	1.49	1.70	1.95	2.26	2.62	3.03	3.49	4.02	4.56	5.11
1	0.81	0.83	0.87	0.92	1.00	1.09	1.20	1.35	1.52	1.73	1.97	2.26	2.59	2.97	3.40	3.87	4.36	4.86
2	0.85	0.87	0.92	0.97	1.04	1.13	1.24	1.38	1.54	1.74	1.97	2.24	2.54	2.89	3.29	3.72	4.17	4.61
3	0.88	0.91	0.95	1.00	1.07	1.15	1.26	1.39	1.55	1.73	1.95	2.20	2.48	2.80	3.16	3.55	3.95	4.36
4	0.90	0.92	0.97	1.02	1.09	1.17	1.27	1.40	1.55	1.72	1.92	2.15	2.41	2.71	3.03	3.38	3.75	4.11
5	0.91	0.93	0.98	1.03	1.10	1.17	1.27	1.39	1.53	1.69	1.87	2.08	2.33	2.60	2.89	3.21	3.54	3.86
6	0.92	0.94	0.98	1.02	1.09	1.16	1.25	1.37	1.50	1.65	1.82	2.01	2.23	2.47	2.74	3.02	3.32	3.61
7	0.91	0.93	0.97	1.01	1.07	1.14	1.23	1.33	1.45	1.59	1.75	1.92	2.12	2.34	2.58	2.83	3.10	3.36
8	0.89	0.91	0.94	0.98	1.04	1.11	1.19	1.28	1.39	1.52	1.66	1.82	2.00	2.20	2.42	2.65	2.88	3.11
9	0.86	0.88	0.91	0.95	1.01	1.07	1.14	1.23	1.33	1.44	1.57	1.71	1.87	2.05	2.24	2.44	2.65	2.86
10	0.82	0.84	0.87	0.91	0.96	1.01	1.08	1.16	1.25	1.35	1.47	1.60	1.74	1.89	2.06	2.24	2.43	2.61
11	0.78	0.79	0.82	0.86	0.90	0.95	1.01	1.08	1.16	1.25	1.36	1.47	1.60	1.73	1.88	2.03	2.20	2.36
12	0.72	0.74	0.76	0.79	0.83	0.88	0.93	0.99	1.07	1.15	1.24	1.34	1.44	1.56	1.69	1.82	1.96	2.10
13	0.66	0.67	0.69	0.72	0.76	0.80	0.84	0.90	0.96	1.03	1.11	1.19	1.28	1.38	1.49	1.61	1.73	1.84
14	0.59	0.60	0.62	0.64	0.67	0.71	0.74	0.79	0.85	0.91	0.97	1.04	1.12	1.20	1.29	1.39	1.49	1.58
15	0.51	0.52	0.53	0.55	0.58	0.61	0.64	0.68	0.73	0.77	0.83	0.89	0.95	1.02	1.09	1.16	1.24	1.32
16	0.42	0.43	0.44	0.46	0.48	0.50	0.53	0.56	0.60	0.63	0.67	0.72	0.77	0.82	0.88	0.94	1.00	1.06
17	0.33	0.33	0.34	0.35	0.37	0.39	0.41	0.43	0.46	0.48	0.51	0.55	0.59	0.62	0.67	0.71	0.75	0.80
18	0.23	0.23	0.23	0.24	0.25	0.26	0.27	0.29	0.31	0.33	0.35	0.37	0.40	0.42	0.45	0.48	0.51	0.53
19	0.12	0.12	0.12	0.12	0.13	0.13	0.14	0.15	0.16	0.17	0.18	0.19	0.20	0.21	0.23	0.24	0.25	0.27

Table 2-4. (*Continued*)

Temperature (°C)	Apparent degree of of alcoholic strength at t°C																	
	0	1	2	3	4	5	6	7	8	9	10	11	12	13	14	15	16	17
21		0.13	0.13	0.13	0.14	0.14	0.15	0.16	0.17	0.18	0.19	0.19	0.20	0.22	0.23	0.25	0.26	0.28
22		0.26	0.27	0.28	0.29	0.30	0.31	0.32	0.34	0.36	0.37	0.39	0.41	0.44	0.47	0.49	0.52	0.55
23		0.40	0.41	0.42	0.44	0.45	0.47	0.49	0.51	0.54	0.57	0.60	0.63	0.66	0.70	0.74	0.78	0.82
24		0.55	0.56	0.58	0.60	0.62	0.64	0.67	0.70	0.73	0.77	0.81	0.85	0.89	0.94	0.99	1.04	1.10
25		0.69	0.71	0.73	0.76	0.79	0.82	0.85	0.89	0.93	0.97	1.02	1.07	1.13	1.19	1.25	1.31	1.37
26		0.85	0.87	0.90	0.93	0.96	1.00	1.04	1.08	1.13	1.18	1.24	1.30	1.36	1.43	1.50	1.57	1.65
27			1.03	1.07	1.11	1.15	1.19	1.23	1.28	1.34	1.40	1.46	1.53	1.60	1.68	1.76	1.84	1.93
28			1.21	1.25	1.29	1.33	1.38	1.43	1.49	1.55	1.62	1.69	1.77	1.85	1.93	2.02	2.11	2.21
29			1.39	1.43	1.47	1.52	1.58	1.63	1.70	1.76	1.84	1.92	2.01	2.10	2.19	2.29	2.39	2.50
30	Deduct		1.57	1.61	1.66	1.72	1.78	1.84	1.91	1.98	2.07	2.15	2.25	2.35	2.45	2.56	2.67	2.78
31			1.75	1.80	1.86	1.92	1.98	2.05	2.13	2.21	2.30	2.39	2.49	2.60	2.71	2.83	2.94	3.07
32			1.94	2.00	2.06	2.13	2.20	2.27	2.35	2.44	2.53	2.63	2.74	2.86	2.97	3.09	3.22	3.36
33				2.20	2.27	2.34	2.42	2.50	2.58	2.67	2.77	2.88	2.99	3.12	3.24	3.37	3.51	3.65
34				2.41	2.48	2.56	2.64	2.72	2.81	2.91	3.02	3.13	3.25	3.38	3.51	3.65	3.79	3.94
35				2.62	2.70	2.78	2.86	2.95	3.05	3.16	3.27	3.39	3.51	3.64	3.78	3.93	4.08	4.23
36				2.83	2.91	3.00	3.09	3.19	3.29	3.41	3.53	3.65	3.78	3.91	4.05	4.21	4.37	4.52
37					3.13	3.23	3.33	3.43	3.54	3.65	3.78	3.91	4.04	4.18	4.33	4.49	4.65	4.82
38					3.36	3.47	3.57	3.68	3.79	3.91	4.03	4.17	4.31	4.46	4.61	4.77	4.94	5.12
39					3.59	3.70	3.81	3.93	4.05	4.17	4.30	4.44	4.58	4.74	4.90	5.06	5.23	5.41
40					3.82	3.94	4.06	4.18	4.31	4.44	4.57	4.71	4.86	5.02	5.19	5.36	5.53	5.71

Table 2-5. Percent alcohol by volume vs. specific gravity at 20°C.[1,2]

Specific gravity	Alcohol (v/v%)	Specific gravity	Alcohol (v/v%)
1.00000	0.00	0.98530	11.00
0.99851	1.00	0.98471	11.50
0.99704	2.00	0.98412	12.00
0.99560	3.00	0.98297	13.00
0.99419	4.00	0.98182	14.00
0.99281	5.00	0.98071	15.00
0.99149	6.00	0.97960	16.00
0.99020	7.00	0.97850	17.00
0.98894	8.00	0.97743	18.00
0.98771	9.00	0.97638	19.00
0.98711	9.50	0.97532	20.00
0.98650	10.00	0.97425	21.00
0.98590	10.50	0.97318	22.00

[1]Specific gravity determined with reference to water at 20°C. To convert the specific gravity with reference to water at 4°C, multiply the above values by 0.99908.
[2]Table based on data in U.S. Bureau of Standards Circular No. 19 (1924).

4. Most analysts prefer to calibrate new alcohol hydrometers against accurately prepared standards.
5. Hydrometers and cylinders must be clean and dry at the time of use to ensure accurate determinations.

Procedure 2-3. Alcohol Determination by Dichromate Analysis

Acidified dichromate can oxidize the alcohol collected by distillation from a wine sample. Excess dichromate is then titrated with iron (II) solution. The ratio of volumes of titrant consumed by the wine sample (V_A) and the blank (V_B) is used to calculate the ethanol content. Careful control of sample temperature, the volume of sample analyzed, and the concentration of dichromate solution are critical for accurate measurements.

I. Equipment
   - Micro-Kjeldahl distillation apparatus (see Fig. 2-9)
   - 50-mL burette
   - Constant-temperature water bath (60–65°C)
   - 50-mL Erlenmeyer flasks (preferably graduated)
   - 500-mL Erlenmeyer flasks
   - Volumetric pipettes (1 and 25 mL)
   - Wash bottle
   - High-intensity light source

II. Reagents (see Appendix II)
   - 1,10-Phenanthroline ferrous sulfate indicator solution
   - Ferrous ammonium sulfate (FAS)

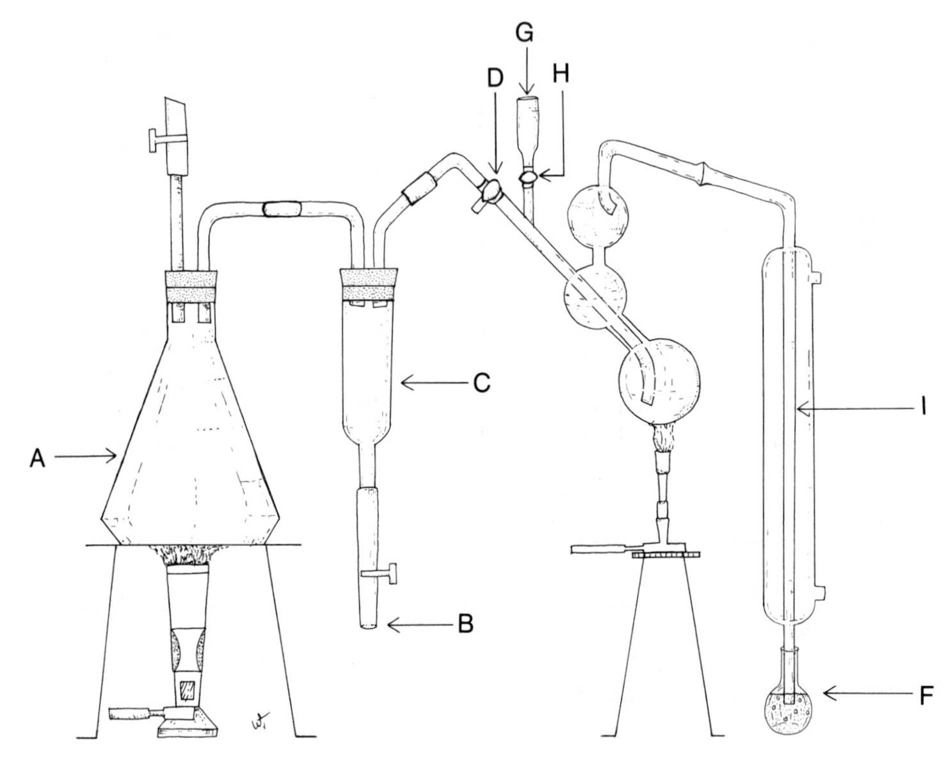

Fig. 2-9. Schematic representation of micro-Kjeldahl distillation apparatus. Lettered components are described in the text.

Potassium dichromate solution
Distilled water

III. Procedure

(a) Preparing the still

1. At the beginning of distillation, water in steam generator *A* should be boiling and condensate passing through condenser exit *I*.
2. Stopcock *D* is turned so that steam from the trap can escape through the side exit.
3. Receiving flask *F* is placed such that the exit tube of condenser *I* is immersed in dichromate solution.
4. Sample stopcock *H* is in the closed position, and funnel *G* contains a small amount of distilled water.

(b) Sample distillation

5. Volumetrically transfer 1.0 mL of wine sample into funnel *G*.
6. Open stopcock *H* to admit wine to the still.
7. Rinse the funnel two or three times with distilled water. Close stopcock *H*.

8. Volumetrically transfer 25 mL of dichromate solution to a 50 mL Erlenmeyer receiving flask, and position it under the condenser. As shown in Fig. 2-9, the tip of the condenser should be immersed in the solution.
9. With stopcock *H* closed, turn stopcock *D* so that steam will flow from the steam trap to the receiver.
10. Position and ignite the microburner. *Caution:* Use only a *very low flame* to prevent breakage.
    (*Note*: The above steps should be carried out in a sequential and continuous manner.)
11. Collect the distillate in the receiving flask to a final volume of approximately 45 mL.
12. When distillation is complete, lower the flask, and rinse the outside of the condenser tip with distilled water.
13. Remove the microburner from its position under the distilling bulb.

(c) Cleaning operation

14. Admit a little distilled water into the steam generator. This will cause dealcoholized residue in the still bulb to siphon into steam trap *C*.
15. Place 10 to 15 mL of distilled water into the still through funnel *G*, and repeat the siphoning operation. Upon completion, leave the funnel full of distilled water.
16. Drain the steam trap at *B*.

(d) Post-distillation sample treatment

17. Stopper the flask and place it in a water bath at 60 to 65°C for 25 minutes.
18. Quantitatively transfer the contents of the flask to a 500-mL Erlenmeyer flask (using distilled water rinses), and titrate it with ferrous ammonium sulfate. When the color of the solution turns to an emerald green, add five drops of indicator. Continue the titration until the color changes (sharply) from blue-green to brown. Some lab personnel prefer to use reflected light to assist in end-point detection.
19. Record the volume of FAS used in the sample titration, and relate it to the volume of FAS used to titrate the "blank," using Equation 2-4. The blank consists of 1 mL of distilled water in 25 mL of dichromate incubated at 60 to 65°C for 25 minutes.

$$\text{Ethanol } (\%\ \text{v/v}) = 25 - 25\,[V_A/V_B] \qquad (2\text{-}4)$$

where:

V_A = titer of wine sample
V_B = titer of water blank

IV. Supplemental Notes

1. Alcohol in the distillate is quantitatively oxidized to acetic acid by an excess of standardized dichromate:

$$3C_2H_5OH + 2Cr_2O_7^{2-} + 16H^+ \rightarrow 3CH_3COOH + 4Cr^{3+} + 11H_2O \quad (2\text{-}5)$$

   Therefore, the time and the temperature of incubation are critical. Unreacted excess dichromate is then determined by titration with standardized FAS.
2. The glass still must be clean for accurate analysis.
3. The pipetting procedure and technique must be followed strictly. Because of density differences between water and dichromate, one may wish to calibrate pipets for accurate delivery of the latter.
4. The temperature of the wine should be the same as that for which the pipet is calibrated. This information is printed on the side of most pipets.
5. In order to monitor the quality/strength of the FAS, it is recommended that one blank titration be run in the morning and another in the afternoon.
6. When the end point is approached, titration should be on a drop-by-drop basis. If overtitration occurs, add 0.5 mL of dichromate solution and again titrate to the proper end point. The additional volume of dichromate used should be added to the initial 25 mL.
7. The coefficient of cubical expansion for potassium dichromate is significant; so this solution should be made to volume at 20°C.

Procedure 2-4. Alternative Procedure of Dichromate Analysis

Some laboratories utilize a modification of the micro-dichromate procedure presented in the previous section. However, instead of titration of the excess dichromate (after heating to 60°C) with FAS, the sample is measured spectrophotometrically at 600 nm. This result is compared to the absorbance of standard solutions of known alcohol levels. The following procedure is that of Caputi et al. (1968).

I. Equipment

Distillation unit (see Procedure 2-3 for setup)
Spectrophotometer

II. Reagents (see Appendix II)

Potassium dichromate solution
Ethanol (200 proof)

III. Procedure

1. Prepare a series of alcohol standard solutions by diluting the volumes indicated in Table 2-6 of 200 proof (100%) ethanol to 250 mL final volume with distilled water at defined temperature.

Table 2-6. Preparation of ethanol calibration standards from 200 proof ethanol.

Volume 200 proof ethanol (mL)	Final concentration (%v/v) in 250 mL volume
50	20
25	10
20	8
10	4

2. Transfer 1.0 mL aliquots of the above standard solutions to individual 50 mL volumetric flasks containing 25 mL acid dichromate. Add approximately 20 mL of distilled water.
3. Stopper the flasks, and incubate them in a 60°C water bath for 20 minutes.
4. Cool to standard temperature, and bring the solutions to volume with distilled water.
5. Mix the solutions thoroughly, and measure the absorbance relative to a blank of dichromate and distilled water at 600 nm.
6. Plot the concentration of each standard versus its respective absorbance at 600 nm.
7. Distill the wine samples, and proceed as in steps 5–11 in Procedure 2.3, comparing the final absorbance to the standard curve.

Procedure 2-5. Gas Chromatographic Analysis of Ethanol

The ethanol in a diluted wine, juice, etc. sample can be separated from other components on a gas chromatographic column. To improve quantitation, 2-propanol (used as an internal standard) solution is used to quantitatively dilute the sample. The peak area ratio for the two chromatographic peaks is compared to the area ratio obtained from injection of a standard ethanol–internal standard mixture.

I. Equipment
   Gas chromatograph equipped with flame ionization detector (FID)
   Electronic integrator (data processor) or strip-chart recorder
   Hypodermic syringe (10 μL)
II. Reagents (see Appendix II)
   Internal standard solution of 2-propanol (0.2% vol/vol)
   Distilled water
   Standard ethanol–water solution (concentration appropriate to samples being run)

Table 2-7. Operating conditions for gas chromatograph (ethanol analysis).

Conditions	Col. a	Col. b	Col. c
Carrier gas	N_2	N_2	N_2
Flow rate (mL/min)	30	15	55
Oven temp., °C	200	105	88
Injector temp., °C	225	150	150
Detector temp., °C	225	150	150

III. Procedure

1. Install one of the following columns in the GC oven:
   a. 6 ft × 2 mm id glass, packed with 80–100 mesh Poropak QS.
   b. 6 ft × 2 mm id glass, packed with 0.2% Carbowax 1500 on 80–100 mesh Carbopack C.
   c. 6 ft × 1/4 in. od copper or stainless steel, packed with 3% Carbowax 600 on 40–60 mesh Chromosorb T.
2. Set the operating conditions given in Table 2-7 for the gas chromatograph (ethanol analysis).
3. Adjust the air and H_2 gas flows to those specified in the instrument operating directions (approximately 300 and 30 mL/min, respectively). These rates may have to be altered somewhat to optimize conditions for the particular carrier gas flow used.
4. Dilute the alcohol standard solution 1 + 99 with the 2-propanol internal standard solution. Inject at least three individual 1.0-μL aliquots into the gas chromatograph, and record the resulting chromatograms. Determine the peak area (integrator) or peak height ratios (recorder) for the alcohol peak to the 2-propanol peak. Calculate the average of the three response ratios (RR').
5. Dilute the sample 1 + 99 with the 2-propanol internal standard solution. Again inject at least three 1.0-μL aliquots into the gas chromatograph, and record the resulting chromatograms. Determine the appropriate response ratios for the alcohol peak to internal standard peak. Calculate the average of the several response ratios (RR). (See Fig. 2-10 for a typical chromatogram.)
6. Calculate the percent alcohol in the wine sample.

$$\% \text{ Alcohol} = (RR \times \% \text{ Alcohol in std})/RR' \qquad (2\text{-}6)$$

IV. Supplemental Notes

1. Oven temperatures may be altered to a limited extent to optimize chromatographic conditions for the specific GC column being used.
2. A calibration curve may be constructed by preparing three or more alcohol standard solutions covering a range of expected sample alcohol concentrations. Follow the procedure listed in steps 4 and 5 above, and

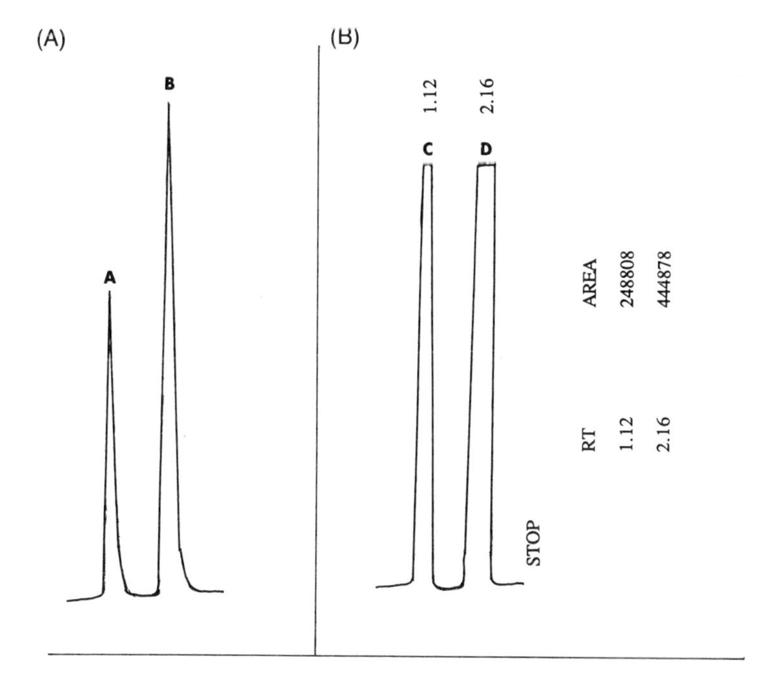

Fig. 2-10. (A) Typical strip-chart recorder and (B) integrator printout of chromatogram of wine sample. The average peak area ratio from three injections is 0.554, whereas the average peak area ratio from three injections of a 11.6% ethanol standard is 0.452. The calculated % ethanol for this sample is 14.2%.

record the resulting chromatograms. The curve is constructed by plotting the average response ratios for each standard chromatographed against the concentration of alcohol in that standard. The unknown values are read directly off the calibration curve.

Procedure 2-6. Gas Chromatographic Analysis of Fusel Oils

The fusel oil content of distillates can be determined by injecting a sample onto a gas chromatographic column, where the individual components are separated. Qualitative information is obtained by comparing elution times for the various components to elution times of components in a mixture of standards. Quantitative information can be obtained by comparison of the standard and unknown chromatographic peak areas.

I. Equipment
   Gas chromatograph equipped with flame ionization detector (FID)
   Electronic integrator or strip-chart recorder
   Hypodermic syringe (10 μL)

II. Reagents (see Appendix II)
 Fusel oil standard stock solution
 Ethanol–water (50% vol/vol) standard solution

III. Procedure

1. Install one of the following columns in the GC oven:
   a. 6 ft × 2 mm id glass, packed with 6.6% Carbowax 20M on 80–100 mesh Carbopack B.
   b. 6 ft × 2 mm id glass, packed with 5% Carbowax 20M on 60–80 mesh Carbopack B.
2. Set the conditions listed in Table 2-8 for operating the gas chromatograph for fusel oil analysis.
3. Adjust the air and H_2 gas flows to those specified in instrument operating directions (approximately 300 and 30 mL/min, respectively). These rates may have to be altered somewhat to optimize conditions for the particular instrument and column used.
4. Prepare a working fusel oil standard by pipetting 5 mL of the standard stock solution into a 100-mL volumetric flask and diluting it to the mark with 50% ethanol–water solution. (See Fig. 2-11 for a typical chromatogram.)
5. Inject 0.5 μL of the working standard into the gas chromatograph. Record the chromatogram and note the retention times of the various components. The components should elute in the order: acetaldehyde, methanol, ethanol, ethyl acetate, *n*-propanol, isobutanol, *n*-butanol, active amyl alcohol, and isoamyl alcohol (Fig. 2-11).
6. Inject 0.5 μL of distillate into the gas chromatograph. Record the chromatogram and note the retention times of the various components. Identify fusel oils present by comparison of retention times with those of the standards. An estimate of the amounts of individual fusel oils present may be made by comparison of the peak areas of the distillate sample to those of the standard solution.

Table 2-8. Operating conditions for gas chromatograph (fusel oil analysis).

Conditions	Col. a	Col. b
Carrier gas	N_2	N_2
Flow rate (mL/min)	40	40
Oven temp., °C	80	80
Program rate, °C/min	4	4
Final temp., °C	200	210
Injector temp., °C	250	250
Detector temp., °C	250	250

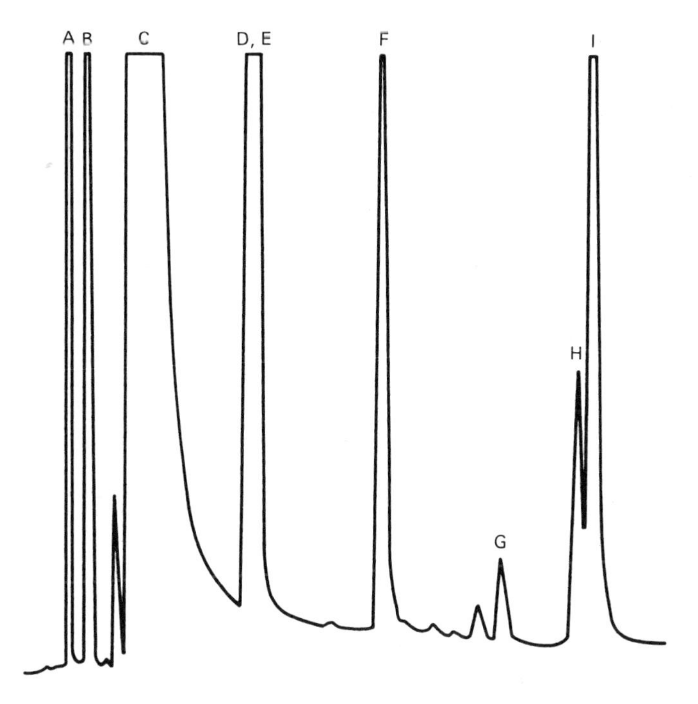

Fig. 2-11. Typical chromatogram of alcohols in 180 proof ethanol. Numbered peaks are: (A) acetaldehyde, (B) methanol, (C) ethanol, (D) ethyl acetate, (E) *n*-propanol, (F) isobutanol, (G) *n*-butanol, (H) active amyl alcohol, (2-methyl-1-butanol), (I) isoamyl alcohol, (3-methyl-1-butanol). The column used was 5% Carbowax 20M on Carbopack B 60/80 mesh.

IV. Supplemental Notes

1. Carbopack columns are capable of resolving complex mixtures of compounds in a variety of alcoholic beverages such as beer and wine. They can also be used to detect acetic, propionic, and other low molecular weight carboxylic acids that are usually adsorbed on other columns.
2. Carbopack columns work best with small or dilute samples. If too large or concentrated samples are used, tailing and poor separation efficiency will result. This is to be expected with the low percent stationary phase used on the packing.
3. Columns should be purchased with the ends packed with phosphoric-acid-treated glass wool. Untreated or silanized glass wool will adsorb acetic acid and inhibit its elution from the column.

Chapter 3

Extract

DEFINITION

Conventionally, extract, as measured by hydrometry, consists of the nonvolatile soluble solids left after dealcoholization of the wine sample (Amerine 1965). The Office International de la Vigne et du Vin (O.I.V.) similarly defines extract as ''the nonvolatile materials in wine,'' but further states that the physical condition for such a definition must be carefully stated. Thus extract includes sugars (or their condensation products), fixed acids, glycerin, 2,3-butylene glycol, and phenols, as well as varying amounts of lactic and acetic acids. Because sugars present in the wine sample are nonvolatile, they contribute to the nonalcohol residue or ''apparent extract.'' The sugar content should therefore be subtracted from the ''apparent extract value'' if one wishes to compare the sugar-free extract contents of different wines.

By international convention, extract is expressed in units of grams/liter (g/L). At present however, data are still frequently expressed as g/100 mL in the United States. Extract levels for dry white wines vary from 20 to 30 g/L (2–3%), whereas dry red wine values often exceed 30 g/L, the increase being attributed to higher levels of phenols. Thus extract content can be used to distinguish light-bodied from heavier-bodied wines. Dry table wines with extract values of less than 20 g/L (2%) often appear thin on the palate, whereas those with 30 g/L (3%) or more have a full-bodied character. Fermentation of low-sugar musts and subsequent fortification can produce wines with low extract values. Furthermore, wines produced by amelioration may also have low extract values.

Extract values depend, in part, upon variations in processing, as well as on the grape variety itself. Extract content may be affected by the type of press used and press pressures. Table 3-1 compares press fractions separated in Methode Champenoise production of Pinot noir cuvees. As can be seen, with increases in pressure there is an increase in extract components such as phenols. Other processing considerations that may increase extract values in wine include the use of pectolytic enzymes. Singleton et al. (1980) note that with increased pomace exposure and fermentation temperature, there are corresponding increases in pH, potassium, protein, and phenol extraction, and therefore extract value. They also note decreased acidity under these conditions.

Table 3-1. California Pinot noir press fractions.

Press fraction	Total phenols (mg/L GAE)	TA (g/L)	pH	Abs. 520 nm	Yield (gal/ton)
1	200	13.0	2.8–3.1	0.25	110
2	250	11.0	3.1–3.25	0.62	20
3	320	9.5	3.3–3.45	1.10	7

SOURCE: Zoecklein (1989).

Processing activities, such as barrel fermentation and storage, increase the extract value as a result of phenol extraction. Storage for one year in new 200-L European oak cooperage may contribute 250 mg/L gallic acid equivalents (GAE) of nonflavonoid phenols (Amerine et al. 1980). By comparison, American white oak contributes about half this level (see Chapter 7).

EXTRACT ANALYSIS

Densimetric Procedures

Extract can be determined from the difference in specific gravity of the original wine sample and that of the alcohol distillate removed from the wine. The specific gravity of the alcohol distillate may also be determined from an accurate value for % ethanol (vol/vol) and an appropriate % alcohol versus specific gravity table (Table 2-5). In another procedure given for extract analysis, the residue from the alcohol distillation is brought back to its original volume with distilled water, and °B is then determined by hydrometer. It should be recalled that °B is defined in the units of g/100 g. To convert to g/100 mL (per U.S. definition of extract), one multiplies the °B by the equivalent specific gravity (Table 1-12). Final results are then expressed in units of g/L.

Several problems may be encountered in carrying out an extract analysis. Most obviously, the physical conditions of analysis (e.g., temperature, size of distillation flask) affect the product(s) formed, lost, or degraded. For example, water, lactic and acetic acids, ethylene glycol, and 2,3-butylene glycol may be differentially lost during distillation. Furthermore, dehydration products resulting from heating of sugars may form, changing the density of the solution formed upon dilution back to the original volume.

Nomographs

In addition to densimetric procedures, nomographs may be used as rapid means of approximating extract values in table and dessert wines. Historically, the Marsh nomograph (Fig. 3-2) for dessert wines has been widely used. However, Vahl (1979) has presented an alternative scale for use with table wines (Fig. 3-1), and he points out that the Marsh nomograph suffers from difficulty of

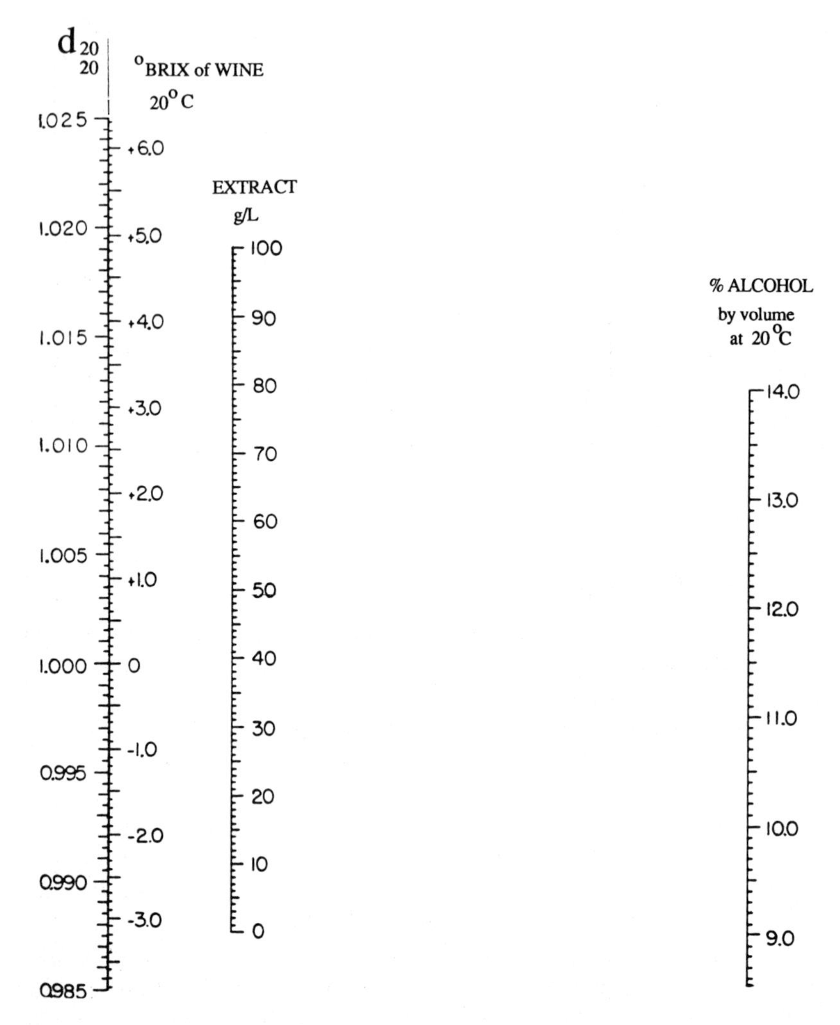

Fig. 3.1. Nomograph for table wine after J. M. Vahl (1979). Measured values of specific gravity (20° / 20°) and % ethanol are used to determine extract (g / L) values for table wines.

interpolation (owing to its relatively wide scale range) as well as scale identification. Specifically, Vahl believes that the scale identified as "Extract" in Fig. 3-2 is, in fact, "Degrees Brix of Extract." In this case, the correct units would be g / 100 g and not g / 100 mL as shown.

As shown in Figs. 3-1 and 3-2, nomographs consist of three scales, "Degrees Brix," "Extract," and "Alcohol"; so if two values are known, the third can

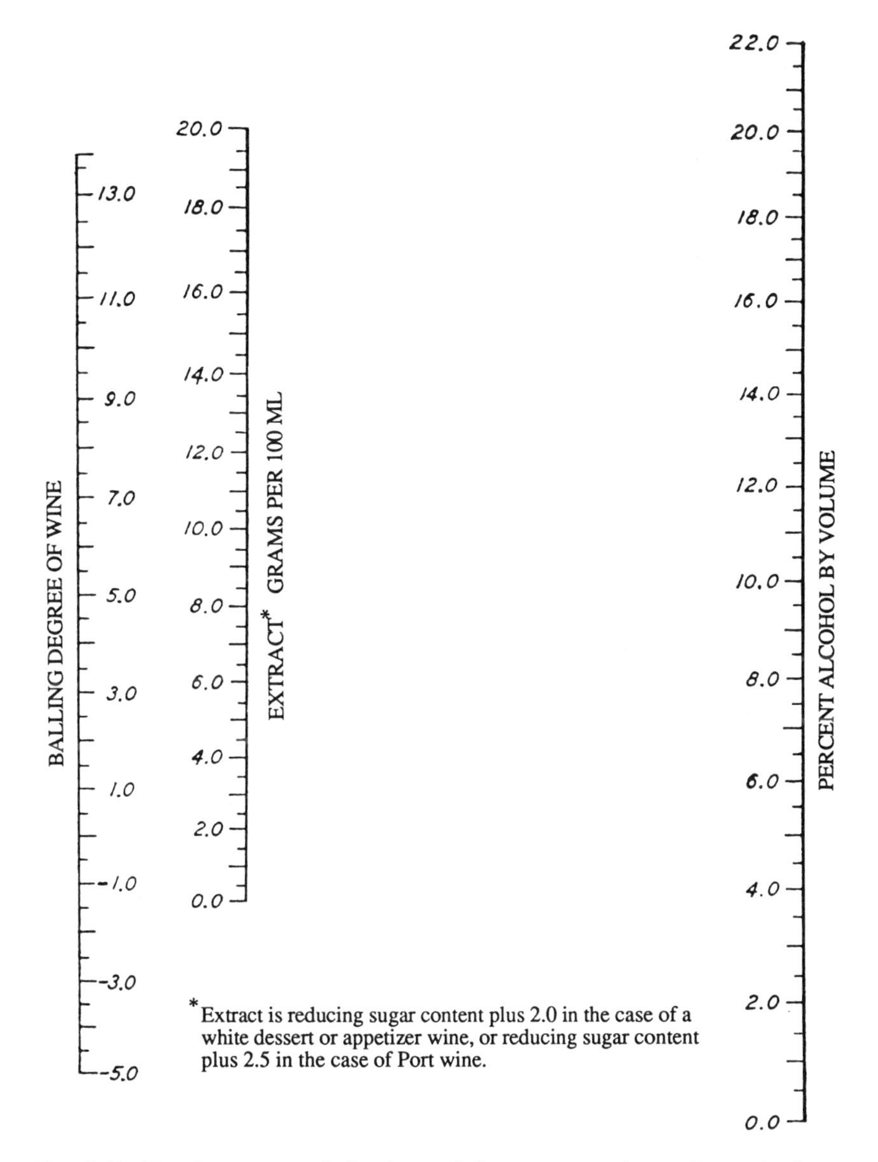

Fig. 3.2. Marsh nomograph for determining extract values of sweet wines.

be obtained directly. As seen in Fig. 3-2, the extract content of a white dessert wine is defined as the reducing sugar content plus 2.0, whereas the extract content of a port-style wine is reducing sugar plus 2.5.

To demonstrate the use of a nomograph, the following two examples are presented using the Marsh scale (Fig. 3-2).

EXAMPLE 1

A winemaker has a sherry with an alcohol of 17.1% (vol/vol) and a reducing sugar content of 2.5 g/100 mL (25 g/L). Determine the °B.

Initially, we know that in the case of white dessert wines, extract is defined as reducing sugar plus 2.0 or, in this case, 4.5 g/100 mL. With a straight edge, align 17.1% on the alcohol scale and 4.5 g/100 mL on the extract scale (Fig. 3-2). Referring to the Brix scale, the intersection value is read as −1.0 g/100 g.

EXAMPLE 2

A winemaker has a port-style wine with an alcohol of 18.5% (vol/vol) and a °B of 6.0 g/100 g. Determine the reducing sugar content.

Aligning the appropriate values of the alcohol and Brix scales in Fig. 3-2 with a straight edge, one can interpolate an extract value of 11.7%. Recalling that the reducing sugar content is defined as extract value minus 2.5 in the case of ports, we arrive at a reducing sugar content of approximately 9.2 g/100 mL (92 g/L).

Procedure 3-1. Extract Determination by Specific Gravity

The *Official Methods of Analysis of the Association of Official Analytical Chemists* (1984) determines extract by the following procedure. In this procedure, extract is determined by taking the specific gravity of the original sample and subtracting the specific gravity of the aqueous ethanol solution distilled from the sample. After adding the specific gravity of water back in, the specific gravity is converted to appropriate units by multiplying by the corresponding °B taken from Table 1-12.

I. Equipment
   - Pycnometer or accurately graduated specific gravity hydrometer
   - Distillation apparatus (see Fig. 2-8)
   - Distilled water

II. Procedure
   1. Determine the specific gravity (20°/20°) of the initial wine sample using a very accurate hydrometer or by pycnometry. One may also measure the °B and convert to specific gravity using Table 1-12.
   2a. Volumetrically transfer 200 mL of the wine sample to the distillation flask. Add approximately 50 mL distilled water.
   2b. Collect approximately 195 mL of distillate. Adjust the solution to volume with distilled water at standard temperature.
   2c. Determine the specific gravity of the distillate as in step 1.
   3. Alternatively, one can determine the alcohol content of the sample by

techniques described in Chapter 2 and convert the measured % Ethanol (vol/vol) to specific gravity using Table 2-5.

4. Calculate extract value according to the following equation:

$$D = S - S' + 1 \tag{3-1}$$

where:

S = Specific gravity of original sample
S' = Specific gravity of aqueous ethanol distillate
D = Specific gravity of dealcoholized sample

5. Using Table 1-12, locate the °B value corresponding to D. Multiply this derived result by the value of D:

$$\begin{array}{c}\text{Extract value}\\ \text{of wine}\\ (\text{g}/100\ \text{mL})\end{array} = \begin{array}{c}\text{Specific gravity X}\\ (20/20)\end{array} \times \begin{array}{c}\text{Corresponding °B}\\ (\text{g}/100\ \text{g})\end{array} \tag{3-2}$$

6. To express the extract value in g/L, multiply the final result by 10.

III. Supplemental Notes

1. The sample distilled in step 2 may be treated with NaOH to eliminate the carryover of volatile acidity, sulfur dioxide, or carbon dioxide. This will not interfere with the measurement of specific gravity of the distillate
2. Extract determinations have been used to detect falsification of wines. Gilbert (1976) proposed the following relationship for calculating a residual (corrected for major acid components) sugar-free extract:

$$E = R - 0.9T - 0.8A - 0.05ET \tag{3-3}$$

where:

E = Sugar-free extract (g/L)
R = Apparent extract − Reducing sugar (g/L)
T = Titratable acidity (g/L)
A = Ash (g/L)
ET = Ethanol (g/L)

Procedure 3-2. Extract Determination Using Brix Hydrometer

Extract is determined by distilling the alcohol out of the wine sample. The nonvolatile residue is brought back to the original volume and temperature and its specific gravity (measured as °B) determined. A unit conversion produces a result for extract in g/L.

I. Equipment

Brix or Balling hydrometer (range: 0–8°B)
200-mL Kohlrausch receiving flask
Hydrometer cylinder
Distilled water

II. Procedure

1. Upon completion of alcohol distillation using 200 mL of initial sample, carefully pour the residue into a clean, dry 200-mL Kohlrausch flask. This step should be followed with several distilled water rinses to ensure complete transfer.
2. Cool to 20°C (68°F), and bring the solution to volume with distilled water. Alternately, temperature corrections may be made using Table 1-11.
3. Mix the solution carefully to ensure sample uniformity.
4. Transfer the contents to a clean, dry hydrometer cylinder.
5. Insert the hydrometer into the solution. When constant level is achieved, record the results as g/100 g of solution (the definition of Brix). To convert results to g/100 mL (as per definition of extract) read the specific gravity corresponding to °B using Table 1-12 and multiply this figure by the recorded °B. To express as g/L, multiply the final result by 10.

III. Supplemental Notes

1. Extract determinations cannot be accurately run on wines that have been neutralized with NaOH prior to distillation, unless one uses the double distillation technique (see Procedure 2-2, Supplemental Note 3). The use of concentrated NaOH (rather than 1–2 N) will minimize volume changes and mass increases.
2. It should be emphasized, and the reader should recall, that the Marsh nomograph (Fig. 3-2) is restricted to dessert wines.
3. Data derived from hydrometric procedures are reported to ±0.1°B; so any report of extract values, using this method, should not exceed this limitation.

Chapter 4

Hydrogen Ion (pH) and Fixed Acids

Hydrogen ion concentration plays a major role in wine production. Its involvements range from physical-chemical and biological concerns to the sensory attributes of wine. Hydrogen ion activity usually is measured in terms of the logarithmic concentration term, pH. Initially, the winemaker is concerned with fruit and must pH. The pH values for white must are often 3.4 or less, whereas slightly higher values usually are observed for red table wines, largely because of extended contact of juice and pomace during fermentation.

It is not uncommon in many grape-growing regions for pH values to rise above optimum levels prior to adequate sugar production, acid reduction, and so on. In addition to the primary effects of climatic and growing conditions, there is evidence that certain varieties typically produce high-pH musts.

INTERACTIONS OF HYDROGEN IONS, TITRATABLE ACIDITY, AND POTASSIUM IONS

Historically, a number of studies have followed the development of the major organic acids and pH in grape berries during maturation (Amerine and Winkler 1942; Hale 1977; Johnson and Nagel 1976). Hydrogen ion concentration (pH) and titratable acidity in mature fruit cannot be explained simply in terms of the concentration of organic acid anions. Investigators have suggested that monovalent metal cations, such as potassium and sodium, enter the cell in direct exchange for protons (H^+) derived from organic acids (Boulton 1980a,b; Hodges 1976). This exchange leads to grapes that have lower titratable acidities and higher pH values than would be expected if one measured the acid anion composition (Boulton 1980a).

Boulton (1980c) proposes a membrane-bound ATPase responsible for potassium uptake (see Fig. 4-1). The evidence for concluding that ATPase-mediated transport occurs in grape berries comes, in part, from observations of the relationship between organic acid levels, titratable acidity, and pH in grape tissue. In grape juice, only about 68 to 74% of the expected protons are found by titration (Boulton 1980d). The missing protons can be accounted for by postulating an ion exchange process in the cell membrane that exchanges potassium and sodium for protons. The discrepancy between the protons expected from

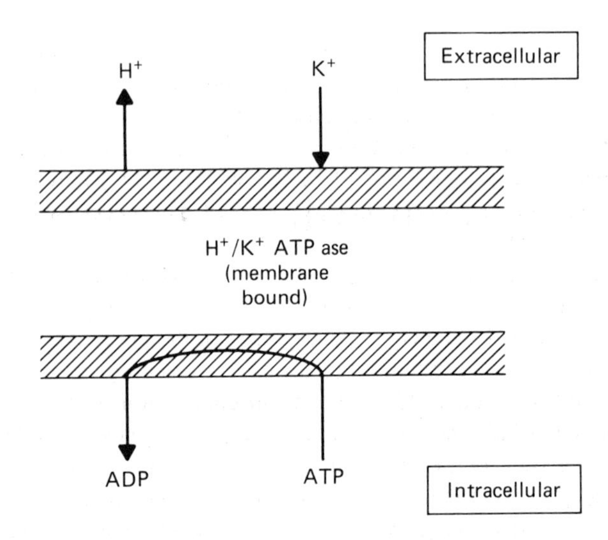

Fig. 4-1. Representation of potassium–hydrogen exchange at the cell membrane.

the organic acid anion content and those found by titration has been shown to be numerically equal to the molar quantity of postassium and sodium ions present (Boultron 1980a,d).

The free proton pool in grape juice generally ranges from 0.001 to 0.01 mol/L, while the concentration of potassium plus sodium ranges from 0.05 to 0.10 mol/L, and the total acid concentration from 0.05 to 0.10 mol/L. As such, the pH value of juice is more sensitive to changes in the concentrations of potassium and sodium than similar molar changes in the principal organic acids—tartaric and malic. At typical juice pH values, an increase of 10% in the potassium and sodium concentrations in the fruit may lead to a shift of approximately 0.1 pH unit. In general, the concentration of potassium in grape berries is about 30 times that of sodium (Amerine and Winkler 1942). Sodium thus usually plays a minor role in ATPase-mediated exchange in grapes (Boulton 1980b).

It has been proposed that potassium is taken up from the soil by hydrogen–potassium ATPase activity in the roots of grapevines. Therefore, the internal levels of ATP in the root tissues, as well as the root system size and the proportion of meristematic tissues in the root system, operate in regulation of potassium uptake.

Once in the roots, cations move by bulk flow to the outer membranes of the leaf and berry cells. The expected presence of an ATPase in the plasmalemma of the cells enables cation transport across the membrane in exchange for internal protons derived from the organic acid pool. Cytoplasmic ATP levels are considered to act as substrate and the primary influence of this transport (Lehninger et al. 1979). Temperature also is considered important in ion transport (Boulton 1980c). The level of monovalent cations, such as potassium and protons in the phloem sap, would be determined by the relative rates of uptake in roots, berries, and other tissues. Potassium uptake by the grape berry may be more rapid than by other tissues such as shoots, petioles, and leaves because of ATP availability (Boulton 1980c).

Freeman and Kliewer (1983) reported that shoots are net exporters of K^+ between anthesis (bloom) and fruit maturity. This export of potassium from the shoots was determined to be equivalent to 43% of the potassium content of the fruit.

The proposed mechanism for potassium accumulation is enzyme-mediated, so exchange is slowed when environmental conditions (i.e., low temperature) do not favor activity of the enzyme. In cool growing regions, the levels of malate in the grape's organic acid pool remain higher than in warmer areas where malate decreases because of respiration. In some instances, conditions may favor greater enzyme activity and uptake, resulting in relatively high levels of berry potassium (>1000 mg/L), high titratable acidity (>10 g/L), and high pH (>3.5) due to the exchange for hydrogen ions from malate. In very cool growing regions, the soil temperature may be sufficiently warm to stimulate enzyme activity and potassium uptake. In these instances, potassium levels and TA are expectedly high (>1000 mg/L and >18 g/L, respectively). However, under such conditions, pH may still remain quite low (2.9–3.2) because of the relatively high concentration of malic acid (Boulton 1985).

In the case of grapes growing in warm regions, enzyme activity is greater because of the temperature, and the acid level is low because of malate respiration. Assuming that potassium is available in the soils and aerobic conditions are present in the root zone, potassium is taken up by the mechanism described and transported throughout the vine, and to the berries, where it exchanges for H^+ arising from malic acid. In this case, pH is high (>3.5), the potassium level is high (>1,000 mg/L), and TA is low (<6 g/L).

Some important parameters in potassium uptake and pH changes include soil, soil exchange capacity, rootstock, vine vigor, leaf shading, cultivar, and crop level, as well as seasonal variations. The rate of potassium uptake by roots appears to be somewhat independent of the concentration of available soil potassium, in that higher concentrations of potassium do not result in greater uptake rates (except when very high). However, if the concentration of potassium is less than that needed for saturation of the hydrogen–potassium ATPase, uptake will be related to soil potassium concentration (Boulton 1985). Waterlogged soils with low oxygen levels allow little respiration to occur in the roots. Thus, not enough energy is generated to activate the enzyme, and little or no potassium uptake occurs.

Shallow soils have more potassium in the upper root system and thus may show greater potassium uptake. Additionally, high-pH soils permit greater potassium uptake. Soils high in magnesium and calcium may cause a reduction in vine potassium accumulation and may lead to potassium deficiency. However, it is difficult to control potassium uptake by soil management.

Devigorating rootstocks, or those that slow vegetative growth, and increased planting density to control vigor may also be utilized to control potassium uptake. Activity that slows vine growth will also slow enzyme activity and thus may affect potassium exchange and pH elevation. Unfortunately, there may not be a good commercially available rootstock for vigor control.

There are differences between different rootstock–scion combinations and hydrogen–potassium ATPase activity (Boulton 1980c). Additionally, enzyme titer may vary between varieties. This is a major reason for the variation seen between pH imbalance and variety.

A higher crop level results in a reduced concentration of potassium principally because of a dilution effect. Higher crop levels in warm climates, however, delay maturity, and, in combination with malic acid respiration, may produce fruit of low TA.

Vines with excessive growth often have a significant amount of leaf shading. Beyond three leaf layers, light exposure is significantly reduced, and shaded leaves are not photosynthetically active. When photosynthesis stops, no sugar is being produced, and ATP may be channeled toward activation of the enzyme for potassium exchange. Thus, additional potassium is pumped into the berry. Malate still may be respired under these conditions, resulting in a decrease in the organic acid pool. As a result of the utilization of malic acid by the plant and uptake of potassium, the fruit has a low TA and high pH.

Unfortunately, attempting to control potassium balance in the vineyard may not be practical. The relationships between leaf shading, vigor, management and fruit and wine chemistry are demonstrated in Tables 1-1 and 4-1.

INFLUENCE OF pH IN WINEMAKING

From the winemaker's viewpoint, must and wine pH levels are important considerations. As with any biological system, pH is involved in a multiplicity of interactions. The opinion of many observers is that pH may be the single most important chemical parameter in premium wine production. Hydrogen ion concentration is known to have significant effects of biological stability, color, oxidation rate, protein stability, bitartrate stability, and overall palatability. Hydrogen ion concentration also plays an important role in microbial activity in must and wine. Generally, the lower the pH, the less likely is the chance for growth of spoilage bacteria and yeast. This phenomenon is not exclusively due to the inhibitory effect of hydrogen ion, but also is related to the concentration of antimicrobial agents, such as sulfur dioxide, sorbic, and fumaric acids, whose active forms are the molecular forms present at lower pH values.

Table 4-1. Expected responses of a ''normal'' pruned and irrigated california vineyard trained on a 2-wire vertical trellis to various cultural practices

Change of Cultural Practices	Light to Cluster	Tartrate (g/berry)	Malate (g/berry)	K (g/berry)	Wine K	Wine Color	Crop Yield	TSS (°Brix)
U-Trellis	Increase	No Effect	Decrease	Decrease	Decrease	Increase	Slight Increase	Increase
Wye-Trellis	Increase	Increase	Decrease	Decrease	Decrease	Increase	Increase	No Effect
GDC Trellis	Large Increase	Increase	Decrease	Decrease	Large Decrease	Increase	Increase	No Effect or Decrease
Severe Pruning	Slight Decrease	No Effect	Increase	No Effect	Decrease	Increase	Decrease	Increase
Topping (Hedging)	Slight Increase	No Effect	Decrease	No Effect	Increase	Decrease	No Effect or Decrease	Decrease
Mod. Water Stress	Increase	Slight Increase	Decrease	Decrease	Decrease	Increase	Slight Decrease	Increase
Removal of Secondary Shoots	Increase	Increase	Decrease	Decrease	Decrease	Increase	Decrease	Increase
Ethephonm (Growth Inhibitor)	No Effect	No Effect	Decrease	No Effect	Large Increase	Large Increase	No Effect To Slight Decrease	No Effect or Slight Increase

SOURCE: Kliewer and Benz 1985.

The dominant wavelength of the wine color also is affected by pH. Increases in the pH of young red wine cause a shift from red to violet or purple hues. Further, high-pH wines tend to be poised toward oxidation. Singleton (1987) points out that twofold increases in the oxidation rate were noted in wines at pH 3.8, compared to wines at pH 3.2.

Hydrogen ion concentration plays an important role in tartrate and protein stability. With respect to protein, lower pH tends to foster more rapid precipitation of unstable fractions. Further, less fining agent generally is required to remove unstable proteins at low pH than would be required to remove an equivalent amount of protein from the same wine at higher pH.

The distribution of tartaric acid into its component species, bitartrate and tartrate, is pH-dependent; so bitartrate stabilization also is closely tied to pH. As seen in Fig. 13-1, bitartrate anion in aqueous solutions (juices) reaches its greatest concentration at pH ~3.5. At pH values greater than ~3.5, potassium bitartrate precipitation consumes hydrogen ions by reaction with tartrate, resulting in an increase in pH. But at pH less than 3.5, potassium bitartrate precipitation releases protons into the medium, thereby lowering pH. In the case of wine, which contains alcohol, the actual pivotal pH is ~3.65. (For further discussion, see Chapters 13 and 18.)

HYDROGEN ION CONCENTRATION AND BUFFERS

The molar concentration of hydrogen ion (H^+) is fruit juice and wine ranges from 10^{-2} mol/L, in the case of lemon and certain other citrus juices, to slightly less than 10^{-4} mol/L for grape juice and wine produced from grapes grown in warmer climates. These molar concentrations are equivalent to pH 2 and 4, respectively, using the classic relationship of Sorenson:

$$\mathrm{pH} = -\log\,[\mathrm{H}^+] \tag{4-1}$$

$$\mathrm{pH} = \log \frac{1}{[\mathrm{H}^+]}$$

Hydrogen ions originate from the dissociation of parental acids. Chemically, there are two general groups of acids with respect to their potential to liberate free hydrogen ions into solution. These include, first, the strong inorganic acids, such as hydrochloric and nitric, which totally dissociate in aqueous solution:

$$\mathrm{HA} \longrightarrow \mathrm{H}^+ + \mathrm{A}^- \tag{4-2}$$

Such acids presently are of relatively little importance to the winemaker. However, by comparison, weak organic acids, such as tartaric, malic, citric, lactic, and acetic, make up the major acid content of grapes and wine. These weak acids dissociate to only a small extent relative to the dissociation of strong acids:

$$HA \rightleftharpoons H^+ + A^- \quad (4\text{-}3)$$

The degree of dissociation of a weak acid is denoted by the dissociation constant (K). Dissociation constants are generally on the order of 10^{-5} for typical wine acids; so the extent of weak acid dissociation in must and wine amounts to about 1%. That is, 99% of the acid content is in the undissociated form (Planc et al. 1980). These constants are frequently utilized in their logarithmic form (pK). Table 4-2 presents pK values of wine acids.

pH vs. Titratable Acidity

A clear distinction should be made between pH and titratable acidity (TA). The former is a measure of the free proton content of the solution, whereas the latter is dependent upon the concentration of wine acids as well as the extent to which they dissociate. Titratable acidity is measured by potentiometric or indicator titration using standardized sodium hydroxide to the agreed-upon end point of pH 8.2. It thus measures free [H^+] plus any undissociated acids that can be neutralized by base. Hydrogen ion concentration (pH), by comparison, is a measure of free H^+ present in solution and is a measure of ion activity. As such, pH depends not only upon the concentration of acids present in the system, but also on their relative degrees of dissociation or activity. [*Note*: The AOAC

Table 4-2. Dissociation values of organic acids present in wine and must.

Wine acid	Molecular weight	Wine acid	Molecular weight
Tartaric acid	150.09	Acetic acid	60.05
Malic acid	134.00	Succinic acid	118.09
Citric acid	192.14	Lactic acid	90.08

Acid	K_A	pK[1]	pK (20°C)[2] Must	pK (20°C)[2] Wine (12% alcohol)
Tartaric	(1) 9.1×10^{-4}	3.04 (25°C)	2.98	3.14
	(2) 4.26×10^{-5}	4.37	4.11	4.32
Malic	(1) 3.5×10^{-4}	3.46 (25°C)	3.38	3.55
	(2) 7.9×10^{-6}	5.10	4.82	5.05
Citric	(1) 7.4×10^{-4}	3.46 (25°C)	3.06	3.23
	(2) 1.74×10^{-5}	4.76	4.44	4.64
	(3) 4.0×10^{-7}	6.40		
Acetic	1.76×10^{-5}	4.75 (25°C)	4.66	4.79
Succinic	(1) 6.16×10^{-5}	4.21 (25°C)	4.11	4.29
	(2) 2.29×10^{-6}	5.64	5.36	5.56
Lactic	1.4×10^{-4}	3.86 (25°C)	3.79	3.96

[1]From *Handbook of Biochemistry*, The Chemical Rubber Co. (1968).
[2]Calculations courtesy of Dr. J. Vahl. (1978).

procedure for the measurement of titratable acidity appears under the title "Total Acidity." This is different from what Boulton (1982) defines as total acidity, as described earlier in this chapter.]

Buffering Capacity

Containment of pH within narrowly defined limits is critical to the physiological well-being of virtually every living system. The buffering capacity of a juice or wine is a measure of its resistance to pH changes. Essentially, it is a measure of the organic acid pool (malic and tartaric) at winemaking pH. A system with a high buffering capacity requires more hydroxide (OH^-) ions or hydrogen ions to change the pH than one of lower buffering capacity. Thus, buffering capacity can be defined, in practical terms, as the quantity of hydroxide or hydrogen ions needed to effect a change of one pH unit (e.g., from 3.4 to 4.4). An example of buffering capacities in relation to other typical wine parameters is presented in Table 4-3.

Buffers occur as either weak acids and their conjugate bases, or as weak bases and their conjugate acids. The net result of buffering action is to create within the system resistance to changes in pH that otherwise would occur with addition of either acid or base. In the case of base addition, excess OH^- ions are consumed by H^+ ions of the buffer's acid component, to form water, whereas excess H^+ ions are consumed by the anion component:

$$HA \rightleftharpoons H + A^- \quad \text{Base addition} \quad OH^- + H^+ \longrightarrow H_2O \qquad (4\text{-}4a)$$

$$BA \rightleftharpoons B + A^- \quad \text{Acid addition} \quad H^+ + A^- \longrightarrow HA \qquad (4\text{-}4b)$$

An example of the effect of buffering capacity is shown in Fig. 4-2. In this case, the weak organic acids present in grape juice are neutralized with a strong base (0.1 *N* NaOH) to yield water. During neutralization, the ratio of $[A^-]$ to [HA] in the initial dissociation ($HA \Leftrightarrow [H^+] + [A^-]$) varies. Theoretically, the ratio of $[A^-]/[HA]$ for single components of solution (e.g., $R\text{–}NH_3^+/R$–

Table 4-3. Buffering capacity of various Cabernet sauvignon wines in relation to other wine parameters.

Wine no.	% Alcohol (v/v)	Extract (g/L)	pH	TA (g/L)	Buffering capacity* (mL)
1	13.4	23.2	3.42	6.55	15.1
2	11.3	29.3	3.38	7.90	18.3
3	10.3	20.2	3.58	5.49	15.2
4	12.6	30.4	3.43	8.22	17.2

*Buffering capacity determined using 0.1 *N* NaOH.
SOURCE: Zoecklein 1987.

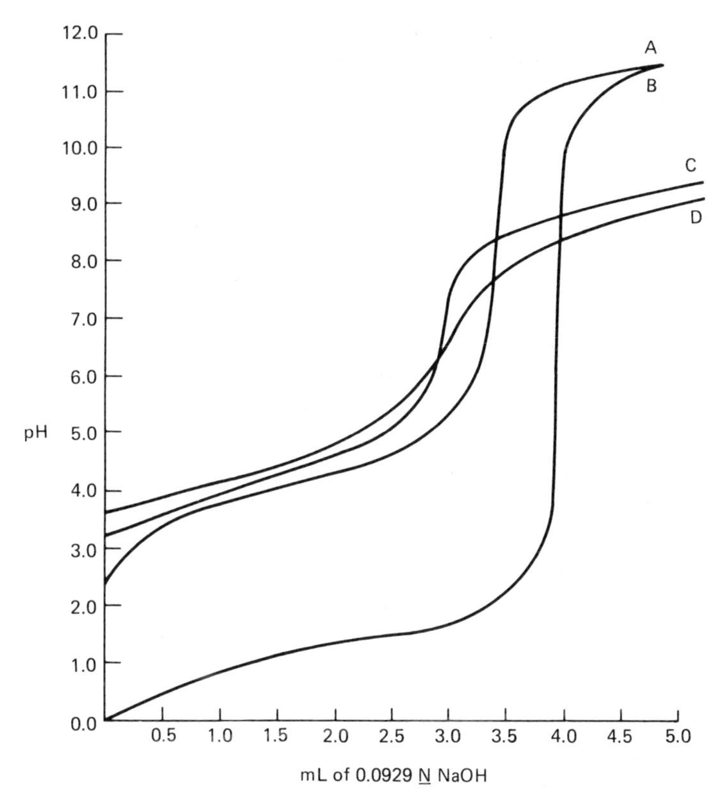

Fig. 4-2. Titration of weak and strong acids with standard base: (A) Acetic acid, (B) Strong acid, (C) Mission grape juice, and (D) French Colombard juice.

NH_2) may be calculated at any pH using the Henderson–Hasselbach relationship:

$$pH = pK_a + \log \frac{[A^-]}{[HA]} \tag{4-5}$$

This relationship may be used to calculate the concentration of acids in undissociated and ionized forms, as well as other components of solution that are capable of ionization and whose dissociation values (pK_a) are known.

ORGANIC ACID CONTENT OF WINE

The acid content of a wine is important from the standpoint of flavor and, indirectly, because of its effect on the pH, color, stability, and shelf life of the product. Titratable acidity in grapes normally runs between 5.0 and 16.0 g/L, although the acid levels obtained are influenced by variety, climatic conditions, cultural practices, and maturity of the fruit.

Table 4-2 lists several organic acids present in wines. In the case of di- and tricarboxylic acids, their several salts also may be present. These organic acids, along with others to be discussed later, are commonly grouped into the categories of "fixed acidity" and "volatile acidity," their sum being "total acidity." Thus, tartaric, malic, lactic, and succinic acids comprise the fixed acid contingent, whereas acetic, propionic, butyric, and sulfurous acids comprise the routinely encountered volatile acids in wine.

The organic acid content of a wine is traceable to three sources. The grape itself contributes tartaric, malic, and, to a much lesser extent, citric acid. By comparison to tartaric and malic acids, which are present at levels of around 0.2 to 1.0% and 0.1 to 0.8%, respectively, citric acid is found in unfermented grapes at 0.01 to 0.05% (Amerine et al. 1980).

Alcoholic fermentation results in formation of lactic, acetic, and succinic acids in addition to very small quantities of other Krebs cycle acids. Lastly, bacterial involvement may produce significant amounts of lactic, acetic, propionic, and butyric acids. Mold growth on the grape may result in gluconic acids levels of 0.15 to 1.0% in the finished wine (McCloskey 1974).

TITRATABLE VS. TOTAL ACIDITY

There is often confusion between the terms titratable acidity and total acid content. The two are used interchangeably by the AOAC to refer to the same measurement. Total acid content may be defined as the concentration of organic acids in the grape (Boulton 1982). It is a measure of the hydrogen ion concentration plus the potassium and sodium ion concentrations. Titratable acidity, on the other hand, is only a measure of hydrogen ions consumed by titration with standard base to a defined end point, usually pH 8.2. Titratable acidity depends on both the amount of acid present and the pH. Titratable acidity is calculated (in the United States) in terms of grams tartaric acid per liter. The equation utilized to calculate this value uses the equivalent weight of tartaric acid (75 g/equivalent) in the formula. Thus this equation is constructed on the assumption that each tartrate species present has two protons to be neutralized by the NaOH titrant. Higher-pH wine or juice samples with some monoprotonated titratable acids present would require less NaOH for the same absolute amount of tartrate or malate. Thus the values obtained by titration [titratable acidity] frequently will indicate less than the absolute concentration of wine or juice acids ("total acid content"), as defined by Boulton.

Acid–Base Titrations

Figure 4.2 compares a typical titration of a strong acid with strong base to the titration of a weak acid with strong base. Grape juice may be considered as a 0.1 *N* acid solution. One may determine the concentration of acid(s) in solution by titration with standardized base. As seen in Fig. 4-2, titration of a weak acid with strong base causes the equivalence point to be displaced to the alkaline side of the pH scale. It is necessary, therefore, to utilize an indicator whose end point is sufficiently basic that it will ensure titration of all available acid groupings. Phenolphthalein, whose end-point color change is in the pH range 8.2 to 10, is the indicator of choice for most organic acid–base titrations.

Interferences in Acid–Base Titrations

Several problems that may be encountered in titrating wine acids are briefly summarized here.

1. Phenolic compounds and amino acids are amphoteric substances, acting as either acids or bases, depending upon the pH of the solution. At low pH values, these compounds act as Bronsted bases, giving up protons and consuming a portion of the added base. The pH dependency of their structures is shown below:

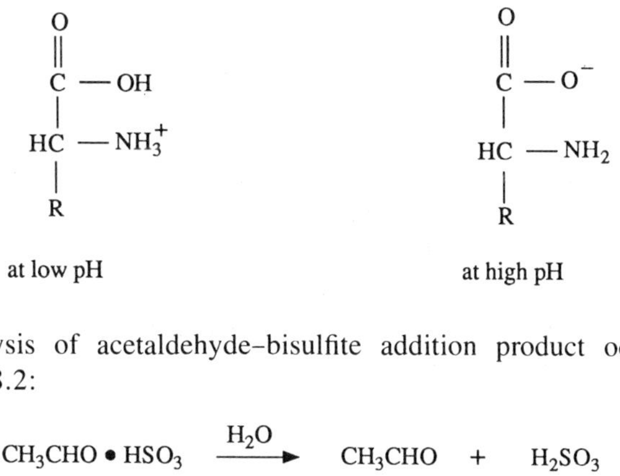

$$\underset{\text{at low pH}}{R-CH(NH_3^+)-C(=O)-OH} \qquad \underset{\text{at high pH}}{R-CH(NH_2)-C(=O)-O^-}$$

2. Hydrolysis of acetaldehyde–bisulfite addition product occurs at pH approaching 8.2:

$$CH_3CHO \bullet HSO_3 \xrightarrow{H_2O} CH_3CHO + H_2SO_3 \qquad (4\text{-}6)$$

The nature of this addition product is more fully developed in Chapter 9.

3. Sugar oxidation begins at pH approaching 8.2:

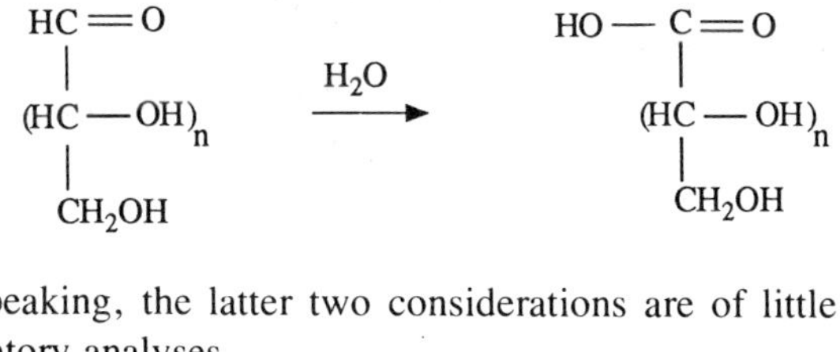

$$HC{=}O-(HC-OH)_n-CH_2OH \xrightarrow{H_2O} HO-C{=}O-(HC-OH)_n-CH_2OH \qquad (4\text{-}7)$$

Practically speaking, the latter two considerations are of little importance in routine laboratory analyses.

4. Carbon dioxide within the sample is titratable as carbonic acid and, as such, can contribute significantly to analytical error:

$$CO_2 + H_2O \longrightarrow H_2CO_3 \xrightarrow{\text{pH 4-6}} HCO_3^- \xrightarrow{\text{pH 8}} CO_3^= \quad (4\text{-}8)$$

There are several procedures for removal of carbon dioxide discussed in the literature, but a commonly utilized measure calls for the addition of a large excess of neutralized distilled water. If the laboratory worker elects to drive carbon dioxide from solution by heating, caution should be taken to avoid loss of volatile acids from prolonged heating. Such losses are reflected in undefinable decreases in apparent levels. Alternately, wine samples may be degassed under vacuum.

5. End-point detection frequently is a major source of error. Especially in red wines, the pink phenolphthalein end point is difficult to detect. Although the problem may be somewhat overcome by dilution of the sample with boiled distilled water prior to titration, the authors recommend the routine use of a pH meter to monitor the progress of titration. By U.S. convention, the sample should be titrated to pH 8.2. The O.I.V., however, suggests a titration end point of pH 7.0. Wong and Caputi (1966) reported 99.5% neutralization of hydrogen ions at pH >7.7; so an end point of 7.0 may represent incomplete titration.

SAMPLE PREPARATION AND REPORTING OF RESULTS

Many constituents of the fruit are not uniformly distributed throughout the berry; so standarized berry sample preparation is an important consideration. The magnitude of these differences is readily seen in Table 1-6. For detailed information on sample preparation, the reader is referred to Chapter 1.

In the United States, results of titratable acid analysis routinely are expressed in terms of the dominant acid contained in the sample, a value achieved by inclusion of the equivalent weight for the respective acid in Equation 4-13 (see Procedure 4-2). The titratable acidity of grape wines in the United States is expressed in terms of tartaric acid. Use of tartaric acid for the expression of titratable acidity is not a standard procedure throughout the world wine community. In France, for example, titratable acidity is expressed in terms of sulfuric acid. By comparison, the German definition of TA is the volume of 0.1 *N* base needed to fully titrate 100 mL of sample.

In analyses of wine or musts, where tartaric is not the principal acid component and one wishes to express results in terms of this acid, the appropriate equivalent weight must be taken into account. (See Table 4-4, for the appropriate values.) Therefore, if it is necessary to express results in terms other than as tartaric acid, one need only use the appropriate equivalent weight of the acid in question.

Table 4-4. Equivalent weight conversions for wine acids.

Acid	Milliequivalent weight (g/meq)
Tartaric	0.075
Malic	0.067
Citric	0.064
Sulfuric	0.049
Lactic	0.090
Acetic	0.060

ADJUSTMENTS IN TITRATABLE ACIDITY AND pH

Adjustment of the acid content or pH of a must or wine may be necessary to produce a well-balanced product. Several of the most commonly employed methods of adjustment are discussed briefly below.

Organic Acid Addition

Several food-grade organic acids currently are approved as wine additives. A complete list of the chemicals that may legally be added to wine is found in Section 240.1051 of the Bureau of Alcohol, Tobacco and Firearms publication *Wine* (Part 240 of Title 27). There may be restrictions on acids approved for different wine types. The most commonly used acidulants in grape wine production are food-grade tartaric, malic, and citric acids. Fumaric acid is sometimes used for control of malolactic fermentations (see Chapter 12). As noted, the TA of grapes and grape wines is expressed as tartaric acid. Apple wines are expressed in terms of their dominant acid (malic), whereas all other fruit wines are expressed in terms of citric acid.

Many winemakers prefer to make necessary acid additions to the must rather than solely to the wine. Additions at this stage may help to maintain a low pH during fermentation, enhance color extraction, and generally produce a more desirable product. At this stage in wine production, tartaric acid generally is used. Several arguments can be offered to support this decision. First, tartaric is a relatively nonmetabolizable acid when compared with malic and citric. Second, tartaric acid is a stronger acid than either malic or citric; so, one achieves a greater pH adjustment per pound of acid used. Third, when such additions can achieve a pH drop to less than 3.65, subsequent precipitation of potassium bitartrate during bitartrate stabilization produces one hydrogen ion per molecule of potassium acid tartrate formed:

$$K^{+} + H_2T \rightleftharpoons KHT + H^{+} \qquad (4\text{-}9)$$

Thus, bitartrate stabilization can be achieved without increases in pH. A reduction in titratable acidity of as much as 2 g/L may occur. Thus the winemaker has achieved the goal of pH reduction without an excessively high TA, remaining.

Some winemakers feel that acid additions are best made using combinations (ratios) of malic and tartaric acids that approximate those of grapes grown in their areas.

For additions to finished wine, citric acid also has been employed. It has the advantages of not forming insoluble precipitates with potassium or calcium in alcoholic solution, as tartatic acid does. Futhermore, citric acid is active in iron chelation and thus helpful in preventing or reducing ferric phosphate casse formation. (The topic of metal casse formation is discussed in Chapters 14 and 15.) Individuals wishing to express citric acid additions in terms of other acids are referred to Table 4-4. As a widely used rule of thumb, 1 pound of citric acid per 1000 gallons raises the TA by 0.13 g/L, expressed as tartaric acid. This relationship results from the eqivalent weights of citric and tartaric acids.

Addition of citric acid to unfermented must should be avoided. Besides slowing the onset of alcoholic fermentation because of the inhibition of EMP pathway enzymes, wine spoilage microbes—especially lactic acid bacteria—may rapidly metabolize citric acid, yielding acetic acid (Peynaud 1984). Furthermore, because citric acid is not a dominant acid in grapes, large additions may be expected to significantly alter the ratio of acids and thus the wine character. Typically, large quantities of this acid, when added to grape wine, result in what many would regard as citrus-like flavor.

Amelioration

In certain winemaking regions of the world, it is permissible to add dry or liquid sugar, water, or a combination of sugar and water, to reduce the acidity and/or increase the degrees Brix of grape must or wine. In California, this practice is permitted only in the production of sparkling wines, special naturals, and fruit wines. If the winemaker wishes to increase the °B in low-sugar musts or to adjust the final residual sugar level of table wines, grape juice concentrate is used. In some cases, amelioration of table wine musts may be permitted, but the resultant product may not carry a California appellation. Such wines may be designated as "American." Federal regulations define the limits of amelioration based upon titratable acidity of the must or wine, provided that the final fixed acid is not lowered to less than 5.0 g/L, expressed as the predominant acid. If the acidity of the must does not exceed 5 g/L, no addition of water is allowed except that amount needed to reduce the soluble level to 22° B. In any case, the maximum level of amelioration permitted in grape and most fruit wine is 35% (vol/vol). In the case of grape and apple wines, the major acids are tartaric and malic, respectively. For all other fruit wines, citric is considered the predominant acid.

Amelioration of must or wine with water has the obvious economic advantage of increasing the yield. However, the process also dilutes aroma, flavor, extract, color etc. In certain intensely flavored varieties, this may be desirable. The expected reduction in extract values for ameliorated wines may be perceived as reduction in body, and so may detrimentally affect the character of certain wines. Reduction in titratable acidity due to amelioration varies with variety as well as level of amelioration. As a result of the unpredictable outcome of this technique, it is often an inappropriate means of deacidification.

Carbonate Deacidification (see Chapter 18)

Neutralization of wine acidity by the use of carbonates is employed in both the U.S. and European wine communities. In both cases, the practice is restricted to potassium and calcium carbonate and potassium bicarbonate. Because of the relatively high pK values for malic acid and the solubility of reaction products, tartaric is the only wine acid to be neutralized under normal conditions of use. The net reaction involves carbonate neutralization of the two hydrogen ions available on tartaric acid and formation of carbon dioxide and water:

$$\underset{\text{carbonate}}{CO_3^{=}} \xrightarrow{H^+} \underset{\text{bicarbonate}}{HCO_3^-} \xrightarrow{H^+} \underset{\text{carbonic acid}}{H_2CO_3} \longrightarrow CO_2 + H_2O \qquad (4\text{-}10)$$

Successful use of a particular carbonate depends upon juice/wine chemistry. For example, high-acid/high-pH juices and wines are best deacidified with calcium carbonate (because of high tartrate ion concentrations), whereas high-acid/low-pH juices and wines are best treated with potassium-containing salts (taking advantage of potassium bitartrate precipitation) (Mattick, 1984). The addition levels needed to achieve the desired reduction in acidity vary for each carbonate. A reduction in titratable acidity of 1.0 g/L requires additions of 0.90 g/L of potassium bicarbonate and 0.62 g/L of potassium carbonate, respectively.

Potassium and calcium components of the salt may become involved in secondary precipitation reactions with bitartrate and tartrate anions, yielding potassium bitartrate and calcium tartrate, respectively. In the case of carbonates containing potassium, precipitation of the salt is improved by chilling the wine or must in advance of the addition. In the case of calcium carbonate additions, however, chilling has little, if any, effect on the rate of precipitation. There is evidence that lower temperatures may favor long-term reduction (Clark et al. 1988). Further, calcium tartrate precipitation is particularly troublesome because, calcium tartrate salts may take weeks to months to form. Unless corrected during post-fermentation processing, calcium tartrate precipitation may well occur in bottled wine (see Chapters 13 and 18).

Double Salt Deacidification

Addition of calcium carbonate to grape juice or wine results in the formation of calcium salts of tartaric and malic acids. Although calcium malate usually is soluble at cellar temperatures, calcium tartrate precipitates from solution, creating an imbalance. This problem may be overcome by the addition of proprietary compounds such as Acidex to the must. The reaction mechanism, as elaborated by Steele and Kunkee (1978), is presented below (Equations 4-11a,b).

Theoretically, the ratio of tartaric and malic acids remaining in solution is constant. This represents the potential advantage of double salt precipitation over the use of calcium carbonate. The mechanism of reaction involves utilization of calcium carbonate to raise the solution pH to a level at which precipitation of the double salt is favored:

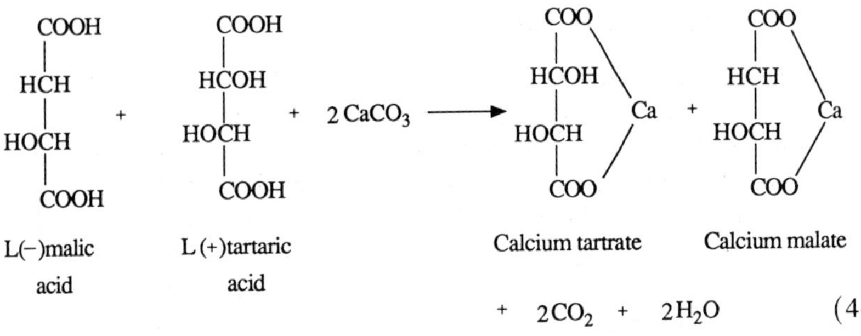

L(−)malic acid (COOH–HCH–HOCH–COOH) + L(+)tartaric acid (COOH–HCOH–HOCH–COOH) + $2\,CaCO_3$ → Calcium tartrate (COO–HCOH–HOCH–COO, Ca) + Calcium malate (COO–HCH–HOCH–COO, Ca) + $2CO_2$ + $2H_2O$ (4-11a)

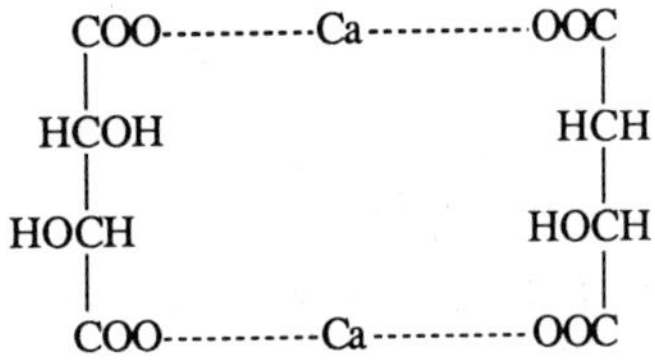

COO----------Ca----------OOC
HCOH HCH
HOCH HOCH
COO----------Ca----------OOC (4-11b)

"Double Salt"

To enhance precipitation, Munz (1960, 1961) included within the carbonate addition approximately 1% crystals of the double salt to serve as nuclei for subsequent crystal formation. In order for the above reaction to operate, the hydrogen ion concentration must be above pH 4.5, or calcium tartrate formation will prevail. At this point, both acids are present as their dicarboxylate anions:

$$R(COOH)_2 + 2OH^- \rightleftharpoons R(COO^-)_2 + 2H_2O \qquad (4\text{-}12)$$

pKa Tartaric acid = 3.1, 4.3

pKa Malic acid = 3.6, 5.1

Following addition of the Acidex to musts, there is an expected rapid increase of pH, followed by slight increases during the storage period. Steele and Kunkee (1978) reported an average decrease in titratable acidity in wines of 3.7 g/L and pH increase of 0.16. In red wines, by comparison, titratable acidity decreased approximately 4.5 g/L with increases in pH of 0.17. However, these workers were unable to demonstrate equimolar precipitation of malic and tartaric acids in their study, a situation they attributed to the lower initial concentration of malic acid in the musts. In this case, tartaric was the principal acid removed. Double salt deacidification is usually limited to grape musts, owing to the lower percentage of free acid found in most grape wines. When this technique is used for deacidification of grape musts (and wines), fixed acid levels may not be reduced to less than 5.0 g/L as tartaric acid (Part 240 of Title 26, 1974).

Ion Exchange (see Chapter 18)

Ion exchange technology has found application in acid reductions in wine. In the case of anion exchange, weakly bound hydroxyl groups on the resin are exchanged for acid anions present in the wine. The hydroxyl groups combine with the dissociated proton from the acid to yield water. Because the acid anion remains bound to the resin, the wine is lower in acidity after the exchange. The fixed acid level of such grape wines may not be reduced to less than 4.0 g/L for reds and 3.0 g/L for whites (White Part 240 of Title 26, 1974). The process, although successful in achieving reductions in total acidity, generally results in decreased overall wine quality. For more specific information regarding ion exchange theory and its applications, consult Chapter 18.

Malolactic Fermentation (see Chapter 12)

Although the pathways and end products may vary, the overall result of this bacterial fermentation is the decarboxylation of L-malic acid, forming L-lactic acid and carbon dioxide. As malolactic fermentations result in conversion of diprotic malic acid into monoprotic lactic acid, the process may result in significant reductions in the titratable acidity and increases in pH of 0.20 or more (Bousbouras and Kunkee 1971). These changes depend upon the species or strain of lactic bacteria used and the concentration of L-malic acid in the wine.

Many winemakers believe that this bacterial conversion contributes complexity and flavors that are not attainable by yeast activity or aging. If the winemaker desires a malolactic fermentation in an already high-pH must, he or she should anticipate the need for subsequent pH adjustment. It is easiest to make such adjustments in pH in the juice or immediately following the completion of the bacterial fermentation. Tartaric acid addition prior to bitartrate precipitation has the advantage of causing pH reduction and acid precipitation in the form of KHT.

Use of Selected Yeasts

Yeasts of the genus *Schizosaccharomyces* are known to utilize L-malic acid as a carbon source during alcoholic fermentation. Thus, these organisms have been employed as a means of must deacidification. It has been noted that, unless controlled, overdeacidification may result from their sole utilization. Common members of the group (*Sch. pombe*) are susceptible to competition from other wine yeasts. Taking advantage of this fact, Snow and Gallander (1979) utilized a secondary inoculation with *Sacch. cerevisiae* near desired titratable acidity levels.

Wines produced from musts deacidified by *Schizosaccharomyces* often have 'odd' sensory properties and pronounced imbalances. At present a good deal of sensory work is needed to properly evaluate the possibilities of such dual fermentations.

The utility of individual strains may become more important as the field of genetic engineering develops. Such potential includes inclusion of the fully expressed gene(s) for the malolactic fermentation within selected strains. Initial work in this area has been successful in implanting the appropriate genes, but the level of expression is far less than that needed for it to be of practical value to the winemaker.

Blending

Blending can be used to obtain proper acid balance by combining high-acid wines with those of lower acidity. Blending operations that result in pH shifts may affect both the biological and the chemical stability of the wine (see Sections II and III of this text).

LEGAL CONSIDERATIONS

California requirements for the fixed acid content of grape wines are presented in Table 4-5. Citric, tartaric, and malic acids may be used to correct deficiencies in grape must acidity, provided that the final titratable acidity, expressed as

Table 4-5. Legal restrictions for total acid content of grape wines.

Wine type	Minimum tartaric acid level (g/L)
Red table wine	4.0
White table wine	3.0
All others	2.5

Adapted from California Administrative Code (Title 17, Section 17,005).

tartaric acid, does not exceed 8.0 g/L. On occasion, fumaric acid is used to inhibit malolactic fermentations (see Chapter 12). Limits for addition of this acid are 25 pounds per 1000 gallons. The fumaric acid concentration in finished wine may not exceed 3.0 g/L. Other pertinent regulations regarding such additions may be obtained by reference to cited government publications.

SENSORY CONSIDERATIONS

Although the acidic character of wine is due to hydrogen ion concentration, both pH and acidity play important roles in the total sensory perception of this stimulus. Amerine *et al.* (1965) noted that at equivalent levels of acidity, the order of perceived sourness of the common wine acids is malic, tartaric, citric, and lactic. Threshold levels for these acids are presented in Table 4-6. Ethanol is effective in increasing acid thresholds, and the increase is even more dramatic with the inclusion of sucrose. Phenols may also be active in increasing minimum detectable acid levels.

SUMMARY OF MAJOR INVOLVEMENTS OF pH IN WINEMAKING

Hydrogen ion concentration (pH) may well be the single most important parameter in winemaking. Its involvements bridge the areas of microbiological, physical, and chemical stability as well as sensory concerns. To emphasize its importance, we have summarized the major areas of interest among winemakers.

Microbiological Activity

Hydrogen ion concentration determines the activity of antimicrobial additives used in winemaking. One of the clearest examples of this is the fraction of free sulfur dioxide in the molecular, or antimicrobial, form. With decreases in pH, the percentage of the molecular form increases dramatically. Thus, the amount of sulfur dioxide needed to inhibit microbial activity in a low-pH must or wine is much lower than that needed for its high-pH counterpart (consult Chapter 9).

Table 4-6. Threshold levels for wine acids in water.

Acid	Threshold level (g/L)
Citric	0.70
Malic	0.50
Tartaric	0.50

Adapted from Berg et al. (1955).

The antimicrobial action of sorbic or fumaric acids also depends largely upon pH. Again, lower pH is associated with greater activity and, thus, decreased levels required to inhibit microbial activity.

Potassium Bitartrate Stability

The ionization and the distribution of tartaric acid into its respective forms, bitartrate and tartrate, are pH-dependent. Hence, tartrate stability is critically dependent upon pH.

Protein Stability

The charge distribution on proteins is related to pH; so its role in protein stability, as well as the level of bentonite needed to achieve stability, represents a major area of involvement.

Oxidation

Hydrogen ion concentration influences the rate of oxidation in red and white wines. Seemingly minor increases in pH may significantly increase the rate of oxidation. The rate of polymerization of phenolics in must and wine is also affected by pH.

ANALYSIS OF WINE ACIDS

Organic acids present in juice or wine may be determined chromatographically by HPLC (Procedure 4-3) or by thin-layer (TLC) techniques. In addition, one may perform chemical or enzymatic analyses for the individual acids present. For example, tartaric acid may be chemically determined by the metavanadate procedure (Chapter 13); malic, citric, and some other acids may be determined by enzymatic techniques. Analysis of L-malic acid is presented in Procedure 4.4.

Estimations of Tartrate/Malate Concentrations

Norton and Heatherbell (1988) have reported development of a titrametric procedure for estimating the tartrate and malate concentrations of juice samples. The procedure involves titration with 0.1 N NaOH over the pH ranges 2.70 to 3.00 and 4.50 to 4.80 using a pH meter capable of being read to the nearest 0.01 pH unit. Estimates of tartrate and malate are made from comparisons with standard curves. The accuracy of this analysis was reported to be 9% and 15% for malate and tartrate, respectively. It may prove to be a useful inexpensive means for monitoring acid levels in maturing fruit.

Procedure 4-1. Hydrogen Ion Concentration (pH)

Hydrogen ion concentration generally is expressed as pH. It is measured using a commercially available meter and glass electrode setup. Standard buffer solutions are used to calibrate the pH meter, and sample pH values are then read by placing the electrode in an undiluted portion of juice, must, etc.

I. Equipment
   pH meter (analog or digital unit) equipped with pH and reference electrode or a combination electrode
   50-mL beaker

II. Reagents (see Appendix II)
   Buffers pH 7.0 and pH 4.00 (pH 3.55 can be prepared from a saturated solution of potassium bitartrate)
   KCl internal filling solution
   Cleaning solution (75% methanol)
   Distilled water

III. Standardization of Meter
   1. Consult the operator's manual for standardization using two buffer solutions.
   2. Rinse the beaker with the sample. Place enough fresh sample in the beaker to cover electrode junctions.
   3. Place electrode(s) in the sample.
   4. Allow the meter reading to stabilize, and record value.

IV. Supplemental Notes
   1. *Never* remove electrode(s) from buffer or sample solutions while instrument is responding. During changing of samples, the meter should be in the standby mode.
   2. When not in use and between measurements, store the electrodes according to the manufacturer's recommendations.
   3. The reference electrode should be checked routinely to ensure an ample supply of KCl filling solution. The level should not be allowed to drop much below the filling port. This port should be uncovered during use.
   4. Electrodes should be periodically cleaned to ensure proper operation. This may be achieved by immersing them in a solution of 75% methanol followed by several rinses in distilled water. Protein contamination is most easily removed by soaking electrodes in a solution of pepsin and 0.1 *N* HCl.
   5. Ceramic junctions may be cleaned by soaking the electrode in hot (not boiling!) distilled water.
   6. If single buffer standardization is used for routine work a saturated potassium bitartrate solution or a buffer close to the pH of material being analyzed is preferred. Two buffer standardizations normally are used where a higher degree of accuracy is required.

7. Using digital meters, results should be reported to three significant figures (e.g., 3.45), as compared with conventional meters with two-place accuracy (pH 3.4). In these cases, the questionable figure may be rounded up or down according to established procedures.

Procedure 4-2. Titratable Acidity

Titratable acidity (TA) is a measure of the organic acid content of the juice, must, or wine sample being analyzed. The titration determines the amount of organic acids that can be neutralized by a dilute alkali solution at pH 8.2. Standardized sodium hydroxide (alkali) is dispensed from a burette until the acids in the sample are neutralized.

I. Equipment
   pH meter
   Magnetic stirring table and stir-bar
   50-mL burette
   5- and 10-mL volumetric pipettes
II. Reagents (See Appendix II)
   0.067 *N* or other standardized NaOH of less than 0.1 *N*
   1% Phenolphthalein
   pH 7.00 and 4.00 (or 3.55) buffers
   Boiled distilled water
III. Procedure 1 (Production Modification)
   1. Standardize the pH meter per operators manual.
   2. Add approximately 100 mL of boiled and cooled distilled water to a 250-mL beaker. Place the stir-bar in the beaker, and position it on the stirring table.
   3. *Carefully* immerse electrode(s) in solution so that they are away from the stir-bar.
   4. Add 2 to 3 mL of sample. (*Note:* This is not a quantitative transfer.) Rapidly titrate with standard base to pH 8.2.
   5. Transfer 5.0 mL of sample into solution using a volumetric pipette. Titrate to pH 8.2. Record the volume of base used for this titration.
   6. Calculate results using Equation 4-13:

$$\text{Titritable acidity (g/L tartaric acid)} = \frac{(\text{mL base})(N\ \text{base})(0.075)(1000)}{\text{mL sample}} \qquad (4\text{-}13)$$

IV. Supplemental Notes
   1. In practice, up to four samples may be run before preparing fresh solution at step 2.
   2. Carbon dioxide in must or sparkling wine, as well as some table wines,

serves as a major interference and source of error. Where CO_2 is known to be present, the authors recommend holding approximately 10 mL of sample in a 60°C water bath for several minutes. After the gas has been removed, cool the solution to a defined temperature before volumetrically transferring 5 mL to the titration beaker. Alternately, one may hold the sample under vacuum for several minutes prior to analysis. Carbon dioxide present in the water diluent may also present problems in end-point detection. In this case, the end point will appear to "fade" (as a result of slow hydration of the CO_2). Therefore, the use of boiled (and cooled) distilled water for sample preparation is recommended.

3. Practically speaking, the volume of water used in diluting the wine sample is not important. This should not be taken to mean, however, that one can make extraordinary dilutions of a red wine to aid in visual detection of a phenolphthalein end point.
4. Sodium hydroxide is not a primary standard; carbonate present in solution acts as a buffer and, as such, alters the actual concentration. Therefore, solutions of NaOH must be standardized against some primary standard. The effects of carbonate interferences can be reduced/eliminated by preparing a saturated solution of NaOH and allowing the carbonate to precipitate over a period of several days. The carbonate-free supernatant then can be diluted to the approximate concentration desired and standardized. Standardization normally is carried out by titration of a known volume of the NaOH with a primary standard acid of accurately determined concentration. Routinely, potassium acid phthalate (KHP) is used for this purpose. Only one proton per molecule of KHP is acidic, and the reaction proceeds accordingly:

$$OH^- + HP^- \longrightarrow H_2O + P^= \qquad (4\text{-}14)$$

where:

HP^- = hydrogen phthalate anion

$P^=$ = phthalate anion

Because titration of KHP is not complete until approximately pH 8.2, phenolphthalein is the indicator of choice. KHP is available in either powder or liquid form, the latter being prepared at known concentrations. Refer to Appendix II for KHP standardization of NaOH.

5. With reference to the normality of NaOH used in the titration, one should avoid using concentrations greater than 0.1 *N*. In readings made from burettes, small volume differences reflect larger relative errors than do larger volume differences. Where production and speed are of the essence, NaOH may be prepared at 0.0667 *N*. At this concentration,

the quantity of base used yields total acidity in the appropriate units. Thus a total volume of 5.0 mL of base (at 0.0667 *N*) corresponds directly to a TA of 5.00 g/L.

6. For routine work, one may expect a probable error of $\pm 0.01\%$. Results of TA analyses usually are expressed to two decimal places (e.g., 7.50 g/L), with rounding-off to the hundredths place.

V. Procedure 2 (AOAC)

The currently recognized procedure of the AOAC is presented below. Reagents utilized are the same as presented in Procedure 1.

1. Add 1 mL phenolphthalein indicator to 200 mL of hot, boiled distilled water. Neutralize to pink with standard NaOH.
2. Volumetrically transfer 5.0 mL of degassed wine or must sample (see Procedure 4.2, Supplemental Note 2).
3. Titrate to same endpoint with standardized NaOH according to prior instructions.
4. Calculate the results according to Equation 4-13.

Procedure 4-3. HPLC Analysis of Organic Acids

Individual organic acids can be separated on an HPLC column and quantitated. Sample preparation involves a preliminary ion exchange separation of sugars that would otherwise co-elute with the organic acids. The summation of amounts of individual organic acids present gives a measure of the "total acid content" of the sample.

I. Equipment

High performance liquid chromatograph with variable-wavelength UV detector
Bio-Rad Organic Acids column (or equivalent)
Guard column
Disposable polypropylene sample preparation columns

II. Reagents (See Appendix II)

Citric acid internal standard solution (2% wt/vol)
Formic acid internal standard solution (2% wt/vol)
Concentrated ammonium hydroxide
H_2SO_4 mobile phase (0.01 *N*)
H_2SO_4 (1 + 4)

III. Procedure

1. Install the column and an appropriate guard column in the high performance liquid chromatograph mobile phase flow stream.
2. Begin pumping H_2SO_4 mobile phase through the column at a flow rate of 0.6 mL/min.

3. Set the column oven temperature at 25°C, and allow the system to stabilize.
4. Turn on the variable-wavelength UV detector, and allow it to warm up. Set the wavelength dial to 210 nm.
5. Prepare and chromatograph the sample as follows:
   a. Volumetrically transfer the following to a capped vial 2.00 mL of wine or grape juice:
      (1) 0.10 mL of a 2% citric acid internal standard solution
      (2) 0.10 mL of a 2% formic acid internal standard solution
      (3) 0.2 mL of concentrated ammonium hydroxide
   b. Mix the contents of the vial, and keep it closed until you are ready to place the contents on the anion exchange resin.
   c. Prepare an ion exchange column by slurrying 1 g Bio-Rex 5, 100–200 mesh (Cl^- form) resin in 3 mL distilled water, and pour it into a disposable polypropylene sample preparation column. Drain off the water, and wash the column with an additional 5 mL distilled water. Do not allow the column to dry out prior to use.
   d. Add the sample plus internal standards to the column, and allow the solution to drain through. Wash the sugars off the column with distilled water (9 mL), and collect them in a 10-mL volumetric flask for subsequent analysis, if desired. (This solution should be brought to the mark with distilled water and mixed well prior to injection into the column.)
   e. Carefully add 2 mL of (1 + 4) sulfuric acid and enough distilled water to collect 10 mL of eluent containing the acids. (Again, mix this sample well prior to injecting it onto the column.)
6. Qualitative analysis may be accomplished by preparing standard solutions of the various acids expected in the sample. Compare observed retention times of peaks in the sample chromatogram with those in the standards chromatogram. (See Fig. 4-3 for chromatograms of typical juice and wine samples.)
7. Standard solutions of various acids at known concentrations can be mixed in the same proportions as above with the internal standards and chromatographed. Peak height (recorder) or peak area (integrator) ratios are compared to those from wine samples, and the unknown concentrations calculated as follows:

$$\text{Acid conc. (g/L)} = (RR \times \text{acid conc. in std, g/L})/RR' \quad (4\text{-}15)$$

where:

RR = response ratio of unknown acid to internal standard
RR' = response ratio of standard acid to internal standard

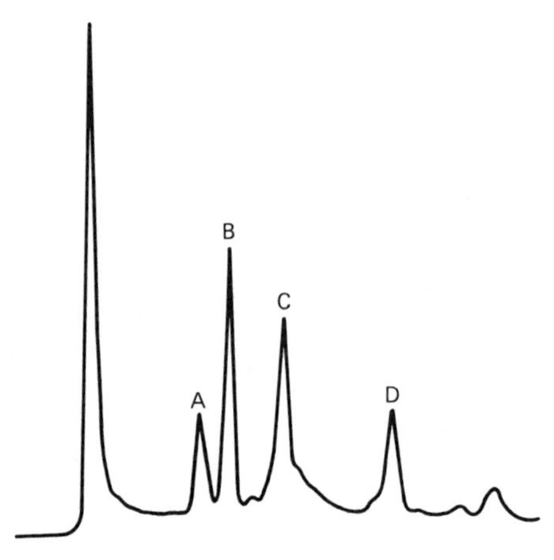

Fig. 4-3. Chromatogram of Fume Blanc wine. Labeled peaks are (A) Citric acid, (B) Tartaric acid, (C) Malic acid, and (D) Succinic acid.

IV. Supplemental Notes

1. Citric acid is the first acid to elute from the column. Often it slightly overlaps the tartaric acid peak, and when present, it causes a definite base-line shift that provides a starting point for the integrator to begin measuring the area of the tartaric acid peak. If no (or very little) citric acid occurs naturally in the wine sample, then the citric acid peak may be used to quantitate the other acid peaks. For those occasions where significant unknown amounts of citric acid are present in a wine sample, the formic acid peak may be used for quantitation.
2. A similar method to that presented above has been developed by Gump and Kupina (1979) for the analysis of gluconic acid in juice or wine. Gluconic acid can be used as a marker compound for the occurance of *Botrytis cineria* infections. Analysis of other fungal metabolites is given in procedure 1-2, and 1-5.

Procedure 4-4. Enzymatic Analysis of L-*Malic Acid*

Enzymatic analysis procedures for the determination of L-malic acid have been developed. The procedures rely on the coupled enzymatic reactions:

$$\text{L-malate} + \text{NAD}^+ \xrightleftharpoons{\text{LMDH}} \text{oxaloacetate} + \text{NADH} + \text{H}^+$$

$$\text{oxaloacetate} + \text{L-glutamate} \xrightleftharpoons{\text{GOT}} \text{L-aspartate} + \alpha\text{-ketoglutarate}$$

LMDH = L-malate dehydrogenase

GOT = glutamate-oxaloacetate transaminase

In Reaction A, the equilibrium is strongly in the direction of the reactants. In the procedure of Mayer and Busch (1963), this reverse reaction is prevented by inclusion of alkaline hydrazine solution. Alternatively, the coupled reaction of oxaloacetate and L-glutamate (Reaction B) can be used to drive the equilibrium toward the products. As seen from Reaction A, formation of reduced NADH is stoichiometrically related to oxidation of L-malate, and thus the concentration of the latter may be determined by measurement of the NADH reduced. Kits of reagents for conducting this procedure are now available to the analyst.

Rather than present a detailed procedure here, the authors refer the reader to the product sheet that accompanies the malic acid kit of reagents, for up-to-date information and a procedure specifically designed for the kit obtained. In general, the malic acid procedure will require the following:

I. Equipment

A spectrometer capable of reading absorbance at 340, 334, or 365 nm plus suitable cuvettes

Micropipettes (10-, 100-, and 200-μL), and adjustable volumetric pipettes capable of delivering 1.00 to 1.5 mL.

II. Reagents

The reagents used may be obtained from Boehringer Mannheim or other suppliers. They are sold in the form of a test kit of sufficient material for approximately 20 L-malic acid determinations.

III. General Procedure

1. Using a reflux apparatus, a wine sample is refluxed with 2 *N* NaOH for 30 minutes (with mixing) to hydrolyze any bound malic acid.
2. The diluted sample is then mixed with reagents (except L-malic dehydrogenase), and the absorbance values are measured.
3. The malate dehydrogenase is then added, and the absorbance again is read after the reaction is complete.
4. A blank is run along with the samples, and absorbance values are corrected for the blank reading.
5. The difference in the corrected absorbance values taken before and after the addition of L-malic dehydrogenase is then used to calculate the concentration of L-malic acid.

IV. Supplemental Notes

The above procedure, as presented, measures spectrophotometrically the change in concentration of NADH at 340 nm (the absorption peak for NADH). Where spectrum-line photometers (equipped with mercury vapor lamp) are used, measurements are made at 334 nm or 365 nm. The appropriate adsorption coefficients are given in the product handouts.

Chapter 5

Volatile Acidity

The total acidity of a wine is the result of the contribution of nonvolatile or fixed acids such as malic and tartaric plus those acids separated by steam volatilization. A measure of volatile acidity is used routinely as a indicator of wine spoilage (see Table 5-1). Increases are usually attributed to microbial production of certain low molecular weight acids and their esters. Although generally interpreted as meaning acetic acid content (in grams per liter), a traditional volatile acid analysis includes all those steam-distillable acids present in the original wine sample. Thus, significant contributions to the volatile acid content may be made by carbon dioxide (as carbonic acid), sulfur dioxide (as sulfurous acid), and, to a lesser extent, lactic, formic, butyric, and propionic acids. In addition, sorbic acid (added to wine as potassium sorbate), used as a fungal inhibitor, is also steam-distillable and should be taken into consideration when appropriate. The contributions of CO_2, SO_2, and sorbic acid interferences can be seen in Table 5-2.

MICROBIOLOGICAL FORMATION OF ACETIC ACID

The volatile acidity of a sound, newly fermented dry table wine may range from 0.2 to 0.4 g/L (Ribereau-Gayon 1961). Increases beyond this level, however, may signal microbial involvement and potential spoilage. Most commonly, increases in volatile acidity in stored wines are attributed to acetic acid bacteria,

Table 5-1. Legal limits of volatile acidity in wines.

Wine type	Federal (g/100 mL)[1]	California (g/100 mL)[2]
Red	0.140	0.120
White	0.120	0.110
Dessert	0.120	0.110
Export (all types)	0.090	—
Late Harvest[3]	0.140	—

[1]From Code Federal Regulations, "Wine" Part 240 of Title 26 (1970).
[2]From California Administrative Code 17,000–17,135.
[3]The U.S. government currently recognizes 0.140 g/100 mL as the limit for late-harvest-type wines (Gahagan 1988).

Table 5-2. Potential interferences in volatile acidity titrations.*

Interferent	Level	Equivalent titration volume in 10-mL wine sample
SO_2	100 mg/L	0.33 mL
Sorbic Acid	200 mg/L	0.18 mL
CO_2	350 mg/L	0.79 mL
	3.92 g/L (upper limit for still wine)	8.90 mL

*At federal legal limit for VA, a 10-mL wine sample will require 2.00 mL (white wine) or 2.33 mL (red wine) of 0.1 *N* NaOH for titration.

but certain strains of malolactic bacteria as well as aerobically growing yeasts may also contribute to overall increases in volatile acids.

Acetic Acid Bacteria

Acetic acid bacteria utilize enthanol, oxidizing the alcohol in the formation of acetic acid. As seen in the following reaction, the alcohol is oxidized first to an intermediate aldehyde and subsequently to the acid:

$$\text{Ethanol} + O_2 \longrightarrow \text{Acetaldehyde} \longrightarrow \text{Acetic Acid} \qquad (5\text{-}1)$$

The above reaction, believed to be the principal source of bacterial acetic acid in wine, operates in the temperature range 20 to 44°C (68 to 111°F), with an optimum temperature of 30 to 32°C (86 to 90°F) (Greenshields 1978). However, specific temperature optima may vary with each species of bacteria. (See Chapter 12 for more detail on bacterial acetification.)

Spoilage Yeasts

In some cases, high levels of volatile acidity may result from growth of both acetic acid bacteria and spoilage yeasts capable of producing the acid. The role of yeast strain and production of volatile acidity has been studied (Rankine 1955). Results indicate that there is considerable variation in production of acetic acid and other by-products among wine spoilage yeasts as well as strains of *Saccharomyces* sp.

Those yeasts involved in wine spoilage include *Brettanomyces* (*Dekkera*), *Pichia*, *Hansenula*, and *Kloeckera*. With regard to acetic acid production, *Saccharomyes*, *Brettanomyces* and its ascospore-forming counterpart, *Dekkera*, are known to produce relatively large amounts. Wang (1985) reports acetic acid production by *Brettanomyces* in white wine after 26 days of incubation (28°C) to increase from 0.31 to 0.75 g/L. The impact of *Dekkera* on volatile acid formation is apparently less than this; under identical conditions, Wang reports formation of 0.62 g/L acetic acid when compared to controls.

Acetic acid is a normal by-product of yeast growth and has its origin primarily in the early stages of fermentation. As mentioned, the volatile acidity of a sound, newly fermented, dry table wine may range from 0.20 to 0.40 g/L (Ribereau-Gayon 1961). However, the activities of spoilage yeast as well as wine yeast growing under stressed conditions may bring about increases in the volatile acid level.

Several extrinsic factors also may affect the formation of acetic acid. These include the pH and sugar content of the must, and fermentation temperatures, as well as interactive affects of other microrganisms present on the grapes and in the must. Dessert wines produced from botrytized grapes often have higher levels of volatile acid than wines made from sound fruit. Ribereau-Gayon et al. (1979) report that mold extracts from *Botrytis cineria* may have a major impact on yeast and bacterial production of acetic acid during alcoholic fermentation. Addition of the extract was found to stimulate production of acetic acid and glycerol.

The effect of high osmotic pressure resulting from high-sugar musts on volatile acid formation also has been reported (Nishino et al. 1985; Rose and Harrison 1970). Such fermentations typically have a longer lag phase with reduced cell viability and vigor. Cowper (1987) also reported higher acetic acid levels in high-sugar musts fermented at low temperature. In his study, Cowper found acetic acid levels ranging from 0.6 to 1.0 g/L in musts ranging from 32° to 42° B when compared to control juice at 22° B. (The acetic acid content of the latter was reported as 0.4 g/L.) Morphologically, yeast cells growing under conditions of high osmotic pressure appear stressed. Generation time (budding) also is delayed. Concomitant with increased production of acetic acid are proportionate increases in glycerol. This is not unexpected when organisms are grown in high-solute (high-°B) environments.

Fermentation temperature is also known to affect the levels of acetic acid produced by wine yeasts. Rankine (1955) examined acetic acid formation in several strains of *Saccharomyces cerevisiae* grown at different temperatures. He found that volatile acid formation was higher in fermentations conducted at 25°C than in fermentations carried out at 15°C. Significant differences between strains also have been noted. Using two stains of *Sacch. cerevisiae*, Shimazu and Watanabe (1981) noted that formation of acetic acid was maximum at 40°C in one case, whereas maximum formation occurred at 10°C in the second strain.

Unless controlled, the temperature of fermentation may rise to a point at which it becomes inhibitory to wine yeast. In practice, inhibition may be noted at temperatures approaching 95°F (35°C) or higher. Because acetic and lactic bacteria can tolerate temperatures higher than those needed to kill (inhibit) wine yeasts, stuck fermentations often are susceptible to secondary growth due to these organisms. Pressure fermentations also may result in higher-than-expected volatile acid contents (Amerine and Ough 1957), possibly due to selective inhibition of wine yeasts and growth of lactic acid bacteria.

Post fermentation Sources of Volatile Acidity

Cellar practices play an important role in volatile acid formation in stored wines. High levels of VA may result when headspace (ullage) is allowed to develop in barreled wines. In this case, the combination of oxidative conditions and surface area may support rapid growth of both bacteria and yeast. Because acetic acid bacteria are strictly aerobic organisms, depriving them of oxygen is a viable means of controlling futher growth. However, wood cooperage may not provide the airtight (anaerobic) environment needed to inhibit growth.

Acetics may survive and grow at the low oxygen levels present even in properly stored wines. Ribereau-Gayon (1985) reported viable populations (10^2 CFU/mL) of *Acetobacter aceti* present in properly maintained wines aging in wood cooperage. In this case, survival of low numbers of the bacteria was attributed to slow exchange of oxygen (approximately 30 mg/L/year) from the cellar into the wine. At this level of surviving bacteria, transitory exposure to air, such as may occur during fining and/or racking operations, may be sufficient to stimulate bacterial growth. In the study, even though the exposure was short-term and the wine was subsequently stored properly, incorporation of dissolved oxygen was found to support continued growth of the bacteria. The problem becomes more apparent with increases in cellar temperature and wine pH.

The volatile acidity of properly maintained barrel-aging red wines may increase slightly without the activity of microorganisms. In one case, for example, the source of acetic acid is coupled oxidation of wine phenolics (see Chapters 7 and 8) to yield peroxide, which, in turn, oxidizes ethanol to acetaldehyde and acetic acid:

$$\text{R-C}_6\text{H}_3\text{(OH)}_2 + O_2 \longrightarrow \text{R-C}_6\text{H}_3\text{(=O)}_2 + H_2O_2 \qquad (5\text{-}2)$$

$$H_2O_2 + CH_3CH_2OH \longrightarrow CH_3CHO + 2H_2O$$

During proper barrel storage, a partial vacuum develops within the barrel over time. Both water and ethanol diffuse into the wood and escape to the outside as vapor. In cellars where the RH is less than 60%, water is lost from the wine to the outside environment, and the alcohol content of the wine increases. Conversely, where a higher RH exists, alcohol is lost to the outside environment. Diffusion of water and ethanol through pores in the staves creates a vacuum in the properly bunged barrel. Thus, even though some headspace may develop under these conditions, the oxygen concentration is very low.

Formation of a partial vacuum in the headspace requires tightly fitted bungs. Winemakers using traditional wooden bungs often store their barrel "bung over" or at the "two o'clock" position to ensure that the bung remains moist and, thus, tightly sealed. Silicon bungs can provide a proper seal without being moist, so barrels closed with such bungs usually are stored "bung up." Topping sealed barrels too often results in loss of vacuum and may accelerate both oxidation and biological degradation of the wine.

Although the practice is not recommended, winemakers forced to store wines in partially filled containers often blanket the wine with nitrogen and/or carbon dioxide. Nitrogen is the preferred blanketing gas because of its limited solubility in wine. Sparging of wines with carbon dioxide is also used, as upon standing the gas escapes slowly from solution and, because of its density, remains at the wine's surface.

ACETATE ESTERS

The volatile character or "acetic nose" is not exclusively the result of acetic acid. Acetate esters, most specifically ethyl acetate, contribute significantly to this defect. Nordstrom (1963) proposes the following enzyme-mediated reaction mechanism for volatile ester formation by yeasts:

$$\text{RCOOH} + \text{ATP} + \text{CoA{\sim}SH} \longrightarrow \text{RCO{\sim}SCoA} + \text{AMP} + H_2O \quad (5\text{-}3a)$$

$$\text{RCO{\sim}SCoA} + \text{R'OH} \rightleftharpoons \text{R—C(=O)OR'} + \text{CoA{\sim}SH} \quad (5\text{-}3b)$$

According to Daudt and Ough (1973), factors that can influence the formation of acetate esters include yeast strain, temperature of fermentation (maximum formation over the range 50–70°F), and sulfur dioxide levels. In their study, temperature was reported as being most important factor relative to the final concentration of ethyl acetate produced.

Although high acetic acid content and the presence of ethyl acetate generally are associated with each other, they may not always be produced to the same extent. Amerine and Cruess (1960) report that ethyl acetate levels of 150 to 200 mg/L impart spoilage character to the wine. This finding follows Peynaud (1937), who suggested that a maximum ethyl acetate level of 220 mg/L be used rather than traditional analyses based solely on acetic acid. These suggestions are based on the fact that in some cases high acetic acid content does not confer a spoilage character to the wine. Although levels vary with the individual wine, those with a volatile acid content of less than 0.70 g/L seldom possess a spoilage character, and, in combination with low concentrations of ethyl acetate may contribute to overall wine complexity.

Acetic acid and ethyl acetate concentrations in unfermented must also have been examined as indicators of spoilage in grapes (Corison et al. 1979). The levels identified for "rejection," based upon ethyl acetate in white and red musts, were 60 mg/L and 115 mg/L, respectively. Corresponding levels in wine were identified at 170 mg/L and 160 mg/L. By comparison, acetic acid levels needed for rejection in white and red musts were 1190 mg/L and 900 mg/L. In white and red wines, rejection levels were lower: 1130 mg/L and 790 mg/L. (See Chapter 1 for a discussion of monitoring grape quality. Details of a gas chromatographic analysis for ethyl acetate are presented in Procedure 2-6.)

REDUCTION OF VOLATILE ACIDITY

Federal government restrictions have been imposed to regulate levels of volatile acidity (expressed as acetic acid) in domestic wines offered for sale. In some instances (e.g., California), more restrictive state regulations have been written. (See Table 5-1.)

Reduction of high volatile acidity in wines is difficult. Attempts to lower volatile acid levels by neutralization generally yield undesirable results because of a concomitant reduction in the fixed acid content. Similar problems are encountered in the use of ion exchange technology in the removal of excessive volatile acidity. The use of yeasts for VA reduction has been studied; the application takes advantage of the fact that yeasts use acetic acid as a carbon source. Utilization of acetic acid by active yeasts has led some winemakers to add high-volatile-acid wine to fermenting musts to lower VA levels. However, such a practice runs the risk of contaminating the entire lot, and it may well have a detrimental impact on wine quality. Judicious blending probably is the best procedure to use in lowering volatile acid content in the case of borderline wines.

ANALYTICAL METHODS FOR VOLATILE ACIDITY

Steam distillation frequently is used in volatile acid analyses. In this procedure volatile acid components are steam-distilled out of the sample. This step is followed by titration with standardized sodium hydroxide, and the results are reported as acetic acid (g/L). Although this is probably the most common procedure for VA determination, enzymatic, gas chromatographic, and high performance liquid chromatographic procedures (Chapter 1) are becoming more popular. Several interfering substances may be encountered in the use of distillation procedures, including carbon dioxide, sulfur dioxide, and lactic, succinic, and sorbic acids (see Table 5-2), in addition to the normally minor components, formic, butyric, and propionic acids. A brief discussion of these interferences is presented in the introduction to Procedure 5-1.

Quantitative recovery of acetic acid from wine using the steam distillation technique is claimed to be relatively poor, with results ranging from 52 to 72% of theoretical (McCloskey 1976, 1980). This problem has been overcome with the development of an enzymatic analysis (McCloskey 1976, 1980; Boerhringer-Mannheim 1980) where, compared to steam distillation, recovery of acetic acid is 100% of theoretical (McCloskey 1980).

The enzymatic procedure originally reported by McCloskey (1976) involves the coupled reactions:

$$\text{Acetate} + \text{ATP} \xrightarrow{(1)} \text{acetyl phosphate} + \text{ADP} \tag{5-4a}$$

$$\text{ADP} + \text{phosphenolpyruvate} \xrightarrow{(2)} \text{pyruvate} + \text{ATP} \tag{5-4b}$$

$$\text{pyruvate} + \text{NADH} + H^+ \xrightarrow{(3)} \text{lactate} + \text{NAD}^+ + H_2O \tag{5-4c}$$

(1) Acetate kinase

(2) Pyruvate kinase

(3) Lactate dehydrogenase

Changes in the concentration of NADH are monitored spectrophotometrically at 340 nm following a 1-hr period of reaction. In a follow-up paper, McCloskey (1980) described an "improved enzymatic procedure" that shortens the time of reaction and includes addition of PVPP or gelatin to remove phenolic interferences that may be encountered in red wines. The shortened time of reaction results from inclusion of the enzyme phosphotransacetylase, which catalyzes conversion of acetyl phosphate:

$$\text{Acetylphosphate} + \text{CoA} \xrightarrow{(4)} \text{acetyl-CoA} + PO_4^{-3} \tag{5-5}$$

(4) phosphotransacetylase

The procedure, as developed by Boehringer-Mannheim (1980), calls for a coupled enzymatic reaction:

$$\text{Acetate} + \text{ATP} + \text{CoA} \xrightarrow{(1)} \text{Acetyl-CoA} + \text{AMP} + \text{pyrophosphate} \tag{5-6a}$$

$$\text{Acetyl-CoA} + \text{Oxaloacetate} + H_2O \xrightarrow{(2)} \text{Citrate} + \text{CoA} \tag{5-6b}$$

$$\text{Malate} + \text{NAD}^+ \xrightarrow{(3)} \text{Oxaloacetate} + \text{NADH} + \text{H}^+ \quad (5\text{-}6c)$$

utilizing:

(1) Acetyl-CoA synthetase

(2) Citrate synthase

(3) Malate dehydrogenase

Reaction 5-6c in the above sequence is required to generate the oxaloacetate present in Reaction 5-6b. Because of the dependence of this intermediate step, NAD^+ reduced to $NADH + H^+$ is not linearily related to the concentration of acetic acid. Thus a correction factor must be incorporated into the final calculation.

Acetic acid has been determined successfully using gas chromatographic techniques (see procedure 5.3). Special inert, carbon-based packing materials are utilized to avoid the problem of acetic acid adsorption on the packing. Quantitative measurement of the acetic acid peak is accomplished by using standards of known concentration. Procedure 1-5 (chapter 1) describes an HPLC method for acetic acid determination.

Sampling Considerations

In the collection of lab samples from tanks where large volumes of wine are involved, stratification of volatile acids may occur. Therefore, the sampling technique is critical. A comparison of volatile acidity samples from a single tank can be done accurately only if the samples are taken from the same location within the tank. Preferably, samples should be taken at the top and the bottom of tanks.

Procedure 5-1. Volatile Acid Determination by Steam Distillation (by Cash still)

In this procedure, steam distillation of the sample is followed by titration with standardized sodium hydroxide, with the results reported as acetic acid (g/L) (see above, under "Analytical Methods for Volatile Acid Determination"). Several interferences may be encountered in using this procedure, as summarized below.

1. Carbon dioxide, titrated in the distillate as carbonic acid, may occur as a major interference and source of error. Several methods have been suggested for its removal from samples. Some laboratories suggest vacuum agitation for several minutes prior to sample withdrawal. Others recommend heating an aliquot to incipient boiling under a condenser and cooling it immediately. On occasion, the authors have found purging the distillate for 30 sec with low-

pressure nitrogen gas, diffused through an air stone, to be useful in driving off carbon dioxide.

2. Sulfur dioxide can be distilled from a sample and titrated in an acid–base analysis. Routinely, one would correct for sulfur dioxide only if the volatile acidity were approaching legal limits, or if the levels of sulfur dioxide in the sample were found to be extraordinarily high. The procedures for free and total sulfurous acid corrections are provided below.

3. Sorbic acid, whose role as a mycostatic agent has resulted in its wide application in the wine industry, is almost completely volatilized by steam distillation. Where sorbic acid is known (or thought) to be an additive in a wine, and uncorrected volatile acidities approach legal maxima, correction for its presence may be essential. A correction for the contribution of sorbic acid in the distillate may be calculated using the following relationship:

$$1 \text{ g sorbic acid} = 0.535 \text{ g acetic acid} \quad (5\text{-}7)$$

4. Lactic acid levels in wine range from 0.2 g/L (0.02%) in the case of wines not having undergone malolactic fermentation to as high as 5 g/L (0.5%) after bacterial conversion (Thoukis et al. 1965). Because lactic acid resulting from malolactic fermentation is steam-distillable only to the extent of approximately 2% (Pilone 1967), it generally is not a major source of interference in analysis of volatile acidity using preliminary sample distillation. Lactic acid is not considered to be a spoilage component per se. Methods for monitoring lactic acid formation are considered in chapter 12.

5. Other volatile fatty acids indicative of microbial spoilage include formic, butryic, and propionic acids. When present in wines, they generally occur at very low levels. Formic acid has been reported at levels of 23 to 89 mg/L in Austrian wines (Amerine 1954). It was thought to result from decomposition of the amino acid leucine upon prolonged storage of wine on lees. Butyric acid plays a minor role among the spoilage acids in wine, its reported levels ranging from 10 to 20 mg/L. However, in wine vinegars the acid may be present at levels of up to 200 mg/L (Amerine 1954).

I. Equipment
   Cash volatile acid still assembly (Fig. 5-1)
   10-mL volumetric pipette
   250-mL Erlenmeyer flask graduated in 50-mL increments
   50-mL burette

II. Reagents (See Appendix II)
   Standardized sodium hydroxide of 0.1 *N* or less
   1% Phenolphthalein indicator
   0.02 *N* Iodine or other dilute standardized iodine reagent
   1 + 3 Sulfuric acid

1% Starch indicator
Distilled water

III. Procedure (see Fig. 5-1)

(a) Distillation

1. Turn on condenser cooling water
2. By releasing clamp *B*, fill boiling chamber *A* with distilled water to the approximate level indicated in Fig. 5-1. When it is filled, make certain that clamps *B* and *C* are secure!
3. With stopcock *G* positioned so that sample will be delivered to the inner chamber through channel *D*, volumetrically transfer 10 mL of wine to funnel *E*. Rinse the sample into inner chamber *F* with distilled water.

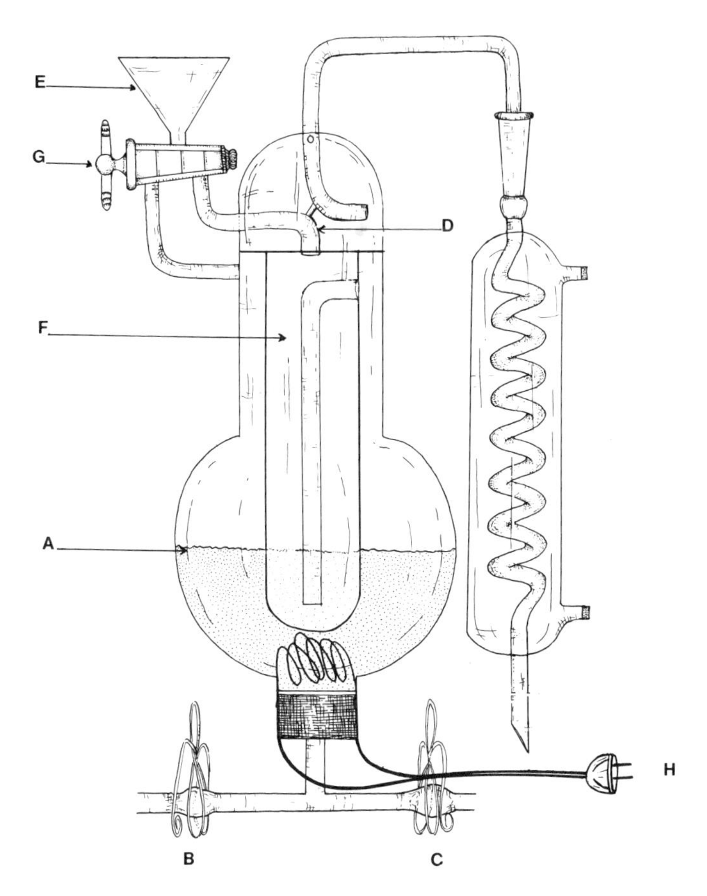

Fig. 5-1. Cash volatile still assembly for volatile acid determination. Lettered components are described in Procedure 5-1.

4. Return the stopcock to the closed position.
5. Plug the heater unit, *H*, into a regulated power moderate supply (Variac). Turn it on, and bring the water in chamber *A* to moderate boiling.
6. Collect 100 mL of distillate in a receiving flask.
7. Immediately upon completion of distillation, *turn the heater unit off.*
8. With the stopcock positioned to deliver to the inner chamber, add approximately 15 mL of distilled water to the funnel for a "self-cleaning operation." Repeat this at least two times.
9. Release clamp *C*, allowing water to drain from the boiling chamber.
10. Refill the boiling chamber with distilled water as previously described.

(b) Titration of distillate

11. Add 2 to 3 drops of phenolphthalein indicator to the distillate and titrate, using previously standardized NaOH, to a permanent pink and point lasting 15 to 20 sec.
12. Record the volume of NaOH used in titration and calculate the volatile acidity (VA in g/L) according to the following equation:

$$\text{Volatile acidity (g/L)} = \frac{(\text{mL NaOH}) \times (N\ \text{NaOH}) \times (0.060) \times (1000)}{\text{mL wine}} \quad (5\text{-}8)$$

(c) Sulfur dioxide correction

Governmentally regulated limits for the volatile acid content of a wine are listed as "exclusive of SO_2." Therefore, in cases where the volatile acidity approaches legal limits, it is necessary to correct results for the contribution of SO_2.

Free sulfur dioxide:

13. Immediately upon completion of VA titration, cool the sample.
14. Add approximately 1 mL of starch indicator and 1 mL of 1 + 3 sulfuric acid.
15. Titrate with standardized iodine solution to a faint blue-green end point.
16. Calculate the free sulfurous acid equivalent (F.S.A.E.) according to the following equation:

$$\text{F.S.A.E. (g acetic acid/L equivalent to } SO_2 \text{ present)} = \frac{(\text{mL } I_2)(N\ I_2)(32)(2)(60)(1000)}{(1000)(\text{sample vol})\ 64} \quad (5\text{-}9)$$

Bound sulfur dioxide:

17. To the same flask, add 10 mL *N* NaOH and boil the solution 3 minutes. Cool it quickly.
18. Add 5 mL (1 + 3) sulfuric acid and 1 mL starch indicator.
19. Titrate with standardized iodine solution to a faint blue-green end point.
20. Calculate the bound sulfurous acid equivalent (B.S.A.E.), again using Equation 5-9.
21. Combining the contributions of "free" and "bound" sulfurous acid and subtracting the result from uncorrected VA yields the actual VA content (g/L) free of sulfurous acid interferences.

IV. Supplemental Notes

1. The Cash still should *never* be operated without water in the boiling chamber! During operation, water levels in the boiling chamber should reflect the approximate level depicted in Fig. 5-1.
2. Carbon dioxide may be present in water used to fill the boiling chamber, so some laboratories recommend venting steam through funnel *E*. This requires that stopcock *G* be in the open position for 10 to 15 sec prior to closure.
3. Direct line current to the heating coil may result in exceptionally vigorous boiling. If this problem is noted, a transformer (Variac) may be necessary in the line from the 110-volt outlet to the heating coil.
4. The connection between the condenser and the distillation unit should be checked routinely to ensure proper seating.
5. Where the unit is run continuously throughout the day, funnel *E* may become warm, introducing the possibility of premature volatilization of the sample. It is recommended that the funnel be filled with distilled water during operation.
6. Workers should be aware of possible interferences, especially from CO_2 and SO_2 in the sample. For correction, refer to Table 5-2 and appropriate sections of this chapter.
7. Results should be reported as ±0.05 g/L.
8. The bound sulfur dioxide correction is included to account for the presence in the distillate of SO_2-containing compounds that are volatilized before hydrolysis.
9 Table 5-3 is presented to facilitate the interpretation of equations used for calculating VA values and corrections. Inserting the appropriate units into Equation 5-9, for example, gives the following:

F.S.A.E. (g acetic acid/L equivalent to SO_2 present) =

$$\frac{(\text{mL I}_2)(N\ \text{I}_2)\dfrac{32\text{ mg SO}_2}{\text{meq}}\ \dfrac{2\text{ mmol CH}_3\text{COOH}}{\text{mmol SO}_2}\ \dfrac{60\text{ mg CH}_3\text{COOH}}{\text{mmol}}\ \dfrac{10^3\text{ mL}}{\text{L}}}{\dfrac{10^3\text{ mg}}{\text{g}}(\text{mL sample})\dfrac{64\text{ mg SO}_2}{\text{mmol}}} \quad (5\text{-}10)$$

Table 5-3. Factors used in VA calculations.

Equation	Factor	Significance
5-8	0.060	Milliequivalent weight of acetic acid.
5-8	1000	Conversion to appropriate units (g/L).
5-9/5-10	32	Equivalent weight of SO_2 in the reaction.
5-9/5-10	2	The iodine titration utilizes a two-electron transfer in the formation of products (see Chapter 9).
5-9/5-10	64	Molecular weight of SO_2.

10. In collection of tank samples for VA analysis, lab personnel should be aware that stratification of volatile acids may occur within the tank. Therefore, sampling techniques are critical (see text).

Procedure 5-2. Acetic acid determination by enzymatic assay

Quantitative recovery of acetic acid from wine using the steam distillation technique is said to be relatively poor (see above, under "Analytical Methods for Volatile Acid Determination"). Further, other steam-distillable acids may serve as major interferences in the subsequent acid–base titration, a problem that has been overcome with the development of an enzymatic analysis [see above discussion of the McCloskey (1976) procedure].

Rather than present a detailed procedure here, the authors suggest that the reader consult the product sheet accompanying the acetic acid kit of reagents for the most recent information and a procedure specifically designed for the kit obtained.

In general the acetic acid procedure will require the following:

I. Equipment

A spectrophotometer capable of reading absorbance at 340, 334, or 365 nm.

Micropipettes (10, 200 μL)

Volumetric pipettes (1.0, 2.0 mL)

II. Reagents

The reagents used in this analysis are those of Boehringer Mannheim and are sold in the form of test kits of sufficient volume for approximately 25 analyses.

III. Procedure

The general thrust of the procedure is as follows:

Samples either are used without any preparation (red wines with acetic acid levels of more than 0.1 g/L and white wines) or are treated with PVPP, filtered, pH-adjusted, and diluted twofold.

The samples are mixed with reagents (except the malate dehydrogenase and acetyl-CoA-synthetase) in a cuvette and the absorbance values measured. Malate dehydrogenase is added and the absorbance again read. Finally acetyl-CoA-synthetase is added and the final absorbance measurement taken. A blank is run with the samples.

Reaction 5-6c is needed to generate the oxaloacetate used in Reaction 5-6b, therefore, the NAD^+ reduced to $NADH + H^+$ is not linear when monitored spectrophotometrically. Thus it is necessary to calculate corrected absorbance measurements for the sample and the blank. The corrected absorbance values are used to calculate the concentration of acetic acid.

Procedure 5-3. Gas Chromatographic Analysis of Acetic Acid

Acetic acid has been determined successfully using gas chromatography. Special inert, carbon-based packing materials are utilized to avoid the problem of acetic acid absorption on the packing. Quantitative measurement of the acetic acid peak is accomplished by using standards of known concentration.

I. Equipment
   Gas chromatograph equipped with flame ionization detector
   Electronic integrator or strip-chart recorder
   Hypodermic syringe (10 μL)
II. Reagents (See Appendix II)
   Acetic acid–water stock standard (100 g/L)
   n-Pentyl alcohol internal standard solution
   Ethanol (10% vol/vol)
   Distilled water
III. Procedure
   1. Install one of the following columns in the GC oven:
      a. 6 ft × 2 mm id glass, packed with 6.6% Carbowax 20M on 80–100 mesh Carbopack B.
      b. 6 ft × 2 mm id glass, packed with 5% Carbowax 20M on 60–80 mesh Carbopack B.
   2. Set the operating conditions for the gas chromatograph shown in Table 5-4.
   3. Adjust the air and H_2 gas flows to values specified in the instrument's operating directions (approximately 300 and 30 mL/min, respectively). These may have to be altered somewhat to optimize conditions for the particular instrument and column used.
   4. Prepare acetic acid calibration standards of 0.1, 0.2, 0.4, 0.7, 1.0, and 1.5 g/L in 100-mL volumetric flasks as indicated in Table 5-5. Store the standard solutions in a refrigerator.
   5. Dilute each standard and wine sample 1 + 1 with the internal standard.

Table 5-4. Operational parameters for gas chromatographic separation of acetic acid.

Carrier gas	N_2
Flow rate, mL / min	30
Oven temp, °C	160
Injector temp, °C	200
Detector temp, °C	250

Table 5-5. Preparation of calibration standards from 100 g / L acetic acid-water stock solution.*

Volume acetic acid stock (mL)	Final conc. (g/L) in 100 mL volume
1.0	0.10
2.0	0.20
4.0	0.40
7.0	0.70
10.0	1.00
15.0	1.50

*Final dilution made with 10% (v/v) ethanol.

6. Inject 1.0 μL of each of the standards into the gas chromatograph. Record the chromatogram, and note the retention times of the various components. Determine the peak area ratios (integrator) or peak height ratios (recorder) for the acetic acid peak relative to the *n*-pentyl alcohol peak. Calculate the respective response ratios (RR').
7. Inject 1.0 μL of the prepared wine sample into the gas chromatograph. Record the chromatogram, and note the retention times of the various components. Identify acetic acid by comparison of its retention time with that of the standard. Determine the appropriate response ratio for the acetic acid peak to the internal standard peak (RR).
8. Calculate the percent acetic acid in the wine sample:

$$\% \text{ Acetic acid} = (RR \times \% \text{ Acetic acid in std})/RR' \quad (5\text{-}11)$$

IV. Supplemental Notes

1. Carbopack columns are capable of resolving complex mixtures of compounds in a variety of alcoholic beverages such as beer and wine. They can also be used to detect fusel oils and other low molecular weight carboxylic acids that are usually adsorbed on other columns.
2. Carbopack columns work best with small or dilute samples. If too large

or concentrated samples are used, tailing and poor separation efficiency will result. This is to be expected with the low percent stationary phase used on the packing.

3. Columns should be purchased with the ends packed with phosphoric-acid-treated glass wool. Untreated or silanized glass wool will absorb acetic acid and inhibit its elution from the column.
4. Some workers recommend saturating the carrier gas with formic acid vapor to reduce tailing and ghosting of the acetic acid peak. Others have not found this step to be necessary.
5. GC supply houses constantly are improving and upgrading chromatographic packing materials. If the above-listed columns are not available, check with your supplier for current recommendations.

Chapter 6

Carbohydrates: Reducing Sugars

Carbohydrates are polyhydroxy aldehydes, ketones, and their derivatives, composed of carbon, hydrogen, and oxygen in the ratio $C_n(H_2O)_n$. On a molecular basis, carbohydrates exist as monosaccharides, such as glucose and fructose, disaccharides, such as sucrose, and long-chained forms, the polysaccharides. Polysaccarides may be hydrolyzed to di- and trisaccharides and, ultimately, to monosaccharides. Examples of polysaccharides that are of potential importance to the winemaker include pectin and starch as well as the alginates used in fining. Other compounds that qualify as carbohydrates include deoxy- and amino sugars, sugar alcohols, and acids.

Reducing Sugars (Hexoses)

To the enologist, the most important carbohydrates are the six-carbon sugars, glucose and fructose, utilized by yeast in alcoholic fermentation. These two sugars also are referred to as reducing sugars. Reducing sugars may be operationally described as those sugars containing functional groups capable of being oxidized and, in turn, bringing about reduction of other components under specific analysis conditions (copper, as Cu II, used in their analysis). Thus, certain pentoses also are classified as reducing sugars, even though they are unfermentable by wine yeasts.

Glucose and fructose may be differentiated on the basis of the location of their respective functional carbonyl group. As seen in Fig. 6-1, the carbonyl group of glucose is located on the first carbon and thus is defined as an aldogroup. In fructose, the carbonyl function is located on the second carbon; thus fructose is an example of a keto-sugar. Intramolecular bond angles create molecular structures for these sugars so that they normally do not exist as straight-chained molecules but rather in cyclic configurations called hemiacetals (glucose) or hemiketals (fructose).

Cyclization does not involve the gain or loss of atoms by the sugar molecule. Thus the straight-chained and cyclic forms are isomers, with the cyclic form representing the more important (prevalent) configuration. Glucose, for example, exists both in solution and in crystalline form almost entirely as the cyclic hemiacetal (Fairley and Kilgour 1966). The fact that sugars display most of the reactions considered typical of aldehydes is the result of an equilibrium established between the open-chained and cyclic configurations present in solution.

Cyclization introduces another structural consideration into the chemistry of sugars. In solution, sugars can occur in rings composed of four carbons and one oxygen or five carbons and one oxygen. The former is termed a furanose ring and the latter a pyranose ring (see Fig. 6-1).

In grapes, glucose and fructose occur in approximately equal concentrations, each contributing approximately 10 g/100 g to juice (Amerine et al. 1972). The disaccharide sucrose is the third most abundant sugar, accounting for 0.2 to 1.0 g/100 g (Hawker et al. 1976). Although glucose and fructose normally are present in a ratio of 1 : 1 in the mature fruit, the proportions may vary significantly. Climatic conditions during the growing season may affect the glucose–fructose ratio; Kliewer (1967a) found that it decreased in warmer seasons and increased during colder periods. Amerine and Thoukis (1958) reported ratios ranging from 0.71 to 1.45 in California's 1955 vintage, whereas Kliewer (1967a, b) cited ratios of 0.74 to 1.05 in *Vitis vinifera* wine varieties. During maturation, the ratio of glucose to fructose usually decreases.

In their review of wine microbiology, Kunkee and Amerine (1970) cite differential utilization of glucose and fructose by yeast. At must reducing sugar levels of 17 to 20%, glucose was reported to be fermented faster, whereas at higher reducing sugar levels (>25%) the rate of fructose utilization was greater. Between 20 and 25% reducing sugar levels, both sugars fermented equally well. Peynaud (1984) notes that the ratio of glucose to fructose declines during fermentation from near 0.95 at the start to 0.25 near the end of fermentation. Thus, it can be seen that fructose usually is present in greater amounts than glucose. As fructose is nearly twice as sweet as glucose, the cited ratios explain the observation that wines sweetened with grape concentrate or muté appear less sweet than wines with the same ***analytical concentration*** of reducing sugar produced by arresting the fermentation.

Reducing sugar analyses play multiple roles in wine processing considerations. The winemaker needs to know the quantity of fermentable sugar remaining in the wine to determine if the fermentation is complete. This may be important

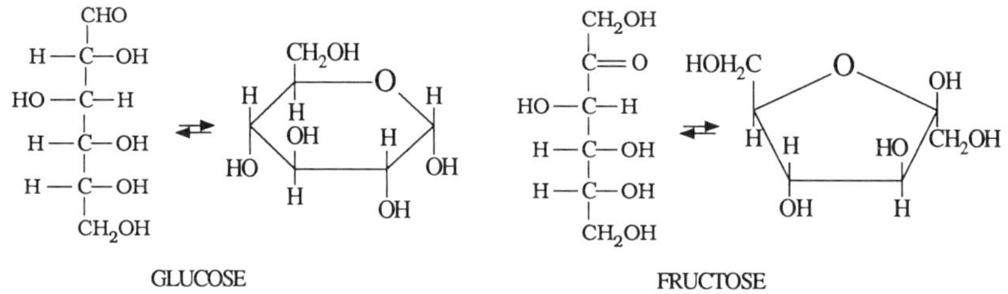

Fig. 6-1. Structures of the two primary sugars in grapes.

so that provision can be made for dealing with microbial stability as well as potential blend preparations. Additionally, monitoring the fermentable sugar content in pomace, distilling material, and so on, is of concern in overall plant efficiency. Traditionally, one attempts to obtain a measure of the residual fermentable sugar by analyzing for all remaining reducing sugars. Thus, although one might expect "dry" table wines to have close to zero residual sugar upon completion of fermentation, typical analytical reducing sugar results are higher because of the contributions of nonfermentable pentoses. As a result, dry wines traditionally have been defined as having reducing sugar levels of 2.0 g/L (0.2%) or less. In contrast, McCloskey (1978) defines the sugar content of a "dry" wine as ranging from 0.15 to 1.5 g/L (when determined by enzymatic assay specific for glucose and fructose). Because the primary reducing sugar content in a dry wine is attributed to pentoses which are not fermentable by yeast, a dry wine ($< 0.02\%$ reducing sugar) generally is considered stable with respect to yeast refermentation.

SUCROSE

The disaccharide sucrose serves as an important energy storage compound in most plants and vegetables. Although sucrose itself is unfermentable, the products of its hydrolysis, glucose and fructose, are utilized readily. In the case of grapes, upon translocation to the berry, hydrolysis by invertase enzymes yields glucose and fructose. Thus, sucrose levels in grape berries, at maturity, range from 0.2 to 1%. Because yeasts produce their own invertase enzyme, chaptelization of sugar-deficient musts with sucrose does not cause problems relative to fermentability.

PENTOSES

The five-carbon monosaccharides, commonly referred to as pentoses, may comprise approximately 28% of the reducing sugar content of a dry table wine (Esau 1967). Among the pentoses present in wine, arabinose is reported to occur in highest concentrations, at 0.40 to 1.3 g/L, followed by rhamnose at less than 0.50 g/L (Amerine et al. 1972). Analytically, this group of sugars is not easily separated; so it is included in traditional analyses of reducing sugar. As to their enological involvements, early evidence suggested that certain pentoses may serve as energy sources in the malolactic fermentation (Doelle 1975). A recent paper by Davis et al. (1988) reports major strain variation among lactic bacteria with respect to pentose utilization. Sugar utilization profiles may play major role(s) in selection of appropriate bacterial strains for induced malolactic fermentations. Pentoses may also be an important source of furfural in baked sherry (Amerine et al. 1972).

POLYSACCHARIDES

The polysaccharide component of must and wine carbohydrates exists as pectins, dextrans, and gums. Because of their size and colloidal nature, these macromolecules present problems in clarification and filtration.

Pectins are naturally occurring plant polymers of galacturonic acid linked via alpha-1.4 bonds. The carboxyl group of acid monomers may exist either in acid form or as the methyl ester. In musts treated with pectic enzymes, the latter may be enzymatically cleaved to yield methanol (see Chapters 2 and 17). Gums present in wine are polymeric mixtures of arabinose, galactose, xylose, and fructose (Peynaud 1984).

Other important polysaccharides in musts and wine exist principally as polymers and shorter-chained oligomers of glucose. These include the glucans, which arise from growth of *Botrytis cineria* on the grape berry. These polymers exist as branched chains of glucose linked via beta-1,3 bonds; the branching results from beta-1,6 linkages. Two glucans are associated with the growth of *Botrytis*. The first, of molecular weight 900,000 is believed to be important in juice and wine clarification difficulties. The second, a heteropolysaccharide of approximately 40,000 is believed to inhibit yeast alcoholic fermentation.

Extraction of glucans from *Botrytis*-infected fruit depends upon grape-handling techniques. For example, careful pressing of whole clusters and press fraction segregation, rather than crushing, minimizes extraction of glucans into the juice, and thus may reduce subsequent clarification and filtration problems. Clarification also may be enhanced by prefermentation utilization of glucanase enzymes. The latter are available as mixed enzyme preparations that have activity toward beta-1,3 and beta-1,4 bonds of the glucan polymer (see Chapter 17).

Other polysaccharides, of bacterial origin, occasionally may be noted in wine. The lactic bacterium *Leuconostoc mesenteroides*, occasionally may present problems in fruit and other wines where sucrose is used in chaptelization. When sucrose is used as a carbon source, this organism produces an extracellular polysaccharide (dextran) that may impede wine clarification. Additionally, some acetic acid bacteria also may produce extracellular polysaccharides when growing on glucose-containing media, including dextrans as well as polymers of mixed sugars.

ANALYSIS OF REDUCING SUGARS

A reducing sugar may be defined as one that contains a free aldehyde or an alpha-hydroxy ketone capable of being oxidized. Thus, sugars in the free aldo- or keto-form or in equilibrium with these forms fit into this category.

Analytically, reducing sugars may be determined by chemical, enzymatic, and high performance liquid chromatographic techniques. The chemical methods generally involve the reaction of reducing sugars with copper (II) in alkaline solution. As seen in the following equations, reducing sugars such as glucose and fructose reduce copper (II) to copper (I) oxide under alkaline conditions:

$$\text{Reducing Sugar} + 2Cu^{+2} \longrightarrow \text{Oxidized Products} + Cu^{+1} \qquad (6\text{-}1a)$$

$$2Cu^{+} + 2OH^{-} \longrightarrow \underset{\text{(yellow)}}{2CuOH} + \underset{\text{(red)}}{Cu_2O} + H_2O \qquad (6\text{-}1b)$$

In alkaline solution, sugars undergo decyclization to yield corresponding aldo- and keto-forms (Joslyn 1950a). This reaction is followed by rearrangement and subsequent degradation. The color change associated with the reaction is believed to be due to enolization, with resulant double-bond formation producing color. For example, glucose in alkaline solution yields 1,2-enediol, 2,3-endiol, and 3,4-enediol. Rupture at the double bond produces a variety of degradation products such as aldehydes and acids. Dependent variables for this reaction include the type and concentration of sugars present, the concentration of alkali, and the temperature and time of reaction. Low temperatures and relatively high concentration of alkali favor formation of the characteristic red Cu_2O precipitate. Sodium-potassium tartate is included in the reagent mix to facilitate separation of Cu_2O precipitate while maintaining sugar oxidation products and unreduced copper in solution. The copper (II) tartrate complex formed is stable even at the high temperatures of reaction. Reduced copper, however, does not complex with tartrate and readily precipitates from solution (Joslyn 1950a).

Once the reaction between sugar and copper (II) is complete, potassium iodide and sulfuric acid are added to the cooled reaction mixture (Rebelein method; see Procedure 6-1). Iodide reacts with the remaining Cu^{2+} ion to produce an equivalent amount of iodine (Equation 6-2a), which is subsequently titrated with standard sodium thiosulfate (Equation 6-2b):

$$2Cu^{+2} + 2I^{-} \longrightarrow 2Cu^{+} + I_2 \qquad (6\text{-}2a)$$

$$I_2 + 2S_2O_3^{-2} \longrightarrow 2I^{-} + S_4O_6^{-2} \qquad (6\text{-}2b)$$

McCloskey (1978) has reported an enzymatic procedure for the analysis of reducing sugars. The reaction sequence is presented in Procedure 6-3. Kits now are commonly available for this determination, which requires the use of a spectrophotometer and the ability to pipette small volumes accurately.

High performance liquid chromatographic procedures have been developed for the analysis of reducing sugars (see Procedure 6-5). These procedures generally are constructed around a speciality HPLC column (sold by one or more manufacturers). Because column technology is improving rapidly in the HPLC area, one should contact an HPLC supplier for columns and procedures.

RAPID DETERMINATION OF REDUCING SUGARS

Rapid reducing sugar measurements may be run routinely by employing a variety of reducing sugar kits originally developed for use by diabetics. By reference to a color code, the reducing sugar content of a measured volume of wine is determined within a range of 0 to 1%. As previously mentioned, the presence of pentoses will prevent the reducing sugar level of wine from reaching zero. Because these pentoses are not fermentable by yeast, a dry wine ($\leq$0.02% reducing sugar) generally is considered stable with respect to yeast refermentation.

In the case of proprietary products such as Clinitest, sensitivity levels are reported as 0.05% (Ames Company 1978). The mechanism of reaction generally is the same as that presented earlier in the chemical method except that the heat required is provided internally by the neutralization reaction of NaOH and critic acid. The major limiting factor in the use of reducing sugar kits is that the sugar level must be less than 1.0%. Thus, these kits are primarily helpful in determining completion of fermentation.

BRIX VS. REDUCING SUGAR VALUES

Confusion may exist between the concepts of °B and reducing sugar. Although this relationship may be indirect, reducing sugar is defined in terms of wt/vol (g/L), whereas °B is defined as % wt/wt ($g/100\ g$). A reducing sugar analysis, then, measures the amount of grape sugar in wine or must directly without interferences from alcohol, carbon dioxide, and so on. On the other hand, °B relates the density of the entire sample (assumed to be g sucrose/100 g sample) to that of pure water at 20°C. Because of the presence of other suspended solids, °B readings, will not be truly accurate measures of the sugar levels in must. In wine or fermenting must samples, the presence of alcohol and carbon dioxide will decrease the specific gravity of the sample below that due to the sugar–water content and yield low °B values. Therefore, hydrometers cannot be used to measure sugar content accurately in these samples (see Chapter 1).

Procedure 6-1. Rebelein Method for Reducing Sugars

In the Rebelein procedure, the excess copper remaining after reaction with reducing sugar subsequently is reduced with excess iodide ion to produce an equivalent amount of iodine. The iodine is then titrated with sodium thiosulfate. To avoid having to standardize reagents, a distilled water blank determination (no reducing sugar) is run. The result then is calculated by comparing the sample titration to that of the blank.

I. Equipment
   Volumetric pipettes (10 mL)
   Erlenmeyer flasks (200 mL)
   Burner
   Burette (50 mL)
   Glass boiling beads

II. Reagents (see Appendix II)
   Copper sulfate ($CuSO_4 \cdot 5H_2O$) solution
   Sulfuric acid (1 *N* and 16% vol/vol)
   Alkaline Rochelle salt (sodium potassium tartrate) solution
   Standard sodium thiosulfate ($Na_2S_2O_3 \cdot 5H_2O$)
   Starch indicator solution
   Potassium iodide
   Distilled water

III. Procedure
1. If reducing sugar is to be determined on a sweet wine, the sugar level first must be reduced, by volumetric dilution, to less than 2.8% (28 g/L).
2. Volumetrically, transfer 10 mL of the copper sulfate solution into a 200-mL Erlenmeyer flask.
3. Add 5 mL alkali Rochelle salt solution, two or three boiling beads, and 2.0 mL wine sample.
4. Bring the solution to a rapid boil for 1.5 minutes. Cool it quickly.
5. When it is cool, add:
   10 mL potassium iodide solution
   10 mL sulfuric acid solution (16%)
   10 mL starch solution
6. Mix this solution, and titrate it with sodium thiosulfate to a cream-white end point.
7. Calculate reducing sugar using the following relationship:

$$\text{R.s. (g/L)} = 28 - 28\,(V_A/V_B) \qquad (6\text{-}3)$$

where:

V_A = Volume titrant used with sugar sample
V_B = Volume titrant used for blank sample

Determination of a blank titration value is accomplished by using the protocol presented in steps 2 through 6 above using 2.0 mL distilled water instead of the wine sample.

IV. Supplemental Notes
1. A blank titration value should be determined each time a new batch of samples is run.

2. Red wines should be decolorized by pretreatment with decolorizing carbon or PVPP (''Polyclar AT'') prior to analysis.
3. Once the potassium iodide, sulfuric acid, and starch are added, the thiosulfate titration should be run as quickly as possible.
4. In this method of chemical analysis of reducing sugar, several variables must be carefully controlled if success is to be achieved, and the calculation (Equation 6-3) is to be valid.
    a. Reagents used for sugar analyses, as ordinarily prepared, consists of two solutions, one containing $CuSO_4$ and the other alkaline Rochelle salt (sodium potassium tartrate). Classic procedure requires that the two be mixed immediately before use. Mixed reagents cannot be kept satisfactorily because of the ease with which the mixture decomposes. The component solutions themselves eventually will decompose over prolonged periods of storage, and it is recommended that the copper sulfate solution be stored in the refrigerator until time for use.
    b. Temperature and duration of heating: The reduction reaction is accelerated with increases in temperature up to 90°C. At temperatures greater than 100°C, however, significant autoreduction of the copper in the reagents may occur, resulting in appreciable error.
    c. The concentration and the composition of alkali affect not only the rate of copper reduction but also the nature of the Cu_2O precipitate formed. Joslyn (1950a) in his review of the subject, noted that superior precipitates are produced by the use of sodium hydroxide rather than potassium hydroxide or carbonates. The quantity of copper reduced is directly related to increasing concentrations of alkali, with best results reported when the solution is 1.6 *N* with respect to NaOH.
    d. The concentration of copper sulfate also affects reduction. Ideally, the ratio of alkali (OH^-) to copper should be 5 to 1. Beyond this, reduction decreases, and the possibility of autoreduction increases.
    e. Increasing the concentration of tartrate beyond recommended levels has no effect on the reaction.

Procedure 6-2. Reducing Sugars by Modified Lane-Eynon Procedure

The Lane-Eynon procedure calls for determining the quantity of standard glucose solution (titrant) that is required to react with a known volume of alkaline copper sulfate under specified heating conditions. One milliliter of wine then is added to a second volume of alkaline copper sulfate, and the quantity of standard sugar solution required to complete the reduction determined. The difference in volumes needed to titrate the wine sample and the blank is related directed to the reducing sugar content of the sample.

I. Equipment
500-mL wide-mouth Erlenmeyer flask
50-mL side delivery burette
Volumetric pipettes (1.0 and 10 mL)
100-mL graduated cylinder
Electric burner
High-intensity light source
Glass boiling beads

II. Reagents (See Appendix II)
Fehling's A solution
Fehling's B solution
0.50% Dextrose (glucose) solution
1% Methylene blue indicator
Distilled water

III. Procedure

(a) Blank determination

1. Add 70 mL of distilled water and several glass boiling beads to a 500-mL wide-mouth Erlenmeyer flask.
2. Volumetrically transfer 10 mL of Fehling's A solution to the flask.
3. Add 10 mL of Fehling's B solution, and mix the resulting solution well. *Note:* This addition need not be volumetric and can be achieved with the use of any pipette.
4. Place the flask on a *preheated* electric burner, and immediately titrate with approximately 18 mL of 0.5% dextrose solution.
5. When the soluion comes to a rapid boil, add five drops of methylene blue indicator and titrate drop by drop until the end point is reached. The end point is detected as the first disappearance of blue indicator and appearance of red in the solution. *Note:* This titration should not take more than 3 minutes to complete. Theoretically, the blank titration should use 21.8 mL of 0.5% dextrose. However, this is not always the case.
6. Blank titrations should be repeated until end points vary by no more than 0.2 mL.
7. Record the volume of dextrose required for the blank (B).

(b) Sample determination:

8. Proceed as in the blank determination, adding 1 mL of wine after the addition of Fehling's B solution in step 3.
9. Titrate rapidly until approximately 2 mL before the estimated end point. At full boil, add five drops of methylene blue, and continue the titration dropwise until the end point is reached.
11. Record the volume of dextrose required for the sample titration (W). Calculate reducing sugar using the following relationship:

$$\text{R.s. (g/L)} = \frac{(B - W)(0.005\ \text{g/mL})(1000\ \text{mL/L})}{\text{sample volume (mL)}} \qquad (6\text{-}4)$$

where:

B = Volume of 0.5% dextrose solution required to titrate blank
W = Volume of 0.5% dextrose solution required to titrate sample

Using 1-mL sample volumes and the outlined procedure, the above equation can be simplified:

$$\text{R.s. (g/L)} = = 5(B - W) \qquad (6\text{-}5)$$

IV. Supplemental Notes

1. Some laboratories elect to decolorize highly pigmented wines prior to analysis. However, this step generally is not necessary when the procedure outlined above is used. Wines with a heavy suspended-solids content should be clarified prior to analysis.
2. During titration, the tip of the burette should be positioned within the neck opening of the flask to reduce oxygen contact. Continuous evolution of steam during titration also helps in reducing air access. Interruptions in steam flow from the titration flask may introduce errors.
3. Several practice titrations may be necessary for consistent end-point detection. Because different people may interpret end points differently, it is essential that the same person run the blank and the samples.
4. End-point detection is the disappearance of methylene blue indicator (formation of the leuco-form). This is best observed in the bubbles and around the edges of the flask by use of a high-intensity light source.
5. Wines with a residual sugar content of more than 5% should be volumetrically diluted prior to analysis.
6. In dry wine types, accuracy is reported to $\pm 0.05\%$. Routinely, one should strive for a reproducibility of ± 0.2 mL between replications.
7. When not in use, Fehling's A solution and dextrose should be stored in the refrigerator. These solutions should be at standard temperature when used.
8. Refer to the Rebelein procedure (supplemental notes) for precautionary comments concerning the use of Fehling's-type solutions.

Procedure 6-3. Enzymatic Analysis for Reducing Sugars

McCloskey (1978) reports an enzymatic analysis for the two major reducing sugars (glucose and fructose) in wine. This method does not detect the pentoses included in traditional analyses of reducing sugar. The results at very low levels

(<1.0 g/L) are superior to those obtained by the Lane-Eynon procedure. McCloskey's procedure utilizes premeasured enzyme reagents sold under the name Glucose Stat-Pak (Calbiochem). Because fructose is not an active substrate, a third enzyme, phosphoglucose isomerase, is necessary. The reaction sequence is:

$$\text{Glucose} + \text{ATP} \xrightleftharpoons{\text{HK}} \text{G-6-Phosphate} + \text{ADP} \qquad (6\text{-}6a)$$

$$\text{Fructose} + \text{ATP} \rightleftharpoons \text{F-6-Phosphate} + \text{ADP} \qquad (6\text{-}6b)$$

$$\text{F-6-Phosphate} \xrightleftharpoons{\text{PGI}} \text{G-6-Phosphate} \qquad (6\text{-}6c)$$

$$\text{G-6-Phosphate} + \text{NADP} \xrightleftharpoons{\text{G6P-DH}} \text{Gluconate-6-Phosphate} + \text{NADPH} + \text{H}^+ \qquad (6\text{-}6d)$$

where:

HK = hexose kinase

PGI = phosphoglucose isomerase

G6P-DH = glucose-6-phosphate dehydrogenase

In lieu of a detailed procedure here, the reader is referred to the product sheet that accompanies the carbohydrate kit of reagents, for the most recent information and a procedure specifically designed for the kit obtained.

In general the carbohydrate procedure will require the following:

I. Equipment
   Spectrometer capable of reading absorbance at 340, 365, or 334 nm
   Single- or multiple-range micropipettes (25 μL and 100 μL)
   Water bath (22–30°C)

II. Reagents
   The reagents used in this analysis are most conveniently purchased in kit form. Such kits may be acquired from Calbiochem (Glucose Stat-Pak) and Boehringer Mannheim (Glucose/Fructose).

III. Procedure
   The procedure usually follows these steps:
   1. Samples generally are diluted 1 + 9 to keep sugar concentration in the 0.1 to 1 g/L range.
   2. An aliquot is mixed with reagents (except hexokinase/glucose-6-phosphate dehydrogenase and phosphoglucose isomerase), and the absorbance is read.

3. Hexokinase/glucose-6-phosphate dehydrogenase is added to initiate the reaction, and the absorbance again is read upon completion.
4. Phosphoglucose isomerase is added, and the absorbance again is read after the reaction ceases.
5. Glucose and fructose levels are calculated by taking differences in the various absorbance levels.

Procedure 6-4. Invert Sugar Analysis

The addition of sugar to fermenting and finished wines is permitted in certain instances. In California, these additions are restricted to formula wines such as champagnes and special naturals. In such cases, the sugars used are generally in the form of concentrates, dry dextrose, sucrose, or syrups of the latter two. Sucrose creates difficulties from an analytical point of view in that it is not a reducing sugar. Thus where such analyses are required, a preliminary sample treatment is necessary. This is usually accomplished by acid hydrolysis of the nonreducing disaccharide to its component reducing monosaccharides glucose and fructose, which are reducing sugars. In an acid medium such as wine, this hydrolysis proceeds normally with time. In like manner, yeast-elaborated enzymes (invertase) bring about the hydrolysis in fermenting must. Thus, in time the acid nature of wine or the fermentative action of yeast will result in the required hydrolysis, making laboratory hydrolysis unnecessary. Once hydrolysis has occurred, one may proceed with the reducing sugar analysis by one of the procedures above, keeping in mind that there will be a correction factor for dilution of the sample.

I. Equipment
   Water bath ($60 \pm 2.0°C$)
   50-mL volumetric flask
   Ice water bath
   Timer
   25-mL volumetric pipet
II. Reagents (See Appendix II)
   Ammonium hydroxide solution (1 + 1.5)
   HC1 (1 + 1)
   Distilled water
III. Procedure
   1. Volumetrically transfer 25 mL of sample into a 50-mL volumetric flask.
   2. Add approximately 10 mL of 1 + 1 HCl.
   3. Place the flask in the 60°C water bath for 15 minutes.
   4. After 15 minutes, cool the flask in the ice bath to 20°C.
   5. Carefully add 10 mL of 1 + 1.5 NH_4OH with mixing.

6. Bring the solution to volume with distilled water at 20°C.
7. Mix the solution thoroughly. Determine reducing sugar using one of the above procedures.
8. Using the appropriate equation, calculate reducing sugar (g/L). Keep in mind the 1:2 dilution factor required in the hydrolysis step.

Procedure 6-5. HPLC Analysis of Carbohydrates

Sugar components can be separated and quantitated by using an appropriate chromatographic column and mobile phase. The method thus allows the analyst to determine all the sugars present in the sample, both fermentable and nonfermentable. In the following procedure, a preliminary separation of the sugars and fixed organic acids in the sample is utilized. A typical HPLC chromatogram may be seen in Fig. 6-2.

I. Equipment
   HPLC with refractive index or variable wavelength UV detector
   Bio-Rad HPX-87H carbohydrate column (or equivalent) plus appropriate guard column
   Disposable polypropylene column
   Pipettes (1 mL) and volumetric flasks (10 mL)
II. Reagents (See Appendix II)
   Mobile phase (0.01 N H_2SO_4)
   Bio-Rex 5, 100–200 mesh (Cl^- form) resin
   Standard solution (must analysis) consisting of 10 g/L glucose, 10 g/L fructose, and 0.5 g/L sucrose.

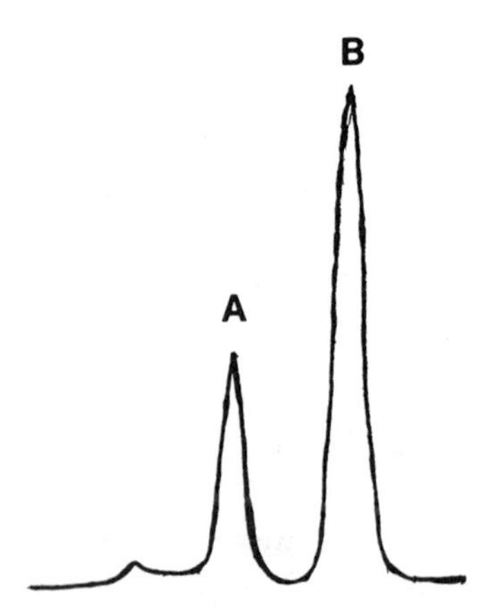

Fig. 6-2. A typical HPLC chromatogram of (A) glucose and (B) fructose in sweet white wine.

Standard solution (wine analysis) containing 0.1 g/L glucose, 0.1 g/L fructose, and appropriate amounts of other sugars of interest

III. Procedure

1. Install a Bio-Rad HPX-87H carbohydrate column (or equivalent) in the high performance liquid chromatograph. This column should be preceded by an appropriate guard column in the mobile phase flow stream.
2. Being pumping 0.01 N H_2SO_4 mobile phase through the column at a flow rate of 0.6 mL/min.
3. Set the column oven temperature at 25°C, and allow the system to stabilize.
4. Turn on the variable-wavelength UV detector, and allow it to warm up. Set the wavelength dial to 210 nm. Alternatively, turn on the refractive-wavelength detector, and allow it to warm up.
5. Prepare and chromatograph the sample as follows:
   a. Pipette 2.00 mL of wine or grape must and approximately 0.2 mL of concentrated ammonium hydroxide into a capped vial. Mix the contents of the vial, and keep it closed until ready to place contents on the anion exchange resin.
   b. Prepare the ion exchange column by slurrying 1 g Bio-Rex 5, 100–200 mesh (Cl^- form) resin in 3 mL distilled water, and pouring it into a disposable polypropylene sample preparation column. Drain off the water, and wash the column with an additional 5 mL of distilled water. Do not allow the column to dry out prior to use.
   c. Add 1 mL of sample to the column, and allow it to drain through. Wash the sugars off the column with distilled water (9 mL) and collect them in a 10-mL volumetric flask. Bring the solution to the mark with distilled water, and mix it well prior to injection onto the column.
   d. If it is desired that the wine acids also be analyzed, carefully add 2 mL of 20% sulfuric acid and enough distilled water to collect 10 mL of eluent containing the acids. Again mix this sample well prior to injecting it onto the column.
6. Qualitative analysis may be accomplished by injecting standard solutions of the various sugars expected in the sample. Compare observed retention times of peaks in the sample chromatogram with those in the standards chromatogram.
7. Standard solutions of various sugars at known concentrations can be injected and chromatographed. Peak heights (recorder) or peak areas (integrator) are compared to those from samples, and the unknown concentrations calculated using the following:

$$\text{Sugar conc (g/L)} = \frac{A}{A'}\text{(sugar conc in std, g/L)} \qquad (6\text{-}7)$$

where:

A = response of unknown sugar
A' = response of standard sugar

IV. Supplemental Notes

1. There are a number of resin columns currently available for carbohydrate analysis. One should consult column manufacturers' literature for columns suitable for the specific analysis desired.
2. The column listed in the above procedure also can be used for the determination of organic wine acids. For this determination, see Procedure 4-3.

Chapter 7

Phenolic Compounds and Wine Color

Grapes and wine contain a large array of phenolic compounds. Derived from the basic structure of phenol (hydroxybenzene), the term ''phenolic'' refers to any compound with a phenol-type structure. Singleton (1980) describes the classes of phenolics in terms of chemical structures that range from relatively simple to complex. One phenolic group is composed of the nonflavonoids, which are derivatives of cinnamic and benzoic acid. Members of the group include hydroxycinnamates and benzoates (and their derivatives), as well as tyrosol. Additional group members are other lower molecular weight phenolics similar to those described, but probably arising from higher molecular weight forms as a result of degradation. These compounds include the volatile phenolics, hydrolyzable tannins, and benzene derivatives. A second large group of phenolics includes the flavonoids. This group encompasses catechins, flavonols, and anthocyanins. A third group is recognized in wine, and represents a composite of phenolic–protein–polysaccharide complexes. Representative structures of these major classes of phenolics are presented in Fig. 7-1.

The phenolic content commonly is measured in terms of equivalent amounts of gallic acid necessary to produce the same analytical response. Results are expressed in terms of gallic acid equivalents (GAE) per weight of fruit or per volume.

Sources for phenolic compounds present in wine include the grape itself, as well as tissues extracted during the course of alcoholic fermentation (see Table 7-1). The total phenol content of wine is less than the level present in the fruit from which it is made. Microbial activity also may result in increased concentrations of certain phenolic compounds. Extraction from oak cooperage during aging as well as from flavorings used in production of special naturals provides additional sources of phenolics. Lastly, degradation of larger-molecular-weight components may be important in the formation of some phenolic compounds.

Phenols in wine contribute significantly to wine character. For example, they are responsible for red wine color, astringency, and bitterness, in addition to contributing somewhat to the olfactory profile of the wine. Phenols also serve as important oxygen reservoirs and as substrates for browning reactions.

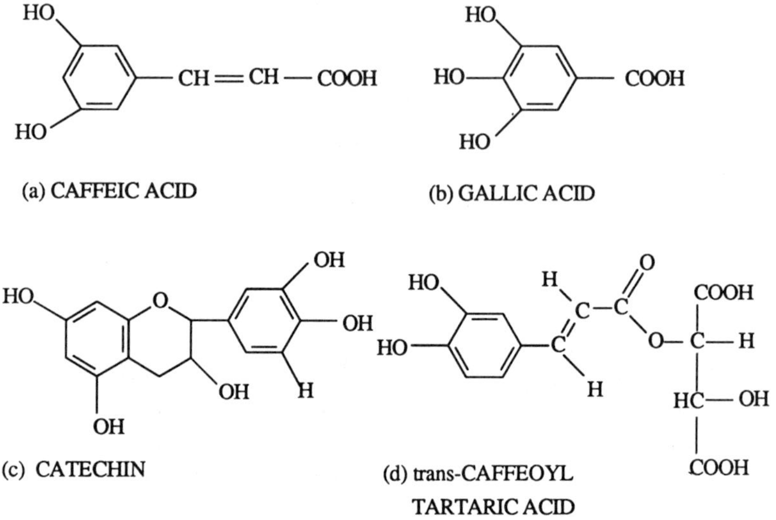

(a) CAFFEIC ACID

(b) GALLIC ACID

(c) CATECHIN

(d) trans-CAFFEOYL TARTARIC ACID

Fig. 7-1. Representative examples of nonflavonoid (a, b, and d) and flavonoid phenolics (c) found in grapes and wine: (a) caffeic (dihydroxycinnamic) acid; (b) gallic (trihydroxybenzoic) acid; and (c) catechin. Structure 7-1d is *trans*-caffeoyl tartaric acid (the most common form in grapes).

NONFLAVONOID PHENOLICS

The aromatic compound cinnamic acid is the parent compound for this group. The structures of cinnamic acid and *trans*-caffeoyl tartaric acid (the most common form in grapes and wine) and its derivatives reported to be in grapes and wine are presented in Fig. 7-1a,d. Other simple phenolic compounds found in wine, which may serve as building blocks for more complex species, include benzoic acid and its derivatives. Ribereau-Gayon (1965) reported seven benzoic acid derivatives in wine (Fig. 7-2).

Singleton (1985) reports that the levels of benzoic acid and its derivatives in red wines range from 50 to 100 mg/L. In whites the level is considerably

Table 7.1. Total phenol levels in Vitis Vinifera grapes.*

Component	Red grapes	White grapes
Skin	1859	904
Pulp	41	35
Juice	206	176
Seeds	3525	2778
Total	5631	3893

*GAE mg/kg berries
From Singleton and Esau (1969).

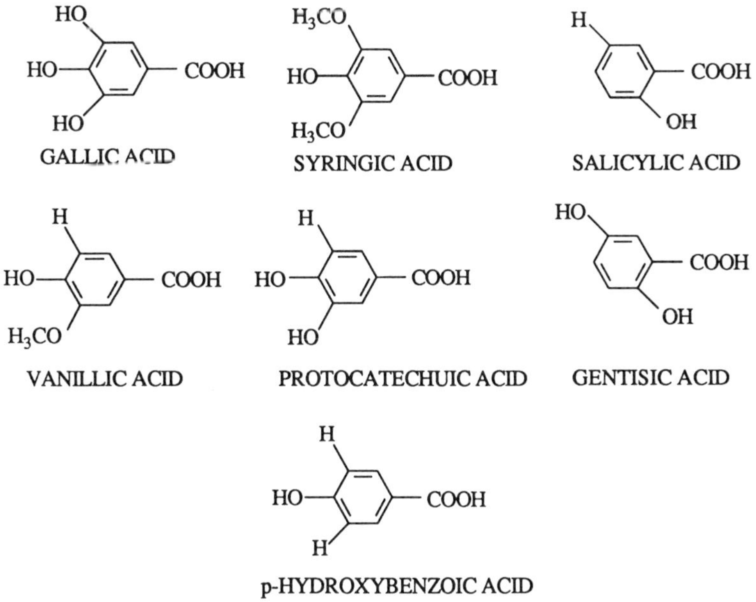

GALLIC ACID SYRINGIC ACID SALICYLIC ACID

VANILLIC ACID PROTOCATECHUIC ACID GENTISIC ACID

p-HYDROXYBENZOIC ACID

Fig. 7-2. Benzoic acid derivatives reported in wines.

lower, on the order of 1 to 5 mg/L; salicyclic acid is present at less than 2 mg/L, and other derivatives are present only in trace amounts.

Most nonflavonoid phenols are present at levels that individually are below their sensory threshold. Collectively, however, members of the group may contribute to bitterness and harshness. The nonflavonoid component arises principally from the juice and extraction during fermentation and secondarily from post-fermentation activity, including exposure to oak cooperage.

Nonflavonoid Content of Juice

The phenol content of juice is largely nonflavonoid (Kramling and Singleton 1969). Hydroxycinnamate derivatives comprise the majority of this class of phenolics in both white and red wines, where they serve as the primary substrate for polyphenol oxidase activity (Singleton et al. 1985). This reaction, involving incorporation of glutathione, yields products resistant to further polyphenol oxidase activity and browning (Nagel and Graber 1988).

Hydroxycinnamates may be present in juice and wine as the free acids and ethyl esters as well as in the form of tartrate or tartrate-glucose esters (Hennig and Burkhardt 1960). Ong and Nagel (1978) noted the presence of three hydroxycinnamic acid–tartaric acid esters in seven white and five red varieties of *Vitis vinifera*. Using high performance liquid chromatography (HPLC) separation techniques, these workers report mean values of monocafeoyltartaric acid, mono-*p*-coumaroyltartaric acid, and monoferuloyltartaric acid as 107.6, 16.3, and 6.1 mg/L, respectively. In the five red varieties they studied, these values increased to 170.5 mg/L for the monocaffeoyl ester and to 25.4 mg/L and 6.3 mg/L for mono-*p*-courmareoyl- and monoferuloyltartaric acid esters. They also proposed the presence of two hydroxycinnamic acid triesters of caffeic acid-tartaric acid-glucose and *p*-coumaric acid-tartaric acid-glucose. Somers (1987) found that the *trans*-isomer of these derivatives is the principal form found in juice.

Nonflavonoids Derived from Fermentation

During alcoholic fermentation, slow (incomplete) hydrolysis of the ester occurs, resulting in the presence of both free acid and ester forms in wine. Caftaric and similar acylated phenols are hydrolyzed to varying degrees, yielding the corresponding free cinnamic acids. The latter may be involved in the formation of microbially produced end products such as 4-ethylcatechol. This transformation involves decarboxylation of the acid to yield 4-vinylcatechol and subsequent reduction to 4-ethylcatechol. Similar microbially induced transformation of benzoic, shikimic, or quinic acids to yield catechol and protocatechuic acid has been reported (Whiting and Coggins 1971). Tyrosol is produced by yeast during fermentation from tyrosine. Somers (1987) reported that over the course of fermentation total hydroxycinnamates decreased by nearly 20% from levels found in juice. They attributed the decrease to adsorption by yeasts.

In general, the levels of nonflavonoid phenols are relatively constant in red and white wines, because of their ability to be extracted from grape pulp. The only appreciable source of nonflavonoids from grape solids is the hydrolysis products of anthocyanins (hydroxycinnamic acyl groups) (Singleton and Noble 1976).

Nonflavonoid Component Arising From Oak Aging

In wines not exposed to oak cooperage, the nonflavonoid phenol fraction is about the same as in the juice from which the wine was produced. Phenols extracted from oak are present almost entirely as hydrolyzable nonflavonoids. Vanillin, sinapaldehyde, coniferaldehyde, and syringaldehyde are reported to be the major species present in barrel-aged wines (Singleton 1985). The type of oak used appears to affect the phenol content of barrel-aged wines. Singleton et al. (1971) report that new wine aged for one year in 200-L French oak barrels may contribute 250 mg/L phenols (GAE). By comparison, wine stored in American white oak barrels had approximately one-half the levels reported from French oak storage. In alcoholic extracts of French and American oak the phenols extracted include ellagitannins as the major component. The conversion of bitter esculin to the less bitter form, esculetin, is thought to be a major reason why wood staves used in cooperage production may improve with seasoning.

The question of whether these compounds are present as oxidation products resulting from laboratory extraction or are initially present in grapes (and wine) has not been completely resolved. It would seem reasonable that the presence of certain members of the group such as vanillin (an aldo-derivative of benzoic acid) is secondary, resulting from cellar practices such as oak aging.

VOLATILE PHENOLS

Degradation of wine flavonoids with age increases the production of more volatile phenols (Singleton and Noble 1976). These low molecular weight hydroxybenzoates and hydroxycinnamates are noted for their odors. Levels of volatile phenols in young white wines are reported to range from 1.5 to 3.2 mg/L (GAE). Higher levels have also been reported. Although the levels of individual volatile phenols in young wines usually are below threshold, cumulative effects may contribute to the perception of smokiness and bitterness.

FLAVONOIDS

Flavonoids represent a large group of over 2000 phenolic compounds associated with vascular plants. Many of these compounds are brightly colored and found in leaves and flowers. Grape flavonoid phenolics are found in skins, seeds, and tissue portions of the fruit. Most flavonoids are believed to occur naturally as glucosides, with their base structure (the aglycone) consisting of two aromatic rings, A and B, joined via a pyran ring. The flavonoid base structure and standard numbering system are seen in Fig. 7-3a. Focusing on the pyran ring, changes in the oxidation state result in variations in hydrogen, hydroxyl, and ketone groups associated with carbons 2, 3, and 4 (Fig. 7-3b), leading to different members of the flavonoid family.

Much of the structure and color associated with wine is due to this group of compounds. Ribereau-Gayon and Glories (1987) view the group of flavonoid phenolics as composed of two reactive members, the anthocyanins and "procyanidins," which serve as precursors to polymeric compounds called tannins. As seen in Fig. 7-3b, "procyanidin" can be used to describe any of three groups of flavonoids: catechins, leucoanthocyanidins, and flavonols.

A main feature of flavonoids is reactivity. Monomeric forms of procyanidins and/or anthocyanins react during the winemaking process to yield dimeric and larger forms. The result is an array of heterogeneous polymeric structures. Polymeric flavonoids make up the major fraction of total phenolics found in all stages of winemaking. Polymerization, either oxidative or nonoxidative, yields tannins and condensed tannins, respectively (see Fig. 7.4). Further polymerization eventually leads to a form that may precipitate from wine. Phenols contribute to astringency, bitterness, and tactile characteristics, as well as wine color.

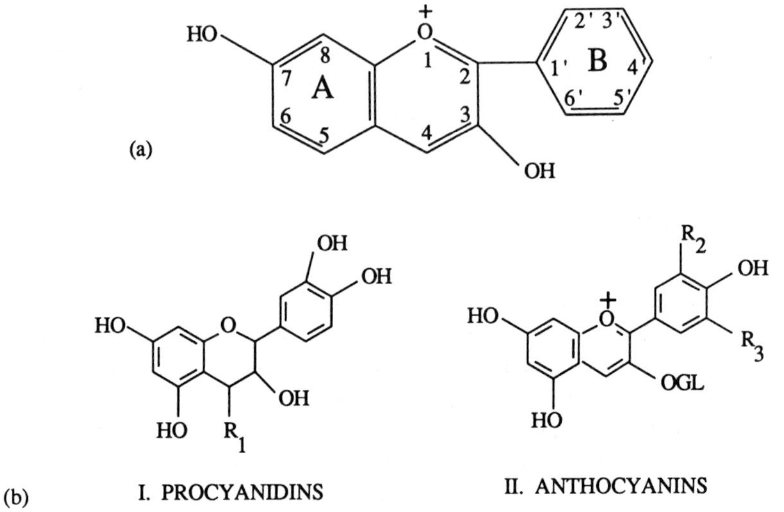

(a)

(b) I. PROCYANIDINS II. ANTHOCYANINS

Fig. 7-3. (a) Base structure of typical flavonoid showing rings and numbering system. (b) Parent structures of wine flavonoids. I. Procyanidins: R = H (catechins); R = OH (leucoanthocyanidins); R = =O (flavonols). II. Anthocyanins: R_2, R_3 = OCH_3 (malvidin); R_2 = OH, R_3 = H (cyanidin); R_2, R_3 = OH (delphinidin); R_2 = OCH_3, R_3 = OH (petunidin); R_2 = OCH_3, R_3 = H (peonidin).

Processing protocol will significantly affect the phenolic composition of wine. Because flavonoid phenolics are localized mainly in the skin, increasing the skin contact time and temperature as well as the extent of berry breakage and extraction, would lead to an increase in phenol content. The distribution of phenols in a red wine with 1400 mg/L (GAE) is seen in Table 7-2. In terms of percentage, flavonoid phenols account for 80 to 90% of the phenolic content of red wine.

Catechins (Flavan-3-ols)

The basic structure for the family is *d*-catechin (Figs. 7-1c and 7-3b). Because of asymmetrism arising at carbons 2 and 3, four stereoisomers may occur: *d*- and *l*-catechin, as well as *d*- and *l*-epicatechin. Hydroxylation at the 5′ position yields *d*-gallocatechin and its isomer, *l*-epigallocatechin. Of these forms, *d*-catechin, *l*-epicatechin, and *l*-epicatechin gallate are the principal catechins found in grapes (Su and Singleton 1969). Singleton (1980) points out that, compared to catechins, gallocatechins are present in much lower concentrations in mature fruit and wines made from it.

Catechins may exist in dimeric and larger groupings, the latter displaying

Table 7-2. Phenolic levels in Red Wine.*

Phenol type	Concentration (mg/L)
Nonflavonoids	200
Flavonoids:	
Anthocyanins	150
Condensed tannin	750
Other flavonoids	250
Flavonols	50
*Total (mg/L GAE)	1,400

Source: Singleton and Noble (1976).

many of the characteristic reactions associated with tannins (see below, under "Tannins"). Catechins are precursors to browning in white wines and to browning and bitterness in red wines. In white wines, catechins contribute significantly to the flavor profile (Singleton and Noble 1976). Sulfur dioxide additions to must increases the extraction of flavonoid phenols such as catechin (Singleton et al. 1980), and such extractions may contribute to the perception of bitterness in white wines. In white wines made with limited skin contact, catechins account for most of the flavonoid phenols. The taste threshold of catechins in white wine is near 200 mg/L (GAE). In contrast, catechins have a reported bitter threshold of 20 mg/L (Dadic and Belleau 1973) in a 5% ethanolic model solutions.

Leucoanthocyanidins (Flavan-3,4-diols)

Leucoanthocyanidins differ from leucoanthocyanins in that the latter contain a sugar attached through a glucosidic linkage. They also differ from catechins in that they have an additional hydroxyl group at carbon 4 (see Fig. 7-3). Three sites of asymmetry (C_2, C_3, and C_4) result in formation of eight possible isomers. Upon heating in acid solution, leucoanthocyanidins are converted to the colored anthocyanidin form. These compounds serve as precursors to larger polymeric forms.

In considering the structure of tannin polymers, Ribereau-Gayon (1974) suggested covalent bond formation between carbon 4 of the flavan-3,4-diol and either carbons 6 or 8 of a second reactive form. Depending upon the nature of that second flavon component, several polymeric species are possible (Fig. 7-4). If carbon 4 is available (Fig. 7-4a), the condensation reaction may continue, yielding a final polymer of three to four or more members (Somers 1987). If reaction at C_4 is blocked (Fig. 7-4b), the condensation is stabilized at the dimer stage. Occasionally, this group is important in pinking reactions seen in the making of white wine.

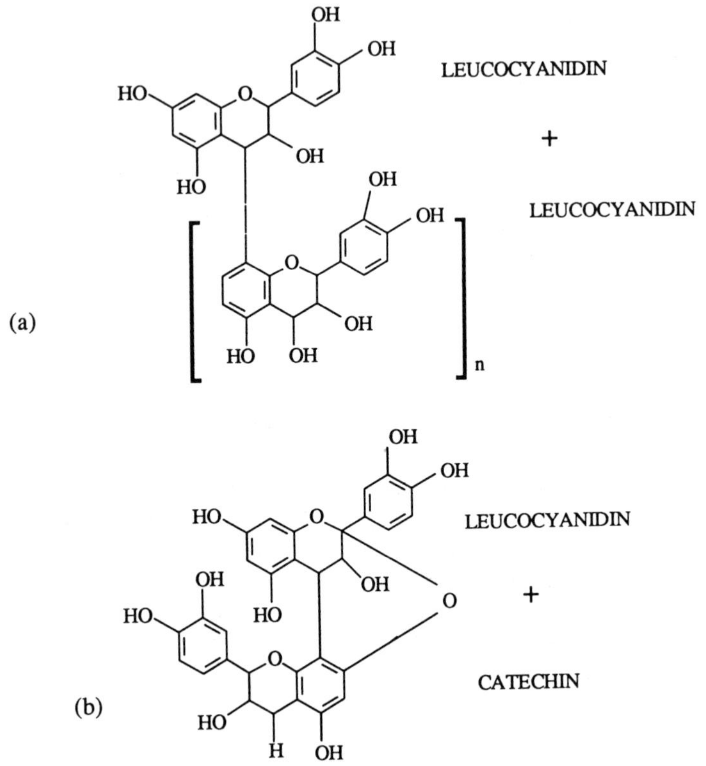

Fig. 7-4. Typical polymeric and dimeric flavan structures found in wine. (a) Polymeric structure with C_4 available and capable of further polymerization; (b) with C_4 blocked and stabilized at the dimer stage.

Flavonols

In grapes, flavonols are localized in the skin fraction, occurring in glycosidic forms. The most commonly encountered sugar moiety in these flavonols appears to be glucose. Ribereau-Gayon (1965) reports glucosides of kaempferol, quercetin, and myricetin, present in the 3-glucoside form.

In wines, Jurd (1969) reports that hydrolysis of the glucose moiety occurs so that the only forms identified are the respective aglycones (Fig. 7-5). Singleton (1982a, 1982b) reports only trace flavonol concentrations in white wines. In young red wines, he reports concentrations of up to 50 mg/L, and in older red wines approximately 10 mg/L. Quercetin comprises the majority of the flavonol fraction. In wines produced without skin contact, flavonols were reported at zero or only trace concentrations.

The bitterness threshold for kaempferol in 5% aqueous ethanol solution is reported to be 20 mg/L. Quercetin and myrecetin made up in the same manner

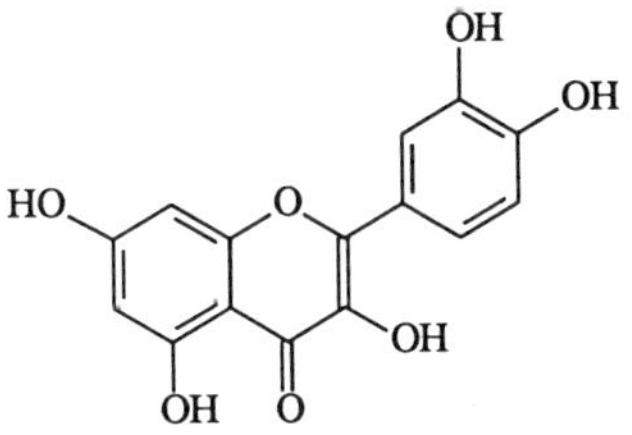

QUERCETIN

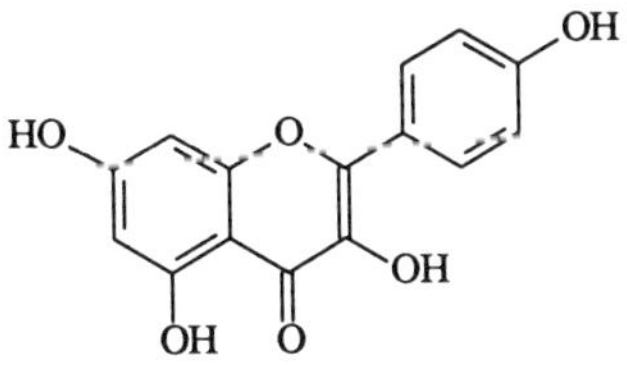

KEMPFEROL

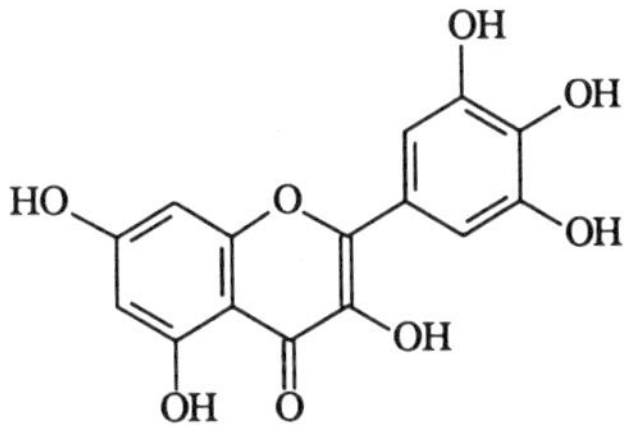

MYRICETIN

Fig. 7-5. Flavonols (aglycone-form) found in wines.

have bitterness thresholds of 10 mg/L (Dadic and Belleau 1973). The glucosides ordinarily are more bitter than the aglycones.

Tannins

Reactions of catechins, flavan-3,4-diols, and flavonols among and between themselves yield larger molecular weight compounds that possess astringent and/or bitter characters. These are generally referred to as tannins.

Tannins are classified as "hydrolyzable" or "condensed." Hydrolyzable tannins exist as esters and, as such, can be degraded or hydrolyzed. More closely related to the class of compounds important in winemaking are the condensed tannins. Wine tannin is composed largely of polymers of leucoanthocyanidins and catechins (Ribereau-Gayon 1965), such as those seen in Fig. 7-4.

Classically, tannins exhibit the following behavior: formation of blue-color complexes upon reaction with Fe^{3+}, as well as reaction with protein. Further, tannins usually are described as astringent. Phenolic compounds eliciting such reactions characteristically have molecular weights ranging from 500 to 3000. Because the molecular weight of dimeric flavonoids exceeds 500, they may be considered tannins. In young wines, tannins exist as dimers or trimers. With time, further polymerization occurs, and tannins may contain 8 to 14 flavonoid units with molecular weights ranging from 2000 to 4000.

In the grape, phenolics are extracted from skin, stems, and seeds. Because of the protective waxy coating of grape seeds, the seed tannin may not comprise the major portion of the final tannin content in wine. When this waxy coating is broached during fermentation, seeds can contribute up to 50% of the final phenolic content of wine. During aging, tannin levels in wines decrease as a result of oxidation interaction with yeast cells, and precipitation with protein.

Tannins also play a role in physical stability. Under conditions of high pH and high tannin levels, iron (Fe^{3+}) may combine with tannin after exposure to air, producing an insoluble colloidal instability known as ferric tannate or "blue casse." (This topic is covered in greater detail in Chapter 15.) Tannins also can combine with high levels of protein, again producing instability. (See Section IV and Chapter 16 for more details.)

The tannin content of most young red wines is well above the flavor threshold. Large polymeric phenols account for astringency in wines, the most astringent components being flavonoid tetramers (Lea and Arnold, 1978). Lower molecular weight phenols generally are perceived as bitter, as discussed earlier. The volatile phenols (lowest molecular weight forms) may possess distinct odors. By comparison to the tannin fractions, however, these forms are less apparent.

As a result of changes in winemaking style, wines produced today tend to be lower in tannins. Singleton and Noble (1976) report the average values for total phenols in California red table wines to be 1400 mg/L, whereas white table wine values average around 250 mg/L.

Astringency masks the perception of bitterness in wine. As wines age or undergo the addition of protein fining agents and tannin precipitation continues, there may be an unveiling of bitterness. (See Chapter 17 for further discussion of this concern.)

Wines deficient in tannin may be either blended with higher tannin stock or supplemented by addition of commercial tannic acid. In white wines, government regulations state that tannin additions must be restricted so that the wine does not contain more than 0.80 g/L GAE after the addition. In red grape wines, the final tannin levels must not exceed 3.0 g/L GAE (Part 240 of Title 27 of the Federal Regulations). Commercial tannic acid often is used for such additions. Tannic acid differs chemically from the grape seed tannin. Berg and Akiyoshi (1956) report increased browning in wines following additions of tannic acid, whereas no such deterioration was seen in samples where grape seed tannin was used. It would appear from these results that grape seed tannin is the compound of choice for increasing wine astringency.

A white wine with a total of 200 mg/L total phenols as GAE can be expected to contain about 100 mg/L caffeoyl tartrate and related cinnamates, 30 mg/L tyrosol and small derivative phenols, and 50 mg/L of flavonoids such as catechins and tannins (Singleton 1980).

The catechin content accounts for most of the total flavonoids found in white wines made with little skin contact. Such wines contain few leucoanthocyanidins. As pomace contact time and temperature is increased, both catechins and anthocyanin fractions increase, with the polymeric tannins increasing faster than the catechins.

Skin contact of crushed white grapes prior to pressing is a standard procedure in producing certain white wines. Seeds, cap stems, and skins provide equal quantities of phenol extractives at cool temperatures (DuPlessis and deWet 1968 and DuPlessis and Uys, 1968). Ough and Berg (1971) reported that increased temperature of the juice during skin contact resulted in more deeply colored whites and higher pH, potassium, proline, and total phenols. For temperature increases of 5 to 15°C, Pallotta and Cantarelli (1979) reported these increases: catechins, 160%; leucoanthocyanins, 58%; tannins, 31%: and total phenols, 42%. Elevated temperatures during skin contact (greater than 10°C) produce wines of deeper color, increased oxidative sensitivity, and coarser character. Also, these wines mature more rapidly during barrel aging (Ramey et al. 1986). These authors also demonstrated increased retention of volatile components in must when pomace temperatures did not exceed 19.5°C.

Anthocyanins/Anthocyanidins

The color of grapes is attributed largely to the presence of members of a large group of plant pigments, the anthocyanins. These compounds are present as glycosides (primarily glucosides) in which specific hydroxyl functions are in combination with sugar residues. Glycosidation may occur as single sugar residues attached to multiple hydroxyls, as di- and trisaccharides, or as any combination of these forms. In varieties of *Vitis vinifera*, pigments are present in the 3-glucoside form, whereas in other species, 3,5-diglucosides may occur (Singleton and Esau 1969). Except for Pinot noir, grape pigments are acylated with acetic, caffeic, or *p*-coumaric acids (Fig. 7-6).

Five anthocyanins generally are found in red grapes: malvidin-, delphinidin, peonidin-, cyanidin-, and petunidin-3-*d*-glucoside (Fig. 7.3b). Of these, malvidin (as the 3-glucoside) is the most common pigment in varieties of *Vitis vinifera*.

Upon acid hydrolysis (Fig. 7-6), anthocyanins yield one or more moles of sugar and a parent anthocyanidin (also referred to as the aglycone). Aglycones are reactive and susceptible to change (polymerization, etc.). The color of these pigments and their stability is a function of pH, the presence of various metallic cations, and other factors, which are discussed, in summary form, under several headings in the following paragraphs.

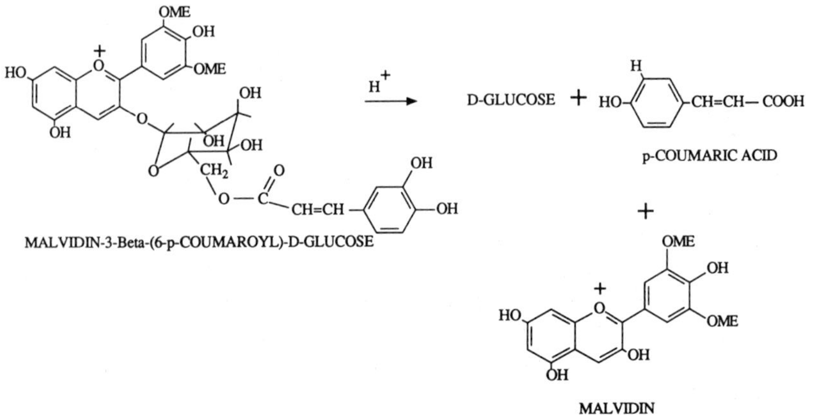

Fig. 7-6. Hydrolysis of typical grape anthocyanin yielding the aglycone (anthocyanidin), glucose, and coumaric acid.

pH

Anthocyanins are amphoteric substances, so the color of these pigments in solution is primarily pH-dependent. The reversible, pH-dependent equilibria that exist for malvidin-3-glucoside are seen in Fig. 7-7; the flavylium ion (red-colored) is in equilibrium with its violet-colored quinone form as well as with its colorless carbinol base form. At lower pH values ($pH < 4.0$), the main equi-

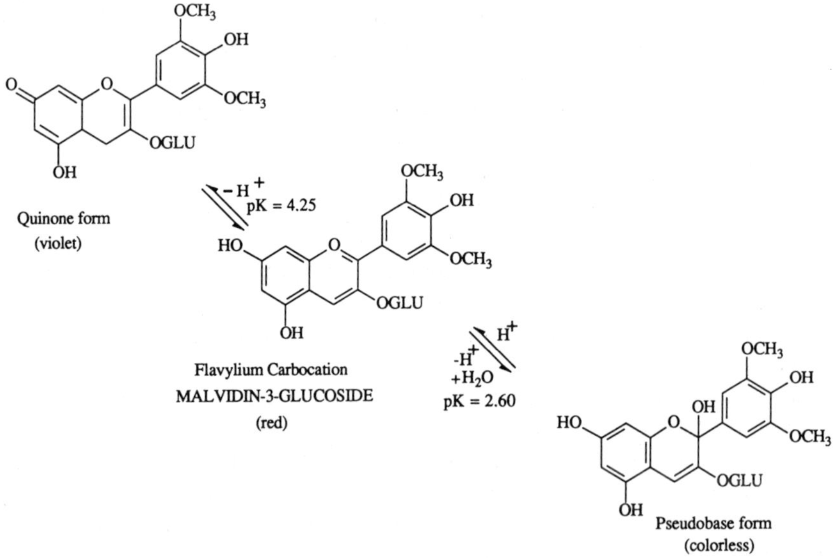

Fig. 7-7. pH dependency of malvidin-3-glucoside pigment.

librium forms are the red-colored flavylium ion and its colorless pseudobase. The pK for equilibrium between these forms (pK = 2.6) favors the colorless form. Thus, at wine pH values greater than 3.0, less than 50% of the potential red color is visible.

Sulfite Bleaching

The well-known observation that SO_2 addition results in temporary color reduction is based upon a reversible reaction presented in Fig. 7-8 (Jurd 1964). Because the site of sulfite binding (carbon 4) is also the point at which reaction with other phenolics can occur, polymerized pigments or tannin-pigment polymers (with linkages between C_4 or C_6 and C_8) are resistant to decolorization by sulfur dioxide (see Fig. 7-4).

Temperature

In general, an increased temperature of wine storage is reported to result in a increased red (or brown) color presumably due to accelerated rates of polymerization and other reactions occurring at the higher temperature.

Polymerization

Even in very young wines, a majority of the anthocyanidin pigment is incorporated into dimeric or larger units. Several possible condensation-polymerization reactions have been studied by Timberlake and Bridle (1976) and Ribereau-Gayon (1973). Copolymerization may take place between pigment and tannin species present in solution. In this case, bond formation may occur between C_4 of the anthocyanidin and either C_6 or C_8 of the tannin to yield a red-colored dimer. Because C_4 of the anthocyanidin is involved in bond formation, such polymers are not reactive toward sulfur dioxide, nor are they responsive to changes in pH. By comparison to monomeric anthocyanidins, polymeric pigments show increased stability toward other reactions.

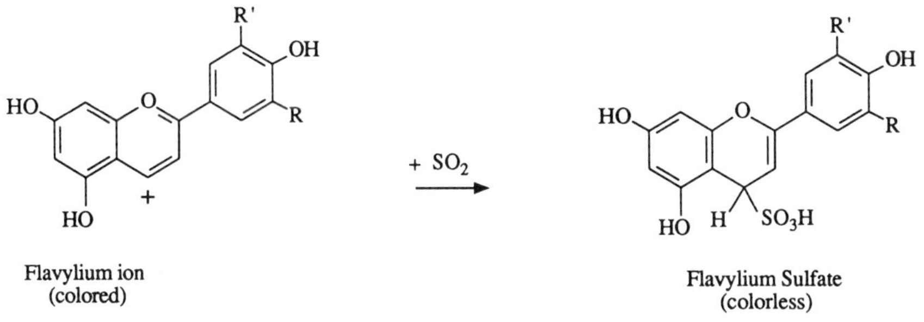

Flavylium ion (colored) + SO_2 → Flavylium Sulfate (colorless)

Fig. 7-8. Sulfur dioxide bleaching of pigments.

In the presence of acetaldehyde (as a carbonium ion), rapid addition may occur between anthocyanidin and tannin precursors (Fig. 7-9). The extent to which these molecules already are polymerized determines the stability of the product. Addition to an already highly polymerized tannin leads to instability and precipitation (and decreased color). By comparison, in cases where polymerization has not reached this state, a stable complex results that is more highly colored than the anthocyanidin pigment. Direct condensation without involvement of acetaldehyde is slow.

Observations of enhanced color after the addition of aldehydic spirits or sugar to red wines may be explained in light of the reactions discussed (Singleton and Guymon 1963; Singleton et al. 1964). The increased color may result from formation of pH-resistant polymers involving anthocyanidins and acetaldehyde. When the anthocyanidin component is present in the colored oxonium ion state

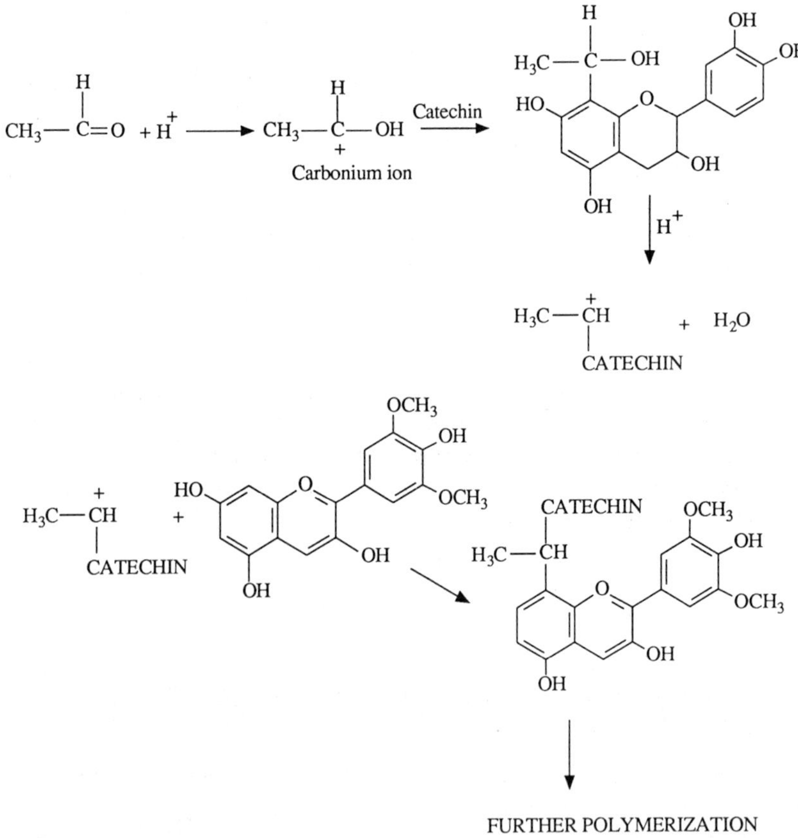

Fig. 7-9. Condensation-polymerization reactions extending from acetaldehyde to polymers with potential for precipitation.

(responsive to changes in pH), increased spectral color results from polymerization. Once formed, such polymers are resistant to changes in spectral color with decreases in pH. Conversely, oxidation of anthocyanidin monomers or precipitation of polymeric species may result in decreased color (Berg and Akiyoshi 1975).

In the case of sugar additions, increases in color also occur. These are thought to involve formation of pH-resistant polymers, but not by the above mechanisms. Instead, sugars are theorized to operate as polymerizing agents, either by hydrogen bonding or by binding water (Berg and Akiyoshi 1975).

GRAPE PHENOLS

Although the amount and the qualitative distribution of phenols within the fruit are determined primarily by variety, climate (macro, meso and micro) may exert an important influence on the former. Flavonoids appear to vary more than nonflavonoids in regard to site, vintage, and climatic variations. For example, the anthocyanin content will be greater for a particular variety grown in a cool region when compared with the same variety harvested at the same maturity in a warmer region under the same viticultural condition. In this case, for example, the relative ratios of pigments are the same, but what varies is the total anthocyanin content.

Maturity effects minor qualitative changes in grape phenols (Singleton 1985). From veraison to harvest, there is a net increase in grape berry phenols. However, this increase is countered by increases in berry size. The net result is that with the exception of anthocyanins, there is a tendency toward lower phenolic content. Overripe and shriveled fruit lose phenols, including anthocyanins, possibly by conversion to unextractable oxidation products (Singleton 1985).

Singleton and Esau (1969) report that on average red grapes have 5500 mg GAE/kg, whereas the average white variety has around 4000 mg GAE/kg. Table 7-1 compares the relative quantities of phenols in the grape berry. Expressed in terms of gallic acid equivalents, 46 to 69% of the total berry phenols are located in the seeds, 1% in the pressed pulp, and 5% in the juice. The balance, 50% for reds, and 25% for whites, is located in the grape skin. Average phenol levels will differ in various red varietal wines, as may be seen in Table 7-2.

WINE COLOR: EXTRACTION, DEGRADATION, AND PRODUCTION CONSIDERATIONS

Upon examination of a wine, one is most immediately impressed with its color, and from this first impression, initial opinions (or biases) often are formed. Thus color becomes one of the most important attributes of a wine and as such is of prime concern to the enologist. In the production of proprietary blends, for example, standardization of color becomes an important quality control measure. Likewise, monitoring color changes in a wine may be used as an index of maturation or deterioration.

Factors Contributing to Wine Color

White musts often contain traces of chlorophyll, carotene, and xanthophyll (Amerine et al. 1972). Occasionally, certain varieties of white grapes, when grown under cool climatic conditions, retain some of their green color, evidently from chlorophyll derivatives (Singleton and Esau 1969), and impart it to the wine. Thus white wines may have a trace of green coloration in addition to the normal range of almost colorless to amber. White wine color considerations are primarily preventive, as the emphasis is on maintenance of desirable color and prevention or delay of browning.

In red wine production, an initial consideration is color extraction. Color in red wines varies from very pale rose to nearly black. Hue is defined as the dominant wavelength, and it is dependent upon such factors as pH, the presence of metal ions, and the degree to which browning overtones have developed in the wine.

The extent to which the red color is extracted into the juice and the wine depends on the grape variety, the region in which it was grown, the nature of the growing season, the degree of maturity, and the physical condition of the fruit (namely, presence of mold, raisining, etc.), and processing. Winemaking practices such as whole cluster pressing, degree of crushing, temperature of fermentation, cap management, SO_2 levels, and alcohol level at dejuicing also affect color extraction. Grape varieties such as Rubired, Royalty, and Salvador have their principal anthocyanins present in the diglucosidic form. This enhances the extractability of these colored compounds. In addition these three varieties are teinturiers with their color pigments present in the pulp, which also makes them more accessible. These grape varieties are used in the production of highly colored wine for blending.

In warm areas such as the Central and Southern San Joaquin Valley, grape pigments tend to be less soluble, and red color development is often poor. Grapes degraded by mold species such as *Botrytis*, *Penicillium*, and *Aspergillus*, and so on, are rich in oxidative enzymes. Unless inactivated or removed, these fungal metabolites may catalyze oxidation of phenolic substrates, producing brown or tawny overtones in red wines and a yellow to amber shift in white wines. Sulfur dioxide can significantly affect the development of color in grape must and wine. In small amounts, it aids in the extraction of color from the pigment-containing cells, the etioblasts. However, Ribereau-Gayon et al. (1976) report that SO_2, at levels used in fermentation, has little effect in this regard. It probably plays a more important role, at regularly encountered use levels, in inactivating oxidase enzymes that catalyze prefermentation oxidation. In the case of mold-damaged fruit, sulfite may play only a limited role in the inactivation of microbially produced oxidative enzymes (e.g., laccase; see Chapter 1).

Color Extraction

In red wine production, color development in nonteinturier varieties is dependent upon extraction of pigments located in the skins, as well as other phenolics, into the fermenting must. Therefore, the handling of the solids portion of a red wine fermentation is of major concern to winemakers. Provisions must be made for ensuring adequate mixing and contact of skins and juice portions of the fermentation during this period.

A number of processing decisions can affect phenol extraction and thus red wine palatability and style. These include the degree of berry breakage, and stylistic variations such as whole cluster, berry, and stem return. The fermentation temperature, as well as the length of time the juice remains in contact with the grape skins, is also an important consideration. Further, the configuration of the fermenter (height to width ratio) affects the extraction process.

Crushing

Adjustments made to the crusher can achieve the desired degree of berry breakage and thus phenolic extraction. Because seeds contain relatively high levels of the low molecular weight phenols that are perceived as bitter, it is imperative to avoid seed scarring and breakage. Bitterness is known to be influenced by the addition of sulfur dioxide during primary processing. Depending upon the level of sulfite added to the must, one may increase the extraction of bitter flavonoid phenols. (Oszmianski et al. 1986) As a result, some winemakers are reducing the amounts of sulfites used in prefermentation processing. This causes increased oxidation, polymerization, and precipitation of phenols and moderates the phenolic content in the wine.

The practice of whole cluster return coupled with light crushing of fruit can be used effectively in the production of red wines with light color, lower astringency, and intensified fruit character. It should be noted that adding back whole berries, instead of clusters, may modify the above generalizations. The presence of stems in whole cluster additions contributes to overall tannin content, possible bitterness, and spiciness. The practice of stem return is used principally in the production of wine from low-phenol varieties such as Pinot noir, and as a stylistic tool. In general, the technique produces wines of lower color, increased pH, and increased spicy character.

Cap Management Concerns

Cap management methodology can have a significant effect on phenol extraction and thus wine palatability and color. Typically, it includes pumping-over or punching-down and may, in some cases, involve specially designed systems to wet the cap (i.e., sprinklers), submerged cap fermentations, and specialty tanks (rototanks), as well as thermovinification.

Periodic mixing of fermenting juice with the cap by pumping-over tends to cause juice to infiltrate unevenly through the cap. The usual result is that much of the pomace is incompletely mixed during a single pump-over. In Bordeaux, manipulation of the cap generally is limited to pumping-over. This results from the belief that Bordeaux varieties such as Cabernet Sauvignon and Merlot have sufficient color to overcome the problems of uneven extraction from skins.

The cap is the chief source of microbial spoilage (acetification) during latter stages of red wine fermentations; so any system that keeps the cap wet not only aids in color extraction and reduces excessive heat buildup, but also potentially reduces the risk of microbial spoilage. Submerged cap fermentations overcome the problems of microbial spoilage associated with the presence of a cap. Further, this technique eliminates the need for pumping-over, thereby saving time. However, production results are less clear with regard to the effectiveness of color extraction.

Generally, the smaller the tank is, the more manageable the cap. Tanks larger than 1500 gallons may have a cap that is very thick and difficult to manage by hand. This is especially true if the height dimensions greatly exceed the tank diameter. For many premium varieties, such as Cabernet Sauvignon, a small height-to-diameter tank is desirable. In these cases the usual industry practice is to select a height-to-diameter ratio on the order of 1.0 to 1.3.

Another consideration relative to primary fermentation is the utilization of open versus closed topped fermenters. Red wine fermentations traditionally have been carried out in open fermenters. Such a tank design may facilitate the removal of hydrogen sulfide as well as the relative ease of cap management. In the case of varieties such as Cabernet Sauvignon, the use of open-topped versus closed fermenters continues to be debated. Some winemakers believe that closed-top fermenters contribute to the retention of complexity.

Specialty tanks such as rototanks have the advantage of being automated mixing systems that maximize juice-to-skin contact. Although phenol extraction usually is greater with their use, the expense involved in the construction of such tanks may not be warranted.

Fermentation Temperature

Within the acceptable temperature range for premium wine production, increases in temperature generally enhance the extraction of phenols and color. Many producers of Cabernet Sauvignon, for example, ferment this variety at temperatures up to 31°C (88°F). With elevated temperatures, however, there is a greater risk of having the fermentation subside prior to complete utilization of the available sugar, and the problem of a stuck fermentation is often complex. Stuck fermentations generally are difficult to restart, and secondary microbial activity during the interim may quickly cause deterioration in the wine. (See Chapters 12 and 16 for more details.)

Duration of Skin Contact

Depending upon variety, and assuming other conditions are not limiting, maximum color extraction in red musts generally is complete within three to five days after the start of active fermentation. However, drawing-off may take place at any time. This will depend upon the initial condition of the fruit, the rate and extent of color extraction, the temperature of fermentation, and production needs relative to the desired color and astringency levels in the wine. Berg and Akiyoshi (1957) demonstrated that color extraction in Carignane and Zinfandel was greatest at 3 to 6% alcohol. In the case of Cabernet Sauvignon, however, Ough and Amerine (1962) found that the process was not complete until the alcohol content reached 12%.

Rosé wines can be produced by drawing-off early in the course of fermentation. Use of pectinolytic enzymes enhances extraction of color during this shortened period of skin contact (Chapter 17). The shortened extraction period yields wines of significantly lower total phenols and color. Singleton and Noble (1976) found that fermentation on the skins for one day produced a wine with 500 mg/L total phenols (GAE) and approximately 10% of the color produced from the same red variety fermented on the skins for three to five days. In the latter case, the additional pomace contact resulted in a total phenol level of 1900 mg/L (GAE).

The practice of extending pomace contact beyond dryness (''maceration'') may produce a wine lower in color, higher in tannins (''soft tannins''), and with increased body. With maceration, the concentration of higher molecular weight phenols also increases. A reduction in bitterness as a function of extended pomace is a result of phenolic polymerization during and after the contact period. Working with Cabernet Sauvignon, Bourzeix et al. (1970) reported that maximum phenol content with maceration, the concentration of higher molecular weight phenols also increases.

Alternative Techniques for Color Extraction

Pectolytic Enzymes

Winemakers often wish to produce red wines with desirable color but reduced astringency. This generally requires shortening the time period for skin contact. Pectolytic enzyme additions may be utilized to enhance color extraction. By comparison to wines produced using conventional processing, and dejuiced at near 12°B, enzyme-treated lots exhibit increased color with less extraction of astringent phenolics. Yield increases of from 3 to approximately 9% have been reported with the use of enzymes. (Refer to Chapter 17 for more details.)

Thermal Vinification

Various processes employing heat, collectively termed thermal vinification, have been used to maximize color extraction in red wine production. Depending upon the format, composite mixtures of skins and juice may be heated together, or the juice alone may be heated and then mixed with crushed grapes. After pressing, the juice is cooled and usually vinified in the manner of standard white wines. The heating time and temperature parameters may vary with the vintner and the geographical location. The net result of thermal vinification is increased color intensity in must and wine. Joslyn and Amerine (1964) reported that heating at 165°F for a period of only 1 minute was sufficient for color extraction in Carignane and possibly most red varieties.

Most thermally vinified wines initially are deeper in color than their traditionally fermented counterparts. Clarification may be slower than with conventionally fermented wines because of increased phenolic extraction, which results in higher concentrations of negatively charged species in solution. Thermal vinification may have an oxidizing or accelerated aging effect on red wine resulting from condensation and precipitation of phenolic constituents. Thus the harshness or astringency associated with young red wines is overcome.

Changes in Red Wine Color

Factors affecting the rate of color change include the phenol composition and concentration, levels of oxygen and sulfur dioxide, temperature of fermentation and storage, types and concentration of metals, and pH. Following fermentation, some workers report a decrease in red color, with losses of up to 33% (Van Buren et al. 1974). These changes are believed to result from pigment adsorption by yeast and colloidal particles as well as by potassium bitartrate crystals. Processing operations such as ion exchange, filtration, centrifugation, cold stabilization, and fining reduce red wine color and may eventually increase browning.

During malolactic fermentations, decreases in red wine color often are noted. The upward shift in pH associated with malolactic fermentation decreases the observed intensity of red wine color in addition to increasing the potential for oxidation (see Fig. 7-7).

The red wine color may vary, depending upon the quantities of pigment present as well as shifts in the equilibrium between colored and leuco-forms of those pigments. During aging, red wine pigments undergo a slow shift from monomeric to polymeric form, and as the process continues, pigments become large enough to precipitate from solution. During prolonged aging, much of the initial color may be lost as precipitate, and its spectral characteristics may resemble those of an oxidized white wine. In such cases, a yellow-brown color corresponding to wine tannins becomes visible and more pronounced as oxidation continues. The color loss may be diminished somewhat by the presence of SO_2. As anthocyanidin pigments are subject to photooxidation, practices are employed that protect the wine from light, such as bottling in colored glass. Despite these measures, changes in pigments (perceived color) will occur.

Several changes in the integrity of pigments, resulting in color loss, are believed to occur during fermentation and aging. These include hydrolysis of the glucose moiety (rendering the pigment unstable), demethylation, and condensation (polymerization) of the pigment(s). However, some winemakers have observed an increase in red wine color following several months of aging, one possible source of this phenomenon being reoxidation of anthocyanins reduced to corresponding flavene forms during fermentation.

Enzymatic Oxidation of Musts

The extent and rate of color change (deterioration) depend upon many parameters, including pH, the amount and type of phenolic substrate present, the temperature, and the amount of dissolved oxygen in the wine. The major viticultural parameters affecting browning in wines include variety, growing region, maturity, and condition of the fruit at harvest. During processing, contact with air and the presence of metal ions, most notably copper and iron, may be principal causes for accelerated browning rates in wine.

Enzymatic oxidation occurs primarily in freshly crushed fruit. The group of enzymes responsible for catalyzing oxidative reactions are the polyphenoloxidases, also referred to as phenolases. Polyphenoloxidases catalyze oxidation of dihydroxyphenols to their corresponding quinones (Fig. 7-10a). The substrates for this reaction are the nonflavanoids (hydroxycinnamates and their derivatives). Oxidized quinones in wine may form polymers of the type presented in Fig. 7-10b. Bond formation occurs between carbon 6 of the first and either carbon 6 or 8 of the second quinone. As condensation continues, color intensifies from an initial yellow to brown (Hathaway and Seekins 1957).

The activity of this group of enzymes is greatest in the skin, their concentration in grape juice depending upon grape variety and condition at harvest as well as the period of skin contact. The addition of SO_2 at crush inhibits polyphenoloxidase activity. Because of its general activity in binding proteins, bentonite addition prior to fermentation may reduce levels of the oxidative enzymes as well as enhancing clarification.

OXIDATION OF WINES

In white wines, as compared with juice, browning is due largely to oxidation of phenolic compounds such as catechins and leucoanthocyanidins; and generally proceeds by chemical means. The following three chemical mechanisms have been proposed to account for wine oxidation.

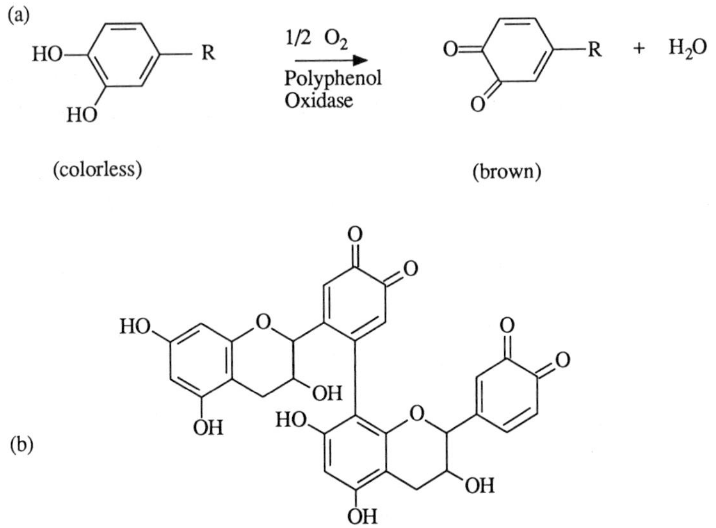

Fig. 7-10. (a) Enzymatic oxidation of nonflavonoid species; (b) polymerization of oxidized flavonoids, as would occur in wine.

Caramelization

Upon heating of hexose sugars in acid solution, rapid decomposition (dehydration) takes place (Fig. 7-11). The reaction does not require proteins or amino acids (see below, Maillard reaction), and may occur in the absence of oxygen.

Caramelization occurs during production of baked sherry. In the United States, shermat with approximately 2% sugar is heated at 120 to 140°F to accelerate the process. As a by-product of the reaction, a dark caramel color is produced. The same reaction takes place during excessive pasteurization of sweet wines. Similarly, browning overtones derived from caramelization may be present in red and white wines vinified from raisined berries.

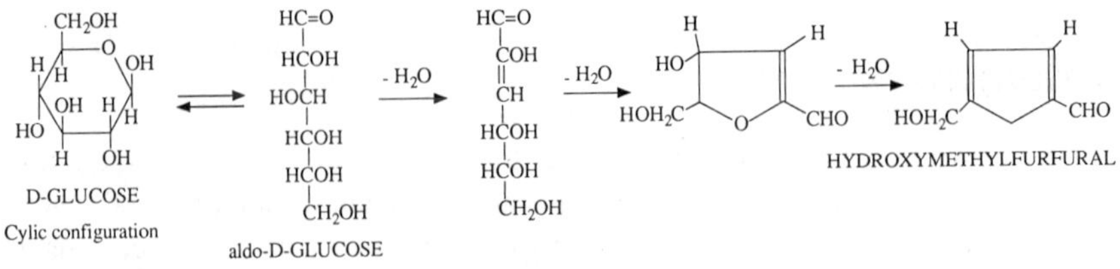

Fig. 7-11. Dehydration of hexose sugars.

Maillard Reaction

This reaction in foods involves condensation of sugars with amino acids and proteins. As seen in Fig. 7-12, active carbonyl groups, as would be present in reducing sugars, aldehydes, and ketones, are necessary for the condensation reaction to occur. Thus, blockage of these active groups, due to prior reaction with SO_2 (yielding a hydroxysulfonic acid), renders the compound unreactive and serves to block the browning reaction sequence.

There has been little supportive evidence for the occurrence of the Maillard reaction in white wine browning (De Villiers 1961; Caputi and Peterson 1966). Nevertheless, because the necessary reactants are present in wine (amino acids, proteins, and reducing sugars), the possibility of its occurrence should not be overlooked.

Direct Phenolic Oxidation

The major nonenzymatic reaction causing browning in wine involves reaction of susceptible phenolic derivatives with molecular oxygen (see Chapter 8). Metals may act as catalysts in this oxidation (see Chapters 14 and 15).

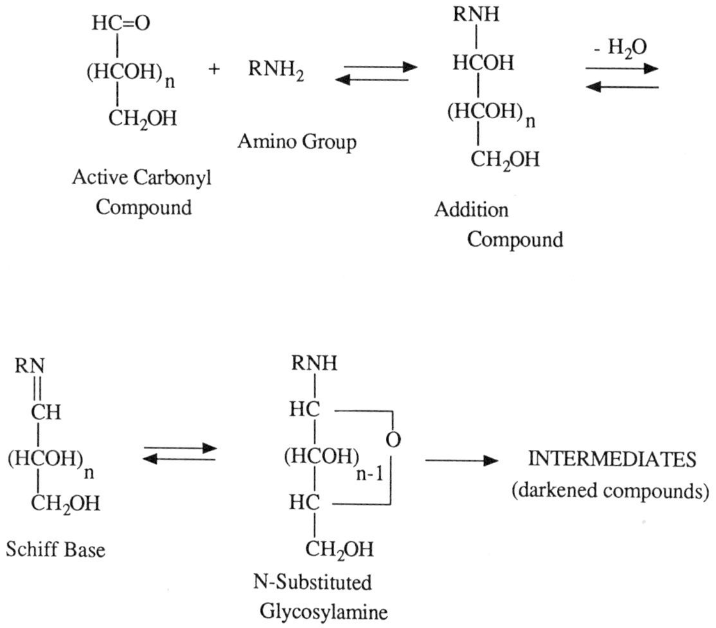

HC=O | $(HCOH)_n$ | CH_2OH + RNH_2 ⇌ RNH | HCOH | $(HCOH)_n$ | CH_2OH ⇌ ($-H_2O$)

Active Carbonyl Compound — Amino Group — Addition Compound

RN || CH | $(HCOH)_n$ | CH_2OH ⇌ RNH | HC—O—HC (ring) $(HCOH)_{n-1}$ | CH_2OH → INTERMEDIATES (darkened compounds)

Schiff Base — N-Substituted Glycosylamine

Fig. 7-12. Involvement of carbonyl groups in browning reactions.

SECONDARY BROWNING REACTIONS

Several nonenzymatic secondary browning reactions may accompany initial quinone formation, yielding still darker products. These include coupled oxidations, condensation-polymerization reactions, and complexation of amino groups.

Coupled Oxidations

As has been pointed out, oxidation of susceptible phenolics to corresponding quinones usually is accompanied by a darkening of color. The oxidized quinone formed, however, may bring about oxidation of other components such as anthocyanidins.

Condensation-Polymerization Reactions

Exposure of susceptible wine phenols to molecular oxygen during aging creates corresponding quinones, as discussed above. Concomitant with this oxidation reaction, hydrogen peroxide is formed (Wildenradt and Singleton 1974), which subsequently oxidizes ethanol to acetaldehyde (Fig. 7-13). The acetaldehyde formed combines with tannins and red wine pigments (see Fig. 7-9).

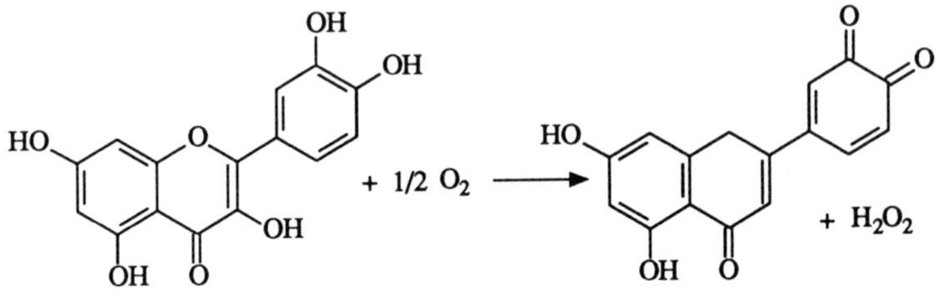

$+ 1/2\ O_2 \longrightarrow$ $+ H_2O_2$

Step I Step II

$$H_2O_2 + C_2H_5OH \longrightarrow CH_3CHO + H_2O$$

Step III

Fig. 7-13. Coupled oxidation of phenolic groups producing various oxidation products (acetaldehyde and acetic acid).

Pinking in White Wines

Development of a red blush in white wines, a reaction called pinking, is observed occasionally. Where there is an overall reduction in oxygen exposure, pinking is reported to be the result of rapid conversion of flavenes to the corresponding red flavylidium salts. Referring to Fig. 7-14, one can see that flavenes are formed in an acidic medium by slow dehydration of corresponding leucoanthocyanins.

In the presence of oxygen, flavenes and leucoanthocyanins are converted to browning pigments. Under reducing conditions, however, accumulations of flavenes may occur. Subsequent rapid exposure of wine to air converts flavenes to their red flavidium salts, which confer a pink blush to the wine.

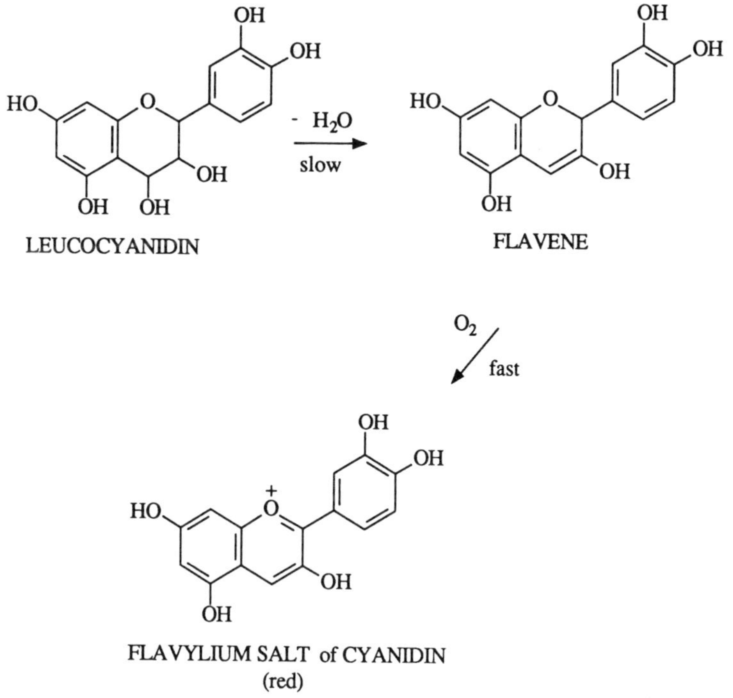

LEUCOCYANIDIN

$- H_2O$

slow

FLAVENE

O_2

fast

FLAVYLIUM SALT of CYANIDIN
(red)

Fig. 7-14. Conversion of leucocyanidins to flavylium salts (colored).

EVALUATION OF COLOR BY SPECTROPHOTOMETRY

Preparation of transmission and/or absorption spectra for wines reveal a great deal about their character. White wines are characterized (Fig. 7-15) by transmission over a broad range of wavelengths. The absence of any strong absor-

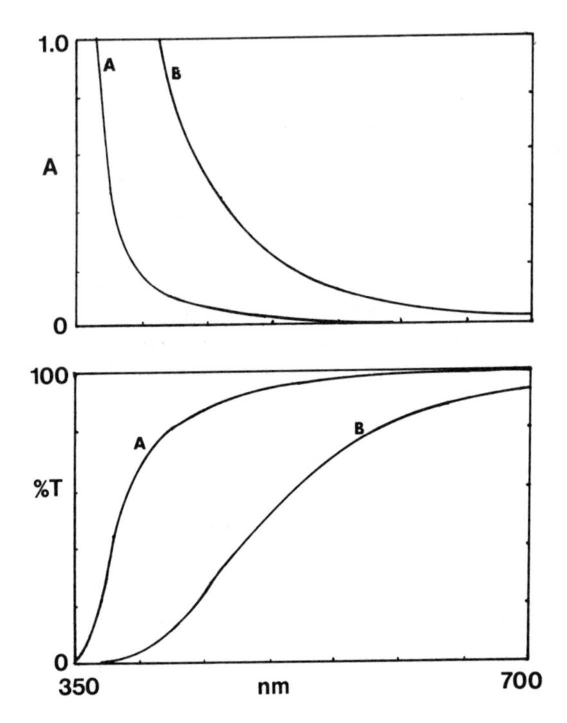

Fig. 7-15. Transmittance and absorbance spectra of (*A*) Chardonnay and (*B*) sherry.

bance bands in the visible spectral region accounts for their absence of color. With sherries, one observes lowered transmittance (more absorbance) in the 400 to 600 nm range, accounting for yellowish tones. The large amount of transmittance in the 600 to 780 nm range produces some greenish color, which when mixed with the yellow yields a characteristic brown tone. With age or oxidation the spectrum of white table wine (Fig. 7-15) shifts toward that of the sherry. Upon excessive oxidation of white wines, the absorption maximum continues to shift toward the ultraviolet range. Hence there is, again, an increase transmittance in the area of the spectrum where brown overtones are seen.

Most young red wines have a transmittance minimum (absorbance maximum) at 520 nm and a transmittance maximum (absorbance minimum) at 420 nm (Fig. 7-16a). These spectral characteristics are also common to blush style (Fig. 7-16b) and port wines (Fig. 7-17). As the wine matures, there is a shift in the absorption maximum to between 400 and 500 nm, usually near 450 nm. With excessive oxidation of red wines, the spectral characteristics approach those of a sherry; the characteristic red pigments are polymerized and precipitated.

The presence of metal ions or complexes can degrade perceived color. The degree to which this becomes a problem is dependent upon the pH of the

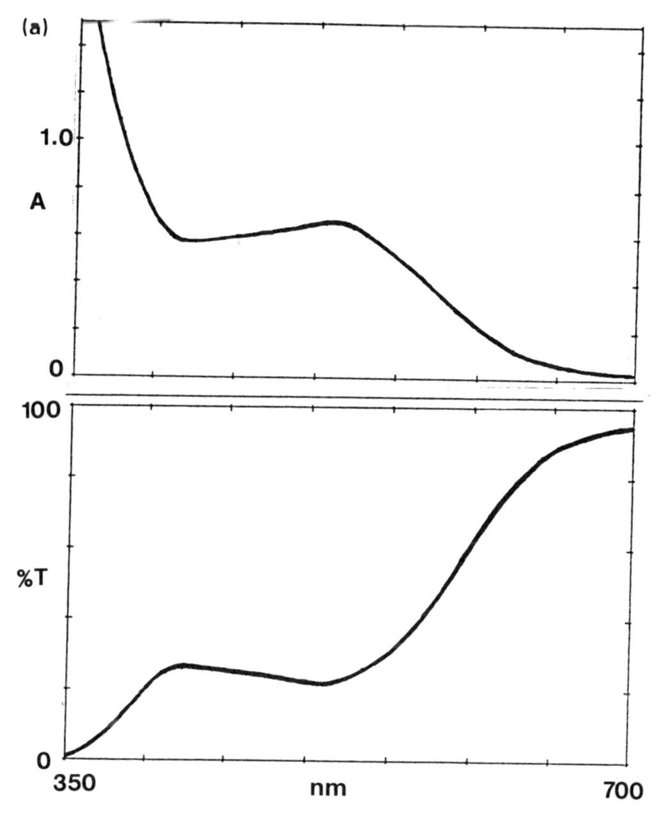

Fig. 7-16. (a) Transmittance and absorbance of 1981 vintage Charbono. Note absorbance maximum at ~520 nm and absorbance minimum at ~420 nm. (b) Transmittance and absorbance spectra of (*A*) white Zinfandel and (*B*) non-vintage rosé wine. Note absorbance maxima at 520 nm and minima at 420 nm are not so distinct as those for red wines seen in Fig. 7-16a.

medium, as well as upon the presence of competitive complexing (chelating) agents such as citric acid (Jurd and Ansen 1966). In the presence of metal ions such as aluminum, copper, and iron, complexes that are formed with anthocyanins and other phenolics result in modification of the red wine color so that there is a shift toward the ultraviolet region of the spectrum. This shift is perceived by the eye as an increase in blue. It has been reported that these anthocyanin–metal complexes may be responsible for the blue coloration of certain native American grape varieties (Jurd and Ansen 1966).

Humans perceive color as a characteristic of the wavelengths and intensity of light being reflected off the surface or being transmitted through an object. This occurs within the wavelength range of 380 to 770 nm of the electromagnetic spectrum, the part that is visible to humans. Color may be completely

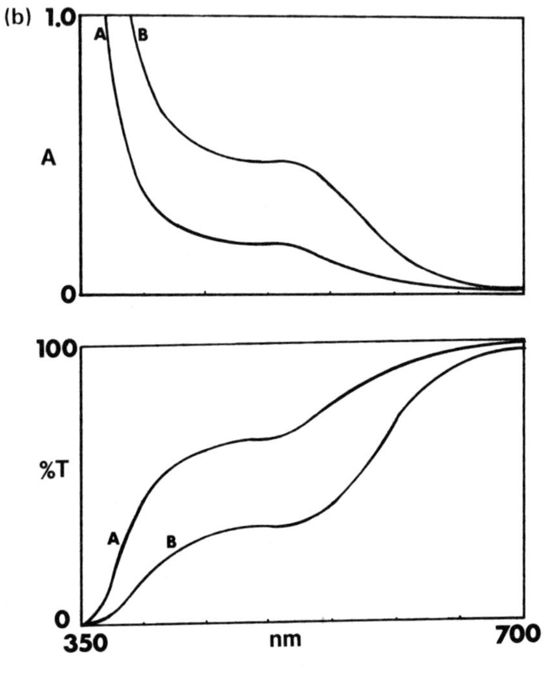

Fig. 7-16. *(Continued)*

defined in terms of three fundamental attributes: (1) the dominant wavelength (hue), such as red, yellow, green, or blue; (2) brightness (luminescence), which defines the amount of gray in the color ranging from white to black; and (3) purity (saturation), which is a measure of the divergence of the color from gray (the percent hue in a color).

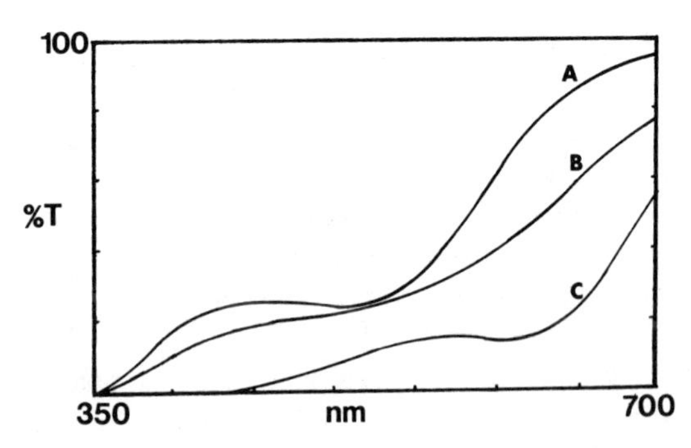

Fig. 7-17. Percent transmittance of diluted (1 : 5) red port wine at several pH values: (A) pH = 3.35 (original pH of wine), (B) pH = 6 (adjusted), and (C) pH = 10 (adjusted).

Values for a particular wine color frequently are assigned by simple measurement of absorbance readings at different wavelengths. For white wines, absorbance usually is determined at 420 nm. In the range 400 to 440 nm, increases in brown coloration of white wines are detected readily. For red wines, absorbance data are collected at varying wavelengths (depending upon the winery). In most cases, however, samples are diluted, and relative absorbances measured at 420 nm and 520 nm (Sudraud 1958). When a dilution procedure is employed, the wine is diluted with distilled water adjusted to the pH of the original wine. The extent of dilution is a function of the original intensity of the wine color, and may range from 1 + 3 to 1 + 9. The importance of adjusting the pH of dilution water can be seen in Fig. 7-17. All particulate matter must be removed from the sample prior to analysis, a step usually accomplished by filtration through a 0.45-μm membrane filter. A measure of color intensity of relative brightness can be achieved by summation of the absorbance readings at 420 nm and 520 nm, whereas hue is measured as the ratio of absorbance readings at these wavelengths.

The standard system for color measurement, against which all other systems are compared, is that of the CIE (Commission Internationale de l'Eclairage). This system specifies color in terms of three quantities, X, Y, and Z, called tristimulus values. These values represent the amounts of the three primary colors, red, green, and violet, required to match the color of the object under consideration. The system further defines chromaticity coordinates, x, y, and z, which are determined by dividing the appropriate tristimulus value by the sum of the three values [for example, $x = X/(X + Y + Z)$]. Chromaticity coordinates give the proportion of the total color stimulus attributed to each primary color.

To specify the color of a wine using this system, the percent transmittance of the sample is measured over three series of wavelengths (516–656 nm for red, 507–640 for green, and 424–508 nm for violet). Ten $\%T$ (% transmittance) values in each series are summed and multiplied by a normalizing factor to obtain the tristimulus values X, Y, and Z. The chromaticity values are then calculated as specified above and indicate the respective fractions of red, green, and violet in the sample. One major advantage of this system is that it specifies color in terms of what one sees, the *transmitted* electromagnetic radiation. The main disadvantage is that one is required to make 30 transmittance measurements in order to calculate the chromaticity values (see Proc. 7-5).

Procedure 7-1. Phenolics by Folin-Ciocalteu

Juices and wines contain a large variety of phenolic compounds, and there is concern about the ability to differentiate between total phenols, one class of phenols, and a specific phenol. Although significant progress has been made in developing high performance liquid chromatographic techniques to accomplish the latter, the most common analyses performed usually are for total phenols.

Like many other "general" analytical methods, these show a lack of specificity. Ultraviolet and visible absorbance measurements suffer from the fact that different phenolic substances can have significantly different molar absorptivities. In addition, nonphenolic substances can be responsible for the major fraction of the absorbance of white wines (Somers and Ziemelis 1972).

The reduction of phenolic substances by the Folin-Dennis reagent (more recently, Folin-Ciocalteu reagent) with subsequent spectrophotometric comparison at 765 nm, is likewise rather broad in interpretation. The procedure utilizes gallic acid as a standard reference compound, and results correspondingly are expressed as gallic acid equivalents (GAE). Folin-Ciocalteu reagent also reacts with monohydroxy phenols and other readily oxidized substances (including ascorbic acid, sulfur dioxide, and aromatic amines). In sweet wines and musts a more serious interference occurs with reducing sugars.

I. Equipment
   Spectrophotometer (visible range)
   10-mm cuvettes
   Glass reflux apparatus
   Pipettes (1, 2, 3, 5, 10 mL)
   Volumetric flasks (100 mL)

II. Reagents (See Appendix II)
   Phenol stock solution (5000 mg/L GAE)
   Folin-Ciocalteu reagent (*Note*: This can be purchased commercially.)
   Sodium carbonate solution

III. Procedures
   1. Prepare calibration standards of 0, 50, 100, 150, 250, and 500 mg/L from phenol stock solution as shown in Table 7-3.
   2. Into a series of 100-mL volumetric flasks, pipette 1 mL of each standard, *or* 1 mL of each white wine sample, or 1 mL of each 1 + 9 diluted red wine sample (10 mL red wine diluted to 100 mL in a volumetric flask).
   3. To each flask add 60 mL distilled water, and mix the solution; add 5 mL of Folin-Ciocalteu reagent, and mix well again for 30 sec; and add

Table 7-3. Preparation of calibration standards from phenol stock (5000 mg/L GAE).

Volume phenol standard (mL)	Final concentration (mg/L GAE) in 100-mL volumetric flask
1	50
2	100
3	150
5	250
10	500

15 mL sodium carbonate solution (add after the 30 sec mixing but before 8 minutes). Mix the solution again, and bring it to the mark with distilled water.

4. Allow standard and sample solutions to sit for 2 hr at 20°C, and then measure the absorbance of each at 765 nm against the blank.
5. Construct a calibration curve from absorbance values of the standards. Read concentrations of white wine samples directly from the curve. Read concentrations of red wine samples from the curve, and multiply by the dilution factor (100/10) to obtain the phenol concentration in the original sample.

IV. Supplemental Notes

1. Folin-Ciocalteu reagent uses lithium sulfate to reduce precipitation problems in the reagent. Lithium salts are more soluble than those of other common cations.
2. Sodium carbonate is used to make the reaction mixture alkaline. The reduction of Mo(VI) and W(VI) requires the presence of the phenolate anion. The reduced heteropoly molybdenum and tungsten molecules are blue; the unreduced molecules are yellow.
3. Reducing sugars (R.s.) are capable of reducing the heteropoly molecules in alkaline media. For R.s. values from 1.0 to 2.5 g/100 mL, divide the total phenol values by 1.03 to correct for this interference. For R.s. values from 2.5 to 10.0 g/100 mL, divide the total phenol values by 1.06. No correction is necessary for dry wines.
4. This method is nonspecific and measures the number of —OH (potentially oxidizable phenolic groups) present in the sample. Different tannins will exhibit different responses. The values obtained are in gallic acid equivalents, as that is the calibration curve standard used.
5. **Nonflavonoid Phenolic Fraction:** One can treat a filtered sample of wine with formaldehyde, and quantitate the remaining phenolics by the Folin-Ciocalteu procedure. Subtracting this value from the original total phenolics value provides a measure of the flavonoid phenolic fraction. The procedure involves treating 10 mL of filtered wine (0.45 μm) with 10 mL of HCl (1 + 3) and 5 mL of formaldehyde solution (8000 mg/L), allowing it to stand for 24 hr, refiltering, and measuring the remaining nonflavonoid phenols (Folin-Ciocalteu). When this value is subtracted from that obtained on a second aliquot of the original wine, the difference is a measure of the flavonoid phenols (also reported in mg/L gallic acid).

Procedure 7-2. Polyphenol Index Determination (Permanganate Index)

Some winemakers perform a quick screening test to aid in determining when to dejuice red fermenters, etc. The permanganate index test that follows provides a crude evaluation of the phenolic content of wines. Low-phenolic wines have index values from 35 to 50. Wines with a high phenol content may reach index values of 100 or above.

I. Equipment
   500-mL Erlenmeyer flasks
   50-mL volumetric pipette
   2-mL volumetric pipette
   50-mL burette
II. Reagents
   Potassium permanganate (0.01 *N*)
   Indigo carmine solution
III. Procedure
   1. Pipette 50 mL of indigo carmine solution into a 500-mL Erlenmeyer flask.
   2. Titrate with potassium permanganate to an end-point color change of blue to straw or orange. Record the volume of permanganate used as volume *A*.
   3. Pipette 50 mL of indigo carmine solution into another 500-mL Erlenmeyer flask; add 20 mL wine to the flask.
   4. Titrate with potassium permanganate to the same end point. Record the volume of titrant used as volume *B*.
   5. Calculate the permanganate index as follows:

$$\text{Permanganate Index} = 5\,(B - A)$$

   6. A permanganate index vaiue of 25 is approximately equivalent to 320 mg/L GAE. A permanganate index value of 95 is equivalent to 2390 mg/L GAE.

Procedure 7-3. HPLC Determination of Pigments.

Analytical methods for specific phenolic compounds and groups of compounds generally are based on one or more liquid chromatographic techniques. The following determination is an example of these methods. Successful HPLC separation of anthocyanidins can be accomplished using a reverse phase column and a gradient elution program. Detection of the various glucosides and diglucosides is accomplished with a UV-VIS detector operated at 520 nm.

I. Equipment
   HPLC capable of running programmed mobile phase gradients and having a UV-VIS detector monitoring absorbance at 520 nm
   Analytical reverse phase (C_{18}) column
   Guard column (C_{18})

II. Reagents (See Appendix II)
Mobile phase I: Acetic acid–water (15:85)
Mobile phase II: Water–acetic acid–methanol (65:15:20)
Anthocyanidin standards (1 mg/mL)

III. Procedure
1. Turn on the HPLC pumps and detector. Set the mobile phase flow rate at 0.2 mL/min using a mixture of 95% mobile phase I plus 5% mobile phase II.
2. Filter a sample of juice through a 0.45-μm filter.
3. Inject 20 μL of the filtered juice sample onto the column. Chromatograph the sample with a mobile phase gradient that increases the percentage of mobile phase II from 5% to 100% over a 20-minute period. This gradient should be nonlinear in that the increase in % mobile phase II is less during the first 10 minutes of the run and greater during the last 10 minutes.
4. Note retention times or volumes as the various pigment peaks elute from the column and detector. For identification purposes, compare these times/volumes to those of standards run under the same conditions.

Procedure 7-4. Wine and Juice Color Specification by Hue

Values for a particular color frequently are assigned by simple measurement of absorbance readings at different wavelengths. For white wines or juices, absorbance usually is determined at 420 nm. In the range 400 to 440 nm, increases in the brown coloration of white wines are readily detected. For red wines or juices, samples frequently are just diluted and relative absorbances measured at 420 nm and 520 nm. When a dilution procedure is employed, the sample must be diluted with distilled water adjusted to the pH of the original. The extent of dilution is a function of the original intensity of color, and may range from 1 + 3 to 1 + 9. Micro cuvettes can be used to avoid dilution.

All particulate matter must be removed from wine samples prior to analysis. This is usually accomplished by filtration through a 0.45-μm membrane filter. A measure of color intensity or relative brightness can be achieved by summation of absorbance readings at 420 nm and 520 nm. By comparison, hue is measured as the ratio of absorbance readings at these wavelengths.

I. Equipment
Single beam visible spectrometer capable of measurements at 420 and 520 nm
Matched cuvettes
pH meter

II. Reagents (See Appendix II)
Concentrated sulfuric acid
Distilled water

III. Procedure

1. Switch on the instrument, and allow appropriate time for it to warm up.
2. Before placing a cuvette in the sample chamber, adjust to zero transmittance.
3. Set the instrument to the desired wavelength.
4. Place a clean cuvette, marked *R* for reference and filled with distilled water, in the cuvette well. Make certain that the cuvette is completely inserted, and close the cover. Adjust the meter to read 100% transmittance (zero absorbance).
5. Remove the reference cuvette, and replace it with filtered sample in cuvette *S*. *Note*: Red wines (juice) should be quantitatively diluted with water adjusted to the wine pH or micro cuvettes employed. Record absorbance (or % transmittance).
6. For each sample, record absorbance at 520 nm and 420 nm. Where necessary one may record these data in $\%T$ and convert to absorbance using the relationship:

$$A = \log \frac{100}{\%T} \tag{7-1}$$

7. A measure of the color intensity or relative brightness can be determined by summation of the absorbances at the two wavelengths:

$$I = A_{420} + A_{520} \tag{7-2}$$

8. The ratio of absorbance at 420 nm and 520 nm is a measure of hue:

$$H = (A_{420})/(A_{520}) \tag{7-3}$$

IV. Supplemental Notes

1. For intensely colored red wines, Ribereau-Gayon (1974) recommends the use of microcell cuvettes (0.1 cm thickness) rather than dilution procedures to obtain absorbance readings within the range of the spectrophotometer. Table 7-4 summarizes experimental data supporting this

Table 7-4. Comparison of wine color data from diluted vs. nondiluted wine samples in regular vs. micro-cuvettes.

Wine	Diluted sample (1 + 9 using 1-cm cuvettes)			Undiluted Sample (using 0.1-cm microcell cuvettes)		
	$A_{420\text{ nm}}$	$A_{520\text{ nm}}$	Hue	$A_{420\text{ nm}}$	$A_{520\text{ nm}}$	Hue
Red wine *A*	0.147	0.250	0.588	1.295	2.076	0.624
Red wine *B*	0.315	0.377	0.836	1.775	1.900	0.934
Red wine *C*	0.248	0.270	0.919	1.675	1.670	1.000

procedure. As per the procedure, dilutions were made using pH-adjusted distilled water.

Procedure 7-5. The Ten-Ordinate Method for Color Specification

Although 30 selected ordinates generally are used in food products that exhibit greater spectral diversity, a ten-selected-ordinate system will suffice in products such as wine that have relatively simple absorption spectra (see Figs. 7-15 through 7-17). The three series of ten wavelengths at which transmittance measurements are taken using this ten-ordinate system are defined in Table 7-5. Transmittance data are summed and totals multiplied by their corresponding factors to yield tristimulus values.

I. Equipment
Recording spectrophotometer covering the range of wavelengths from 400 to 660 nm
Matched cuvettes
pH meter

II. Reagents (See Appendix II)
Concentrated sulfuric acid
Distilled water

III. Procedure
1. Using a membrane-filtered (0.45-μm) wine sample, collect transmittance data for the three sets of ten wavelengths shown in Table 7-5. (*Note*: Red wines should be quantitatively diluted with deionized distilled water that has been adjusted to the wine pH.)

Table 7-5. Measurement of CIE color values using the ten-selected ordinate procedure and iluminant A (Tungsten filament lamp at 2854 K).

Ordinate number	X_{nm} (Red)	Y_{nm} (Green)	Z_{nm} (Violet)
1	516.9	507.7	424.9
2	561.4	529.8	436.0
3	576.3	543.7	443.7
4	587.2	555.4	450.5
5	596.5	566.3	456.8
6	605.2	576.9	462.9
7	613.8	587.9	469.2
8	623.3	600.1	476.8
9	635.3	615.2	487.5
10	655.9	639.7	508.4
Normalizing factors	0.10984	0.10000	0.03555

See steps 1 and 2 of Procedure 7-5 for directions for the use of this table.

2. Using the following format, total the collected $\%T$ values in each column, and then multiply the individual column totals by their corresponding normalizing factors:

$$X = \text{Total } \%T \text{ (for } X) \times 0.10984 =$$
$$Y = \text{Total } \%T \text{ (for } Y) \times 0.10000 =$$
$$Z = \text{Total } \%T \text{ (for } Z) \times 0.03555 =$$

$$\text{Total } X + Y + Z =$$

$$x = \frac{X}{X + Y + Z} \tag{7-4a}$$

$$y = \frac{Y}{X + Y + Z} \tag{7-4b}$$

$$z = 1 - (x + y) \tag{7-4c}$$

IV. Supplemental Notes

1. The number of selected ordinates chosen will depend upon the complexity of the spectral curve measured. In the case of wine, the goal of color specification can be accomplished by use of ten selected ordinates. This reflects the relatively simple nature of white and red wine spectra.
2. Transmittance data are collected at different wavelengths for X, Y, and Z tristimulus values. This procedure reflects directly upon the standard observer curves, which report human visual perception of color, *not* objective monitoring.

Procedure 7-6. Spectral Evaluation of Juice and Wine

The various phenolic constituents in grapes are known to absorb radiation in the UV and visible regions of the electromagnetic spectrum. Absorbance of a red juice or wine at 280 nm has long been used as an index of flavonoids present. This is primarily due to the influence of flavonoids extracted from skins. Somers and Evans (1977) and later Somers and Ziemelis (1985) developed more sophisticated spectrophotometric methods for characterizing phenolic compounds in young red and white wines, respectively.

One can determine the relative amounts of simple and polymeric pigment forms by measuring the absorbance of a wine at 520 nm, before and after addition of excess SO_2. Anthocyanins decolorize upon addition of SO_2, while polymeric pigment forms do not. Treatment of young red wines with acetaldehyde releases anthocyanins that are bound to SO_2. The resultant increase in absorbance is a measure of the amount of total anthocyanins previously bound (and decolorized) with SO_2.

By shifting the pH of a young red wine to values < 1.0, anthocyanins are converted entirely to their highly colored flavylium form yielding a large increase in absorbance at 520 nm. Since polymeric pigments are much less affected by low pH, this technique permits one to estimate the amount of monomeric anthocyanins present.

In white juices and wines broad absorbance maxima occur in the ranges 265–285 nm and 315–325 nm. These absorbance bands are produced by flavonoids (280 nm maxima) and hydroxycinnamate esters (320 nm maxima), other non-phenolic compounds absorbing around 265 nm, and sorbic acid (if present) absorbing at 255 nm. The latter interference can be eliminated by extracting the wine with iso-octane to remove the sorbic acid. Estimates of the presence of non-phenolic interferences can be made by treating wines with polyvinyl pyrrolidone to remove phenolic compounds and remeasuring absorbances at 280 and 320 nm.

In doing this work Somers and Ziemelis demonstrated that the majority of white wines gave corrected absorbance values equal to (A_{280} − 4) and (A_{320} − 1.4), as measures of total phenolic components and total hydroxycinnamates, respectively. Phenols extracted from skins and seeds, may lead to coarseness and bitterness in a wine, as well as a tendency for oxidative browning. The flavonoid content may be calculated by further correcting the measured absorbance at 280 nm by the amount caused by the presence of hydroxycinnamates (that also absorb to a limited extent at 280 nm).

I. Equipment
   UV − VIS spectrophotometer covering wavelength range from 280 to 520 nm
   Quartz cuvettes 1, 2, 5, and 10 mm pathlength
   volumetric pipets
   micropipets covering range of 20 to 200 μL
   membrane filters (0.45 μm)
   centrifuge and tubes

II. Reagents (see Appendix II)
   sodium metabisulfite solution [20% (wt/vol)]
   aqueous acetaldehyde solution [10% (wt/vol)]
   hydrochloric acid 1 *M*
   polyvinyl pyrrolidone or Polyclar AT

III. Procedure
   1. Centrifuge juice and hazy wine samples at 4000xg for 10 minutes. Bring samples to 25°C in a water bath.
   2. Prepare a blank from a saturated solution of potassium hydrogen tartrate with 11% ethanol (wine) or 20% glucose (juice). Treat blank same as samples.

3. For red juice or wine samples:
   a. Place sample in 1, 2, or 5 mm pathlength quartz cuvette. Measure absorbances at 280, 420, and 520 nm against prepared blank.
   b. Add 20 μL sodium metabisulfite solution to same sample, mix by inversion for 1 minute, and remeasure absorbance at 520 nm.
   c. Add 20 μL acetaldehyde solution to a fresh sample of juice or wine. Let sit for 45 min at 25°C, then measure absorbance at 520 nm.
   d. Dilute 100 μL sample (200 μL, if light colored) to 10.0 mL with 1 *M* HCl. Let sit for 3 to 4 hr at 25°C, then measure absorbance at 520 nm in a 10 mm pathlength cuvette. Correct absorbance value for dilution by multiplying reading by 10.0 mL/0.10 mL or 10.0 mL/0.20 mL.
   e. Correct all absorbance readings to a 10 mm pathlength cuvette.
4. Calculate red wine color parameters as follows:
   a. Intensity of wine color $= A_{(420\text{ nm})} + A_{(520\text{ nm})}$
   b. Wine hue $= A_{(420)}/A_{(520)}$
   c. Total phenolics (absorbance units) $= A_{(280)} - 4$
   d. Measure of total anthocyanins (mg/L)

$$= 20[A^{\text{HCl}}_{(520\text{ nm})} - (5/3)A^{\text{SO2}}_{(520\text{ nm})}]$$

   e. Measure of colored (ionized) anthocyanins (mg/L)

$$= 20[A_{(520\text{ nm})} - A^{\text{SO2}}_{(520\text{ nm})}]$$

   f. Percent of total anthocyanins in colored (ionized) form

$$= \frac{A_{(520\text{ nm})} - A^{\text{SO2}}_{(520\text{ nm})}}{A^{\text{HCl}}_{(520\text{ nm})} - (5/3)A^{\text{SO2}}_{(520\text{ nm})}} \times 10^2$$

   g. Measure of extent that total anthocyanins have been decolorized by binding with SO_2

$$= A^{\text{CH3CHO}}_{(520\text{ nm})} - A_{(520\text{ nm})}$$

   h. Percent total anthocyanins present in colored form corrected for effect of SO_2 upon wine color

$$= \frac{A^{\text{CH3CHO}}_{(520\text{ nm})} - A^{\text{SO2}}_{(520\text{ nm})}}{A^{\text{HCl}}_{(520\text{ nm})} - (5/3)A^{\text{SO2}}_{(520\text{ nm})}} \times 10^2$$

   i. Chemical age factors (degree to which polymeric pigment forms have replaced monomeric pigment forms)

$$A^{SO2}_{(520\ nm)}/A^{HCl}_{(520\ nm)}$$

5. For white juice or wine samples:
   a. Measure absorbance of undiluted samples at 280 and 320 nm in 1 or 2 mm cuvettes against prepared blank.
   b. Mix 5 mL sample with 5 g polyvinyl pyrrolidone (pvp) in a centrifuge tube. Let stand 30 min and clarify by centrifuging at 4000xg for 10 min. Membrane filter (0.45 μm) supernatent and remeasure absorbance values at 280 and 320 nm.
   c. Correct all absorbance readings to 10 mm pathlength.
6. Calculate white wine color parameters as follows:
   a. Total phenolics (absorbance units) = $A_{(280\ nm)} - 4$
   b. Total hydroxycinnamates (absorbance units)

$$= A_{(320\ nm)} - A^{pvp}_{(320\ nm)}$$

   c. Total hydroxycinnamates as mg/L caffeic acid equivalents (mg/L CAE)

$$= \frac{A_{(320\ nm)} - A^{pvp}_{(320\ nm)}}{0.90} \times 10\ mg/L$$

   d. Total flavonoids (absorbance units)

$$= [A_{(280\ nm)} - 4] - (2/3)[A_{(320\ nm)} - A^{pvp}_{(320\ nm)}]$$

IV. Supplemental notes
   1. Spectral measurements of color intensity and hue may be compared to values obtained by measuring tristimulus values (see Procedure 7-5).
   2. The total phenolics measurement includes a correction for the 280 nm absorbance of non-phenolic compounds present in juice and wine. This correction has been determined from studies of a number of different vinifera wines and juices.
   3. At low pH simple anthocyanins are in their colored flavylium form causing an increase in absorbance. When treated with SO_2, monomeric anthocyanins are decolorized. Since polymeric pigments also contribute to a limited extent to increased color at low pH, a correction factor of 5/3 is used to get a more accurate measure of monomeric anthocyanin content in the juice or wine.
   4. At wine pH the difference between absorbance values without and with SO_2 provides a measure of the amount of colored (flavylium ion) anthocyanins. Since polymeric pigments are more resistent to bleaching upon addition of SO_2, no correction for these components is made.

5. Acetaldehyde binds more strongly with SO_2 than do anthocyanins. Treatment of a wine with acetaldehyde should therefore release those anthocyanins decolorized by any SO_2 treatments during fermentation, and give a better measure of anthocyanins present.
6. The ''chemical age'' factors presented should increase as a wine ages and the anthocyanins are lost and polymeric pigment forms increase.
7. White wines with abnormally high absorbance values at 280 nm may have sorbic acid present as a preservative. Sorbic acid can be completely removed by extraction with iso-octane (see Procedure 11-3) to eliminate this interference.
8. Hydroxycinnamates are the major non-flavonoid components of white wines and juices. They absorb radiation at 320 nm, as do a number of non-phenolic compounds. Polyvinyl pyrrolidone (pvp) removes hydroxycinnamates and catechins from juice and wine samples and allows one to obtain a spectral correction for UV absorbing non-phenolics.
9. Approximate measurements of the flavonoid content of a white wine or juice requires a correction factor for the 280 nm absorbance of the hydroxycinnamates whose absorbance at 280 nm is approximately 2/3 of that measured at 320 nm.
10. Recent work by Tryon et al. (1988) using Vidal blanc has suggested difficulty in carrying out the above procedure in white wines.

Chapter 8

Oxygen, Carbon Dioxide, and Ascorbic Acid

OXYGEN

Oxygen contact with must and wine is a concern to the winemaker throughout the winemaking process. In some instances (e.g., in juice processing, during barrel aging, etc.), selected and controlled exposure to oxygen may play an important and beneficial role in wine quality. In situations such as bottling, oxygen levels should be as low as possible to prevent premature deterioration.

Since the early 1970s, results of research as well as commercial wine production have suggested that limited oxygen contact of the must prior to fermentation may not be as detrimental as once thought (Muller-Spath et al. 1978; Long and Lindbloom 1986). Studies indicate that oxidative browning occurring in juice may be reversed during fermentation. Oxidative polymerization of phenols in white juice reduces the phenolic content of the subsequent wine and aids in buffering the wine against further oxidative degradation.

Derivatives of hydroxycinnamates are the primary phenols responsible for enzymatic browning in juice. Utilizing these substrates, polyphenoloxidases catalyze oxidation leading to browning. S-Glutathionyl caftaric acid is the major product formed, and is itself resistant to further attack by polyphenoloxidases and browning (Nagel and Graber 1988). The rate and extent of browning depend on the concentration of hydroxycinnamates. However, the reaction is intimately related to all the conditions existing in the must and fermenting wine (pH, oxygen levels, increasing alcohol content, etc.).

There is mounting general concern regarding medical ramifications of sulfur dioxide usage. As a result, emphasis has been directed toward reducing use levels and, eventually, eliminating sulfur dioxide in processing altogether. In addition to medical concerns, proponents of sulfite elimination in wine processing point to several other areas of concern relative to its use (Muller-Spath et al. 1978): (1) Total phenolics are higher in musts receiving prefermentation additions of sulfur dioxide. (2) sulfur dioxide reacts rapidly with several compounds present in juice and wine, chief among them acetaldehyde. The latter reaction product is rather stable and has no or questionable activity in prevention of oxidation or inhibition of most microorganisms. Further, formation of the addition compound reduces the levels of free molecular sulfur dioxide present, thereby creating a need for further additions in order to achieve the desired levels of molecular sulfur dioxide needed for microbiological and oxidative stability. Because of the reactivity of SO_2 and acetaldehyde, the timing of sulfite additions has a major influence on final concentrations of both components in wine. Wines produced from sulfited juice, or which have had sulfite additions made during fermentation, have significantly higher levels of acetaldehyde than do those produced with no added sulfite. (3) Where a malolactic fermentation is considered desirable, high levels (> 50 mg/L) of total sulfur dioxide used in juice processing may inhibit the potential for microbial growth.

To reduce the levels of sulfur dioxide used while maintaining high standards of wine quality, careful control of virtually every facet of wine production is necessary, from the vineyard to the bottled product. Concern regarding the potential for excessive oxidative degradation begins in the vineyard. Relevant factors include grape variety, climatological conditions, harvest maturity, and fruit temperature at harvest, in transport, and during processing.

Grape chemistry and integrity also play roles in oxidation. High-pH musts and wines tend to oxidize at a faster rate than low-pH lots. Winemakers forced to deal with mold-damaged fruit expect to observe more oxidation than is seen in must produced from sound fruit. In these cases, the use of higher levels of sulfur dioxide may be needed to control further deterioration by bacteria and wild yeast. In some instances, microbially produced oxidases may not be inhibited by sulfur dioxide used at reasonable levels. During its growth, *Botrytis cinerea* produces laccase that is relatively insensitive to sulfur dioxide and alcohol (Ribereau-Gayon 1988). (A procedure for laccase analysis is presented in Chapter 1.) One technique for dealing with oxidase activity employs prefermentation juice fining (see Chapter 17). Although a variety of enzymes are present in sound fruit, fermentation and processing reduce their activity substantially.

Polyphenoloxidase activity in wine is not considered to play a major role in further oxidative degradation, although laccase activity may continue (see Chapters 1, 7, and 12). Browning of wines thus is mainly a function of nonenzymatic oxidation of phenolic compounds.

The color of wine is one of its most important characteristics, so a departure from what is generally accepted as the "normal" color of a wine can be a serious problem. The potential for color changes leading to browning in wines may be closely tied to grape variety, because of differences in concentration and activity of polyphenoloxidase. Grapes grown in warmer regions tend to darken faster than the same variety grown in cooler areas. Also, maturity seems to affect the tendency toward browning. It would appear that controlled oxygen contact with the must plays an important, and probably beneficial, role in wine quality.

IMPORTANCE OF OXYGEN IN YEAST METABOLISM

Oxygen levels in must are important with respect to the potential for oxidation and the rational use of sulfur dioxide. Additionally, oxygen plays important roles in the physiological status of yeast.

Although fermentation traditionally has been described as growth in the absence of oxygen, small amounts of oxygen may stimulate fermentative activity in certain yeast species. Called the "Custers" or "negative Pasteur effect," stimulation of fermentation (at sugar levels of less than 3 to 4%) by oxygen in trace amounts is seen in some strains of *Saccharomyces*, as well as in wine spoilage yeasts *Brettanomyces* and *Dekkera*.

Molecular oxygen is required in the synthesis of the lipids and steroids (ergosterol, 24,28-dehydroergosterol, and zymosterol) needed for functional cell membranes. Steroids play a structural role in membrane organization, interacting with and stabilizing the phospholipid component of the membrane. It has been shown that yeast propagated aerobically contain a higher proportion of unsaturated fatty acids and up to three times the steroid level of conventionally prepared cultures. This increase correlates well with improved yeast viability during the fermentative phase.

As fermentation begins, oxygen initially present in the must is rapidly consumed, usually within several hours. After utilization of the initial oxygen present, fermentations become anaerobic. As anaerobic yeast growth continues, cell division results in a redistribution of the steroids and unsaturated fatty acids needed for membrane synthesis. Because the cells are unable to synthesize membrane components in the absence of oxygen, existing steroids must be redistributed within the growing population. Under such conditions, yeast multiplication usually is restricted to four to five generations, in part because of diminished levels of steroids, lipids, and unsaturated fatty acids.

The need for oxygen supplementation may be overcome by addition of steroids and unsaturated fatty acids. The lipid present in grape cuticle, oleanolic acid, may serve as an exogenous source during anaerobic fermentation (Radler 1965). This fatty acid has been shown to replace the yeast requirement for ergosterol (major steroid produced by yeast) supplementation under anaerobic conditions (Brechot et al. 1971). There is also evidence that exogenous addition of yeast "ghosts" or "hulls" may overcome the oxygen limitation, possibly by providing a fresh source of membrane components. Yeast hulls refers to the cell wall-membrane complex of yeasts.

Labatut et al., (1984) studied the toxic effects of octanoic and deconoic acids on fermenting wine yeast. They found that in the presence of these fatty acids sugar consumption by yeast was inhibited as well as the yeast ability to produce desirable volatile compounds. The addition of 0.2 g/L yeast hulls 48 hours after the onset of fermentation increased the yeast population and prolonged their growth phase. The addition of yeast hulls appears to prolong viability of nonmultiplying yeast and thus stimulates fermentation. Yeast hulls appear to act like carbon particles in their ability to bind these inhibitory fatty acids. As such, yeast hulls are employed as an aid in stuck fermentations, in juice produced from mold-degradated fruit, and occasionally in heavily clarified juices.

The methodology of starter propagation is important with respect to subsequent oxygen requirements. When propagation is conducted in an environment in which yeasts are in contact with a relatively high oxygen titer, subsequent growth and fermentation proceed at a faster rate than with starters prepared without aeration (Wahlstrom and Fugelsang 1987). Yeast deprived of oxygen are known to undergo autolysis more rapidly during subsequent storage. Additional information on yeast growth is provided in chapter 12.

REDOX POTENTIALS OF WINE SYSTEMS

The reduction (redox) potential of a chemical system (such as wine) is a measure of the tendency of its molecules or ions to gain electrons. Molecular oxygen is an example of a compound with a high (increasingly positive) redox potential; oxygen will readily accept electrons and thus is reduced to water in the process. Similarly, a lower redox potential (or negative value) is given to molecules or ions with increasing tendencies to lose electrons. An example of the latter is ascorbic acid, which loses electrons to become oxidized to dehydroascorbic acid (Fig. 8-1).

The various components in wine can exist as mixtures of their oxidized and reduced forms (called redox pairs). Wine then is a complex system made up of many redox pairs. Examples of some of these redox pairs may be found in Table 8-1 along with their standard reduction potentials (as described above, a measure of the tendency of the oxidized form of the compound to be reduced).

A component with a large positive reduction potential (e.g., oxygen) will react to produce the reduced form of that component (e.g., water). Conversely, a component with a negative potential value (e.g., sulfur dioxide or sulfite ion) will more likely react to produce the oxidized form (e.g., sulfate ion). Thus

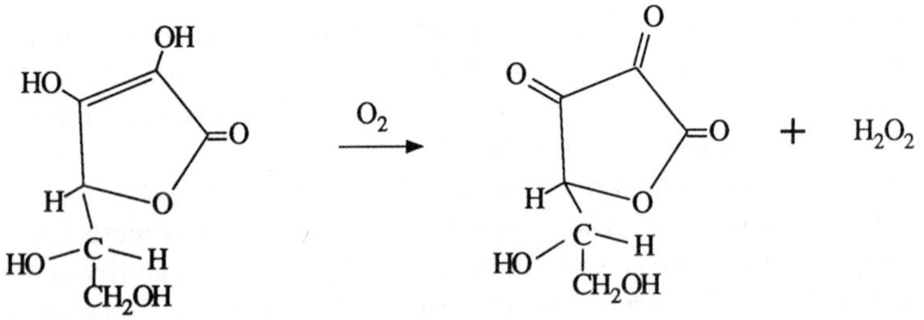

Fig. 8-1. Oxidation of ascorbic acid to dehydroascorbic acid and hydrogen peroxide.

Table 8-1 Typical redox pairs of compounds found in wine.*

Half-reaction	Standard Potential (volts)	
	at pH 3.5	at pH 7.0
$\frac{1}{2}O_2 + 2H^+ + 2e^- \rightarrow H_2O$	1.022	0.816
$Fe^{3+} + 1e^- \rightarrow Fe^{2+}$	0.771	0.771
$O_2 + 2H^+ + 2e^- \rightarrow H_2O_2$	0.475	0.268
Dehydroascorbate + $2H^+ + 2e^- \rightarrow$ Ascorbate	0.267	0.06
Fumarate + $2H^+ + 2e^- \rightarrow$ Succinate	0.237	0.030
$Cu^{2+} + 1e^- \rightarrow Cu^+$	0.158	0.158
Oxaloacetate + $2H^+ + 2e^- \rightarrow$ Malate	0.105	−0.102
Acetaldehyde + $2H^+ + 2e^- \rightarrow$ Ethanol	0.044	−0.163
Pyruvate + $2H^+ + 2e^- \rightarrow$ Lactate	0.027	−0.180
Acetyl-CoA + $2H^+ + 2e^- \rightarrow$ Acetaldehyde + CoA	−0.203	−0.41
$SO_4^{2-} + 4H^+ + 2e^- \rightarrow H_2SO_3 + H_2O$	−0.244	−0.657
Acetate + $2H^+ + 2e^- \rightarrow$ Acetaldehyde	−0.39	−0.60

*Standard conditions are unit activity for all components with the exception of H^+, which is 3.2×10^{-4} *M* or 10^{-7} *M*; the gases are at 1 atm pressure.
Recalculated from Florkin, M. and T. Wood, *Unity and Diversity in Biochemistry*, Pergamon Press, London, 1960.

reduction of one component causes oxidation of another until a final equilibrium point is reached (net amount of reduction equals net amount of oxidation). Because pH affects the values of some redox potentials (see Table 8-1), the position of the final equilibrium point in a wine is highly dependent on the pH.

Oxygen is known as an oxidizing agent, a compound that causes other components to be oxidized (lose some of their electrons). A reducing agent, on the other hand, is some molecule or ion that causes other components to be reduced. The main reducing agents found in wines are sulfur dioxide, ascorbic acid, and the various phenols. These compounds can react or bind with oxygen and lower the overall redox potential of the system (i.e., a lower oxygen level yields a lower redox potential for the system). As can be seen in Table 8-1, the reducing power of these agents is also a function of pH: The higher the pH, the more negative the redox potential of many compounds, and thus the better these compounds act as reducing agents.

The reaction speeds of different reducing agents are highly variable. Ascorbic acid acts very fast, sulfur dioxide is much slower, and phenols are still slower. As an example, after bottling, the oxygen content, and thus the redox potential, of a wine, will be lowered over a three- to five-day time period because of relatively slow reactions of oxygen with sulfur dioxide and phenols. Ascorbic acid is a more efficient antioxidant than sulfur dioxide because it reacts more rapidly with oxygen. However, the reaction of ascorbic acid and oxygen generates hydrogen peroxide (Fig. 8-1), which can react with ethanol producing acetaldehyde. The latter binds with the free sulfur dioxide, so that it is unavailable as an antioxidant. Hence, more susceptible substrates (phenols) are oxidized.

A wine may be buffered with respect to oxidation by the presence of phenols and, to a lesser extent, sulfur dioxide (in the form of SO_3^{2-}), which can bind with the oxygen present. A poorly buffered wine at oxygen saturation has a high redox potential due to the lack of oxygen binding compounds. A well-buffered wine, having a high capacity for oxygen, will have a much lower redox potential than a poorly buffered wine.

Therefore, winemaking practices may influence and enhance a wine's buffering capacity with respect to oxidative changes and, hence, the rate at which the particular wine ages. The Burgundian practice of extended lees contact (*sur lie*) is a frequently used stylistic tool in production of certain white wines. The technique utilizes storage of the newly fermented wine on yeast lees for varying periods of time. During the yeast cell death phase, intracellular enzymes (including proteases) bring about hydrolysis of cytoplasmic components (including protein). Upon autolysis, these components are released into the wine, where they no doubt contribute to the pool of oxidizable material that plays a part in the redox (buffering) system of the wine.

In addition to improved aging potential, winemakers using extended lees contact report enhanced complexity. Typically, this involves conversion of fruity characters to more muted vinous aromas.

The redox state of wine has an effect on the volatility of some components. For example, sulfur compounds (hydrogen sulfide, mercaptans) exist in molecular and ionic forms in equilibria that are affected by the redox potential. It is the molecular form that is volatile and responsible for aroma (ions themselves being nonvolatile). A shift in the redox potential of a wine may produce the more volatile form of a number of compounds, causing a change and/or an increase in the aroma. The opposite may also occur, with aromas being masked or eliminated. Because pH affects the redox potential (see Table 8-1), it also is involved in determining the equilibrium state of a wine and the relative volatility of some aroma compounds.

ACETALDEHYDE

As wines age, acetaldehyde levels increase because of chemical oxidation of ethanol and, in the case of improperly stored wines, growth of film yeast at the wine's surface. In flor sherry production, for example, levels of acetaldehyde may exceed 500 mg/L. Acetaldehyde also is an intermediate in the bacterial formation of acetic acid. Under low-oxygen conditions and/or alcohol levels above 10% (vol/vol), acetaldehyde tends to accumulate instead of being oxidized to acetic acid. In the case of alcohol inhibition, it is reported that aldehyde dehydrogenase is less stable than ethanol dehydrogenase (Muraoka et al. 1983).

Other parameters that reportedly result in higher levels of acetaldehyde include high pH and fermentation temperatures. Oxygen uptake such as may occur during bottling may result in oxidation of ethanol to acetaldehyde. The muted varietal character of newly bottled wines, especially those with low sulfur dioxide levels, reflects this transitory oxidation and accumulation of acetaldehyde. Although some winemakers believe that sulfur dioxide additions at bottling will limit short-term oxidation, the binding rate of SO_3^{2-} is a very slow process (see Chapter 9). Acetaldehyde may be analyzed chromatographically (procedure 2-6) or enzymatically (procedure 8-3)

At levels normally encountered in newly fermented table wine (<75 mg/L), acetaldehyde is of little sensory importance. Berg, et. al. (1955) report its sensory threshold in table wines to range from 100–125 mg/L. At levels above threshold, the compound has been variously described as having an aroma resembling ''overripe bruised apples'' as well as ''nutty'' or ''sherry-like.''

CARBON DIOXIDE

The carbon dioxide (CO_2) present in table wines may arise from several sources. Alcoholic and malolactic fermentations represent the important biological origins, whereas use of the gas in post-fermentation processing and bottling represents the principal abiotic source. The presence of residual CO_2 arising from yeast fermentation may play an important protective role in minimizing contact with oxygen at the wine surface. Retention of residual CO_2 in wine depends largely on temperatures of fermentation and subsequent storage, as well as reducing sugar levels and final alcohol content. At equivalent temperatures, CO_2 retention is inversely related to alcohol levels. That is, more gas is retained in wines of lower alcohol than in corresponding higher-alcohol lots. Because the malolactic fermentation usually occurs in the later stages of alcoholic fermentation or during storage of young wines, the presence of naturally produced CO_2 may be attributed to this source. Another source of carbon dioxide in wines is refermentation. Unstabilized sweet wines may, on occasion, referment when conditions favorable for yeast growth are present. Gas formation in these cases usually is much greater than that attributed to the malolactic fermentation. The common causative species involved in secondary yeast fermentations is usually *Saccharomyces* (see Chapter 12).

Carbon dioxide retention is favored by fermentation and storage of wine at low temperatures. Unless care is taken to prevent loss, pumping, nitrogen sparging, and filtration may be expected to reduce the amount present, whereas the use of CO_2 in purging oxygen from tanks and lines may contribute to increases. Nitrogen gas should not be used to sparge sparkling wine cuvees.

Wineries may elect to sparge wines with CO_2 during storage and prior to bottling. At present, CO_2 additions are permitted in still wines provided that not more than 3.92 g/L is present at the time of sale. The reader is referred to Section 240.531 of Title 27 of the Federal Regulations for further details as they may apply. Sensory detection and interpretation of CO_2 in wine depends upon several variables. Capt and Hammel (1953) report threshold CO_2 levels of 0.7 g/L in table wines.

ASCORBIC ACID

Ascorbic acid, also known as vitamin C, can be isolated from a variety of fruits, and ranges in concentration from 500 mg/kg in oranges to over 3000 mg/kg in guavas. The ascorbic acid content in grapes is low compared to other fruits, ranging from 5 to 150 mg/kg of fruit (Amerine and Ough 1974).

As can be seen in Fig. 8-2, the compound exists as a monobasic acid with lactone ring formation occurring between carbons 1 and 4. Because of asymmetrism at C_4 and C_5, four stereoisomers may occur: D- and L-ascorbic, D-isoascorbic (erythorbic), and D-erythro-3-ketohexuronic acids. Of these, D- and L-ascorbic and erythorbic acids are of interest to winemakers.

Modes of Action of Ascorbic Acid

Ascorbic acid appears to have no significant antibacterial or fungal properties, nor does it seem to play a major role in limiting enzymatic browning (Amerine and Joslyn 1970). Oxidation is greatly accelerated in the presence of trace metals, especially copper, and the polyphenolase enzyme, ascorbic acid oxidase. Ascorbic acid readily undergoes autooxidation in the presence of flavone oxides, the latter being derived from oxidation of flavonoids in solution (Braverman 1963) See Chapter 7. The coupled oxidation mechanism (Fig. 8-3) may

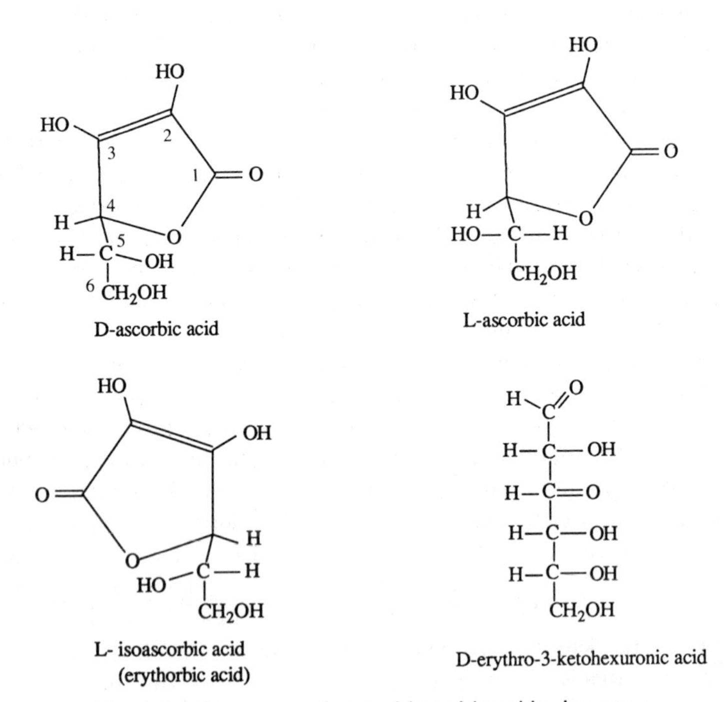

Fig. 8-2. Structure of ascorbic acid and its isomers.

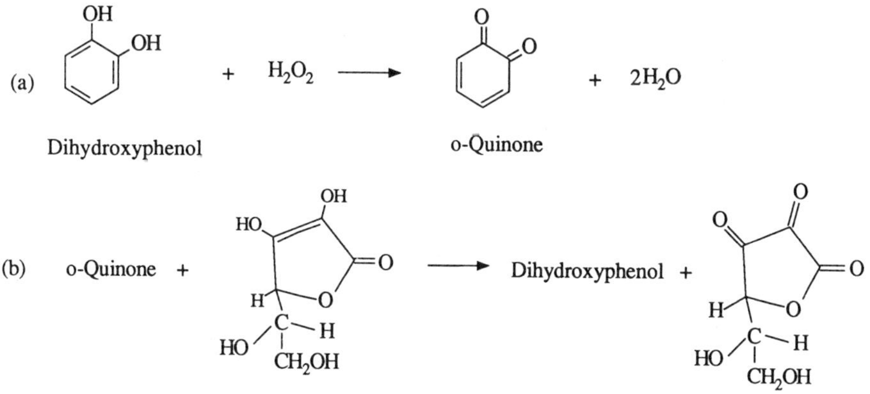

(a) Dihydroxyphenol + H_2O_2 ⟶ o-Quinone + $2H_2O$

(b) o-Quinone + [ascorbic acid] ⟶ Dihydroxyphenol + [dehydroascorbic acid]

Fig. 8-3. Coupled oxidation of dihydroxyphenol to corresponding quinone and subsequent oxidation of ascorbic acid to dehydroascorbic acid.

occur under anaerobic conditions and after enzyme inactivation has taken place. Although its intended action is that of an antioxidant, Kielhofer (1960) demonstrated that ascorbic acid may not be effective and may, in fact, catalyze oxidation of some wine constituents. The reaction of ascorbic acid and oxygen generates hydrogen peroxide (Fig. 8-1), which can directly oxidize sulfur dioxide or react through a coupled oxidation with ethanol, producing acetaldehyde. The latter binds with the free sulfur dioxide making it unavailable as an antioxidant. As a result, more susceptible substrates (phenols) are oxidized. Thus, the effectiveness of ascorbic acid as a wine additive is disputed. In any event, because of its reactivity with molecular oxygen, the use of ascorbic acid should be restricted to additions at bottling. Ascorbic acid and its optical isomer, erythorbic acid, first received governmental approval for use in wine and fruit juice production in 1956 and 1958, respectively. In countries of the European Economic Community, the use of ascorbic acid is permitted at levels not exceeding 150 mg/L (Peynaud 1984). At present, no maximum limit has been established by U.S. government.

Because of ascorbic acid's ability to catalyze oxidation of sulfurous acid, it should be used only in wines with low levels of SO_2. Fessler (1961) suggests that with the use of 0.5 to 1.5 pounds of ascorbic or erythorbic acid per 1000 gallons (60–182 mg/L), the total SO_2 level should be less than 100 mg/L. Ascorbic acid interferes with SO_2 determinations by Ripper titration (see Procedure 9.1).

Procedure 8-1: Carbon Dioxide Determination by Titrimetric Analysis

The procedure of Caputi and Walker (1987) calls for converting all CO_2 in the sample to the carbonate ion (CO_3^{2-}) by adjusting the pH of the sample to 11 to 12 using 50% (wt/wt) NaOH, and adding carbonic anhydrase. In earlier work, without the use of anhydrase, it was not possible to recover theoretical amounts of CO_2 present in standardized samples. The accuracy was at least partially dependent upon the speed at which the analysis was performed. This suggested a slow hydration of localized CO_2:

$$CO_2 + H_2O \longrightarrow CO_2 \cdot H_2O \longrightarrow H_2CO_3 \quad \text{(slow)} \qquad (8\text{-}1)$$

The solution to this problem was to add the enzyme, carbonic anhydrase, immediately after the addition of NaOH. Carbonic anhydrase catalyzes the rapid hydration of CO_2 (to H_2CO_3) (Equation 8-2) so that it can be quantitatively converted into carbonate ion in the presence of strong base (Equation 8-3):

$$CO_2 + H_2O \xrightarrow{\text{Carbonic anhydrase}} H_2CO_3 \quad \text{(fast)} \qquad (8\text{-}2)$$

$$H_2CO_3 + 2OH^- \longrightarrow CO_3^{2-} + 2H_2O \quad \text{(instantaneous)} \qquad (8\text{-}3)$$

This latter reaction is quite rapid. The sample is then titrated with acid to bring the pH down to 8.6, a step that neutralizes the excess hydroxide ions (Equation 8-4) and converts 99.5% of the carbonate, CO_3^{2-}, present to bicarbonate, HCO_3^-, (Equation 8-5):

$$Na^+ + OH^- + H^+ \longrightarrow H_2O + Na^+ + Cl^- \qquad (8\text{-}4)$$

$$CO_3^{2-} + H^+ \longrightarrow HCO_3^- \qquad (8\text{-}5)$$

Subsequent titration to pH 4.0 with standardized acid converts all of the bicarbonate to carbon dioxide and water (carbonic acid):

$$HCO_3^- + H^+ \longrightarrow CO_2 + H_2O \qquad (8\text{-}6)$$

Anions other than bicarbonate are present in wine samples, so a degassed (CO_2-free) separate blank must be run for each sample analyzed. The difference in volumes of standardized acid used to titrate the degassed blank and its corresponding effervescent counterpart over the pH range 8.6–4.0 is directly related to the quantity of CO_2 (in mg/L) present in the sample.

I. Equipment
   Burette (25 mL)
   pH meter (pH 4.00 and 10.00 standard buffers)
   Magnetic mixer and stir-bar
   Vacuum flask and source (28 torr)
   Volumetric Pipette (10 mL)

II. Reagents (See Appendix II)
   NaOH (50% wt/wt)

Carbonic anhydrase (0.1 mg/mL) (stored under refrigeration until use)
Standardized H_2SO_4 ($\leq 0.100\ N$)

III. Procedure

(a) Blank titration

1. Transfer a 50-mL aliquot of the wine sample to a vacuum flask (with stir-bar). Stopper, place on stirrer plate, and degas the sample under a vacuum ($\geq$ 28 inches) for 5 minutes with maximum mixing.
2. Transfer 10 mL of the degassed sample to an appropriate-sized beaker. Add 40 mL of recently boiled (and cooled) distilled water. Insert the pH electrode.
3. Using NaOH (50% wt/wt), adjust the pH to 10–11.
4. Add three drops of carbonic anhydrase, and mix the sample.
5. With mixing, adjust the pH to 8.6 using standardized H_2SO_4. Record the burette volume.
6. When the meter reading is stabilized at pH 8.6, begin titration with standard H_2SO_4.
7. Titrate to a final pH of 4.0. Record the volume of acid required to bring the pH of the blank from 8.6 to 4.0, and report it as the "blank titer."

(b) Sample titration

8. Carefully open the sample bottle, avoiding unnecessary agitation.
9. To a 375-mL sample bottle add 5 mL 50% NaOH. To a 750-mL sample bottle add 10 mL 50% NaOH. To larger volumes, corresponding amounts of base may be added.
10. Quickly stopper the bottle, and mix the solution thoroughly.
11. After mixing, transfer a 10-mL aliquot to a beaker containing 40 mL of recently boiled (and cooled) distilled water, add three drops of carbonic anhydrase, and proceed as for the blank titration.

(c) Calculations

Calculate the carbon dioxide content of the wine sample in units of mg/L using the following equation:

$$CO_2\ (\text{mg/L}) = \frac{(A - B)\ (N\ H_2SO_4)\ (44)\ 1000}{\text{Sample volume}} \times 1.01 \qquad (8\text{-}7)$$

where:

A = milliliters H_2SO_4 needed to titrate sample from pH 8.6 to 4.0
B = milliliters H_2SO_4 needed to titrate degassed blank from pH 8.6 to 4.0

IV. Supplemental Notes

1. The factor of 1.01 in the above calculation takes into account the dilution resulting from the addition of base to the samples.

2. If one uses titrant of 0.0682 *N* and a 10-mL sample volume, 1 mL of titrant is equivalent to 30 mg CO_2/100 mL. This conversion does not take into account dilution due to the addition of NaOH.
3. As noted, a *separate* blank titer must be determined for *each* wine sample run.
4. In the original paper it was reported that the differences in CO_2 between precooled samples and those run without cooling were not significant for routine control purposes.

Procedure 8-2. Determination of Oxygen Using a Dissolved Oxygen Meter

Oxygen in juice, must, and wine can be determined readily with an electrochemical measuring system. A typical oxygen measuring system is that produced by YSI (Yellow Springs Instrument Co.), consisting of an oxygen probe (sensing device) connected to a signal processor/readout unit (oxygen meter).

The YSI probe is a functioning polarographic cell containing two electrodes with sufficient potential applied between them to reduce any oxygen present. This polarographic cell is isolated from the outside environment by a thin membrane that is permeable to gases. When the probe is placed in a liquid sample, oxygen in the sample passes through the permeable membrane at a rate proportional to the amount of oxygen in the sample. This oxygen is rapidly reduced at the cathode electrode in the cell, producing a current that is measured and displayed by the signal processing unit in terms of oxygen concentration.

I. Equipment
   YSI dissolved oxygen meter or equivalent
   YSI 5700 Series probe or equivalent

II. Reagents (See Appendix II)
   Half-saturated KCl with two drops of Kodak Photo Flo per 100 mL of solution.

III. Procedure
   1. Following the manufacturer's instructions, fill the probe body with KCl electrolyte, and cover the probe end with a Teflon membrane.
   2. Follow instructions provided for calibrating the dissolved oxygen meter.
   3. Place the probe in the sample, allow the sample to equilibrate, and read the dissolved oxygen level off the readout meter.
   4. Store the sensing probe according to the manufacturer's instructions.

IV. Supplemental Notes
   1. Membranes will last indefinitely if properly installed and treated with care during use. Poor membrane application or membrane damage results in erratic readings. Erratic behavior can be caused by loose, wrinkled, or fouled membranes, or by bubbles in the probe from electrolyte loss. The gold cathode in the probe may become plated with

silver (from the anode) if the sensor is operated for extended periods of time with a loose or wrinkled membrane.

2. Several gases are known interferents (SO_2, CO, H_2S, halogens, and neon). Suspect oxygen readings may be due to the presence of one or more of these gases.
3. The gold cathode always should be bright and untarnished. Some gases contaminate the sensor, as evidenced by discoloration of the gold. Clean the gold cathode by vigorous wiping with a soft cloth, lab wipe, or hard paper. Do not use abrasives or chemical cleaners!

Procedure 8-3. Enzymatic Analysis for Acetaldehyde

During fermentation, acetaldehyde is produced as the immediate precursor of ethanol by decarboxylation of pyruvate. Other sources of acetaldehyde in wine include its formation, and under certain circumstances accumulation, as an intermediate in the bacterial formation of acetic acid (see Chapters 5 and 12) as well as in coupled oxidation reactions.

The presence of acetaldehyde may be used as an indicator of oxidation in sparkling wine cuvees as well as table wines. Acetaldehyde may be analyzed by gas chromatography—see procedure 2-6. The enzymatic assay for acetaldehyde utilizes the enzyme aldehyde dehydrogenase to catalyze the quantitative oxidation of acetaldehyde to acetic acid. During the course of the reaction, the coenzyme nicotinamide-adenine dinucleotide (NAD) is reduced to NADH; and the amount of reduced NADH formed in the reaction is measured using a spectrophotometer. Prepared kits containing the necessary components for the analysis are commercially available.

Rather than present a detailed procedure here, the authors refer the reader to the product sheet accompanying the acetaldehyde test kit, for the most recent information and a procedure specifically designed for the kit obtained.

I. Equipment

In general, the acetaldehyde procedure will require the following equipment:

Spectrophotometer capable of reading absorbance at 334, 340, or 365 nm
Single or multiple-range micropipettes (0.05–3.00 mL)

II. Reagents (See Appendix II)

Acetaldehyde Enzymatic Assay Kit (Boehringer-Mannheim)
Distilled water

III. Procedure

1. Follow directions on the product information sheet for proper dilution of the sample.
2. Mix an aliquot of the sample in a cuvette with all the reagents provided *except* the aldehyde dehydrogenase solution. Read the absorbance of the mixture.

3. Add the aldehyde dehydrogenase solution to the cuvette, and mix the sample. After the reaction is complete, again read the absorbance of the mixture.
4. The acetaldehyde concentration in the sample is calculated from the difference in the two measured absorbances.

IV. Supplemental Notes

1. Each supplier of enzyme kits will include various precautions and suggestions on the detailed product sheet. It is advisable to read this carefully prior to beginning the analysis.
2. Because standard acetaldehyde solutions are not used for the calibration, the accuracy of the method depends on the care with which one delivers the precise volumes of reagents and samples specified.
3. Acetaldehyde may also be measured by gas chromatographic procedures (See procedure 2-6 and figure 2.11)

Section II

MICROBIAL STABILITY

Microbial activity plays a major role in the chemical and physical properties of wine. Changes occurring at every step in the process from vineyard to bottled wine may dramatically affect the sensory properties of that wine. Section II deals with processing considerations that impact the potential for microbial activity. Beginning with a discussion of sulfur dioxide (Chapter 9) and its application(s) in preventing or slowing oxidative changes, relevant supportive analyses provide the winemaker-chemist with several methods for determination. Other sulfur-containing compounds, namely hydrogen sulfide and mercaptans, are also considered in this section (Chapter 10). Antimicrobials, aside from sulfur dioxide, are discussed in Chapters 11 and 12. These include sorbic, benzoic, and fumaric acids, as well as dimethyldicarbonate. Chapter 12 presents a review of wine microbiology addressing the role(s) of bacteria, yeast and other fungi in grape growing and winemaking. The intent of this chapter is to provide relevant information of an applied rather than academic nature. Included among the procedures discussed are cell counting techniques and routine diagnostic schemes for bacterial and yeast isolates.

Chapter 9

Sulfur Dioxide

Sulfur dioxide (SO_2), in its several commercially available forms, is widely used in wine and related food industries as a chemical preservative and an inhibitor of microbial activity. It is active as an antioxidant in reducing chemical and enzymatic browning.

About a century ago, the Pasteur Institute announced a 10,000-franc prize to anyone able to identify a substance that could reproduce the desirable characteristics of sulfur dioxide and yet have fewer detriments than SO_2. The fact that the prize remains unclaimed underscores the effectiveness of sulfur dioxide and the difficulty of finding a viable substitute.

Although historically sulfites have been considered ''generally recognized as safe'' (GRAS), the U.S. Food and Drug Administration (FDA) recently has determined that the presence of unlabeled sulfites in foods and beverages poses a potential health problem to a certain class of asthmatic individuals. As a result, the Bureau of Alcohol, Tobacco and Firearms has implemented regulations requiring a declaration in labeling of sulfites present in alcoholic beverages at a level of 10 or more mg/L (ppm), measured as total sulfur dioxide, by any method sanctioned by the Association of Official Analytical Chemists (AOAC).

MODES OF ACTION OF SULFUR DIOXIDE

Historically, three modes of action have been suggested to account for the activity of SO_2 in wine and juice. These theories are discussed in the following paragraphs.

The Antioxidant Theory

This theory proposes that SO_2 is effective because of its antioxidative activity, being oxidized in preference to other compounds whose oxidation is accompanied by undesirable changes. It has been well established that hydroxyphenol in the presence of oxygen will be oxidized to produce quinone and hydrogen peroxide (H_2O_2) (Equation 9-1a). The peroxide thus formed subsequently may oxidize sulfite (SO_3^{2-}) to sulfate (SO_4^{-2}) (Equation 9-1b), instead of reacting with substrates whose oxidation is accompanied by undesirable color changes.

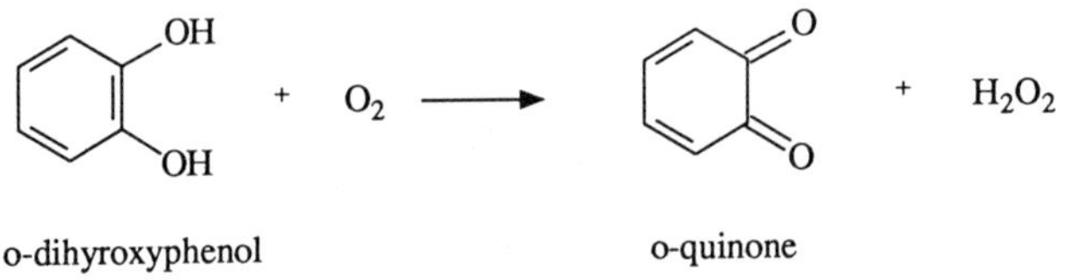

$$\text{o-dihyroxyphenol} + O_2 \longrightarrow \text{o-quinone} + H_2O_2 \quad (9\text{-}1a)$$

$$SO_3^{-2} + H_2O_2 \longrightarrow SO_4^{-2} + 2H_2O \quad (9\text{-}1b)$$

Alternatively, SO_2 may be active in the direct reduction of oxidized quinones, as shown in the following reaction sequence:

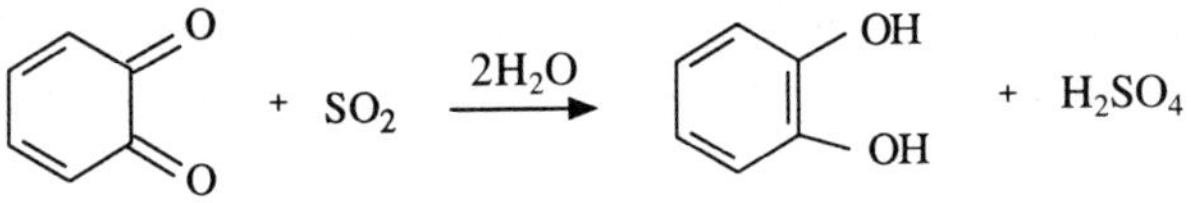

$$\text{o-quinone} + SO_2 \xrightarrow{2H_2O} \text{o-dihydroxyphenol} + H_2SO_4 \quad (9\text{-}2)$$

The Bleaching Theory

This theory suggests that SO_2, while failing to inhibit oxidation reactions, does act to bleach any darkened compounds as they are formed (see Chapter 8).

The Addition-Compound or Active-Carbonyl Theory

This theory proposes that SO_2 reacts with active carbonyl groups to form stable compounds not susceptible to browning. Examples of active carbonyl groups are the $-\overset{|}{C}=O$ groups of reducing sugars and aldehydes.

SULFUR DIOXIDE AS AN INHIBITOR OF BROWNING REACTIONS

Browning in juice and wine may result from either chemical or enzyme-catalyzed mechanisms. Three major browning reactions are discussed below, as they relate to the topic of SO_2. For further details, see Chapters 7 and 8.

Role of SO_2 in Prevention of Enzymatic and Nonenzymatic Browning

Juices and wines contain many readily oxidized compounds, including polyphenols. The antioxidative role of SO_2 in wine and must lies in its competition with oxygen for susceptible chemical groupings. As a reducing agent, sulfur dioxide can inhibit some of the oxidation caused by molecular oxygen. Aging may be considered a controlled oxidative process, where some oxidation by molecular oxygen is important in wine development. Thus, Ribereau-Gayon (1933) recommended that oxygen absorption be limited to the rate at which it could be catalytically reduced by oxidizable substances such as flavonoids and sulfur dioxide. Thus in the case of white wines where flavonoid substrates are relatively low, SO_2 may play a more important role in the consumption of oxygen.

The commonly observed browning phenomenon of freshly cut fruit is due to the activity of a group of plant enzymes called polyphenoloxidases. These enzymes, also active in freshly crushed grapes, catalyze the formation of darkened phenolic compounds from previously colorless species:

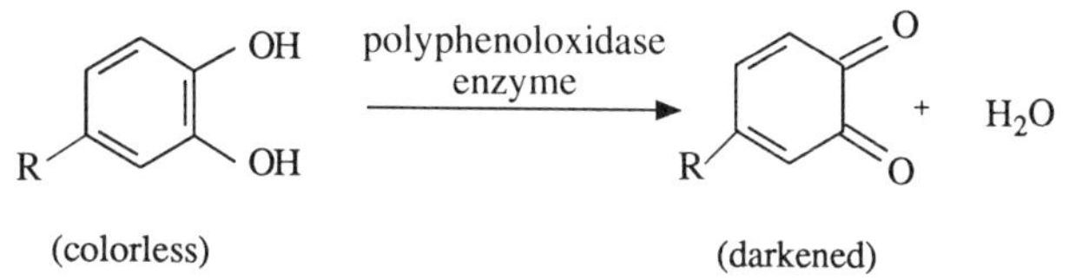

R, OH, OH (colorless) —polyphenoloxidase enzyme→ R, O, O (darkened) + H_2O (9-3)

Although the exact mechanism of inhibition is not fully understood, it is known that SO_2 is active in destabilizing the disulfide bridges that help stabilize enzymes in their "native" or active form. The addition of 35 mg/L of sulfur dioxide to must was found to completely inhibit oxygen uptake by polyphenoloxidase, according to White and Ough (1973).

Enzymatic oxidation during vinification may be significant because of the abundance of oxidative enzymes. In addition to the natural polyphenoloxidase, grapes degraded with *Botrytis cinerea* contain the oxidative enzyme laccase. Laccase is known to cause rapid oxidation of susceptible substrates (Peynaud 1984). This enzyme is reported to oxidize *o*- and *p*-dihydroxyphenols and, unlike polyphenoloxidase, does not hydroxylate monophenols. It is also more soluble and more resistant to sulfur dioxide (see Chapters 1, 7 and 8).

The bisulfite ion species (HSO_3^-) helps protect juices and wines from oxidative browning reactions, as well as to scavenge hydrogen peroxide formed from oxidative reactions. (For more information on browning reactions in wine see Chapter 7.)

COMPOUNDS THAT BIND WITH SULFUR DIOXIDE

Several compounds present in juice and wine are active in binding with SO_2. In red wines, the major binding compounds are acetaldehyde and anthocyanins. Burroughs (1975) reported that at free SO_2 levels of 6.4 mg/L (i.e., 10^{-4} *M*), acetaldehyde present in solution was 98.5% bound. This level of binding was followed by that of malvidin 3,5-diglucoside (63%), pyruvic acid (39%), and alpha-ketoglutaric acid (15%). Additionally, dihydroxyacetone present in microbially deteriorated musts and wines may bind significant amounts of free sulfur dioxide (Lafon-Lafourcade 1985). The latter arises from oxidation of glycerol in grapes and musts by *Gluconobacter oxydans*.

Reactions with Acetaldehyde

Acetaldehyde is the principal aldehyde normally present in wine. It is produced as an intermediate during alcoholic fermentation, the majority of the acetaldehyde being reduced to ethanol during this phase. Upon prolonged storage of wine in the presence of air, nonenzymatic oxidation of ethanol may yield small amounts of acetaldehyde (Kielhofer and Wurdig 1960a). The majority of acetaldehyde formed, however, results from microbial oxidation of ethanol under aerobic conditions (Reed and Peppler 1973). This may take place in the latter stages of alcoholic fermentation (as CO_2 evolution subsides) or in wine stored improperly in partially filled containers. In the latter case, the growth of aerobic film yeasts at the surface of the wine may result in the accumulation of large amounts of acetaldehyde. As an intermediate in the bacterial formation of acetic acid, acetaldehyde may accumulate even under conditions of low oxygen concentration (see Chapters 5 and 12).

Bisulfite reacts with acetaldehyde present in wine to yield the hydroxysulfonate addition product presented below:

$$CH_3-\overset{\displaystyle H}{\overset{|}{C}}=O + HO\underset{\underset{\displaystyle O}{\|}}{S}O^- \underset{K_D = 5\times10^{-6}}{\overset{K_f = 2\times10^{5}}{\rightleftharpoons}} CH_3-\overset{\displaystyle H}{\overset{|}{\underset{\underset{\displaystyle OH}{|}}{C}}}-\overset{\displaystyle O}{\overset{|}{\underset{\underset{\displaystyle O}{\|}}{S}}}O \qquad (9\text{-}4)$$

The small dissociation constant for this reaction ($K_D = 5 \times 10^{-6}$) indicates that the equilibrium favors formation of the product (Burroughs and Whiting 1960). One can compare this to dissociation constants for addition products with alpha-ketoglutaric acid (8.8×10^{-4}), pyruvic acid (4×10^{-4}), and glucose (6.4×10^{-1}). The comparison indicates that the addition product formed with acetaldehyde is the most stable of those listed here. Hydrogen ion concentration plays an important role in the above equilibrium in that it affects the concentration of bisulfite, as opposed to the molecular sulfur dioxide ($SO_2 \cdot H_2O$) present in solution (see Fig. 9-1). Because bisulfite ion is the major species present at wine pH, the formation of addition products is the favored reaction. Winemakers take advantage of acetaldehyde binding by adding additional sulfur dioxide to wines where the "aldehyde nose" is a sensory detriment. Such a procedure may help to freshen up the wine aroma.

At levels normally encountered in newly fermented table wine (<75 mg/L), acetaldehyde is of little sensory importance. Berg et al. (1955) report its sensory threshold in table wines to range from 100 to 125 mg/L. At levels above threshold, the compound has been variously described as having an aroma resembling "overripe bruised apples" or being "nutty" or "sherry-like."

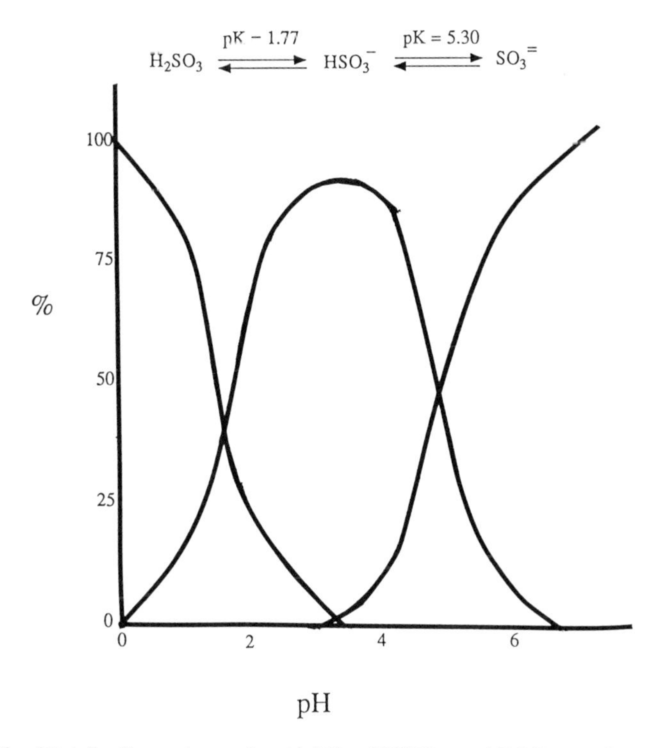

Fig. 9-1. Distribution of species H_2SO_3, HSO_3^-, and $SO_3^=$ as a function of pH in dilute solution.

Reaction with Pigments

Anthocyanin pigments have a strong affinity for bisulfite. However, the reaction of bisulfite and anthocyanin is not the same as the active carbonyl addition demonstrated by aldehydes, ketones, and sugars. Jurd (1964) has shown that the pigment–bisulfite reaction involves addition of the bisulfite anion and the colored anthocyanin cation to form, among others, the colorless species anthocyanin-4-bisulfite:

(colored) $\xrightarrow{SO_2}$ SO_3 Anthocyanin-4-bisulfite (colorless) (9-5)

Timberlake and Bridle (1976) report formation of the anthocyanin-4-bisulfite to be approximately 85% complete after the addition of only 15 mg/L SO_2. Burroughs (1975), however, pointed out that more than half of the red wine color may be attributed to polymeric anthocyanins, which are resistant to SO_2 addition because of prior substitution.

Sulfur dioxide enhances color extraction in red wine fermentations because the addition compound formed by sulfur dioxide and anthocyanidin is more polar and therefore more soluble in water–ethanol than was the anthocyanidin by itself. Furthermore, wines retain color longer in the presence of moderate amounts of SO_2 than in its absence. (This topic is covered in greater detail in Chapter 7.)

Reactions with Sugars

Sugars as a group combine with bisulfite (HSO_3^-) at a much slower rate than do aldehydes and ketones, and the products formed with sugar are less stable than those with aldehydes and ketones. Joslyn and Braverman (1954) reported that in a mixture of acetaldehyde, glucose, and bisulfite, the aldehyde combined with bisulfite first, with 90 to 95% being bound after 2 minutes. Furthermore, addition of acetaldehyde to a glucose–bisulfite solution resulted in replacement of glucose by acetaldehyde. The fact that sugars bind with bisulfite at a slower rate and produce less stable products can be explained by reference to the equilibrium reaction presented in Fig. 9-2. The majority of glucose present in solution is in the unreactive or potentially active pyranose ring form.

Formation of the addition product thus is dependent upon more complex equilibria than those existing between simple aldehydes and ketones and bisulfite. Also, it is contingent upon the dissociation of $SO_2 \cdot H_2O$ in solution

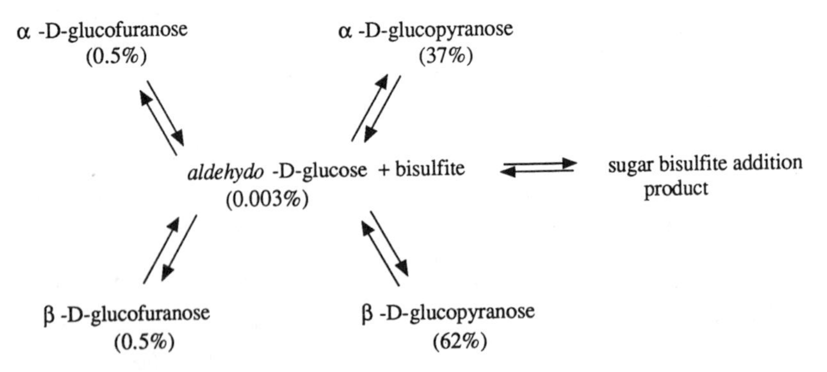

Fig. 9-2. Equilibrium between straight-chained and ring configurations of glucose in solution.

(see below, Equations 9-6a–c). An additional relevant finding is that of Braverman (1963), who reported that the keto-sugar fructose does not form an addition compound with bisulfite ion. This is particularly significant in that fructose comprises approximately half of the reducing sugar content of must, a ratio that increases as the fermentation reaches completion (see Chapter 6).

DISTRIBUTION OF SULFITE SPECIES IN SOLUTION

Upon solution of sulfur dioxide in water, the following equilibria are established:

$$SO_2\,(g) \rightleftharpoons SO_2\,(aq) \qquad (9\text{-}6a)$$

$$SO_2\,(aq) + H_2O \rightleftharpoons SO_2 \bullet H_2O \qquad (9\text{-}6b)$$

$$SO_2 \bullet H_2O \overset{K_1\,=\,1.7\times10^{-2}}{\rightleftharpoons} \underset{\text{(bisulfite)}}{HSO_3^-} + H^+ \qquad (9\text{-}6c)$$

$$HSO_3^- \overset{K_2\,=\,6.3\times10^{-8}}{\rightleftharpoons} \underset{\text{(sulfite)}}{SO_3^{-2}} + H^+ \qquad (9\text{-}6d)$$

In addition, the metabisulfite anion can produce similar sulfite species:

$$S_2O_5^{-2} + H_2O \rightleftharpoons 2HSO_3^- \qquad (9\text{-}6e)$$

The presence of any form(s) of sulfur dioxide in solution is pH-dependent. A distribution-of-species diagram for SO_2 (Fig. 9-1) shows that only the sulfite form occurs at pH levels above ~9.0, and the bisulfite form dominates in the pH range 2 to 7. At the pH of wine, bisulfite (HSO_3^-) and $SO_2 \cdot H_2O$ are the dominant species.

BOUND AND FREE (INCLUDING MOLECULAR) SULFUR DIOXIDE

Sulfur dioxide in wine occurs in two forms, bound (or fixed) and free, their sum equaling total SO_2. The term ''bound SO_2'' refers to formation of addition compounds between bisulfite ion (HSO_3^-) and other substances such as aldehydes, anthocyanins, pectic compounds, proteins, and aldo-sugars. The resultant inactive compound is of the generalized structure:

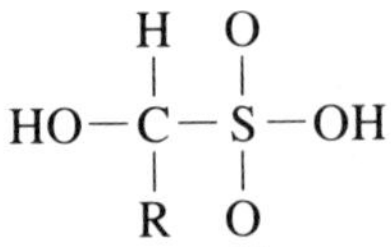

$$
\begin{array}{ccccccc}
 & & H & & O & & \\
 & & | & & | & & \\
HO & - & C & - & S & - & OH \\
 & & | & & | & & \\
 & & R & & O & &
\end{array}
$$

Hydroxysulfonate
addition compound

The above form may represent up to 75% of the SO_2 present in wine (Fornachon 1965).

The undissociated molecular form of free sulfur dioxide is the most important antimicrobial agent. Within the pH range of juice and wine, the amount of free sulfur dioxide in the molecular form varies considerably. Table 9-1 (see p. 206) illustrates these changes. Wine pH thus is a key indicator of the resistance of a wine to both oxidation and microbial degradation. Attempting to control microbial growth through the measurement of total SO_2 or free SO_2 without reference to pH and molecular free SO_2 is of little value.

The proper level of molecular SO_2 for the control of microbial growth will vary with temperature, ethanol content, micronutrient level, and so on. Beech et al. (1979) determined that for white table wines, 0.8 mg/L molecular free sulfur dioxide achieved a 10,000-fold reduction in 24 hours in the number of viable *Brettanomyces* species, certain lactic acid bacteria, and other wine spoilage organisms. The amount of free SO_2 needed to obtain 0.8 mg/L of the molecular form at various pH levels is also given in Table 9-1. Thus, one can see that 30 mg/L free SO_2 may well be an excess of molecular SO_2 in one wine (lower pH) and an insufficient quantity for controlling microbial growth in another (higher pH).

If the goal is microbial inhibition, adding a large quantity of sulfur dioxide less often is preferable to making multiple smaller additions. In the latter case, competitive formation of addition compounds (aldehydes, sugars, etc.,) may decrease the effect of SO_2 on the microbial population in question.

With regard to its activity toward microbes, it is known that free molecular SO_2 is strongly inhibitory toward enzyme systems, especially those containing sulfhydryl (—SH) groups and disulfide linkages, as well as the coenzyme nicotinamide-adenine dinucleotide (NAD^+) (Fig. 9-3). In this respect, SO_2 may be active in preventing or delaying enzymatic decarboxylation of pyruvate in fermentative metabolism (Braverman 1963). During fermentation, yeasts are more resistant to SO_2 because of rapid fixation by aldehydes; so attempting to stop fermentation by the addition of sulfur dioxide often is not effective.

SULFUR DIOXIDE USE IN WINE PRODUCTION

There is an industry-wide trend toward reducing sulfur dioxide use whenever possible. The reasons for the trend include public health concerns, better quality fruit, possible desire for a malolactic fermentation, and the perceived delicacy of wines (lower levels of phenols in white wines) that have not had SO_2 added. The current trend is away from grapey flavors and toward more vinous wines. Reduction of sulfur dioxide levels particularly in the juice is consistent with such stylistic goals. Additionally, wines produced with low levels of sulfur dioxide are thought to be softer on the palate.

A number of sensory characteristics in wines have been directly attributed to the presence of sulfur dioxide. High levels of SO_2 impart a metallic (tinny) and harsh character to wines. Furthermore, excessive levels of free sulfur dioxide add a pungent aroma, a sharpness in the nose, and a "soapy" smell.

Grape Processing Considerations

If sulfur dioxide additions are made at the crusher, the levels added are usually dependent upon the cultivar, maturity, fruit temperature, fruit condition, wine style desired, pH, and overall sanitation. Sulfur dioxide added to grapes at crush

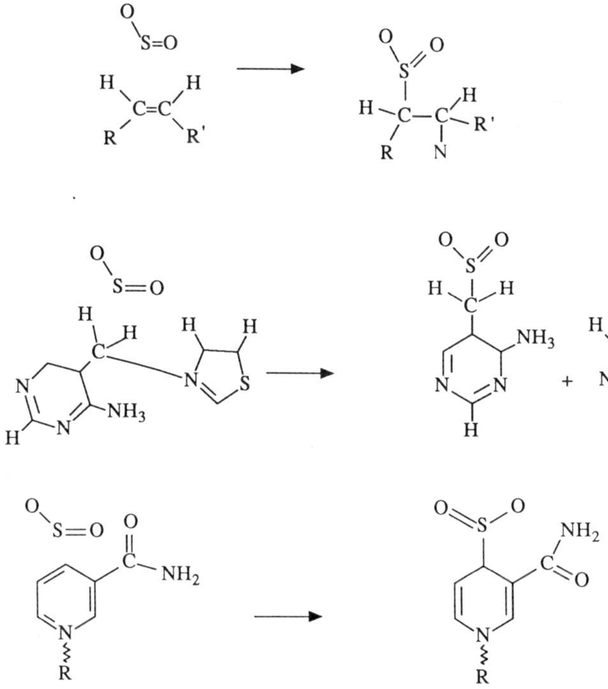

$$R_1{-}S{-}S{-}R_2 + SO_2 \longrightarrow R_1{-}SH + R_2{-}SO_3^- \quad \text{(d)}$$

Fig. 9-3. Potential reactions of sulfur dioxide; (a) addition across a double bond, (b) cleavage of thiamine (vitamin B_1), (c) addition to coenzyme NAD, and (d) rupture of disulfide bonds.

increases the extraction of flavonoid phenols (Singleton et al. 1980), which contribute to both bitterness and astringency. Sulfur dioxide additions to the juice following pressing are common. Must processing in the absence of sulfur dioxide will increase oxidative polymerization, (Ough and Crowell 1987) and precipitation of phenols; so winemakers may choose to limit sulfur dioxide additions prior to fermentation (see Chapter 7).

Cellar Considerations

Although bound sulfur dioxide has little inhibitory effect upon most yeast and acetic acid bacteria, it can affect the growth of lactic acid bacteria. High concentrations of the acetaldehyde–bisulfite complex are known to inhibit malolactic fermentation. Because of this suppression, wines designed to undergo a malolactic fermentation should have only moderate amounts (less than 50 mg/L SO_2) of sulfur dioxide.

The quantity of SO_2 fixed and the rate of binding is a pH-dependent reversible reaction; the lower the pH, the slower the addition. Temperature affects the equilibrium in a similar manner. Monitoring the concentration of SO_2 in free and bound forms following an addition to wine or must generally would show a rapid increase in the levels of the free form initially, followed by a decrease over 2 to 8 hours and corresponding increases in the percentage of the bound form. This is especially true in the case of sweet wines and fortified wines. SO_2 may be regarded as only a temporary preservative in wine, owing to oxidation to sulfate and binding. Lactic fermentations utilizing glucose lowers the pyruvate concentration thus helping to maintain higher levels of free SO_2.

Bottling Considerations

The level of molecular free SO_2 in wine is particularly important at the time of bottling. Most winemakers increase its content immediately prior to this last step in processing in order to take advantage of packaging a product with a higher percentage of the free form. Although amounts may vary, this adjustment generally is calculated to yield molecular free SO_2 levels of 0.8 mg/L. This requires levels from 20 to 30 mg/L of free SO_2, depending on the wine pH (see Table 9-1). A survey of California wineries conducted in 1988 found free SO_2 levels of from 20 to 45 mg/L (average 34 mg/L) for white wines at bottling. The values for total SO_2 reportedly ranged from 30 mg/L to as high as 170 mg/L (average 120 mg/L). In the same report, levels of free SO_2 in red wines at bottling were from 5 to 40 mg/L (average 34 mg/L), with total SO_2 ranging from 20 to 125 mg/L. For red wine, the average total SO_2 level at bottling was reported to be 120 mg/L (Blackburn 1988).

The loss of free SO_2 in wine is proportional to the dissolved oxygen content. In this respect, the headspace in bottled wine may contain up to 5 mL of air, which amounts to 1 mL (1.4 mg) of oxygen available to initiate oxidation. Generally, 4 mg of SO_2 is needed to neutralize the effects of 1 mg of oxygen. Using this relationship, an additional 5 to 6 mg of free SO_2 is needed to reduce molecular oxygen in the headspace. This may represent a rather significant loss of free SO_2 that otherwise would be available as an antioxidant or antimicrobial agent.

There are some important limitations to the capabilities of sulfur dioxide as an antioxidant. Because its reaction with molecular oxygen is very slow, it may be impractical to use SO_2 for scavenging oxygen from wines exposed to significant aeration.

SOURCES OF SULFUR DIOXIDE

Originally sulfur dioxide was obtained by burning sulfur. It is difficult to measure both the amount of SO_2 produced and the quantity dissolved in wine with this procedure. Today, alternative sources of sulfur dioxide are available to the wine industry; potassium bisulfite, potassium metabisulfite (potassium disulfite is the preferred IUPAC name), and compressed sulfur dioxide gas are the most common. Sodium sulfite, potassium sulfite, sodium bisulfite, and sodium metabisulfite (sodium disulfite) are other sulfite compounds allowed by the BATF. Compressed gas has the advantages of both being economical and avoiding the addition of potassium to the juice or wine. Whether one elects to use compressed gas or some salt of sulfur dioxide, it is necessary to establish the correct addition levels. A typical addition of SO_2, using (1) the gas and (2) the salt potassium metabisulfite, is presented in the following example.

A winemaker wishes to increase the amount of total sulfur dioxide in 1000 gallons of wine by 28 mg/L (ppm). This may be accomplished in two ways

(1) *Addition of sulfur dioxide gas* (of 90% activity):
   (a) 28 mg/L = 0.028 g/L
   (b) 1000 gal × 3.8 L/gal × 0.028 g/L = 106 g
   (c) $\dfrac{106 \text{ g}}{454 \text{ g/lb}} = 0.234$ lb sulfur dioxide gas
   (d) It is given that the gas is 90% sulfur dioxide. Therefore, the amount of SO_2 gas is:

$$\frac{0.234 \text{ lb}/1000 \text{ gal}}{0.90} = 0.260 \text{ lb}/1000 \text{ gal}$$

(2) *Addition as potassium metabisulfite:* Potassium metabisulfite, or KMBS, is the water-soluble potassium salt of sulfur dioxide. Theoretically, avail-

able sulfur dioxide makes up 57.6% of the total weight of potassium metabisulfite ($K_2S_2O_5$):

$$
\begin{aligned}
K_2 &= 39.10 \times 2 = & 78.20 \\
S_2 &= 32.06 \times 2 = & 64.12 \\
O_5 &= 16.00 \times 5 = & \underline{80.00} \\
& \text{Total weight} = & 222.32
\end{aligned}
$$

The percentage of the active form, S_2O_4, in the total weight of salt is calculated as follows:

$$\frac{S_2O_4}{K_2S_2O_5} = \frac{128.12}{222.32} = 0.576$$

$$0.576 \times 100 = 57.6\%$$

Thus, the winemaker calculates the 28 mg/L sulfur dioxide addition as follows:

$$\frac{0.028 \text{ g/L} \times 3.8 \text{ L/gal} \times 1000 \text{ gal}}{0.576} = 184.7 \text{ g}$$

$$\frac{184.7 \text{ g}}{454 \text{ g/lb}} = 0.41 \text{ lb KMBS/1000 gal}$$

Some winemakers choose to use solutions of sulfur dioxide in water for additions. Such solutions usually are created by bubbling liquid sulfur dioxide or by weighing KMBS into a measured volume of water, thereby creating $SO_2 \cdot H_2O$. Sulfur dioxide should never be added to juice or wine in dry or highly concentrated form.

During storage, dry forms of sulfur dioxide may lose their potency. Elevated temperature, high humidity, and so on, can contribute to significant loss of strength. Therefore, it is desirable to monitor the potency of these compounds, particularly if they are stored for extended periods.

Cellar workers frequently use solutions of sulfur dioxide prepared in water at pH 3.0 or lower for sanitation purposes. Because of cost considerations, the necessary pH adjustment often is accomplished using citric acid. Active levels of molecular sulfur dioxide may be determined by reference to Table 9-1.

ANALYSIS OF FREE AND TOTAL SULFUR DIOXIDE

Sulfites have been used extensively for many years as preservatives; so there is a long history of analytical methodology for the determination of sulfites in wine and a variety of foods. At present, the official AOAC method for SO_2 in wine is the modified Monier-Williams method (Procedure 9-4). This method involves distillation of SO_2 out of the sample into peroxide, and subsequent titration of the H_2SO_4 formed. The method is cumbersome and gives answers only for total SO_2 content. A variant of the modified Monier-Williams method has been developed for use with wines. This, the aeration oxidation (AO) method (Procedure 9-3), utilizes the same chemistry as the former procedure, but simpler equipment and shorter analysis times. It also can be used to give answers for both free and total sulfur dioxide.

The Ripper method for sulfur dioxide (Procedure 9-1), which is more than one hundred years old, uses standard iodine to titrate the free or total SO_2 in a sample. Although it is universally recognized that this method is somewhat inaccurate, the procedure is so simple (and, in general, is accurate enough for routine monitoring) that it is the most common method employed in the winery laboratory. The Ripper analysis is now available in kit form.

Due to proposed reductions in allowable SO_2 levels, there has been a renewed interest in developing simple, fast, and reliable methods of analysis for sulfur dioxide. Such new methods generally are instrumental rather than chemical, and thus lend themselves to automation. Techniques of current interest involve enzymatic analysis, gas and liquid chromatography (including ion chromatography), potentiometry and polarography, ultraviolet and visible spectrophotometry, atomic absorption, and fluorometric spectrometry (including flow injection analysis). One should stay abreast of the literature if interested in any of the "newer" techniques. All "official" methods are published in the Journal of the AOAC, with a description of necessary equipment and a procedure, as well as some evaluation of the reliability of the procedure.

Procedure 9-1. Sulfur Dioxide by the Ripper Method

In this method standard iodine is used to titrate free sulfur dioxide. The end point of the titration traditionally is monitored using starch indicator solution, but a number of commercial electrochemical detector systems also are available and are in common use. Free sulfur dioxide is determined directly. Total sulfur dioxide can be determined by first treating the sample with sodium hydroxide to release bound sulfur dioxide. Commercial kits for conducting the analyses of sulfur dioxide by the Ripper technique are available.

I. Equipment
   250-mL Erlenmeyer flask (preferably wide-mouth)
   10-mL burette
   25-mL volumetric pipette
   Rubber stopper of appropriate size
   High-intensity light source

II. Reagents (see Appendix II)
- $1 + 3\ H_2SO_4$
- 1 *N* NaOH
- 0.02 *N* or other dilute *standardized* iodine solution
- 1% Soluble starch indicator
- Sodium bicarbonate

III. Procedure: Free Sulfur Dioxide
1. Volumetrically transfer 25 mL of wine or must to a clean 250-mL Erlenmeyer flask.
2. Add approximately 5 mL of starch indicator and a pinch of bicarbonate.
3. Add 5 mL $1 + 3\ H_2SO_4$.
4. *Rapidly* titrate with standard iodine solution to a blue end point that is stable for approximately 20 sec. The use of a high-intensity light source aids end-point detection in red wines.
5. Calculate the free SO_2 concentration (in mg/L) using the following equation:

$$SO_2\ (\text{mg/L}) = \frac{(\text{mL iodine})\ (N\ \text{iodine})\ (32)\ (1000)}{\text{mL wine sample}} \qquad (9\text{-}7)$$

IV. Procedure: Total Sulfur Dioxide
1. Volumetrically transfer 25 mL of wine sample to a clean 250-mL Erlenmeyer flask.
2. Add 25 mL of 1 *N* NaOH, swirl the solution, and stopper the flask. Allow 10 minutes for the hydrolysis reaction(s) to occur.
3. Add approximately 5 mL starch indicator and a "pinch" of bicarbonate.
4. Add 10 mL $1 + 3\ H_2SO_4$.
5. *Rapidly* titrate with standard iodine solution to a blue end point that remains stable for approximately 20 sec.
6. Calculate the total SO_2 concentration (mg/L) using Equation 9-7.

V. Supplemental Notes
1. The analysis for free and total SO_2 is dependent upon the redox reaction:

$$H_2SO_3 + I_2 \rightarrow H_2SO_4 + 2HI \qquad (9\text{-}8)$$

Completion of this reaction is signaled by the presence of excess iodine in the titration flask. The excess iodine can complex with starch (blue-black endpoint) or be sensed with a platinum electrode. A number of commercial electrode detection systems are available (Beckman, Fisher, etc.) to take advantage of the electrochemical properties of iodine.

2. Because the iodine solution will oxidize quickly, the titer should be checked on a regular basis. When not in use, iodine solutions should be stored in amber glass bottles.
3. The standardization procedure for iodine solution is accomplished with standard sodium thiosulfate solution:
   a. This solution can be purchased already made and standardized, or it can be prepared from solid reagent. Then proceed as follows:
   b. Volumetrically transfer 10 mL of thiosulfate solution to a 250-mL Erlenmeyer flask.
   c. Add starch and titrate with iodine to a blue end point.
   d. Calculate the normality of the iodine solution:

$$N \text{ iodine} = \frac{(\text{mL thiosulfate})\,(N \text{ thiosulfate})}{\text{mL iodine solution}} \qquad (9\text{-}9)$$

   The above procedure is based upon the reaction:

$$2Na_2S_2O_3 + I_2 \longrightarrow 2NaI + Na_2S_4O_6 \qquad (9\text{-}10)$$

4. In the analysis for total SO_2, it is necessary first to hydrolyze bisulfite addition compounds. This is accomplished by addition of 1 N NaOH according to the following reaction:

$$R-\underset{OH}{\overset{H}{\underset{|}{\overset{|}{C}}}}-\underset{O}{\overset{O}{\underset{|}{\overset{|}{S}}}}-ONa + NaOH \longrightarrow R-\underset{H}{\overset{O}{\underset{|}{\overset{\|}{C}}}} + NaO-\underset{O}{\underset{\|}{S}}-ONa + H_2O \qquad (9\text{-}11)$$

   The completeness of this reaction is pH-dependent. Tomada (1927) reported less than 5% completion in the pH range 6–8, whereas in the pH range 8–10, 50% of the addition product is hydrolyzed. Beyond pH 12, almost total dissociation occurs. As used in the analysis of total SO_2, this is a timed reaction; so allowing additional time to elapse may introduce error.
5. In the determination of free SO_2, the sample first is acidified to reduce oxidation of wine polyphenols by iodine, and to drive the reaction equilibrium to H_2SO_4 production as shown in Equation 9-8.
6. Based on equilibrium reactions presented above (as Equation 9-6), it is apparent that at low pH SO_2 is in the volatile form. Burroughs (1975)

pointed out that acidification and the resultant drop in pH liberates SO_2 bound to anthocyanins. Therefore, upon addition of 1 + 3 H_2SO_4 (the last step before titration), it is imperative that titration be carried out as rapidly as possible to reduce loss from volatilization. Furthermore, as titration continues (removing SO_2 from solution), further dissociation of anthocyanin–bisulfite occurs. As a result, data for free SO_2 in red wines may be expected to be erroneously high.

7. To aid in end-point detection in red musts and wine, a high-intensity lamp should be positioned such that light is transmitted through the sample.
8. In deeply colored red wines, a quantitative dilution of sample with distilled water often is made. This facilitates detection of the starch end point, but results must be multiplied by the appropriate dilution factor. Furthermore, dilution may affect the equilibrium between free and total SO_2 in solution.
9. The starch indicator solution is subject to rapid microbial decomposition, so it should be stored in the refrigerator when not in use. If the integrity of the indicator is in question, place a few milliliters in a test tube and add a drop or two of iodine solution. The contents should turn an immediate and intense purple. If an amber color is noted, discard the starch.
10. The anticipated error using the procedure outlined is ± 7 mg/L (for total SO_2), which is generally acceptable for routine winery analyses. Results of this analysis should be rounded off to the nearest whole number.
11. In SO_2 determinations for products where ascorbic acid has been added or is naturally present in significant amounts, iodine titration leads to erroneously high results, an effect largely due to competitive oxidation of ascorbic acid and SO_2 by the iodine titrant. Thus, in determinations of SO_2 in such products, it is necessary either to correct for the presence of the acid or to utilize a procedure not involving redox titrations. The Aeration Oxidation method has been used successfully in these cases.
12. To summarize, the Ripper procedure for free and total SO_2 suffers from several deficiencies, including (1) volatilization and loss of SO_2 during titration, (2) reduction of iodine titrant by compounds other than sulfite, and (3) difficulty of end-point detection in colored wines.

Procedure 9-2. Alternative Iodate Reagent Procedure

Schneyder and Vlcek (1977) proposed the use of iodate rather than iodine as a titrant for sulfur dioxide determinations. The advantage of such a substitution is stability of the iodate solution compared to iodine solutions. In this procedure an excess of iodide is added to the juice or wine sample, which is then titrated with iodate as above.

The iodate reacts with the iodide and sulfuric acid in the sample, producing iodine in situ:

$$IO_3^- + 8I^- + 6H^+ \longrightarrow 3I_3^- + 3H_2O \qquad (9\text{-}12)$$

The iodine produced (triiodide ion) reacts with the SO_2 as before. The titration proceeds as in Procedure 9-1, and the results are calculated using Equation 9-7.

I. Equipment
   Same as in Procedure 9-1
II. Reagents (see Appendix II)
   0.02 *N* KIO_3 primary standard solution
   Starch plus KI indicator solution
   (1 + 2) H_2SO_4
III. Procedure
   1. Follow the instructions for Procedure 9-1 for free and total SO_2, substituting the above reagents for those specified therein.
   2. Calculate the results using Equation 9-7.

Procedure 9-3. Aeration Oxidation Procedure for Sulfur Dioxide

Sulfur dioxide in wine or juice is distilled (with nitrogen as a sweeping gas or with air aspiration) from an acidified sample solution into a hydrogen peroxide trap, where the volatilized SO_2 is oxidized to H_2SO_4:

$$H_2O_2 + SO_2 \longrightarrow SO_3 + H_2O \longrightarrow H_2SO_4 \qquad (9\text{-}13)$$

The volume of 0.01 *N* NaOH required to titrate the acid formed to an end point is measured, and is used to calculate SO_2 levels. The apparatus presented in Fig. 9-4 is typical of a number of commercial glassware systems used in this procedure. The efficient condenser and cold coolant water are effective in preventing distillation of volatile acid components in total SO_2 determinations. In routine production operations where approximate values for free SO_2 levels are adequate, glassware setups similar to that described below, but without the condenser, have been used.

I. Equipment
   Recirculating ice bath
   Micro-burner

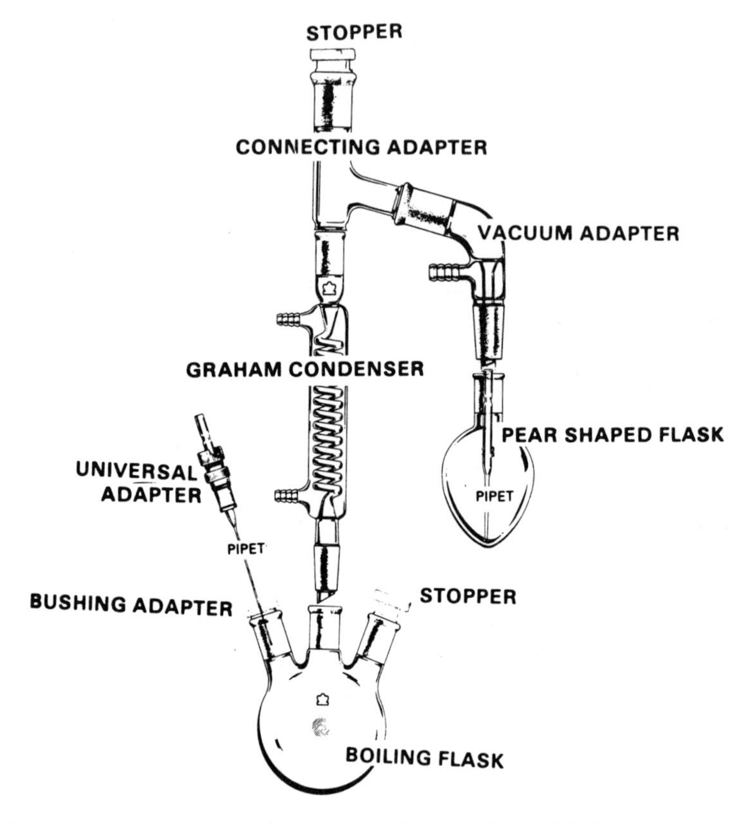

Fig. 9-4. Recommended glassware for aeration oxidation procedure.

Volumetric pipettes (10 and 20 mL)
Glass distillation unit consisting of:

A. Round-bottom flask, 3-neck, 100 mL
B. Graham condenser, 300 mm
C. Distilling adapter, 75° bend
D. Vacuum adapter, 105° bend
E. Pear-shaped flask, 50 mL
F. Thermometer adapter, 14/20
G. Bushing adapter, 24/40–14/35
H. Stopper, 24/40
I. Pasteur pipettes, 9-inch length

II. Reagents (see Appendix II)
Phosphoric acid (1 + 3)
Standardized sodium hydroxide (0.01 *N*)
Hydrogen peroxide solution (0.3%)
Indicator solution

III. Glassware Assembly

Refer to Fig. 9-4 for assembly. Connect the 3-neck 100-mL round-bottom flask at the middle neck to the 300-mm Graham condenser. Place the bushing adapter in one of the outer necks of the round-bottom flask. Place a Pasteur pipette through the thermometer adapter, and connect it to the round-bottom flask at the neck containing the bushing adapter. Adjust the height of the Pasteur pipette so that it comes within 1 mm of the bottom of the flask. Place the stopper in the third neck of the flask.

Connect the distilling (connecting) adapter (75°) to the top of the Graham condenser. Using a short piece of Tygon tubing, connect a Pasteur pipette to the central glass tube of the vacuum adapter. Adjust the length of the pipette so that it will reach to within 1 mm of the bottom of the pear-shaped receiver flask. Connect the vacuum adapter (105°) to the distilling adapter, and connect the pear-shaped flask to the vacuum adapter.

Run a vacuum hose between a water aspirator and the side arm of the vacuum adapter.

Connect a recirculating ice bath to the Graham condenser.

IV. Procedure for Free Sulfur Dioxide

1. Place an ice bath below the round-bottom flask.
2. Rinse all glassware with distilled water, and turn on the pump supplying ice water to the Graham condenser. To the pear-shaped flask add 10 mL 0.3% hydrogen peroxide solution and six drops indicator. Titrate from the initial violet color to a turquoise end point with 0.01 *N* NaOH (usually one drop with a fine-tip burette). Be sure to note the color, as it will be the same color as the final end point of the determination. Connect the pear-shaped flask to the vacuum adapter.
3. Distillation of sample: Remove the sample bottle from the refrigerator and allow to warm to standard temperature. Add exactly 20.0 mL of sample to the round-bottom flask immersed in an ice water bath. Add 10 mL of $1 + 3$ phosphoric acid to the same flask. Replace the glass stopper in the round-bottom flask. Begin aspirating vigorously, and continue aspirating the sample for exactly 10 minutes, taking care that none of the peroxide solution is lost from the pear-shaped flask.
4. Titration: At the end of the 10-minute period, cease the aspiration, and remove the pear-shaped flask. Using distilled water, rinse the inside of the vacuum adapter, and the outside of the Pasteur pipette connected to it, collecting the rinse in the pear-shaped flask.
5. Using a 10-mL burette, titrate the contents of this flask with 0.01 *N* NaOH to the end point noted above. Read the titration volume to the nearest 0.01 mL. Report the sulfur dioxide concentration to the nearest ± 1 mg/L. *Note*: It is preferable to set up and begin running the total SO_2 determination (next section) prior to titrating the "free" SO_2 sample. This first titration can be done during the aspiration of the subsequent run.

6. Calculate SO_2 (mg/L) as follows:

$$SO_2\ (\text{mg/L}) = \frac{\text{mL NaOH} \times N\ \text{NaOH} \times 32 \times 1000}{20\ \text{mL (sample size)}} \qquad (9\text{-}14)$$

V. Procedure for Bound Sulfur Dioxide

1. Initial setup: Follow the procedure listed above for "free" sulfur dioxide, except place a micro-burner below the round-bottom flask.
2. Distillation: Using the same sample that was used for the above free analysis, heat the contents of the round-bottom flask (sample and phosphoric acid) to a boil with the micro-burner. A few drops of an inert antifoam may be necessary to control excessive foaming during distillation. When boiling commences, aspirate the sample vigorously for exactly 15 minutes, taking care that none of the peroxide solution is lost from the pear-shaped flask.
3. Titration: At the end of the 15-minute period, turn off burner and stop the aspiration. Remove the pear-shaped flask. Using distilled water, rinse the inside of the vacuum adapter, and the outside of the Pasteur pipette connected to it, collecting the rinse in the pear-shaped flask. Using the 10-mL burette, titrate the contents of this flask with 0.01 *N* NaOH to the end point noted above. Read the titration volume to the nearest 0.01 mL. Report calculated sulfur dioxide values to the nearest ± 1 mg/L.
4. Calculate the bound sulfur dioxide (mg/L) using Equation 9-14.

V. Calculation of Total Sulfur Dioxide

For the NaOH titration volume, take the sum of the volumes used in the "free" and "bound" titrations. Calculate the mg/L SO_2 as above (Equation 9-14).

VI. Supplemental Notes

1. Potential sources of error in the procedure include:
   a. Carryover of CO_2 and volatile acids resulting from high initial VA, an inefficient cooling condenser, or cooling water that is not cold enough.
   b. For work in a laboratory where sulfur dioxide or carbon dioxide vapors might be a problem, an air scrubber may be necessary. The air scrubber is a vacuum flask containing 0.3% hydrogen peroxide plus 1 mL dilute NaOH, which is stoppered with a one-hole rubber stopper with a Pasteur pipette passing through it (adjust the Pasteur pipette so that it extends down into the hydrogen peroxide solution). The connection from the arm of the vacuum flask to the inlet (Pasteur pipette) on the round-bottom flask is accomplished using Tygon tubing.

c. Acidification of the sample may liberate some SO_2 bound to wine anthocyanins; so it should be the last step in the addition sequence.
d. As increased temperatures accelerate dissociation of bound SO_2, analysis for the free form should be carried out at less than 20°C (hence the need for an ice bath).

2. Nitrogen gas may be used to push the sample through the apparatus; the results will be comparable to those obtained using aspiration.
3. Following completion of the free SO_2 determination, the analyst may prefer to use a new wine sample in the reaction flask. The analytical value obtained in this instance is the total sulfur dioxide level in the sample.
4. Glassware kits designed specifically for this analysis are commercially available.

Procedure 9-4. Modified Monier-Williams Procedure for Sulfur Dioxide

This procedure has been the official AOAC method for total SO_2 for a number of years. The chemistry involved is essentially the same as that utilized in the Aeration Oxidation procedure. SO_2 is distilled out of the sample, oxidized with hydrogen peroxide, and titrated as sulfuric acid against standard sodium hydroxide. When other titratable compounds are distilled over with the sulfur dioxide, a gravimetric precipitation of the sulfate ion with barium can be employed to correct for this interference.

I. Equipment
Refer to a current volume of the AOAC *Official Methods* for appropriate SO_2 distillation glassware.

II. Reagents
Refer to a current volume of the AOAC *Official Methods* for appropriate reagents and their preparation.

III. Procedure (Total Sulfur Dioxide)
Refer to the latest edition of *Official Methods* (AOAC) for specifics of the procedure.

1. Transfer a 100- to 300-mL wine sample to a distillation flask. Dilute it to approximately 400 mL with distilled water.
2. Add 90 mL HCl $(1+2)$ to a separator funnel, and force it into the flask.
3. Start the nitrogen gas flow, and heat the flask to boiling. Reflux for 1.75 hr with a steady flow of bubbles.
4. Titrate the contents of the collector tube with 0.01 *N* NaOH, using methyl red as an indicator. Calculate mg/L SO_2 (ppm) using Equation 9-14.
5. If desired, a gravimetric determination can be carried out by acidifying

the sample and precipitating the sulfate ion with $BaCl_2$ solution. The precipitate is washed, dried, and weighed to determine the amount of $BaSO_4$ collected.

IV. Supplemental Notes

1. The Food and Drug Administration has conducted a collaborative study on an improved version of this procedure. When approved by the AOAC Methods Board, it will be sanctioned for use. The Optimized Monier-Williams method is suitable for determining SO_2 at the 10-mg/L level.
2. The Modified Monier-Williams and the Optimized Monier-Williams methods are certified only for total sulfur dioxide levels in wine.

Table 9-1. Distribution of free SO_2 as a function of pH.

pH	% SO_2 (molecular)	% HSO_3^-	% SO_3^{-2}	Free SO_2 to obtain 0.8 ppm molecular SO_2
2.9	7.5	92.5	0.009	11 ppm
3.0	6.1	93.9	0.012	13
3.1	4.9	95.1	0.015	16
3.2	3.9	96.1	0.019	21
3.3	3.1	96.8	0.024	26
3.4	2.5	97.5	0.030	32
3.5	2.0	98.0	0.038	40
3.6	1.6	98.4	0.048	50
3.7	1.3	98.7	0.061	63
3.8	1.0	98.9	0.077	79
3.9	0.8	99.1	0.097	99
4.0	0.6	99.2	0.122	125

Source: C. Smith. Enology Briefs Feb/March 1982, University of California, Davis

Chapter 10

Sulfur-Containing Compounds

Volatile sulfur-containing compounds are known to impart distinctive aromas to wines—loosely described as rubbery, skunky, or like onion, garlic, cabbage, kerosene, and so on. The objectionable odor of hydrogen sulfide, generally described as rotten-egg-like, also is occasionally observed in wines. If no correction is made for it, hydrogen sulfide may undergo reaction with other wine components to yield mercaptans, which can have detrimental effects on wine palatability that may be difficult to remove. For a determination of hydrogen sulfide vs mercaptans see procedure 10-1. An understanding of the formation and control of sulfur-containing compounds is essential for premium wine production.

HYDROGEN SULFIDE (H_2S)

Sulfur, an essential element for yeast growth, is utilized in the formation of cell components such as protein and vitamins. Available as sulfate (SO_4^{2-}) in grape juice, it can be reduced to hydrogen sulfide by the reaction series presented in Fig. 10-1.

The concentration of sulfate in grape juice varies with the variety, soil type, and year, generally ranging from 100 to 700 mg/L. A sulfate concentration of approximately 5 mg/L is considered sufficient to support yeast cell growth (Eschenbruch 1983). However, before yeast can incorporate and utilize the sulfur available in sulfate, it must be reduced to a lower oxidation state. As indicated in Fig. 10-1, H_2S formation can result from the reduction of sulfates. Thus, it is a normal intermediate in fermentative yeast metabolism and will always be produced to some degree during alcoholic fermentation. Threshold levels of H_2S in wine have been reported to be 1 mg/L (ppm) (Rankine 1963) and as low as 5-10 μg/L (ppb) (Amerine and Roessler, 1976). Because the objectionable compound is perceivable at very low levels, it has a significant negative impact on wine quality.

As H_2S is an integral part of yeast metabolism, it is not possible to prevent its formation. However, vineyard management including selection and timing of spray applications in the vineyard and wine processing techniques, may effectively minimize its detrimental effects. According to Schutz and Kunkee

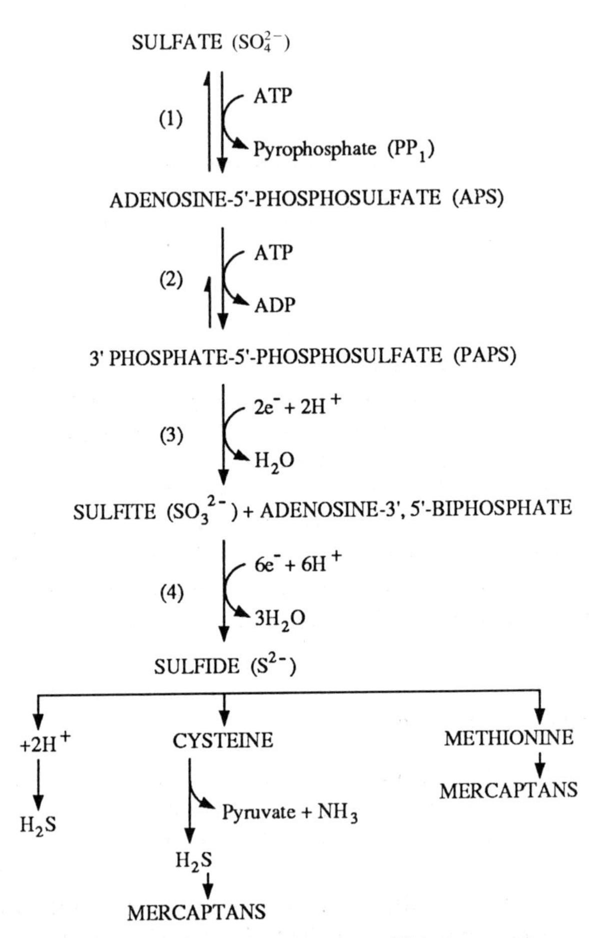

Fig. 10-1. Formation of hydrogen sulfide by yeasts.

(1977), the concentration of hydrogen sulfide produced in a wine is dependent upon: (1) the kind and amount of elemental sulfur present on grapes, (2) the oxidation–reduction state of the must and the wine, (3) the ethanol concentration, and (4) the physiological condition of the yeast during alcoholic fermentation. Of these, elemental sulfur is the most important source of H_2S in winemaking. Details about each of these factors are presented in the following sections.

I. ELEMENTAL SULFUR

Elemental sulfur is used as a fungicide in vineyards throughout the world. Thoukis and Stern (1962) report that elemental sulfur levels of only 5 mg/L in musts are sufficient to produce quantities of H_2S that workers describe as difficult to remove. The shorter the time interval between the last application of sulfur spray and the grape harvest, the greater is the residual sulfur content present in the juice, and this can result in significant levels of H_2S in the resulting wine. Eschenbruch (1983) recommends that sulfur not be applied for at least 35 days prior to harvest. Rankine (1963) reported the quantity of H_2S formed by elemental sulfur to be inversely proportional to particle size. Schutz and Kunkee (1977) compared H_2S formation from musts treated with dusting sulfur as well as wettable, sublimed, precipitated, and colloidal sulfurs. Of the sources of elemental sulfur evaluated, these workers report that colloidal sulfur appeared to bring about the most dramatic increase in H_2S. Wettable and dusting sulfur produced less H_2S, followed by precipitated and sublimed sulfur. Because of increasing awareness of the problems associated with sulfur in winemaking, many viticulturists are using micronized sulfur, which consists of very small particles, ranging from 6 to 8 μm in size, which are readily miscible in water. Suppliers of the product report that it fumes or sublimes at lower temperatures (72°F) than conventional formulations. An advantage of micronized sulfur is that the application rate is less than one-third the normal dusting sulfur rate for the same measure of fungal control. Recommended addition levels for micronized sulfur range from 1 to 2.5 pounds/acre, versus 6 to 8 pounds for wettable formulations and 10 to 20 pounds for dust.

An additional source of elemental sulfur in juice is sulfur candles, used by some vintners to disinfect fermentation and storage containers. These candles may not burn completely, so that unburned sulfur enters the wine or juice. The use of dripless sulfur sticks and/or sulfur cups may effectively overcome this problem.

In their study, Schutz and Kunkee (1977) found that the amount of H_2S produced during fermentation was linear up to at least 100 mg/L of colloidal sulfur present at the start. The addition of colloidal sulfur to must at this level yielded H_2S levels of 1200 $\mu g/L$. The formation of H_2S from elemental sulfur was greatest toward the end of fermentation, a phenomenon attributed to the increasing solubility of sulfur in the ethanol produced.

II. REDOX STATE AND TEMPERATURE

Hydrogen sulfide formation also is a function of the oxidation–reduction (redox) state of the must during fermentation. Rankine (1963) found higher levels of H_2S produced from fermentations carried out in tall-form (height to diameter) tanks. The design of such fermenters is conducive to a rapid drop in redox potential. The fermentation temperature also affects the overall formation of H_2S; generally, less H_2S is produced at lower temperatures (Rankine 1963). However, at lower temperatures, less H_2S is lost through entrainment with carbon dioxide.

III. HYDROGEN SULFIDE FORMATION BY YEAST

Yeasts differ significantly in their ability to form hydrogen sulfide. In addition to variations in alcohol tolerance, one of the major problems associated with the use of wild yeast fermentations is the production of higher than acceptable levels of H_2S. Furthermore, relatively large quantities of H_2S are produced by those yeasts that ferment more rapidly and, hence, bring about a more immediate drop in redox potential (Rankine 1963). (See Chapter 8)

Among commercially available strains. Pasteur Champagne (UCD 595), Epernay 2, and Prise de Mousse are known to be low-H_2S producers, whereas Montrachet (UCD 522) and some strains of Steinberg are known to produce higher levels of H_2S. The reasons for these differences are not clearly understood, but some yeasts are known to have deficiencies in their sulfur metabolism that promotes increased production of H_2S. Such yeasts have an absolute requirement for the vitamins pantothenate and/or pyridoxine. Although grape juices normally are not deficient in these two vitamins, must treatment may result in the depletion of one or both. The addition of methionine tends to suppress sulfate reduction and subsequent H_2S production (Eschenbruch 1974); however, other undesirable products can be formed.

MECHANISMS OF HYDROGEN SULFIDE FORMATION

Eschenbruch (1974) proposed an enzyme-mediated reaction in which H_2S is an intermediate (Fig. 10-1). The pathway utilizes two ATP and NADPH. By comparison, Schutz and Kunkee (1977) speculate that there is direct chemical (nonenzymatic) reaction of the amino acid crysteine and elemental sulfur at the yeast cell surface. Biosynthetic pathways leading to formation of amino acids generally are regulated by the concentration of the final amino acid, so the increased formation of H_2S associated with certain strains of wine yeasts (e.g., Montrachet) may be tied to the organisms' requirements for greater or lesser amounts of cysteine.

Vos and Gray (1979) report that yeast extracellular proteolytic enzymes may attack protein and other nonassimilable nitrogen sources (in nitrogen-deficient musts) to yield assimilable forms. In the process, H_2S is produced from sulfur-containing amino acid degradation products. Figure 10-2 graphically presents the concept. The results of this study suggest that the free amino nitrogen (FAN) component of must plays an important role in subsequent H_2S formation. Specifically, assimilable free amino nitrogen content was found to be inversely related to H_2S levels. When the assimilable must nitrogen content is low, the proteolytic activity of the yeast is stimulated; proteins and large peptides are degraded to assimilable forms in a effort to supplement this deficiency. The result is protein breakdown and liberation of H_2S. In support of their contention, these workers point out that the sulfur content present in must proteins ranges from 0.5 to 2%.

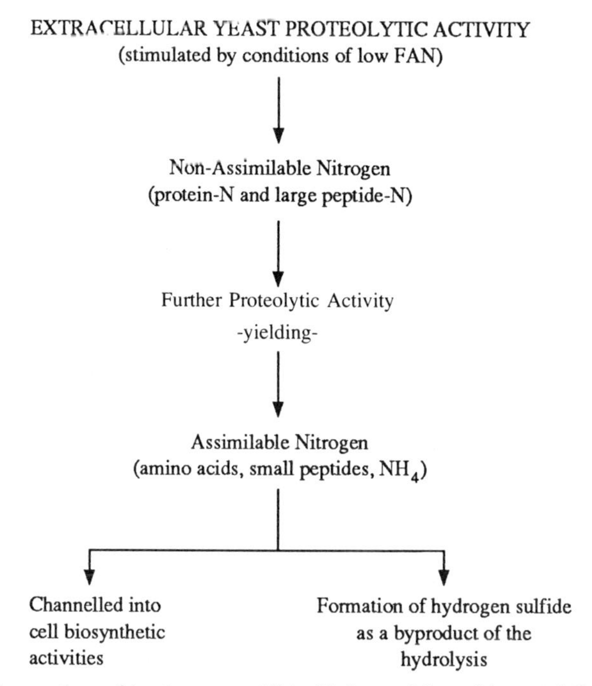

Fig. 10-2. Formation of hydrogen sulfide (Adapted from Vos and Gray 1979).

Centrifugation of juice may remove insoluble proteins, but it may have only a limited effect in reducing hydrogen sulfide formation because insoluble proteins represent only a small fraction of the total protein content. Bentonite addition to juice and its subsequent removal prior to fermentation is a step frequently employed in wine production. This addition is believed to lower H_2S formation in the wine as a result of reduction in juice protein, amino acids, and elemental sulfur (Ferenczi 1966). Complete suppression of H_2S by means of bentonite treatment of juice appears unlikely. Although the activity of bentonite is not specific, it affects mainly protein nitrogen and amino acids. Retarded or incomplete fermentation resulting from prefermentation clarification (and reduction in levels of FAN) using bentonite may be due to lack of assimilable nitrogen.

In their study, Vos and Gray (1979) found that wines fermented in contact with bentonite had increased levels of H_2S. In this regard, bentonite may cause conformational changes in protein, which may include cleavage of protein disulfide bridges, resulting in an increase in sulfhydryl groups. It is presumed that, because of steric changes, the protein molecule is rendered more susceptible to subsequent protease attack (Vos and Gray, 1979).

It should be noted that commercial trials in California (where assimilable nitrogen is added to clarified juice) have not demonstrated increased levels of H_2S. Fermentation in contact with bentonite may be advantageous in obtaining protein stabilization in white wines (see Chapter 17). A comparison of the work of Vos and Gray with that of Eschenbruch suggests that H_2S may arise from catabolic protein metabolism stimulated by insufficient quantities of assimilable nitrogen (FAN) in the must.

SULFITE

As an intermediate in the reaction series forming methionine and cysteine, bound sulfite may be released into the wine, where it is active in contributing to total sulfur dioxide (SO_2) levels. In the reaction mechanism, sulfate reduction is dependent upon (regulated by) the requirement of yeast for the amino acids cysteine and methionine. Thus, if the content of these amino acids in must is sufficiently high, their formation by yeast might be expected to be low, and correspondingly, SO_2 and H_2S production suppressed. Amerine and Joslyn (1970) reported that methionine and cysteine in must range from 1 to 14 mg/L and from 0 to 4 mg/L, respectively, whereas methionine and cysteine in wine are present at concentrations of 4 to 49 mg/L and 17 to 106 mg/L, respectively.

In his study, Eschenbruch (1974) found that 20 out of 250 strains of *Saccharomyces cerevisiae* examined produced more than 25 mg/L SO_2 during fermentation. Of these, five produced from 60 to 70 mg/L SO_2. Weeks (1969) found that *Sacch. oviformis* (currently, *Sacch. cerevisiae*) produced from 35 to 76 mg/L during fermentation. In this study, seven of over 60 strains produced greater than 20 mg/L SO_2 from an initial sulfate concentration of 350 mg/L. The sulfate content of California grapes is reported to range from 70 to 3000 mg/L (Amerine and Joslyn 1970). Sulfate also may be introduced to the must from the addition of $CaSO_4$ in plastering. Relative to its role in potential H_2S formation Thoukis and Stern (1962) state that SO_2 levels in excess of 100 mg/L are necessary to produce noticeable H_2S in wine.

ORGANIC SULFUR-CONTAINING COMPOUNDS

Goniak and Noble (1987) determined the sensory threshold for several important volatile sulfur compounds in white wine. The threshold values obtained for dimethyl sulfide, dimethyl disulfide, diethyl sulfide, diethyl disulfide, and ethanethiol were 25, 29, 0.92, 4.3, and 1.1 $\mu g/L$ (ppb), respectively. De Mora et al. (1987) determined the sensory threshold of dimethyl sulfide in Cabernet Sauvignon to be 60 $\mu g/L$ (ppb).

Many terms have be used informally to describe sulfur compounds, which have been said to suggest garlic, cabbage, onion, rubber, and so on. For example, the aroma of dimethyl sulfide has been variously described as reminiscent of asparagus, corn, or molasses, whereas ethanethiol is considered to be more onion- or rubber-like (Goniak and Noble 1987). Spedding and Raut

(1982) demonstrated that low concentrations of dimethyl sulfide had a beneficial effect on the quality of some wines. In fact, this may be true for other sulfur compounds as well.

The mercaptans are another group of sulfur-containing compounds that are also important from a sensory standpoint. Their name arises from the presence of a terminal —SH moiety. Although only a few such compounds exist in nature, those present are very odorous. For example, *n*-butyl mercaptan is responsible for the objectional odor of the skunk. The odor of ethyl mercaptan is reported to be detectable in air at levels of 50 ppb (English and Cassidy 1956).

In wine, ethyl mercaptan is believed to be formed by the reduction of H_2S with ethanol via the following reaction:

$$\underset{\text{acetaldehyde}}{CH_3CHO} \xrightarrow{H_2S} \underset{\substack{\text{thioacetaldehyde}\\ \text{(intermediate)}}}{CH_3CHS} + H_2O \longrightarrow \underset{\text{ethyl mercaptan}}{CH_3CH_2SH} \qquad (10\text{-}1)$$

Methyl mercaptan, considered to be of greater importance than ethyl mercaptan in wine, is thought to be produced from the amino acid methionine upon storage:

$$\underset{\text{methionine}}{CH_3SCH_2CH_2CHNH_2COOH} \longrightarrow \underset{\substack{\text{methyl}\\ \text{mercaptan}}}{CH_3SH} + \underset{\alpha\text{-aminobutyric acid}}{CH_3CH_2CHNH_2COOH} \qquad (10\text{-}2)$$

Methionine is a direct reduction product of sulfate. Potentially methyl mercaptan formation may occur at methionine levels of 1 to 5 mg/L (Gomes 1973).

In wines exposed to light, the amino acid methionine may undergo decomposition to yield several odorous compounds, including hydrogen sulfide, methanethiol, dimethyl disulfide, dimethyl sulfide, and ethyl methyl sulfide. Commonly referred to as "light struck," these wines are characterized as having a cheese- or plastic-like aroma. The time of light exposure necessary to catalyze the reaction may be very short, even in green glass bottles. Light struck is a particular problem in sparkling wines due to the magnifying effect carbon dioxide has upon the perception of aroma components.

YEAST AUTOLYSIS

Upon yeast cell death, degradation and rupture of cell membranes release cytoplasmic components (collectively referred to as autolysate) into the wine. The products of protease-mediated decomposition include free amino acids, peptides, and polypepetides. Other degradation products include fatty acids, as well as components of the yeast nucleic acids, and vitamins.

Yeast autolysate may play an important role in the character and complexity of wine. However, the process may result in the production of 'off' flavors and aromas, including H_2S and mercaptans (Rankine 1963; Schanderl 1955).

METALS

Copper, manganese, and zinc are components of many metal-containing fungicides used in vineyards. The vineyard management protocol in certain areas calls for late-season application of metal-containing fungicides to the grapes not only to control microbial growth but also in the belief that subsequent H_2S formation during fermentation will be diminished or eliminated. Additionally, it is not uncommon to find winemakers in certain regions of the United States adding Cu (II) sulfate directly to fermenters, hoping to achieve control of H_2S production. Because copper ions serve as powerful oxidants, the authors believe that, in the case of unfermented juice, these practices should be avoided.

Eschenbruch and Kleynhans (1974) report a definite relationship between the use of copper-containing fungicides on grapes and increased incidences of H_2S formation in wines produced from those grapes. In one case, these workers report almost a fourfold increase in copper present in must from fungicide-treated lots (Cu^{2+}) when compared to untreated controls ($Cu^{2+} = 0.98$ mg/L). Hydrogen sulfide formation in resultant wines also was stimulated. In the case of the fungicide-treated lot, the final H_2S level was 52.7 μg/L, as compared with 8.5 μg/L in the control. Rankine (1963) reports similar problems with hydrogen sulfide formation in wine and beer held in contact with these metals, particularly when sulfur dioxide is present.

The question as to how copper can stimulate H_2S formation is not resolved. Copper ions are both constituents of certain enzyme systems such as polyphenol oxidase as well as known inhibitors of respiration. Ashida et al., 1963, suggest that yeast grown in the presence of copper adopt a protective mechanism of H_2S formation and consequently copper sulfide formation. Based upon the above, the advantage of late season copper sprays for hydrogen sulfide control must be questioned.

CONTROL OF HYDROGEN SULFIDE AND MERCAPTANS IN WINE

Formation of excessive H_2S in white wines often can be minimized by settling, centrifuging, or filtering the juice prior to fermentation. These practices all accomplish the same goal, that is, removal of high-density solids along with associated elemental sulfur when present. In the case of red wine fermentation, some winemakers deal with excessive H_2S by aeration at first racking, thus volatilizing the H_2S. Some workers reported increased H_2S production, however, if aeration was carried out too soon after the completion of alcoholic fermentation. In these cases, elemental sulfur was believed to act as a hydrogen acceptor, forming H_2S. Coincidental with H_2S formation were increases in yeast populations arising as a result of transient exposure to oxygen.

Additional techniques for controlling H_2S include sparging problem wines with nitrogen gas several weeks after the completion of alcoholic fermentation. This practice may be relatively effective in eliminating minor quantities of H_2S, but desirable volatile wine components also may be swept away during sparging operations. In cases where methyl mercaptan appears to be the problem, carefully controlled aeration may bring about oxidation of methyl mercaptan to the less objectionable compound dimethyl disulfide (see Fig. 10-3).

The inverse relationship between FAN and H_2S production demonstrated by Vos and Gray (1979) may provide the best control. Suppression of H_2S has been obtained by exogenous addition of nitrogen in the form of ammonia (diammonium phosphate).

Some winemakers remove objectionable H_2S and mercaptans by direct contact with copper. This may be accomplished either by use of "in line" brass fittings or by addition of copper-containing agents. In the case of exposure of wine to copper-containing alloys (brass), resultant copper uptake into the wine is difficult to predict and may lead to oxidation (see Chapter 14). For this reason,

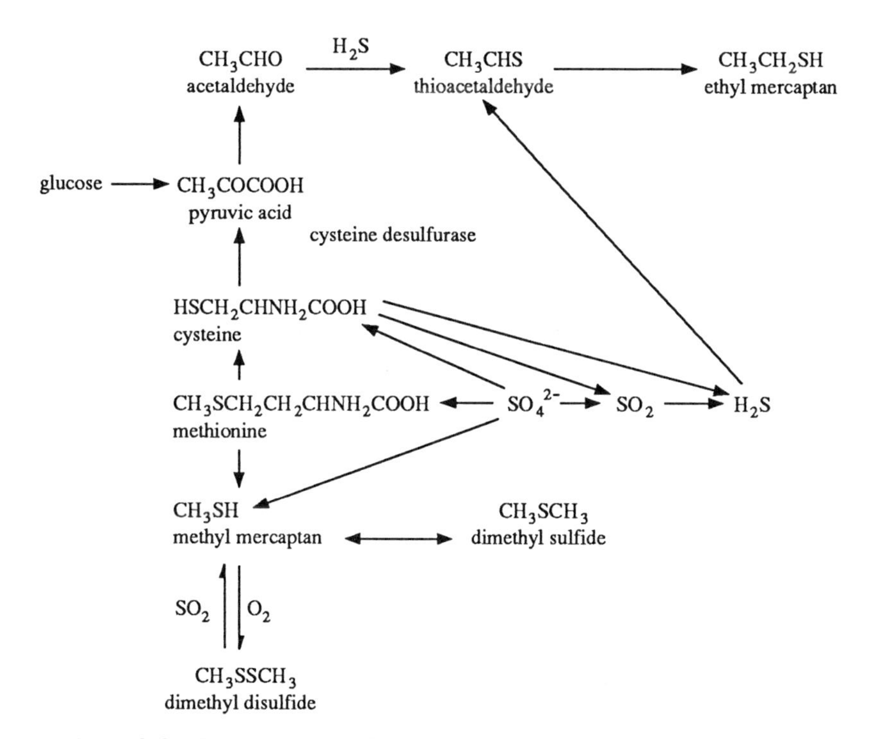

Fig. 10-3. Formation of H_2S during fermentation and storage of wine.

winemakers may elect to use copper-containing salts or fining agents. The addition of 4 g copper (II) sulfate ($CuSO_4 \cdot 5H_2O$) per 1000 gallons raises the copper content by 0.2 mg/L. Although governmental regulations permit additions of up to 0.5 mg/L (as copper), residual levels in the wine cannot exceed 0.2 mg/L (as copper). One should consult state and federal regulations governing the use of copper compounds for such purposes. The results of such procedures are often variable, and careful laboratory testing should precede any additions to wine. If residual copper levels are higher than 0.1 mg/L, it may be necessary to counterfine the wine with Cufex or Metafin (see Chapters 14, 17, 19).

It should be noted that although mercaptans react with copper, dimethyl disulfide does not (see Procedure 10-1). Thus if the wine in question has undergone any oxidation, it may be necessary to reduce dimethyl disulfide back to the reactive species, methyl mercaptan (see Fig. 10-3). This can be accomplished by addition of ascorbic acid. Generally, addition levels of 50 mg/L or

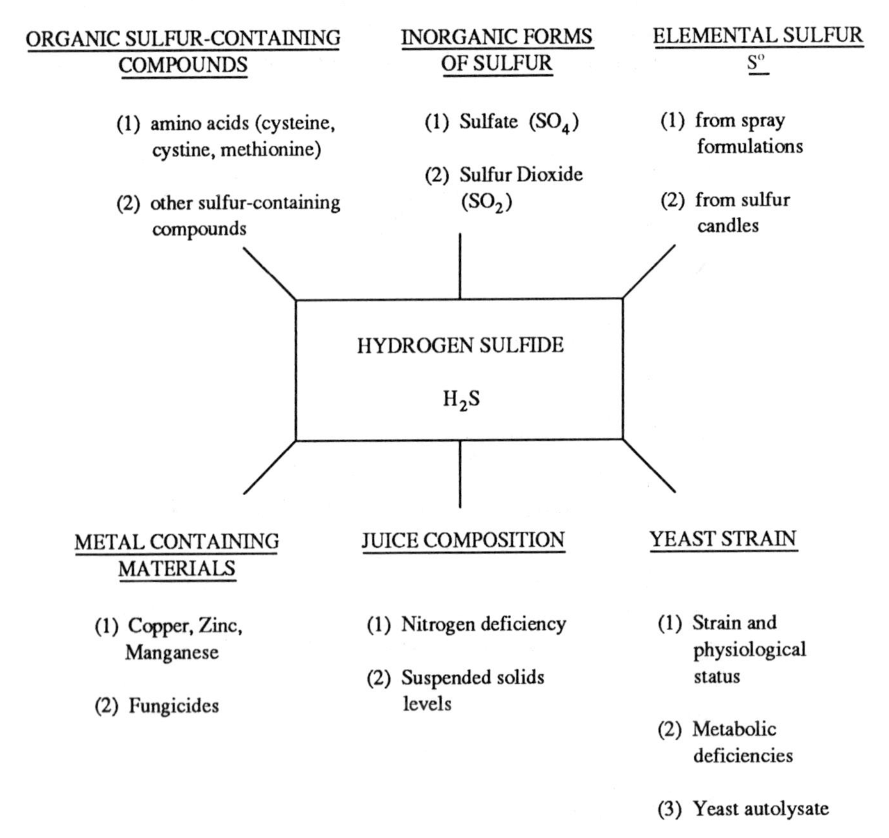

Fig. 10-4. Factors effecting H_2S formation (adapted from Eschenbruch 1983).

more of ascorbic acid are used, and such additions usually are made several days prior to the addition of copper. The reader is cautioned that sulfur dioxide analysis by Ripper titration cannot be performed accurately in wines containing ascorbic acid, as the latter also reacts with the iodine titrant. In any case, copper should not be added until the fermentation is complete and the yeast titer reduced by racking, filtration, and so on. Yeast cells will bind copper ions to cell surfaces and may reduce reactivity with H_2S. (See Figure 10-4 for a review of H_2S formation.)

Lastly, the addition of SO_2 to wines may reduce H_2S levels. The addition results in an SO_2-induced oxidation of H_2S to yield elemental sulfur, which, after precipitation, may be removed by centrifugation or filtration.

$$SO_2 + 2H_2S \longrightarrow 3S^{0} + 2H_2O \qquad (10\text{-}3)$$

Procedure 10-1. Sensory Separation of Hydrogen Sulfide and Mercaptans

The following procedure is a modification of the method suggested by Brenner et al. 1954). The technique involves separation of the two groups of volatiles based upon sensory examination before and after treatment with copper- and cadmium-containing compounds. In the case of copper [as copper (II) sulfate] addition, reaction with both sulfides and mercaptans yields nonvolatile and relatively odorless products. Cadmium salts, by comparison, are chemically reactive toward hydrogen sulfide, but have little effect on mercaptans.

The following technique provides a quick, relatively simple and inexpensive diagnostic examination of wines suspected of having hydrogen sulfide and/or mercaptan problems.

Note: Wines evaluated using this technique *should not be tasted!* This procedure is designed for evaluation by smell only.

I. Reagents (See Appendix II)
 1% wt/vol Copper (II) sulfate (as $CuSO_4 \cdot 5H_2O$)
 1% wt/vol Cadmium sulfate
 10% wt/vol Ascorbic acid

II. Procedure
 1. Fill three clean glasses with 50 mL of the wine to be evaluated.
 2. Mark glass 1 "Control."
 3. Mark glass 2 "Copper." To the 50 mL of wine, add 1 mL of the copper sulfate solution.
 4. Mark glass 3 "Cadmium." Add 1 mL of 1% cadmium sulfate to the 50 mL of wine.
 5. Mix the contents of each glass thoroughly.
 6. Smell the control sample, noting the presence and relative intensity of offensive odors. In like manner, evaluate glasses 2 and 3. Attention should be directed to the decrease or absence of a reduced odor.

Table 10-1. Interpretation of results from sensory evaluation of cadmium- and copper-treated wines.

Glass 1 (Control)	Glass 2 (Copper)	Glass 3 (Cadmium)	Interpretation
Presence of offensive odor	Odor is gone	Odor is gone	H_2S present
	Odor is gone	No change	Mercaptans
	Odor is gone	Odor has lessened but is not gone	Both H_2S and mercaptans
	No change	No change	**

**The objectionable odor does not stem from either mercaptan or H_2S sources, *or* methyl mercaptan has been oxidized to dimethyl disulfide.

7. Interpret the results of the test according to Table 10-1.
8. If odor evaluation results indicate that neither H_2S nor merceptans are responsible for the offensive odor (last line, Table 10-1), further evaluation for the presence of dimethyl disulfide should be carried out (see steps 9–12 below).
9. Transfer 50 mL of untreated wine sample to a fresh, clean glass.
10. Add 0.5 mL of the 10% ascorbic acid solution. Mix and let it stand for several minutes.
11. Add 1 mL of copper (II) sulfate solution. Mix and evaluate for changes in the "nose."
12. If the offensive odor is less apparent than in glass 2, the presence of dimethyl disulfide is suspected. Dimethyl disulfide does not react with copper. It must be reduced back to the reactive species methyl mercaptan (see text).

III. Supplemental Notes

1. The control sample should not change appreciably over the short course of this test. However, if the objectionable character is diminished as a result of oxidative changes, this would suggest a possible cellaring technique for correction.

Chapter 11

Other Preservatives: Sorbic Acid, Benzoic Acid, and Dimethyldicarbonate

SORBIC ACID

Sorbic acid is a short-chained unsaturated fatty acid widely used in the wine and food industry as a preservative. Since this compound is not readily soluble, it usually is sold as the soluble salt, potassium sorbate. Effective primarily as a fungistat, sorbic acid is added to sweet wines (those with a reducing sugar content of more than 1.5 g/L) immediately prior to bottling.

To be effective against microorganisms, sorbic acid must be incorporated into the cell. The antimicrobial activity appears to reside in the undissociated molecule. The relative amounts present in the undissociated form versus the ionized (negatively charged) form will largely determine ease of movement across the negatively charged cell membrane of bacteria and yeast. The pK_a of sorbic acid is pH 4.7 (the pK_a of an acid being the pH at which the ratio of undissociated to ionized forms is 1.0). Further decreases in pH bring about decreasing amounts of the ionized form and proportionate increases in the concentration of the active undissociated form. Thus, as the pH is lowered, the activity of the antimicrobial form of the acid increases, and the concentration needed for inhibition decreases. Once incorporated into the cell, there is evidence that the chemical may be operative against the dehydrogenase enzyme system of yeasts and molds (Desrosier and Desrosier 1977), interfering with oxidative assimilation of carbon.

The additive does not kill yeast, but, if properly utilized, it inhibits yeast growth. It has little inhibitory activity toward acetic and lactic acid bacteria. The activity of sorbic acid is directed toward controlling the growth of fermentative yeast in wine. Its effectiveness is dependent upon several parameters, including:

Wine pH
Concentration of free (molecular) sulfur dioxide
Percent alcohol by volume
Concentration of sorbic acid
Viable yeast cell concentration

As the percentage of undissociated sorbic acid decreases with increases in pH, the effectiveness of the compound is correspondingly reduced. Peynaud (1984) estimates that the functional activity of the preservative decreases by approximately one-half with a pH increase from 3.1 to 3.5. Table 11-1 compares the presence of active (undissociated) sorbic and benzoic acids as a function of pH.

Sulfur dioxide appears to operate in a synergistic manner with sorbic acid. Ough and Ingraham (1960) report inhibition of yeast growth in wine with additions of sorbic acid at 80 mg/L coupled with free sulfur dioxide levels of 30 mg/L. In a study of stability in white and red wines, Auerbach (1959) reported that inhibition in whites was achieved at a combination of sorbic acid at 75 mg/L and sulfur dioxide at 200 mg/L. In table wines, under routine conditions of bottling, 150 to 200 mg/L sorbic acid usually is utilized.

Although a generally effective inhibitor of fermentative yeasts, sorbic acid has little inhibitory activity toward lactic acid bacteria present in wine. The relative inactivity of sorbic acid toward wine bacteria may be due to formation of an addition compound between sorbic acid and sulfur dioxide (Heintze 1976). Reaction between components is thought to occur in a 1 : 1 ratio. This may result in removal of enough free sulfur dioxide from the system to support the growth of wine bacteria. In a review of the subject, Schmidt (1987) points to the importance of adequate levels of sulfur dioxide in addition to sorbate for inhibition of sorbic acid–resistant lactic bacteria.

Table 11-1. Concentration (%) of active sorbic and benzoic acids as a function of pH.

pH	% Undissociated sorbic acid	% Undissociated benzoic acid
2.8	99.0	96.2
2.9	98.9	95.2
3.0	98.4	94.0
3.1	98.0	92.6
3.2	97.5	90.9
3.3	96.9	88.8
3.4	96.2	86.3
3.5	95.2	83.4
3.6	94.1	80.0
3.7	92.6	76.0
3.8	90.9	71.4
3.9	88.8	66.6
4.0	86.3	61.5
4.2	79.9	50.0
4.3	76.0	44.3
4.8	50.0	20.1

Source: Gump (1987).

Alcohol content is known to affect the activity of sorbic acid, and thus is also a consideration in addition levels. As expected, sweet wines of higher alcohol content require less sorbic acid for stabilization than those of lower alcohol levels. Table 11-2 compares wine alcohol content with the concentration of sorbic acid needed for inhibition of wine yeast. The concentration of yeast at the time of addition also is important. Peynaud (1984) states that cell concentrations must be less than 100 CFU/mL (colony forming units/mL).

Internationally, legally permissible levels of sorbic acid in wine vary from zero to several hundred mg/L. Currently, the legal limit for sorbic acid in table wines produced in the United States is 300 mg/L, whereas for wine coolers, addition levels may not exceed 1000 mg/L. In the case of coolers, benzoic acid, as potassium or sodium benzoate is a permissible additive in addition to sorbate. When sorbate is used in combination with benzoates, the total (combined) concentration of both additives may not exceed 1000 mg/L. However, unfavorable sensory responses to this concentration of either or both compounds would preclude their use at this level.

Attempts to establish threshold levels for sorbic acid in wines have resulted in a relatively wide range of values. Ough and Ingraham (1960) reported the threshold levels in wine to be as low as 50 mg/L, with an average threshold of 135 mg/L, among trained panelists. By comparison, Tromp and Agenbach (1981) found the threshold for the additive to be 300 to 400 mg/L. Results of triangle difference analyses revealed that a significant portion of judges were able to detect sorbic acid at levels of 150 mg/L, and those correctly identifying the "odd sample" consistently showed a preference for untreated controls (Fugelsang 1974). Practically speaking, however, the potential presence of objectionable properties must be weighed against the possibility of refermentation in bottling of nonfortified sweet wines.

As sorbic acid may be detected by sensory evaluation at use levels, winemakers contemplating the addition should use only pure, fresh potassium sorbate; the use of old, yellow oxidized sorbate may result in detection at even lower levels than those discussed above. The oxidation of sorbate-treated wines also may lower the sensory threshold for the additive. Therefore, wines treated with the additive should be stored under low-oxygen conditions, with the SO_2 level kept high enough to prevent incipient lactic bacterial growth. For these

Table 11-2. Inhibitory interaction of alcohol content and levels of sorbic acid in wine.

Alcohol content (% v/v)	Sorbic acid (mg/L)
10–11	150
12	100
14	50

Ough and Ingraham, 1960.

reasons, it is recommended that the preservative be added to sweet wine just prior to bottling. The use of potassium sorbate in wines destined for long-term aging is not recommended. De Rosa et al. (1983) reported development of an odor attributed to ethyl sorbate in Charmat-style sparkling wines after one year of storage. The compound was reported to have a celery–pineapple-like odor.

An undesirable odor due to lactic bacterial decomposition of sorbic acid occasionally may be noted in sorbated wines (Edinger and Splittstoesser 1986). This odor, suggestive of geraniums, is commonly referred to as a "geranium tone." Wurding et al. (1974) studied geranium tone and concluded that it resulted from the formation of 2,4-hexadien-1-ol and its lactate esters and acetates. It now appears that a significant portion of the off-character results from formation of an ether, 2-ethoxyhexa-3,5-diene, by rearrangement of hexadienol (Crowell and Guymon 1975):

$$\text{SORBIC ACID (2,4-hexadienoic acid)} \xrightarrow[\text{ETOH}]{\text{pH} < 4.0} \text{ETHYL SORBATE} \quad (11\text{-}1a)$$

$$\downarrow \text{growth of lactic acid bacteria}$$

$$\text{SORBYL ALCOHOL} \xrightarrow{\text{rearrangement}} \text{ETHYL SORBYL ETHER} \quad (11\text{-}1b)$$

$$\downarrow$$

$$\text{3,5-hexadien -2-ol} \xrightarrow[\text{ETOH}]{H^+} \text{2,ETHOXYHEXA-3,5-DIENE} \quad (11\text{-}1c)$$

Probably the most succesful way to deal with the problem of geranium tone is to use blending. Unfortunately, the blend ratios of sound to defective wines are generally high, usually on the order of at least 11 : 1.

Because of the poor solubility of sorbic acid (see above), a more soluble salt, potassium sorbate, is used in the wine industry. Normally, the salt is hydrated in wine or water prior to addition. Uniform mixing of the preservative with the wine may be accomplished using a Y on the suction side of a positive displacement pump or proportioning pump, and so on.

In using potassium sorbate, it is necessary to correct for the differences in molecular weights between the acid and its salt. This relationship is:

$$\begin{array}{c}\text{Wt of}\\ \text{potassium sorbate}\\ \text{required/L}\end{array} = \frac{\text{Molecular wt salt}}{\text{Molecular wt acid}} \times \begin{array}{c}\text{Addition level}\\ \text{of sorbic acid}\\ (\text{mg/L})\end{array} \quad (11\text{-}2)$$

Therefore to obtain a sorbic acid level of 125 mg/L it is necessary to add 167.5 mg/L of potassium sorbate

$$\frac{150.22}{112.13} \times 125 \text{ mg/L} = 167.5 \text{ mg/L of potassium sorbate}$$

DETERMINATION OF SORBIC ACID CONTENT

Several analytical procedures commonly are used for determining the sorbic acid content of wines. These are included in Procedures 11-1, 2, and 5. Because the acid is completely steam-distillable (greater than 99%), preliminary separation by distillation in a Cash volatile acid still is an effective method for removing sorbic acid from the wine matrix. The acid (molecular) form is extractable in isooctane, so preliminary separation from the wine matrix also may be accomplished in this fashion. Once isolated, sorbic acid may be derivatized to form a colored compound, which can be measured spectrophotometrically with a simple visible colorimeter. Molecular sorbic acid, because of its conjugation, also can absorb radiation in the UV region of the spectrum, and this property is the basis for several other common analytical procedures. Finally, samples containing sorbic acid can be diluted and analyzed using HPLC techniques (See Procedure 11.5).

BENZOIC ACID

The use of benzoic acid as the potassium or sodium salt has been restricted to the food industry. The product is not an approved additive for table wines, but recently its use in wine coolers was approved. Although the legal limit of addition is 1000 mg/L, concerns relative to sensory properties dictate much lower use levels. Further, when it is used in combination with sorbic acid, the combined addition levels may not exceed 1000 mg/L.

Like sorbic acid, benzoic acid is antimicrobially active in the un-ionized form. Thus at pH values below its pK_a of 4.2, the percentage of the active form increases significantly, and the amounts needed for inhibition decrease correspondingly (Table 11-1).

In addition to the effects of sulfur dioxide and alcohol in combination with benzoates, carbonation also acts to enhance antimicrobial activity. Schmidt (1987) reports linear decreases in the concentration of sodium benzoate required for inhibition with increases in carbonation levels in soft drinks.

In wine coolers, sodium or potassium benzoate frequently is used in combination with potassium sorbate and, as discussed earlier, sulfur dioxide. The combination of sorbate and benzoate provides the needed level of antimicrobial activity at a concentration that generally is not objectionable.

DIMETHYL DICARBONATE (DMDC)

Dimethyl dicarbonate (DMDC) is the methyl analog of diethyl dicarbonate (DEDC). The latter compound also has been known as diethylpyrocarbonate. Both DMDC and DEDC are active in inhibition of yeast at relatively low levels of addition (less than 250 mg/L). The mechanism of inhibition probably is similar for both. That is, each brings about hydrolysis of yeast glyceraldehyde-3-phosphate dehydrogenase and alcohol dehydrogenase, yielding inactive forms of the enzymes. Within the respective enzymes, the site of reaction appears to be the imidiazole ring of histidine (Fig. 11-1). The carbethoxy- or carbmethoxyhistidine produced from the reaction cannot bond with adjacent cysteine; so the native configuration of the enzyme is disrupted, resulting in inactivation.

DEDC use in wine originally was approved in 1960, and it was revoked in 1972. The removal of DEDC was based on reports that under certain conditions ethyl carbamate (urethane) may be produced upon decomposition of the sterilant in wine. Subsequently, it has been proposed that the dimethyl analog replace DEDC. Because this compound does not contain the ethyl components needed in the formation of urethane, it is believed to be an acceptable alternative.

In both DEDC and DMDC, the temperature and alcohol levels appear to act synergistically with the sterilant. That is, higher temperature and alcohol levels reduce the time needed for kill. In early studies on the effect of DEDC, Splittstoesser and Wilkinson (1973) reported the killing rate for *Saccharomyces cerevisiae* to be 10 to 100 times faster at 40°C when compared to the death rate at 20°C. Using DMDC, Porter and Ough (1982) reported that the yeast cell count was reduced to zero (from 380 cells/mL) in ten minutes upon the addition of 100 mg/L of the DMDC to wines (10% alcohol and 2% reducing sugar) held

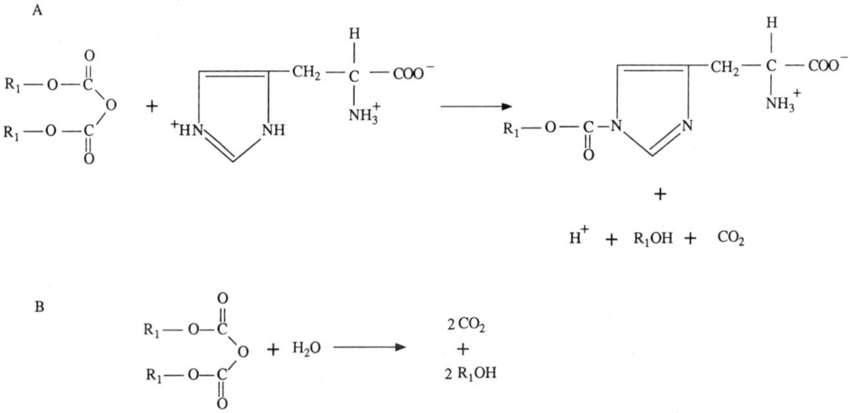

Fig. 11-1. (A) Reaction of DMDC and/or DEDC with histidine. (B) Hydrolysis of DMDC to carbon dioxide and methanol Porter and Ough (1982).

at 30°C. A review of their paper shows that viable cells were recovered only when the reaction was carried out at 20°C.

DMDC is in the latter stages of review by governmental agencies, and approval of its use in wines (not coolers) is pending. When it is approved, maximum use levels will probably be set at 250 mg/L (Gahagan 1989).

Procedure 11-1. Colorimetric Analysis for Sorbic Acid

As developed by Caputi et al. (1974), this procedure follows the initial separation of sorbic acid by steam distillation with its oxidation to the intermediate malonaldehyde. Reaction of the latter with thiobarbituric acid yields a highly colored condensation product that is measured at 530 nm:

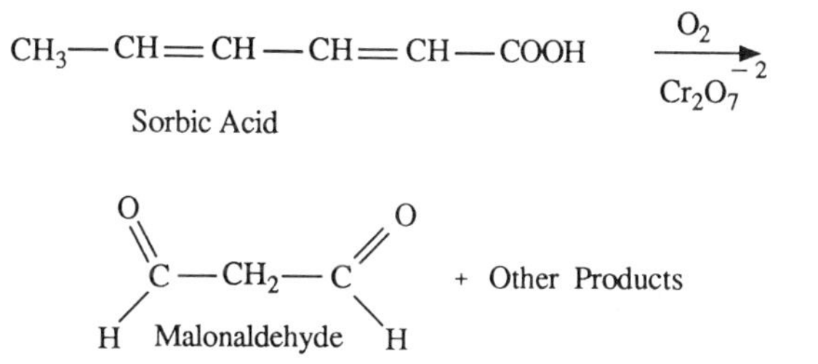

$$\underset{\text{Sorbic Acid}}{CH_3{-}CH{=}CH{-}CH{=}CH{-}COOH} \xrightarrow[Cr_2O_7^{-2}]{O_2} \underset{\text{Malonaldehyde}}{OHC{-}CH_2{-}CHO} + \text{Other Products} \tag{11-3a}$$

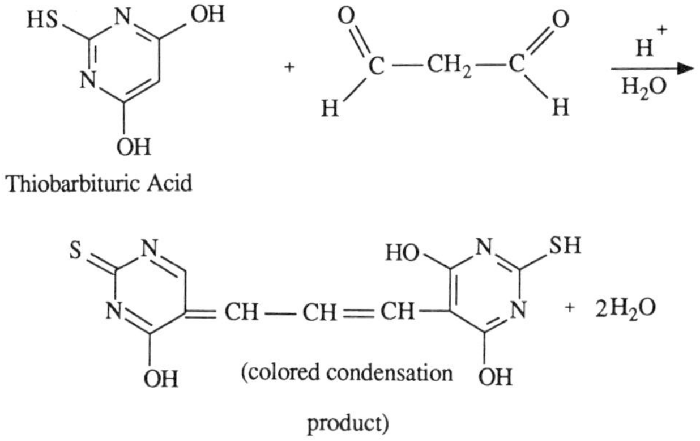

$$\underset{\text{Thiobarbituric Acid}}{\text{HS, OH, OH-pyrimidine}} + OHC{-}CH_2{-}CHO \xrightarrow[H_2O]{H^+} \text{(colored condensation product)} + 2H_2O \tag{11-3b}$$

I. Equipment
   Cash volatile acid assembly
   Bausch & Lomb Spectronic 20
   Boiling water bath
   500-mL volumetric flasks
   100-mL volumetric flasks

Volumetric pipettes (2, 5, 10, 15 mL)
Several 15-mL Pyrex test tubes

II. Reagents (See Appendix II)
0.5% Thiobarbituric acid solution
Potassium dichromate solution
0.3 *N* H_2SO_4
100 mg/L Stock sorbic acid solution
Distilled water

III. Procedure

(a) Standard curve preparation

1. Prepare a series of sorbic acid calibration standards from the 100 mg/L stock solution according to Table 11-3.
2. Volumetrically transfer 2 mL of each standard to separate test tubes. Blank preparation consists of 2 mL of distilled water.

(b) Preparation of wine sample

3. Volumetrically transfer 2 mL of wine to a receiving funnel on the Cash still. Rinse the funnel several times with distilled water to ensure complete transfer.
4. Collect approximately 190 mL of distillate in a 200-mL volumetric flask. Dilute the distillate to volume with distilled water.
5. Volumetrically pipette 2 mL of distillate into a test tube.

(c) Analysis of samples (to be done simultaneously with standards)

6. To each test tube, add 1 mL of 0.3 *N* H_2SO_4 and 1 mL of potassium dichromate solution.
7. Heat the tubes in the boiling water bath for exactly 5 minutes.
8. Remove the tubes and cool them in an ice water bath.
9. Pipette 2 mL of 0.5% thiobarbituric acid reagent into each tube. Return the tubes to the boiling water bath for 10 minutes.
10. Cool the tubes to room temperature. Determine the absorbance of each standard at 532 nm, and plot it against the appropriate concentration (in mg sorbic acid/L).
11. Using the standard curve, determine the concentration of the unknown. Remember, this value must be multiplied by 200 mL/2

Table 11-3. Preparation of standard sorbic acid solutions from 100-mg/L stock.

Volume of sorbic acid standard (mL)	Final concentration (mg/L) in 100-mL volume
1	1
2	2
3	3

mL (the reciprocal of the original 2 mL to 200 mL dilution factor) to yield a result in mg/L.

Procedure 11-2. Ultraviolet Analysis for Sorbic Acid

The distillate produced in Procedure 11-1 can be analyzed directly at 260 nm without forming a colored derivative. Owing to its state of conjugation, sorbic acid absorbs light in the ultraviolet area of the spectrum. Using this property to their advantage, Melnick and Luckman (1954) developed this rapid procedure. It is necessary to acidify the distillate to prevent the sorbic acid from dissociating into its nonabsorbing ionic form.

I. Equipment
   UV spectrophotometer
II. Reagents: (See Appendix II)
   Sorbic acid stock (100 mg/L)
   0.1 *N* HCl
   Distilled water
III. Procedure
   (a) Standard curve preparation
      1. Prepare sorbic acid calibration standards from 100 mg/L stock according to Table 11-3. Before bringing dilutions to volume, add 0.5 mL of 0.1 *N* HCl.
   (b) Preparation of Wine Samples
      2. Volumetrically transfer 2 mL of wine to the receiving funnel of the Cash volatile acid still.
      3. Collect approximately 190 mL of distillate in a 200-mL volumetric flask containing 0.5 mL of 0.1 *N* HCl. Dilute the distillate to volume with distilled water.
   (c) Colorimetric analysis
      4. Determine the absorbance of each standard and sample at 260 nm (using a blank of 0.5 mL HCl in 199.5 mL of distilled water). Plot absorbances of standards against the respective concentrations (in mg/L sorbic acid).
      5. Compare absorbances of samples to the standard curve. Multiply the values from the curve by 200 mL/2 mL (the reciprocal of the dilution factor); express the result in units of mg/L.

Procedure 11-3. Alternative Procedure for Sorbic Acid

Because of potential problems of component loss during preliminary distillation, Zimelis and Somers (1978) have developed a direct extraction procedure. In this procedure, 0.25-mL aliquots of wine sample are extracted in isooctane

(2,2,4-trimethyl pentane), and absorbance is measured at 255 nm relative to an isooctane blank.

I. Equipment
   Micropipette
   UV spectrophotometer
   Quartz cuvettes
   Screw-cap test tubes (25 mL) and glass boiling beads (2 mm)

II. Reagents (See Appendix II)
   Sorbic acid stock (500 mg/L)
   Isooctane (2,2,4-trimethyl pentane)
   Phosphoric acid
   Ethanol (12% vol/vol)

III. Procedure
1. Prepare a series of calibration standards from the 500-mg/L sorbic acid standard according to the scheme presented in Table 11-4. Bring solutions to the mark with 12% (vol/vol) ETOH reagent.
2. Using a micropipette, transfer 0.25 mL of each standard solution to a screw-cap test tube.
3. Similarly transfer 0.25-mL aliquots of wine samples into screw-cap test tubes.
4. Add 0.1 mL H_3PO_4, 10.0 mL isooctane, and two or three glass beads to each standard and sample.
5. Replace the caps, and shake the tubes vigorously for 2 minutes.
6. Allow 3 to 5 minutes for phase separation.
7. Determine the absorbance of each standard and sample against an isooctane blank at 255 nm.
8. Prepare a standard curve by plotting the absorbance of standards vs. concentration (mg/L).
9. Compare absorbance values of samples to the standard curve. *Note:* No dilution factor is involved in the expression of final results.

Table 11-4. Preparation of sorbic acid calibration standards from 500-mg/L stock solution.

Volume of sorbic acid stock (mL)	Final concentration (mg/L) in 100 mL volume
5	25
10	50
15	75
20	100
30	150

IV. Supplemental Notes

1. Isooctane was chosen as the solvent of preference because of its limited extraction properties for wine phenolics. The latter compounds absorb in the ultraviolet. Isooctane has the further advantage of reduced volatility.
2. Phosphoric acid is used to enhance the separation between aqueous and organic phases.
3. Zimelis and Somers (1978) reported sorbic acid recovery ranging from 98.7% to 101.9% of that added to wine samples. For red wines, these workers report a background error (that attributed to interferences) of 5 mg/L, and for white wines, 3 mg/L.
4. Prepared standards are more concentrated than those used in Cash still methods, as the 2 mL sample to 200 mL distillate dilution step (Cash still distillation) is not part of *this* procedure.

Procedure 11-4. Spectrophotometric Determination of Benzoic Acid

Benzoic acid can be determined by measuring its absorbance in the UV region at 272 nm. The acid first is extracted into ether to separate it from other absorbing substances in the sample. Then the absorbance spectrum of the acid is scanned between 265 and 280 nm, and a base line constructed from the two absorbance minimums. Net absorbance is measured from this baseline to the peak of the absorbance maximum (at ~272 nm) and related to the concentration of benzoic acid. As presented, the method is applicable to samples containing from 200 to 1000 mg/L benzoic acid.

I. Equipment

UV spectrometer
Stoppered silica cuvettes
500-mL seperatory funnel
10-mL pipette

II. Reagents (See Appendix II)

Benzoic acid stock standard (1000 mg/L)
Diethyl ether
Saturated NaCl in distilled water
HCl (1 + 999)

III. Procedure

1. Prepare calibration standards of benzoic acid in ether from the 1000-mg/L stock solution according to Table 11-5. Keep the standards tightly stoppered to prevent evaporation.
2. Scan the absorbance spectrum of the 50-mg/L standard over the range 265 to 280 nm. Note the wavelength of maximum absorbance (ca. 272 nm) and wavelengths of adjacent absorbance minimums (ca. 267.5 and 276.5 nm).

Table 11-5. Preparation of benzoic acid calibration standards from 1000-mg/L stock solution.*

Volume of sorbic acid stock (mL)	Final concentration (mg/L) in 100 mL volume
2	20
5	50
7.5	75
10	100

*Dilute to volume with diethyl ether.

3. Measure absorbance values for the remaining standards at the maximum and minimum wavelengths determined in step 2.
4. For each standard, determine the average *minimum* absorbance value. Subtract this calculated average from the maximum absorbance value, and plot this corrected absorbance against the concentration.
5. Transfer 20 mL of the wine sample to a separatory funnel, dilute it to 200 mL with saturated NaCl solution, and acidify the solution (litmus paper) with HCl.
6. Extract the acidified sample sequentially with 70-, 50-, 40-, and 30-mL portions of ether, mixing well to ensure complete extraction. Drain and discard the aqueous phase.
7. Wash the combined ether extracts sequentially with 50-, 40-, and 30-mL portions of HCl, and discard the aqueous washings.
8. Dilute the combined ether extracts to 200 mL with additional ether. Determine the absorbance of the sample in a stoppered cuvette at the above-determined minimum and maximum wavelengths.
9. Average two minimum absorbance values as before, and subtract the result from the maximum absorbance value. Determine the concentration of benzoic acid from the standard curve.
10. Multiply the concentration value by 200 mL/20 mL (reciprocal dilution factor) to obtain the benzoic acid concentration in the original sample. Multiply this value by 1.18 (MW sodium benzoate/MW benzoic acid) to obtain the equivalent concentration of sodium benzoate.

IV. Supplemental Note

1. In some instances an additional purification step may be necessary. Further extract the washed ether extracts from step 7 sequentially with 50, 40, 30, and 20 mL of 0.1% NH_4OH, and discard the ether phase. Neutralize the combined NH_4OH extracts with HCl, and add 1 mL excess. Extract this acidified solution sequentially with 70, 50, 40, and 30 mL of ether. Combine the ether extracts, and proceed with procedural step 8.

Procedure 11-5. HPLC Determination of Benzoic Acid and Sorbic Acid

Benzoic acid, as well as sorbic acid, can be determined by using the technique of high performance liquid chromatography (Fig. 11-2). A commercial organic acid column with an acetonitrile-modified sulfuric acid mobile phase will resolve these two acids. Detection is best accomplished at 233 nm.

I. Equipment
 HPLC unit with variable-wavelength detector
 Organic acid HPLC column and appropriate guard column
 Millipore filters (0.45 μm) or equivalent
 2-mL pipette
 50-mL volumetric flasks

II. Reagents (See Appendix II)
 Sodium benzoate standard stock solution
 Potassium sorbate standard stock solution
 Mobile phase: 85% 0.01 N H_2SO_4, 15% acetonitrile
 Distilled water

III. Procedure
 1. Prepare calibration standards of benzoic and sorbic acids at concentrations of 2, 5, 10, and 20 mg/L from benzoic and sorbic acid stock standards (Table 11-6). Use the mobile phase as the solvent for these standards.
 2. Transfer a 2.0-mL sample to a 50-mL volumetric flask. Dilute it to the mark with the mobile phase.

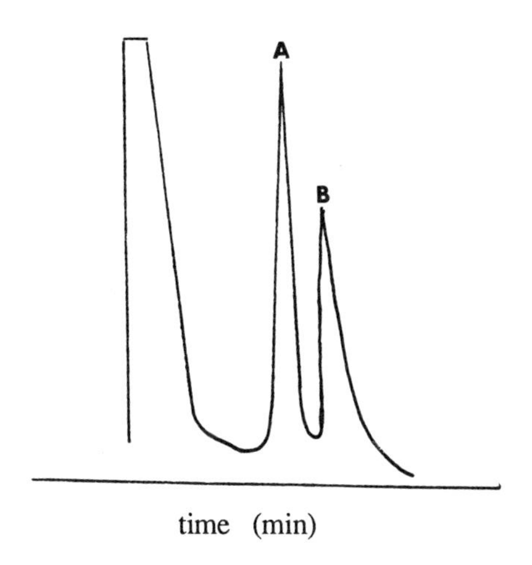

Fig. 11-2. HPLC analysis of sorbic and benzoic acids using "fast acid analysis" column (Bio-Rad) and a mobile phase consisting of 85% (0.01 N) H_2SO_4: 15% acetonitrile. The column is operated at 65°C with a mobile phase flow rate of 1.0 mL/min. Detection is by UV absorbance at 233 nm.

Table 11-6. Preparation of benzoic acid and sorbic acid calibration standards from 100-mg / L stock solutions.

Volume of benzoic acid or sorbic acid stock (mL)	Final concentration (mg/L) in 100 mL volume*
2.0	2.0
5.0	5.0
10.0	10.0
20.0	20.0

*Dilute to volume with the prepared mobile phase.

3. Set up the chromatograph with a 1.0-mL/min flow rate of a freshly sparged (air-free) mobile phase. Bring the column temperature to 65°C, and set the detector to 233 nm. Allow the unit to stabilize.
4. Prior to the injection of standards and samples, filter them through 0.45-μm membrane filter disks.
5. Inject 10 μL of each standard onto the chromatograph. Note the retention times and response factors. If desired, construct a calibration curve.
6. Inject 10 μL of each diluted sample. Record the values provided by the data processing unit, or by reference to the prepared calibration curve. If the sample peak response is greater than that of the most concentrated standard, dilute it by an appropriate amount, and chromatograph again.
7. Calculate the concentrations of the original samples by multiplying the chromatograph result by the appropriate reciprocal dilution factor (e.g., 50 mL/2 mL).

IV. Supplemental Note
1. An alternative procedure for the HPLC analysis of benzoic acid can be found in *Official Methods* (AOAC), section 12.018–12.021 (14th ed., 1984). This procedure utilizes a C_{18} reverse-phase column and a fixed-wavelength (254 nm) UV detector.

Procedure 11-6. Analysis of Dimethyldicarbonate by Gas Chromatography

Dimethyldicarbonate in wine reacts to form breakdown products of methanol, carbon dioxide, and ethyl methyl carbonate. The ethyl methyl carbonate residues in wine can be determined using conventional extraction and gas chromatographic techniques. An estimate of the original addition level of dimethyldicarbonate can be calculated from the ethyl methyl carbonate and percent ethanol analytical results (Stafford and Ough 1976).

I. Equipment
Gas chromatograph equipped with a flame ionization detector and a stainless steel column: 10 feet × 1/8 inch od packed with 10% Carbowax 20M on 100/120 mesh Chromosorb G. Instrument conditions include: helium

carrier flow rate = 20 mL/min, oven temperature = 160°C, injector temperature = 150°C, and detector temperature = 200°C.
10-μL Hamilton syringe
Strip-chart recorder or electronic integrator
1-L and 250-mL volumetric flasks

II. Reagents (See Appendix II)
Carbon disulfide (washed with fuming nitric acid)
Ethyl methyl carbonate (EMC) standard solution (500 mg/L)
Diethyl carbonate (DEC) internal standard (500 mg/L)
95% (vol/vol) ethanol
Distilled water

III. Procedures

1. Prepare calibration standards of ethyl methyl carbonate (EMC) in 12% (vol/vol) ethanol at concentrations of 1, 2.5, 5, and 7.5 mg/L from EMC standard solution (Table 11-7).
2. Add 100 mL of the wine sample *or* working standard plus 1 mL DEC internal standard solution plus 1 mL absolute ethanol to a 250-mL separatory funnel.
3. Extract the sample with 20 mL carbon disulfide for 10 minutes, using a gentle rolling motion. Allow the layers to separate.
4. Transfer 10 mL of the organic phase to a 15-mL conical-bottom centrifuge tube. Centrifuge for 2 to 3 minutes in a clinical centrifuge.
5. Transfer 5 mL of the clarified extract to a glass vial, and store it in a freezer until analysis.
6. Inject 5 μL of the sample or working standard extract into the gas chromatographic column for methyl carbonate (EMC) analysis. Approximate retention times should be 6 to 7 minutes for EMC and 8 minutes for DEC.
7. Determine the peak height ratios for EMC/DMC. Using ratios from injected working standards, prepare a standard curve of EMC/DEC ratios vs. EMC concentrations (in mg/L). Compare the peak height

Table 11-7. Preparation of ethyl methyl carbonate calibration standards from 500-mg/L EMC stock standard.

Volume EMC stock standard (mL)	Final concentration (mg/L) in 1000-mL volume*
2.0	1.0
5.0	2.5
10.0	5.0
15.0	7.5

*Dilute to volume with 12% (v/v) ethanol.

ratios for wine sample extracts to the standard curve to determine their EMC concentrations.

8. Calculate the original dimethyldicarbonate addition level in the wine samples from the following equation:

$$\text{DMDC level (mg/L)} = \frac{\text{EMC conc (mg/L)} \times 100}{(0.39) \times \text{\%ETOH (v/v)}} \quad (11\text{-}4)$$

IV. Supplemental Notes

1. The factor 0.39 comes from linear regression data of the average amount of EMC produced vs. the % (vol/vol) ethanol in various wine samples (both red and white wines) when 100 mg/L of DMDC was added to those wines. The factor for white wines alone is 0.41; the factor for red wines alone is 0.37.
2. At the time of writing (1989) DMDC is in its latter stages of review by governmental agencies, and its approval for use in standard wines (not coolers) is pending. When approved, maximum use levels probably will be set at 250 mg/L (Gahagan 1989).

Chapter 12

Wine Microbiology

This chapter is intended to highlight those groups of microorganisms that are important to the winemaker. The procedures that follow are designed to familiarize the reader with routine laboratory monitoring and diagnostic techniques.

MOLD AND MOLD COMPLEXES ASSOCIATED WITH GRAPES

Molds are multicellular, filamentous fungi. Most molds are saprophytes, requiring a readily available source of exogenous nutrient. Macroscopically, mold growth appears as a mass of threadlike elements, the mycelium. Typically, a mold mycelium is composed of branched elements called hyphae. Depending upon the organism in question, these may or may not be septate.

The usual mode of reproduction among molds is asexual, that is, without the union of sex cells. In most species, asexual elements—the spores—are borne on specialized aerial hyphae termed sporangiophores or conidiophores, depending upon the mold in question. Sporangiospores and condiospores may be differentiated in that the former normally are protected within the body of a sporangium until dehiscence, whereas the latter are borne unprotected at the apices of condiophores.

Molds are important in winemaking because, when conditions permit, they may cause damage to the fruit in the vineyard, thus detrimentally affecting ultimate wine palatability. Common organisms involved in vineyard spoilage include *Penicillium, Aspergillus, Mucor, Rhizopus,* and *Botrytis.*

Mold growth on grapes is considered undesirable except for the association of *Botrytis cinerea* in sweet wine production. In some cases, mold growth and associated degradation of fruit stimulate the activity of wild yeast and acetic acid bacteria. Growth of the latter groups, in turn, produces ethanol (from yeast metabolism) and its resultant bacterially produced oxidation product, acetic acid. Populations of acetic acid bacteria (chiefly *Acetobacter aceti* and *A. pasteurianus)* have been reported to reach levels of near 10^6 cells/g (Joyeux et al. 1984 a,b). In addition, gluconate and ketogluconate may be produced from bacterial oxidation of glucose (De Ley and Schell 1959). Ethyl acetate often is produced as a result of yeast growth on deteriorating fruit. Additionally, keto-acids (e.g., pyruvate) and acetaldehyde, present as intermediates in the fermentation of sugars, and dihydroxyacetone, from the bacterial oxidation of glycerol, serve as important binding substrates for subsequent sulfur dioxide additions. Thus, the potential for the oxidation of musts and the uncontrolled growth of other undesired microorganisms is greatly increased by mold growth.

Botrytis cinerea

The gray mold, *Botrytis cinerea*, generally is found on grapes wherever they are grown. However, under conditions of high temperature and low relative humidity, little or no mold growth occurs. In central European countries, the eastern United States, and other areas where the relative humidity is generally high, *B. cinerea* is common. The time needed for infection varies with temperature (see Table 12-1). As seen in Table 12-2, relative humidity also is related to the extent of infection.

When appropriate environmental conditions occur, spore germination proceeds, and the infection phase begins. Penetration of the berry and growth of the fungus loosen the skin of the berry. After infection, depending upon climate conditions, two different effects may be seen on the fruit:

1. In rainy weather, the infected grapes do not lose water, and the percentage of sugar remains nearly the same and may even decrease. Secondary infection by other microbes (yeast and acetic acid bacteria) may follow. Cold and wet conditions lead to excessive *Botrytis* growth without drying, a condition called *pourriture grise* in the French literature. Such conditions are favorable for the growth of *Penicillium* sp. and Aspergillus sp., as well as other fungi and yeast that may overgrow *Botrytis*. The latter condition is referred to as vulgar rot (*pourriture vulgaire*). Attack upon and breakdown of the grape integument provides a substrate for the growth of native (''wild'') yeasts and acetic acid bacteria present in the vineyard. This continued growth of wild yeast and acetic acid bacteria is called *pourriture acide*, or sour rot.

2. In contrast to the above cases, *Botrytis* infection followed by warm sunny

Table 12-1. Approximate moisture period required for *Botrytis* bunch rot infection at various temperatures.

Temperature	Approximate Moisture Period*
30°C (86°F)	35 hr.
26.5°C (80°F)	22 hr
22.5°C (72°F)	15 hr
15.5°C (60°F)	18 hr.
10°C (50°F)	30 hr.

*Free water (or humidity >92%) at the fruit surfaces. Source: Sall et al., 1982.

Table 12-2. Effect of relative humidity on the proportion of table grapes infected by *Botrytis cinerea*.

R.H. of incubation (%)	Percentage of *Botrytis*-infected berries
79	18
86	24
92	62
95	88
97	100

Source: Nelson 1951.

windy weather causes berries to lose moisture (dehydrate) by evaporation. With dehydration, shriveling occurs, and the sugar levels increase. This condition is known as *pourriture noble*, or noble rot. Growth of the mold and associated acetic acid bacteria consumes a portion of the grape sugar. However, in cases where climatic conditions favor dehydration, the utilization of sugar is countered by increases in sugar concentration due to loss of water. The relatively high sugar content in dehydrated fruit, in addition to mold metabolites, has led to the commercial use of *Botrytis*-infected fruit in the production of sweet white wine in certain areas of Europe and the United States.

As seen in Table 12-3, the concentration of tartaric and malic acids decreases in *Botrytis*-infected fruit. But again, because of the effects of dehydration (resulting in the concentration of nonvolatiles), infected fruit actually may have a higher titritable acidity than uninfected fruit.

Gluconic acid is a product of the metabolic activities of both mold and acetic acid bacteria. In the case of molds, it is produced by the enzyme-mediated oxidation of glucose:

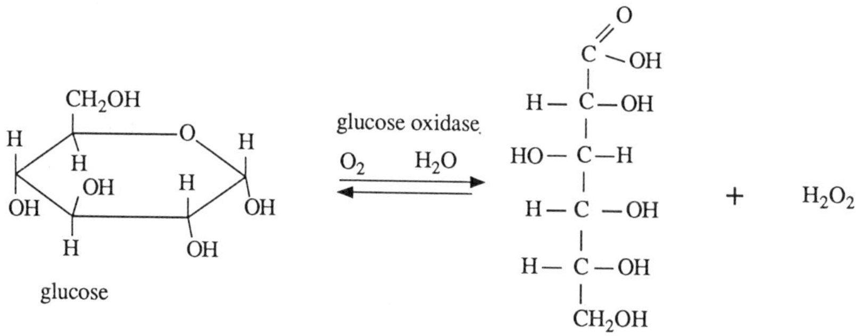

$$\text{glucose} \xrightarrow[\text{glucose oxidase}]{O_2 \qquad H_2O} \text{HOOC–CH(OH)–CH(OH)... : } \begin{array}{c} \text{C(=O)OH} \\ | \\ \text{H–C–OH} \\ | \\ \text{HO–C–H} \\ | \\ \text{H–C–OH} \\ | \\ \text{H–C–OH} \\ | \\ CH_2OH \end{array} + H_2O_2$$

The glucose oxidase that catalyzes the reaction is produced by a number of molds, including *Aspergillus, Botrytis cineria,* and *Penicillium*. In wines, gluconic acid is an indicator of possible fruit degradation. In addition, it may well have a sensory effect. Ribereau-Gayon (1988) reports gluconic acid levels in

Table 12-3. Malic and tartaric acid content in mold-infected fruit (expressed as a percentage of total acidity).

Fruit condition/type of mold infection	% Malic and tartaric acids compared to sound fruit
Healthy fruit	85–90
Noble rot	55–70
Sour rot	30–40

Ribereau-Gayon 1988

'clean' fruit to be trace and in wine from clean fruit to be near 0.5 g/L, whereas in wine produced from fruit infected with *B. cineria* levels increase to 1 to 5 g/L. In the case of sour rot or vulgar rot, where bacterial growth occurs along with the mold growth, levels may reach 5 g/L. The methodology for the analysis of gluconic acid in must is presented in Chapter 4 (procedure 4.3).

Botrytis cineria also produces significant amounts of polyols, of which glycerol is quantitatively the most important. Quantities produced may be as high as 20 g/L (Ribereau-Gayon et al. 1980). Glycerol may be metabolized prior to harvest. Ribereau-Gayon (1988) suggests that the ratio of glycerol to gluconic acid may be used as an indicator for judging the "quality" of the rot. That is, higher ratios indicate the growth of true noble rot, whereas lower ratios suggest sour rot. *Gluconobacter oxydans* and *Acetobacter aceti* are known to carry out the oxidation of glycerol to dihydroxyacetone (see Chapter 2). In this regard, *Gluconobacter oxydans* is a more important consumer of glycerol than *Acetobacter* and is used in the commercial production of dihydroxyacetone. Thus, the presence of these species in significant numbers may impact the final concentrations of glycerol in the must.

In addition to reductions in juice volume in infected fruit, rots have other detrimental effects, generally centering around oxidative changes in grape consituents. The growth of *Botrytis* results in the formation of laccase by the mold. Like grape polyphenoloxidase enzymes, laccase is an oxidative enzyme capable of oxidizing grape phenolic substrates. However, laccase has greater activity than polyphenoloxidase and is relatively resistant to the action of sulfur dioxide.

As described in Chapter 1, the analysis of laccase can be used as an indicator of the extent to which *Botrytis*-degradation has occurred. Major substrates affected by the growth of *Botrytis* are anthocyanins (hence color), as well as other grape and wine phenols and aroma components. In cases where sour rot is present along with *Botrytis*, the oxidation of phenolics may proceed to an even greater extent.

Noble rot often imparts a "honey" or "roasted" component to the aromatic character of wine. Nishimura and Masuda (1983) suggest that the characteristic aroma of wines made from *Botrytis*-infected fruit is due to 3-hydroxy-4, 5-dimethyl-2-furanone.

In the case of sour rot or that of sour rot and *Botrytis* (sour bunch rot), several aroma modifications may occur. Generally, fruitiness disappears with the formation of unpleasant odors that are described as "phenol" and "iodine."

Botrytis also destroys monoterpenes, which are, in part, responsible for the aroma of muscat varieties, Johannesburg Riesling, and Gewürztraminer. Reductions of up to 90% of the primary monoterpenes such as nerol, geraniol, and linalool are reported to occur with the growth of the mold (Ribereau-Gayon 1988). An analysis procedure for grape monoterpenes is given in proceedure 1-6. *Botrytis* also secretes an esterase that hydrolyzes yeast esters.

One of the greatest impacts of *Botrytis* growth is the formation of polysaccharides that create problems in clarification of the wine. Pectins are hydrolyzed by mold-produced polygalacturonase, with the formation of beta-1,3- and 1,6-glucans. Commercially, several glucanases are available to minimize clarification problems arising from this source.

Musts produced from *Botrytis*-infected fruit may be difficult to ferment. Fermentation problems may be due to osmotic pressure in high-sugar musts as well as the formation of antimicrobial compounds during the growth of the fungus. A heteropolysaccharide that possesses some antibiotic activity has been identified in must produced from this fungus.

Penicillium

In its early stages of growth, *Penicillium* appears as a white cottony colony. Later, conidiospores are produced. After early fall rains, *Penicillium* may develop in cracks in grape berries, making the fruit unfit for wine production. *Penicillium* sp. are known to be cold-weather molds, showing abundant growth between 15 and 24°C. *Penicillium expansum* growth on grapes can produce a mold flavor and odor.

Aspergillus

Aspergillus niger is a common vineyard fungus, found growing on damaged fruit. Initially, white mycelial elements of the mold resemble *Penicillium*. With conidial formation, colony growth becomes black. The growth of this mold is more abundant in warm climates than in cooler areas. It can metabolize grape sugars to produce citric acid, resulting in an increase in the acid content of the juice.

Mold Growth in the Winery

Molds may be present in the winery, where they can grow on the surface of cooperage as well as on walls and other porous surfaces. In the case of improperly maintained cooperage, growth may impart a moldy odor and flavor to the wine. Because molds are aerobic organisms and rather intolerant of alcohol, mold growth is not observed in wine. Although it is not uncommon to observe mold fragments in microscopic examination of a wine, they most likely originate from mold growth on grapes processing equipment or from storage vessels. Thus, microorganisms that can grow directly in wine are limited to alcohol-tolerant lactic and acetic acid bacteria and several species of yeast.

YEASTS

Yeast Morphology

Microscopically, most yeasts appear oval to ellipsoidal in shape. Like molds, they usually reproduce asexually; but in this case budding is the usual mode of replication (see Photomicrograph No. 12-1 on p. 286). Some yeasts may undergo a sexual cycle of replication; the mother cell produces ascospores that subsequently germinate, producing new vegetative cells.

Yeast taxonomy is complicated somewhat by the fact that not all yeast species have an identified sexual phase. In those cases, the asexual yeasts are classified among the "imperfects," within the class Deuteromycotina (formerly Fungi Imperfecti). In many instances, however, the imperfect form may closely resemble another ascospore-forming yeast, in which case a dual taxonomy is used. In isolations where ascospores are not observed, the yeast is given the "imperfect" designation (e.g., *Brettanomyces*), whereas another isolate may appear identical except that it produces ascospores on special media. In the above example, the microbiologist would use the "perfect" taxon (i.e., *Dekkera*). Thus, in the laboratory identification of yeast, the mode of reproduction (both asexual and sexual) varies and is an important criterion in identification. However, to the winemaker this discussion is largely academic because both "species" may behave similarly in wine.

In addition to cell morphology (shape), which varies with the age of the culture and type of media on which it is grown, the position of buds provides initial information of value in identification. With the exception of a few genera, buds normally arise on the shoulders and axial areas of vegetative cells. In some genera where the vegetative cell has no apparent axis, buds may arise at any place on the surface. This type of budding is called multilateral budding.

Still other species exhibit restricted bud formation. For example, budding limited to the axial areas of the cell is referred to as polar budding. This type of reproduction is characteristic of the apiculate or lemon-shaped yeast.

As compared with budding, asexual reproduction in *Schizosaccharomyces* sp. occurs via fission in a manner that is visually similar to that seen in bacteria. In these cases, cell division occurs by formation of cross walls or septa without constriction of the original cell wall. When septum formation is complete, the newly formed cells separate.

Several species of yeast typically grow as films at the surface of wines exposed to air. These films, originating from the budding of single cells, proliferate rapidly when growing in low-alchohol wines stored under oxidative conditions. In this case, however, film formation results from incomplete separation upon completion of budding. Although the mother and daughter cells are functionally separate, they tend to adhere and bud further. The result of this type of budding is a spreading type of growth called a "pseudomycelium." The latter differs from a true mycelium in that there is no cytoplasmic continuity between cells comprising the pseudomycelium. This type of growth is seen in the several species of *Candida* and *Pichia* that may occasionally be encountered in wine spoilage.

Wild Yeast Fermentations

A variety of yeasts have been found on grapes and in grape juice and wine. Rankine (1972) has reported finding 18 genera and 147 species of yeasts occurring on grapes and grape products. These range from non- or weakly fermentative species to yeasts that have been cultivated for commercial wine production. Historically, the use of selected commercially available yeast cultures was not permitted in some wine-producing countries owing to the belief that suitable strains of *Saccharomyces* and other yeast flora are present in vineyards.

Typically, unsulfited fermentations utilizing native yeast flora (without inoculation of wine yeast cultures) undergo a succession of populations. Generally, weakly fermentative species that are present on the fruit in greater numbers than *Saccharomyces* strains appear first and initiate fermentative activity. However, these strains are rather susceptible to increasing alcohol levels, and are overgrown by *Saccharomyces*, which then finishes the fermentation.

Research has been conducted on the use of mixed-culture fermentations, especially the use of *Kloeckera apiculata* and *Hansenula anomala* in conjunction with cultures of *Saccharomyces* (Florenzano 1949; Rankine 1955). Although some winemakers believe that mixed cultures contribute to overall complexity and produce wines of superior quality, the outcome of such work is unclear, and more research is needed in this area.

Attributes of Wine Yeasts

Because of the species diversity noted above, most winemakers generally employ pure yeast cultures to carry out alcoholic fermentation. The advantages of selected pure yeast cultures over fermentations by wild strains include: (1) rapid onset of active fermentation and a predictable rate of sugar-to-alcohol conversion, (2) complete utilization of fermentable sugars, and (3) improved alcohol tolerance. Additionally, winemakers may consider other properties, such as: (4) SO_2 production, (5) formation of volatile acidity, acetaldehyde, and pyruvate, (6) tendency to foam, and, upon completion of fermentation, (7) clarification attributes (i.e., flocculation). Other important considerations include (8) the production of larger amounts of desirable by-products (ester formation) with (9) minimal formation of detrimental by-products such as H_2S, and so forth. Methode champenoise producers place very specific demands on yeasts used to carry out the secondary fermentation, such as pressure tolerance, cold tolerance, agglutinating (flocculation) ability, limited lipid production, and so on.

In certain special applications, yeasts also may be selected for their ability to produce or utilize oxidative components (Guerzoni et al. 1981). For example, in flor sherry production, selection is made for the production of acetaldehyde and certain other metabolites important in the sensory profile of that particular wine type. Sherry-flors comprise several strains of *Saccharomyces*. In addition, *Pichia*, *Hansenula*, and *Candida* sp. may be involved in producing the sherry character.

Other specific applications of yeast strains include reduction of total acid content of must and wine. Such reductions involve the utilization of malic acid in the formation of carbon dioxide and ethanol. Fermentative yeasts differ in their ability to reduce the overall acidity of grape must. Rankine (1966a) demonstrated that certain strains of *Saccharomyces* could reduce the L-malic acid content of grape must up to 45% or more. By comparison, selected strains of *Schizosaccharomyces* are capable of decomposing anaerobically all the malic acid present in grape musts (Peynaud et al. 1964).

Claims that various strains of the same species impart notable sensory differences to wine may be overstated. Although some winemakers may note sensory differences in young wines fermented with different yeast strains, these differences are frequently short-lived. However, different species (and strains of the same species) of yeast generally do produce varying concentrations of the same end products (Rankine 1972). For example, *Saccharomyces oviformis* (currently *Sacch. cerevisiae*) has been reported to produce up to 19% (vol/vol) alcohol (Minarik and Nagyova 1966). Thus this organism may find greater application in the production of dry wine types, whereas it may be of less value in sweet wine fermentations. Breeding or hybridization of yeast strains for specific purposes may well be an important future development in the wine industry.

Yeasts of Importance to the Winemaker

Saccharomyces sp.

Strains of this genus represent the most commonly used wine yeast. Morphologically, this group of yeast appears spherical to ellipsoidal in shape with approximate dimensions of 8 × 7 μm, depending upon the species and the growth medium. All species of *Saccharomyces* reproduce by multilateral budding as well as by occasional ascospore formation. Many strains of *Sacch. cerevisiae* can ferment hexose sugars to yield alcohol levels of up to 16% (vol/vol). Use of supplemental or "syruped" fermentations (addition of small amounts of sugar at intervals) may enable the yeast to produce as much as 18%, or more, alcohol.

Among winemakers, there is a good deal of discussion as to which strain might be advisable for any particular varietal and/or style of wine. It may be difficult to discern differences in wine that can be directly attributed to yeast activity versus those differences that result from the composition of the juice, the processing protocol, or the combined effects of each. However, when the variables are adequately controlled, one generally can distinguish (at least in young wines) differences between wines fermented by different strains of yeasts. These differences typically lie in the areas of ester formation and retention of fresh fruit and varietal character of the grape. As a generality, those strains that are capable of fermenting well at less than 15°C tend to produce a fruitier style of wine, whereas wines made with those that typically ferment at higher temperature develop more subtle and complex characters.

Spoilage Yeasts

Saccharomyces may grow oxidatively as a film on the surface of wine exposed to air. One strain, *Sacch. beticus*, is an important species in the flor of Spanish-style sherries. Other strains, including those used in table wine fermentations, may grow on the surface of improperly stored wines where production of aldehydes is less desirable.

On occasion, *Saccharomyces* may be an important contaminant in sweet ($>$1.5 g/L reducing sugar) wine. Unless the wine is bottled under sterile conditions and/or chemically stabilized, it may support yeast growth. In one report, refermentation was observed in wine with an initial inoculation of as few as 5 yeast/mL at bottling (Meidell 1987). The extent of yeast growth in bottled wine may range from a marginally perceivable sediment on the bottom of the bottle to turbidity and sufficient gas to push corks.

Clouding and sediment in wine may result from the growth of a wide range of yeast in addition to *Saccharomyces*. Spoilage may result from either early involvement in fermentation or secondary oxidative deterioration due to surface growth in improperly stored wines. Oxidative (aerobic) alcohol-utilizing spoilage yeast may grow as films on low-alcohol wines exposed to the air. These yeasts include the genera *Pichia* and *Candida*, among others. These organisms usually do not grow in wines containing more than 12% (vol/vol) alcohol, and most are inhibited by lower alcohol concentrations. Members of the group exhibit mainly oxidative metabolism, so their activity may be controlled by maintaining wine in fully topped containers. Low pH and adequate levels of molecular sulfur dioxide also aid in control. The growth of these organisms in wine generally imparts off or detrimental characteristics to the product.

Candida sp.

The nonsporulating (''imperfect'') genus *Candida* includes a variety of yeast from varied sources linked by similar morphology and the absence of observed ascospore formation in the life cycle. In wine, *Candida* typically grows as a surface yeast, producing a chalky white film on the surface of low-alcohol wines. *Candida* is stongly oxidative, using ethyl alcohol in addition to wine acids as carbon sources. Most species of *Candida* are fast growers. Microscopically, these yeasts appear as long cylindrical cells of approximate dimensions 3–10 × 2–4 μm. Asexual reproduction is by budding. The incomplete separation of mother and daughter cells leads to formation of an extensive pseudomycelium, as well as a true mycelium in some instances.

Pichia sp.

Another example of a film yeast, *Pichia* may be found growing at the surface of wines with alcohol contents of up to 13% (vol/vol). However, the majority of species are inhibited by alcohol levels of near 10% (vol/vol). Growth on the wine surface appears as a very heavy, balloon-like, chalky film. The continued growth of this organism imparts an aldehydic character to the wine. The organism is sensitive to free SO_2 but tolerant of bound SO_2 up to 122 mg/L in dry white wine at 10% alcohol, pH 3.35 (Rankine 1966b).

Ethiraj and Suresh (1988) isolated *Pichia membranaefaciens* from spolied mango juice stabilized with sodium benzoate at concentrations ranging as high as 1500 mg/L. Although lower pH was observed to enhance the antimicrobial activity of the benzoate, growth was observed in one lot (pH 3.5, benzoate concentration of 500 mg/L) after eight days.

Actively growing cells appear as short ellipsoids to cylindrical-shaped rods. Reproduction is usually by multilateral budding, which leads to development of an extensive pseudomycelium. On lab media, the organism may produce ascospores, which, when present, usually are round to hat-shaped and range in number from one to four per ascus.

Hansenula sp.

Hansenula is widely distributed and may be represented in significant numbers in fermentations originating from native grape flora. The organism is both fermentative and oxidative, being capable of partially fermenting juice as well as pellicle formation in low-alcohol wines. Members of the genus reproduce by asexual budding, forming buds of varying shapes. Formation of a pseudomycelium and, in some cases, a true mycelium has been observed. Sexual reproduction yields two to four hat-shaped (Saturn-shaped) ascospores per ascus. Members of the group resemble *Pichia* in appearance but are often more strongly fermentative than *Pichia*.

The most common member of the genus found in wine is *Hansenula anomala*, which microscopically, appears as an oval to oblong-shaped cell of approximate dimensions 2.5 × 5–10 μm. *Hansenula* sp. may form large amounts of volatile esters, particularly ethyl acetate. Amerine et al. (1972) reported levels of acetate in wine attributed to *Hansenula* to range from 2.52 to 4.39 g/L. When they are grown in mixed cultures with *Saccharomyces cerevisiae*, formation of these volatile esters has been reported to add flavor and bouquet to the wine (Wahab et al. 1949; Saller 1957). At the wine surface, *H. anomala* forms a wrinkled chalky-white pellicle.

Brettanomyces sp. (sporulating form: *Dekkera*)

These fermentative yeasts occasionally are found in wine spoilage, particularly red wines. The growth of *Brettanomyces* in wine may result in the formation of large amounts of acetic, isobutyric, and isovaleric acids, yielding a pungent, acetate-like character as well as imparting a metallic taste and a "mousy" or "horsey" odor to the product.

Microscopically, the organism resembles *Saccharomyces cerevisiae*, although it is usually somewhat smaller than the latter, with vegetative cells ranging from ogival to ovoidal. Asexual reproduction is by polar budding. In older cultures maintained on laboratory media, many cells may appear elongate. Occasionally, incomplete separation of daughter cells may result in chains of the organism being present. Frequently up to one-third of the cells within a microscopic field are ogival or arch-shaped.

Brettanomyces may be differentiated from many other yeasts genera on the basis of their relative resistance to actidione (cycloheximide), growing in concentrations of 100 mg/L (see Appendix 3). The latter characteristic is used effectively in laboratory isolation techniques from mixed cultures. Reportedly, several strains of *Brettanomyces* are relatively resistant to the action of SO_2 at normally used levels (Schanderl and Staudenmayer 1964).

Brettanomyces principally is a problem in red wines, although its activity has been noted in whites. The outcome of *Brettanomyces* growth appears to vary with the extent to which the organism has grown in the wine. Red wines degraded with *Brettanomyces* often have objectionably high levels of ethyl acetate and a characteristic metallic taste. *Brettanomyces* is particularly difficult to deal with because its presence may go unnoticed until growth has reached the point where the wine is permanently tainted. Industry-wide, this yeast may represent one of the most important spoilage problems. It appears to spread from winery to winery through contaminated wine and/or equipment. Used barrels are particularly important sources of the organism.

Kloeckera (perfect form: *Hanseniaspora*)

These small, lemon-shaped, fermentative organisms occur in abundance during the early stages of natural fermentations. The apiculate shape characteristic of this group arises from repeated budding at both poles.

Yeast Growth Dynamics and Starter Propagation

Following yeast population dynamics from the beginning through completion of fermentation, one can identify the classic stages of the growth cycle.

Lag Phase

Upon transfer of culture to fresh medium, one observes a short period in which cell activity (i.e., budding) appears to be low. Depending upon environmental parameters (e.g., temperature) and the condition of the culture used for the transfer or, in the case of active dry yeast, the technique of rehydration, this period may last from one to several hours.

Eventually, cells begin to bud, producing daughters, and the culture enters an accelerated lag phase. Yeast metabolic activity in this stage of growth is directed toward the production of biomass.

Exponential Growth Phase

Yeast populations growing exponentially replicate at regular intervals. Properly prepared, yeast starters take advantage of the addition of yeasts in this phase of growth. As fermentation is associated with the generation of a considerable amount of heat, the winemaker may wish to control the rate of activity during exponential growth by cooling. Other parameters affecting growth and fermentation rates include pH, osmotic stress (high-sugar musts), and limiting the availability of nutrient.

Stationary Phase

Eventually, populations reach maximum numbers, and, with the presence of increasing concentrations of alcohol and decreasing nutrient levels, budding slows. Deficiency in available nitrogen (FAN) may initiate hydrogen sulfide formation (see Chapters 10 and 16). According to Monk (1986), yeast activity in this stage of growth is directed toward the fermentation of sugar. Up to 50%, or more, of the initial sugar present in must is utilized in stationary growth phase metabolism.

Prolonged stationary growth activity under near-anaerobic conditions affects yeast cell membrane function. Specifically, oxygen-requiring steroid and fatty acid syntheses are blocked. Additionally, toxic effects of ethanol combined with nutrient depletion play a major role in yeast viability. At this stage, the potential for stuck fermentations is a concern.

Death Phase

Assuming that some other factor, such as nitrogen depletion, did not cause the fermentation to stop prematurely, the death phase coincides with the utilization of the available carbon source (fermentable sugar).

Prior to addition to must, yeast must be expanded so the viable cell numbers, upon addition, are on the order of $2\text{–}5 \times 10^6$ cells/ml. In addition to ensuring relatively high cell numbers compared to competitive species originating on the fruit, the proper preparation of yeast starters ensures that, upon inoculation, the majority of the yeast in the must is at, or entering, the exponential phase of growth. This is accomplished by manipulation of the culture from the lag phase to the rapid growth phase, and may include addition of exogenous nutrient as well as mechanical agitation or sparging of the starter to incorporate oxygen needed for growth.

Propagation from Slants

Prior to inoculation into must, laboratory cultures must be expanded, without contamination, to a starter volume that represents from 1 to 3% (vol/vol) of the total volume to which it will be added. Subsequently, the starter tank must be maintained, without contamination, until transfer. Logistically, several loopfuls of the culture initally are transferred from the slant to 100 mL sterile-filtered or heat-sterilized juice (100°C for 15–20 minutes). In the case of sterile-filtered juice, no sulfur dioxide should be present at inoculation, hydrogen ion concentration (pH) should be raised to above 3.5, and nitrogen levels should be supplemented with 2 g/L diammonium phosphate. Once this culture is actively growing, its volume is expanded twofold using sterile juice. From this point, the culture is expanded in a series of five-to-tenfold increases to yield the final yeast inoculum. This volume is used to prepare the larger yeast starters for direct addition to must. The juice used for expansion of laboratory starters also should be sterile.

After addition of the yeast inoculum to the starter tank volume, 24 to 72 hours may be required before cell numbers reach the level at which the starter can be added to must. Ideally, lab personnel should follow the growth microscopically, noting viability as well as percentage of budding cells. Cell viability usually is determined using dyes such as methylene blue, Ponceau S, or Walford's stain. Methods for the use of these dyes and interpretation of the results are presented in Procedure 12-2 and Appendix III. In actively growing starters, budding cells ideally compromise 60 to 80% of the total cell number.

Starter tanks should be constructed of polished stainless steel and provided with sterile air locks so that the tank may be operated aseptically. According to Monk (1986), tanks should have a 3:1 height-to-diameter ratio with a maximum operating capacity of 70% of the total tank volume. Starter tanks must be maintained and monitored constantly to ensure that ample nutrient in the form of nitrogen and carbohydrate is available, and that the levels of alcohol do not become inhibitory. Transfer of the starter culture should occur before sugar levels are fully depleted. Nutrient exhaustion may force growing yeast into a secondary lag phase, significantly reducing cell viability and delaying activity upon transfer.

The practice of holding back a portion of the starter to serve as inoculum for fresh starter tanks, while convenient from the production standpoint, eventually may cause problems. Over the course of the season, this practice may introduce contamination from other yeast and bacteria, thereby making the starter a source for initial introduction of spoilage organisms. Further, prolonged use of this technique may result in depletion of critical cell components. One specific example is reduction in the levels of membrane steroids when the cell is continually grown in near-anaerobic conditions. Such stressed cells may not yield vigorous starters, so that fermentations originating from these starters prove troublesome.

Yeast Culture Maintenance

Maintaining stock cultures throughout the year may cause problems. Cultures held on agar slants at refrigeration temperatures tend to dehydrate and, unless transferred regularly, lose viability. Several techniques are available to extend the longevity and viability of a culture; they have included overlaying the culture with mineral oil as well as other techniques (Pilone 1979). However, even with regular transfers, cultures maintained on agar media for extended periods of time tend to lose, and, in some cases, gain, physiological traits that previously distinguished them from other strains. In some cases, these changes may impact the properties of the yeast to a point where future fermentations may be affected. The problems of strain maintenance and physiological integrity may be overcome by periodically using the yeast to carry out small-scale fermentations and then reisolating it. Needless to say, this requires considerable effort and expense.

Wine-Active Dry Yeast

Before use, active dry yeast must be rehydrated. In some instances, vintners simply spread the yeast pellets over the surface or mix them into the must. However, such a practice results in incomplete rehydration and low viable cell numbers. The correct protocol calls for prior rehydration in warm water.

Physically, rehydration in warm (40°C) water or must tends to disperse the yeast to a much greater extent than is seen when yeast is added to cold must. In the latter case, pellets tend to remain intact and, in fact, clump, resulting in reduced nutrient and oxygen availability and incorporation into the cells. The final outcome of this practice is reduction in the numbers of active yeast cells (Kraus et al. 1981).

Physiologically, the effect of rehydration at cold versus warm temperatures seems to center on the integrity of the cell membrane. Hydration at low temperature results in the reactivation of the membrane to a point where essential soluble cytoplasmic components escape before membrane function can be reestablished. Viability is greatly reduced at low rehydration temperatures. Cone (1987) points out that hydration at 15°C (60°F) may result in up to 50 to 60% cell death. By comparison, activation at warm (40°C) temperatures quickly establishes the membrane barrier and function before any leaching effect can occur. When yeasts are rehydrated in water, they should not be allowed to remain there for more than 30 minutes before transfer to must; longer hydration periods tend to reduce viability. Kraus et al. (1981) reported the optimal temperature range for rehydration to be between 37.8 to 43.0°C. Within this temperature range, 15 minutes was sufficient for rehydration and maximum activity.

Normally, winemakers inoculate fermentations at levels of $2\text{–}5 \times 10^6$ cells/mL. Potentially troublesome fermentations, such as late harvest and/or *Botrytis*-infected grapes, may receive slightly higher addition levels. In the case of starters produced from active dry yeast, budding cell numbers usually are much lower than those inoculated from starter tanks. In this case, budding cells may comprise 2 to 3% of the viable cell count.

Yeast Oxygen Requirements

Yeast need a certain amount of oxygen for activity. Specifically, the synthesis of cell membrane components (fatty acids and steroids) requires molecular oxygen. Prolonged growth under semianaerobic conditions reduces the steroid (ergosterol) content of the cell membrane, thereby making the yeast more sensitive to the effects of alcohol. In the final stages of yeast manufacture, producers generally manipulate growth conditions so that upon rehydration the yeast will have sufficient steroids for four to five generations. Beyond this, budding diminishes important membrane components, resulting in stress and the potential for stuck fermentations. When levels of unsaturated fatty acids and steroids reach approximately one-quarter of that found in aerobically grown cells, budding stops, and there is a decline in protein synthesis (Giudici and Guerzoni, 1982). The situation is readily reversible. Upon provision of oxygen, protein synthesis starts rapidly, along with synthesis of unsaturated fatty acids and ergosterol (Gordon and Stewart 1972). Yeast suppliers recommend oxygen incorporation into starter tanks during the course of expanding volumes. Cone (1987) points out that actively growing yeast starters take up oxygen rapidly before must oxidation occurs. Because simple mechanical agitation or pumping over may not increase the oxygen concentration to the levels needed, it is suggested that compressed air be bubbled directly into the starter tank. The use of pure oxygen should be avoided because of its toxicity (Fugelsang, 1987).

Actively growing yeast cultures should not be transferred to chilled must. Temperature shock may reduce the viable cell count by up to 60% and, in general, result in slow growth and increase the potential for stuck fermentations. Further, sudden drops in temperature may result in hydrogen sulfide production (Monk 1986). When fermentation at lower temperature is needed, starters should be acclimated to growth at the lower temperature prior to inoculation.

When purchasing active dry yeast, a vintner should plan to order only enough to fulfill the needs for that season. Dehydrated yeast, even when stored under ideal conditions, will lose viability over the course of a year. Simpson and Tracey (1986) report decreased viability in active dried yeast stored at several temperatures for 6 months. Generally at storage temperature greater than 4°C, viability was decreased by 15–25% compared with storage at 4°C. In those strains reported, storage at less than 4°C did not improve viability and, in fact resulted in lower viability in several cases. Cone (1987) confirms that at 4°C activity may be reduced by 5% per year, and at 20°C (68°F), activity drops by nearly 20% per year.

Killer Yeasts

Yeast producers now are marketing killer yeast strains. Killer yeasts are species that can kill sensitive members of their own species and, in some cases, those of other sensitive species and genera. Killer toxins have been isolated from several strains of *Saccharomyces cerevisiae* used in wine as well as brewing and sake fermentations. The killer characteristic also is found in yeast from other genera, including *Candida*, *Pichia*, *Hansenula*, and *Torulopsis*.

Killer yeasts secrete a plasmid-coded protein (3.8 KB, 2.5×10^6) toxin (Tipper and Bostian 1984). In the case of *Saccharomyces cerevisiae*, killer protein is encoded by a double-stranded RNA virus-like plasmid (Tipper and Bostian 1984). Killing results from the production of an exocellular protein toxin in the case of *Sacch. cerevisiae* or or glycoprotein in certain other cases. Killing probably occurs in at least two steps. Initially, the toxin appears to bind to the 1,6-β-D-glucan receptor component of the cell wall of sensitive strains. Then, after binding with the receptor site, the toxin interacts directly with protein components of the cell membrane, disrupting the normal state of electrochemical ion gradients.

During the course of fermentation, maximum killer activity appears to be somewhat delayed, by up to three days, decreasing gradually from that point. Shimizu et al. (1985) report continued activity until the end of fermentation.

Toxin stability also diminishes somewhat with increases in temperature. At 20°C, workers reported a slight reduction in activity even after ten days. At 30°C, toxin activity decreased rapidly. The activity of killer toxin in strains studied appears to be stable over the pH range from 2.9 to 4.9. Above pH 5.4, stability decreases rapidly. Optimal activity in cases reported is at pH values above 4.0.

A recent study of stuck fermentations in South Africa points to the activity of killer yeasts toward sensitive strains. In their report, Van Vuuren and Wingfield (1986) cite 5 different stuck fermentations in which 90% of original yeast inoculum (killer-sensitive) were dead. The majority of those viable cells remaining were killer strains.

Several commercially available strains of killer yeast are available for use in the wine industry. These include Prise de Mousse, as well as Montrachet 1107 (Lalvin), Campagne 111B, and Wadenswil 27 (Lalvin).

Immobilized Yeasts

Recent studies have shown the potential for using immobilized yeast in table and sparkling wine (Methode Champenoise) fermentations (Fumi et al. 1988). In these cases, yeasts are immobilized within calcium alginate carrier beads. The beads may be packed into a column of suitable dimensions and a volume of juice pumped over the immobilized yeast. In the case of Methode Champenoise fermentations, the yeast-impregnated beads are placed directly into the bottle with the cuvee. By use of this technique, riddling and disgorging are facilitated.

The utility of individual strains may become more important as the field of genetic engineering develops. There is the potential for development of strains for specific applications as well as inclusion of the fully expressed gene(s) for the malolactic fermentation within selected strains. Initial work in this area has been successful in implanting the appropriate genes, but the level of expression is far less than is needed for it to be of practical importance to the winemaker.

WINE BACTERIA

Acetobacter and *Gluconobacter*

Bacteria of these genera utilize ethanol (and glucose) aerobically in the formation of acetic acid. Thus the growth of acetic acid bacteria in wine as well as in must and on deteriorating grapes may significantly increase the volatile acid content. (For details regarding the analysis of volatile acidity in wine, the reader is referred to Chapter 4.) Often much of the spoiled character associated with the growth of acetic acid bacteria may be the result of ethyl acetate formation. The presence of this ester in wine is perceived as the odor of fingernail-polish remover. Peynaud (1984) points out that ethyl acetate may be sensorily detectable at levels of 160 to 180 mg/L and thus, may be a more reliable indicator of spoilage than using the acetic acid content alone. As evidence, he points to the cellar observation that a wine with lower volatile acidity (as determined by steam distillation) may be judged totally defective, whereas another wine with higher VA but lower concentrations of ethyl acetate may appear less defective, or even sound.

The taxonomy of acetic acid bacteria has undergone considerable revision. The genus *Acetobacter*, as described in *Bergeys Manual of Determinative Microbiology*, 8th edition (Buchanan and Gibbons, 1974), comprised three species with nine subspecies:

1. *A. aceti* (subsp. *aceti*, *orleanensis*, *xylinum*, *liquefaciens*)
2. *A. pasteurianus* (subsp. *pasteurianus*, *lovaniensis*, *estunensis*)
3. *A. peroxydans*

This represented a revision from Bergey's seventh edition (1957), which recognized seven species. The consolidation results from acceptance of the genus *Gluconobacter* with a single species (*G. oxydans*) and four subspecies: *G. oxydans* (subsp. *oxydans*, *industrius*, *suboxydans*, *melanogenes*). In this version, *Gluconobacter* was placed within the family Pseudomonaceae, whereas *Acetobacter* was not given family status.

In Bergey's ninth edition, both *Acetobacter* and *Gluconobacter* are placed in the family Acetobacteraceae. Their placement within a single family recognizes their most apparent similarity (i.e., the ability to oxidize ethanol to acetic acid). Additionally, modern taxonomic tools such as r-RNA and DNA hybridization, and so on, have shown their close affinity (Gillis 1978; Gillis and De Ley 1980). As a result of the reclassification, the present status of the group recognizes two genera, *Gluconobacter* and *Acetobacter*. The genus *Gluconobacter* is comprised of a single species, *G. oxydans*, whereas *Acetobacter* consists of four species: *A. aceti*, *A. pasteurianus*, *A. liquefaciens*, and *A. hansenii*.

Vaughn (1955) reported *A. aceti* and *A. oxydans* (currently *G. oxydans*) to be the two most commonly encountered species of acetic acid bacteria in California wines. *Acetobacter aceti* also was found to be the most commonly isolated species in aging Bordeaux wines (Joyeux et al. 1984a). By comparison, Drysdale and Fleet (1985) reported *A. pasteurianus* to be the most common isolate among aging Australian red wines.

Acetobacter and *Gluconobacter* are gram-negative (becoming gram-variable in older cultures) rods frequently occurring in pairs or chains. Although size varies (depending on growth media), cell dimensions range from 0.6 to 0.9 μm by 1 to 3 μm. Flagellation also may be of some value in morphological separation. *Gluconobacter* exhibits polar flagellation (when present), whereas *Acetobacter* is peritrichous. Endospores are not formed in either group.

Both *Acetobacter* and *Gluconobacter* exhibit totally respiratory (aerobic) metabolism. Thus their growth generally occurs on the surface of wine and is seen as a translucent, adhesive film. This film may separate, resulting in a patchy appearance. The formation of surface films or pellicles may cause the wine to appear hazy or cloudy. It is currently thought that these bacteria may be able to survive at very low (semianaerobic) oxygen concentrations as may be found in stored wines (Drysdale and Fleet, 1989).

The principal physiological similarity of interest to the winemaker between *Acetobacter* and *Gluconobacter* is the ability of both to carry out oxidation of ethanol (and glucose) to acetic acid. However, the extent to which this oxidation occurs varies with the organism involved (De Ley and Schell 1959). In the case of glucose metabolism, both genera utilize the hexose monophosphate pathway in the formation of acetic and lactic acids. An important taxonomic distinction between the two is that *Acetobacter* may carry the oxidation a step further, converting acetic and lactic acids to CO_2 and H_2O via the TCA cycle. For this reason, *Acetobacter* sp. are called "overoxidizers." *Gluconobacter*, by comparison, does not demonstrate this overoxidation. The reasons for the inability of *Gluconobacter* to oxidize acetic acid further lie in the fact that the organism lacks functional key enzymes of the TCA cycle. Specifically, alpha-ketoglutarate dehydrogenase (catalyzing the formation of succinyl-CoA from alpha-ketoglutaric acid) and succinate dehydrogenase (catalyzing the formation of fumaric acid from succinic acid) are not operational (Greenfield and Claus 1972).

With regard to the formation of acetic and lactic acids from glucose, species of *Acetobacter* appear less capable of bringing about the complete oxidation than is *G. oxydans* (De Ley 1958). The major end products resulting from growth of *Acetobacter* on glucose-containing substrates are gluconic and ketogluconic acids. The utilization of glucose in the formation of acetic and lactic acids via the hexose monophosphate pathway is dependent upon *both* the pH of the growth medium and the concentration of sugar. Glucose concentrations of 5 to 15 m*M* and pH levels of less than 3.5 inhibit completion of the oxidation (Olijve and Kok 1979). In this case, gluconic acid accumulates as an end product. Gluconic acid subsequently may be oxidized to ketogluconic acid when glucose levels drop to 10 m*M* and pH rises above 3.5 (Weenk et al. 1984).

In slowly fermenting or stuck fermentations, the growth of *Acetobacter* and *Gluconobacter* may result in oxidation of glucose, and formation of gluconic acid (Vaughn 1938) also may occur. Depending upon must chemistry (pH and sugar concentration), gluconic acid may not be further metabolized and thus will accumulate. Gluconic acid levels up to 70 g/L have been reported in grape musts where growth of *G. oxydans* has occurred (Joyeux et al. 1984b). Production of gluconic and ketogluconic acids also has been reported in grapes where the growth of *Gluconobacter* and *Acetobacter* has taken place. *Acetobacter* reportedly produces lower concentrations than does the growth of *Gluconobacter* (Joyeux et al. 1984b).

Some species of *Acetobacter* and *Gluconobacter* do not oxidize fructose, or do so to a limited extent (Joyeux et al. 1984b). As a result of the relative sweetness of fructose (compared to glucose), disproportionate utilization of glucose by these species may result in a fermenting wine (or stuck fermentation) with a sweet–sour character. In this case, the sweetness is attributed to unoxidized fructose (Vaughn 1938, 1955).

Under oxidative conditions, *G. oxydans* is capable of oxidizing glycerol to dihydroxyacetone (Hauge et al. 1955; Eschenbruch and Dittrich 1986), which may play an important role in the sensory properties of wine. Because the reaction is inhibited by alcohol concentrations of greater than 5% (Yamada et al. 1979), its occurrence in wine is questionable. However, it is likely that glycerol oxidation occurs in infected grapes and musts. The growth of acetic acid species here may produce significant quantities of dihydroxyacetone, which may end up in the fermented wine. In one study, dihydroxyacetone levels of 260 mg/L were reported in must infected with *G. oxydans*. Levels of dihydroxyacetone in wine made from this must contained 133 mg/L of dihydroxyacetone (Sponholz and Dittrich 1985).

Acetobacter is distinguished from *Gluconobacter* on the basis of overoxidation of acetic and lactic acids to carbon dioxide and water via the TCA cycle. Thus it would be expected that *Acetobacter* species are capable of oxidizing other TCA cycle intermediates including citric, succinic, malic, and fumaric acids. *Gluconobacter* lacks two enzymes of the pathway, so it does not oxidize these acids.

Optimal growth temperatures for *Acetobacter* and *Gluconobacter* also vary. Reported growth optima for *A. aceti* range from 30 to 35°C, whereas *A. oxydans* (*G. oxydans*) grows best at around 20°C. Joyeux et al. (1984a) report the growth of *A. aceti* in wine at 10°C.

Control of Acetic Acid Bacteria

The need for oxygen in the survival of acetic acid bacteria is questionable. Although the group are obligate aerobes, evidence suggests that they may survive in wine under relatively low (semianaerobic) oxygen conditions. In one case, viable populations were recovered from wines stored in barrels.

Aside from oxygen, other important parameters affecting the growth of acetic acid bacteria in wine include pH and alcohol. Dupuy and Maugenet (1962) reported little growth in wines of pH less than 3.2 and/or alcohol levels greater than 13% (vol/vol). Subsequently, Dupuy and Maugenet (1963) report *Acetobacter pasteurianus* grew at pH 3.4 in 12.5% alcohol, whereas in 8% ethanol, its survival was noted at pH 3.0. Where pH and/or other parameters are not limiting, maximum alcohol tolerances for acetic acid bacteria are generally reported as 14 to 15%.

The literature cites a relatively wide range of SO_2 levels needed for control of acetic acid bacteria in wine and must. However, most reports cite levels of free and total SO_2 present at isolation without reference to the concentration of molecular SO_2. Inhibition of most species of wine acetics may be achieved at molecular SO_2 levels of 0.8 mg/L. (For details regarding the importance of SO_2, in its various forms, the reader is referred to Chapter 9.) Other factors such as sugar or tannin levels or the addition (elimination) or minor elements may play less of a role in the activity of acetic acid bacteria. However, these parameters have not been reported in detail.

Acetic acid bacteria are ubiquitious, found wherever fermentation has produced a suitable substrate for their growth. The problem may begin with development of the sour rot complex in the vineyard, so that already badly degraded grapes may be transported to the winery at harvest. In the cellar, the most common sites of isolation are puddles of wine under barrels and tanks, as well as under other equipment where wine has been allowed to accumulate with exposure to the air. Partially filled storage containers serve as an excellent and potentially undetected growth site for acetics. Problems associated with the storage of wine in barrels and techniques designed to avoid acetic acid bacterial development are discussed in Chapter 5.

Thus the winemaker must work toward control, rather than elimination, of acetics in the cellar. General sanitation, as well as proper barrel maintenance, is needed to reduce the potential for pockets of contamination within the winery. As a result of improved winery sanitation and processing techniques, spoilage attributed to acetic acid bacteria now is relatively uncommon in the U.S. wine industry.

Bacillus sp.

Normally this organism is found in soil, and secondarily in water. Occasional observations of the isolated colonies in wine sample preparations usually are attributed to contamination from winery water. The nature of its activity in wine is uncertain, and the U.S. literature makes only passing mention of the organism (Vaughn 1955).

Lactic Acid Bacteria

The malolactic fermentation is a catabolic pathway in which L-malic acid is enzymatically oxidized to L-lactic acid and carbon dioxide. Depending upon the species (or strain) of lactic bacteria involved, several by-products may be produced that impact the sensory properties of the wine.

The occurrence of malolactic fermentation is common to all wine-producing areas of the world. Studies have shown that lactic acid bacteria probably originate on the grape, where they may be isolated from the berry surface and grape leaves. Their numbers, however, are rather low—in most instances, less than 100 cells/mL (Lafon-Lafourcade et al. 1983). Other studies suggest that winery equipment is an important source of infection. Contamination, in these cases, is secondary, resulting from improperly sanitized sites where lactics have been allowed to accumulate and proliferate.

Chemically, the most significant changes observed during the course of a malolactic fermentation are increases in pH and corresponding decreases in titratable acidity. Depending upon the concentration of malic acid and the extent of microbial growth, increases in pH of 0.3 unit have been documented (Pilone et al. 1966; Rankine 1977). Decreases in titratable acidity generally are on the order of 0.1 to 0.3%.

Therefore, successful induction of the malolactic fermentation in high-acid, low-pH wines is, potentially, a useful technique for acid and pH adjustment. However, in the case of a high-pH wine, the malolactic fermentation may significantly (and negatively) impact sensory properties of the wine. Changes in flavor and aroma that may result from bacterial conversion include a loss of floral intensity and aroma. Additionally, instability, both chemical and microbial, may occur as a result of increased pH. Such wines tend to be susceptible to subsequent growth of other microorganisms whose activity may be clearly regarded as spoilage. In cases where a malolactic fermentation may occur in low-acid wines, follow-up adjustments in acidity may be desirable. In fact, many winemakers acidify shortly after the completion of a malolactic fermentation to help attain acceptable pH levels. Such adjustments are best accomplished using tartaric acid. (For additional discussion of tartaric acid addition and potential problems, the reader is referred to Chapters 13 and 18.)

Sensory Aspects of the Malolactic Fermentation

In addition to overall increases in pH and reduction in acidity, by-products of this conversion may be important from a sensory point of view and may contribute to complexity (Rankine 1972; Amerine et al. 1972). These products include, principally, biacetyl and, secondarily, acetoin and 2,3-butanediol in addition to acetic acid and its esters. Perceived as buttery in character, biacetyl is produced to an extent that varies with the strain of lactic bacteria involved. At concentrations greater than 5 mg/L, biacetyl may be objectionable (Rankine 1977). At lower levels, however, its buttery character in combination with other wine components may add complexity to the wine. For example, at low concentrations, the buttery overtones of biacetyl may be used as a stylistic tool in the production of certain white wines such as Chardonnay. Because of its importance in sensory properties, the biacetyl content is monitored by some wineries (see Procedure 12-5). The formation of higher concentrations of the metabolite (>5 mg/L) is reported to be associated with the growth of *Pediococcus* sp.

Many California wineries strive to produce white wines with the emphasis on delicacy and balance. The prevalent attitude is that high malic acid levels and low pH add life and freshness consistent with the winemakers' stylistic goals. Malic acid is considered to be an important part of the palate structure. In this regard, however, the malolactic fermentation lowers the acid content and freshness.

Other vintners, however, believe that the malolactic fermentation broadens and lengthens a wine's finish and flavor. Because the malolactic fermentation brings about pH increases and can reduce titratable acidity by as much as one-third (the resulting lactic acid is not as acidic on the palate), the combination contributes to their goals. The perceived benefit of malolactic fermentation is that the wine's flavor is modified by increasing complexity, but this benefit may be largely based on anecdotal evidence rather than experimental results. Many winemakers claim to be able to distinguish wines, particularly white wines, that have gone through a malolactic fermentation. However, several researchers have reported taste panel results in which the tasters could not, statistically, tell whether a wine had undergone a malolactic fermentation (Radler 1968; Rankine 1972; Silver and Leighton 1981), or had been deacidified by some other technique such as potassium carbonate (Castino, et al. 1975).

Lactic populations in wine may reach levels equivalent to yeast populations (10^6–10^8 cells/mL). Aside from their primary transformation, it might be expected that by-products of their metabolism would have a continuing influence on the wine long after the bacterial cells were gone. It is believed that esterases, lipases, and proteases of bacterial origin may play an important sensory role in wines undergoing malolactic fermentation. However, the nature of enzymatic activity is still unclear (Davis, et al. 1988).

Amino acid utilization also changes with the growth of lactic acid bacteria in wine. No consistent trends are reported although in the case of *Leuconostoc oenos* there is an increase in ornithine with corresponding decreases in arginine concentration following the conversion. Lastly, some lactics are known to degrade proteins, but it is unknown if any wine proteins are so affected.

Taxonomy of Lactic Acid Bacteria

Taxonomically, lactic bacteria are placed in the families Lactobacillaceae (*Lactobacillus* sp.), encompassing rod-shaped species, and Streptococcaceae (*Leuconostoc* sp.). The latter appear spheroid, often ranging to lenticular (''coccobacilloid'')-shaped cells. *Leuconostoc oenos* is the species most commonly reported in induced malolactic fermentations. The genus *Lactobacillus* includes gram-positive species that produce D- or L- or DL-lactic acid from glucose.

The amount of lactic acid formed from glucose and the pathway of its formation serve to separate the lactic acid bacteria into two groups, the hetero- and homofermenters. Homofermenters produce primarily (generally more than 85%) lactic acid as the end product of glucose metabolism. Lactics in this group use the Embden-Meyerhoff Parmas (EMP) pathway; instead of decarboxylation of pyruvate to acetaldehyde and subsequent reduction to form ethanol, homolactic bacteria reduce the pyruvate to yield two moles of lactic acid and two ATP per mole of glucose. Heterofermenters, however, lack the aldolase enzyme, which mediates cleavage of fructose-1,6-diphosphate in the formation of dihydroxyacetone phosphate and glyceraldehyde-3-phosphate. This group utilizes the oxidative pentose, phosphate pathway, seen in Fig. 12.1b (Stanier et al. 1976). As can be seen, once glyceraldehyde-3-phosphate is produced by cleavage of ribulose-5-phosphate, the pathway leading to lactic acid is the same as that used by homofermenters. The second product of the cleavage (acetyl phosphate) is either reduced to ethanol via two successive reduction steps involving the coenzyme NADH, or oxidized to produce acetic acid.

Morphologically, the lactobacilli occur as nonmotile asporogeneous rods ranging in shape from coccobacilloid to elongate. Individual cells are less than 4 to 5 μm in length. In some cases, cells may adhere to each other, resulting in filamentation. In the case of *Lactobacillus trichodes* and *L. hilgardii*, filament formation may be extensive, leading to the appearance of mycelia-like growth in the case of the former. This characteristic has led to the descriptive names ''cottony bacillus'' and ''Fresno mold'' for *L. trichodes*. Also, the cell shape may be variable.

Control of Lactic Bacterial Growth

As compared with acetic acid bacteria, the lactics are microaerophilic to facultatively anaerobic, requiring reducing (low oxygen) conditions for normal growth. Further, lactics are nutritionally fastidious microorganisms, requiring complex organic media for growth. The lactics have lost their ability to synthesize many specific compounds required for metabolic activity and require certain preformed compounds such as vitamins and amino acids for growth, in

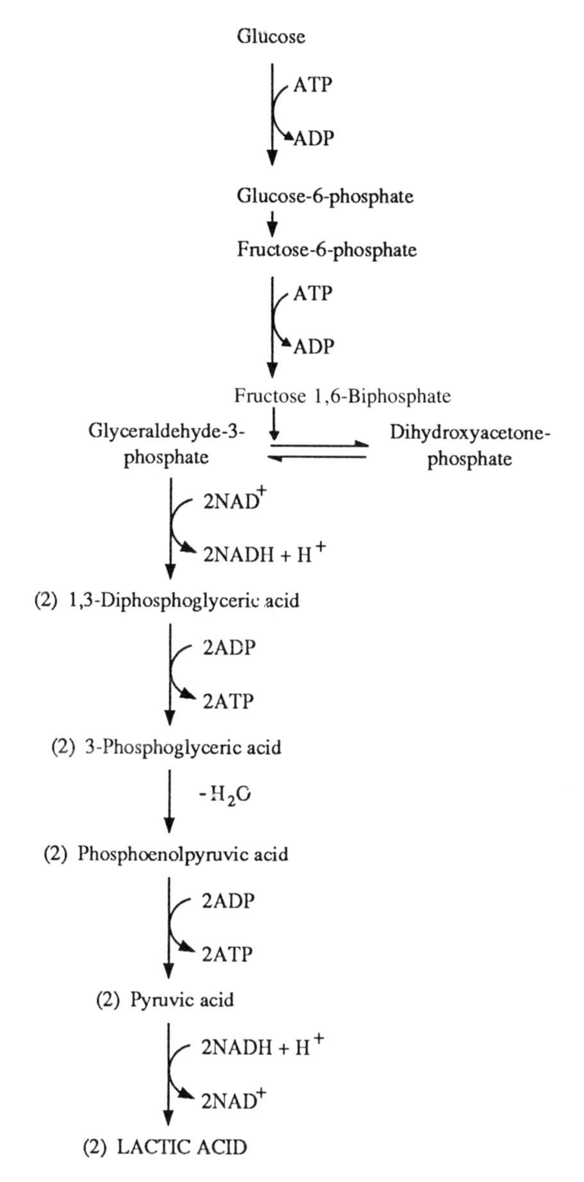

Fig. 12-1*a*. Homolactic Fermentation. Glucose is degraded to pyruvic acid via the EMP pathway. Two moles of lactic acid are produced per mole of glucose yielding a net 2 ATP.

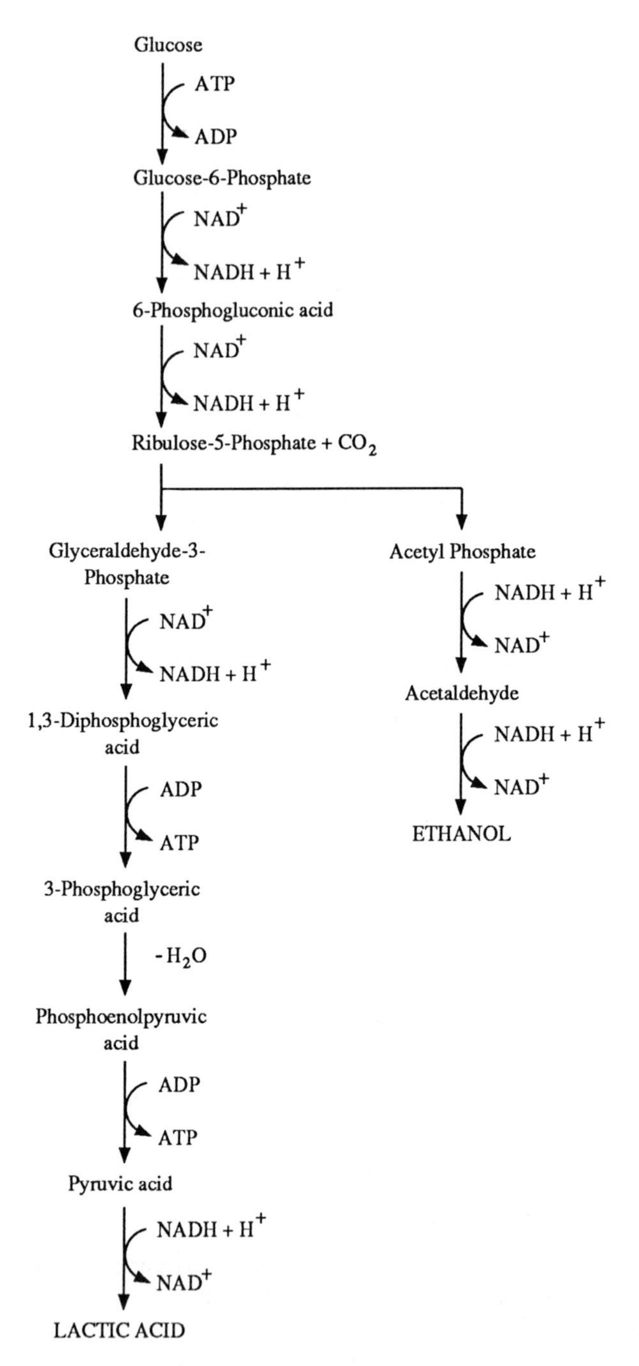

Fig. 12-1*b*. Major endproducts of the heterolactic fermentation become lactic acid, ethanol, and CO_2 (in equimolar amounts). By comparison to homolactic fermentations, the energy yield is only 1 ATP/glucose.

addition to other organic compounds. In laboratory cultivation, many workers find it necessary to augment standard culture media of yeast extract and protein hydrolysates with fruit or vegetable juices. Thus variations in the susceptibility of wines to lactic acid bacteria are partly due to differences in available nutrients and metabolic intermediates, as well as variations in particular lactic strains.

If other factors are not limiting, the addition of small amounts of yeast autolysate may stimulate the growth of these bacteria in wine (Carter 1950). The B-complex vitamins, produced by yeasts, are especially important to the growth of lactic bacteria, and it is believed that the addition of lactic starter cultures during the course of alcoholic fermentation takes advantage of the increased supply of nutrient provided by the yeast. Because wines contain sufficient quantities of required inorganics for the growth of these bacteria, their activity may depend in large measure on the availability of essential "growth factors." This assumes that other factors such as pH, alcohol, and SO_2 levels are not inhibitory.

Hydrogen ion concentration is the most important wine parameter controlling the growth of lactic bacteria. The hydrogen ion concentration establishes whether lactics will grow in wine, and selects which species will grow and their growth rate, as well as the concentration of metabolites produced during the growth cycle. For example, Bousbouras and Kunkee (1971) report completion of malolactic fermentation using *Leuconostoc oenos* in 14 days at pH 3.83. The same wine with the pH adjusted to 3.15 required 164 days for completion. A similar pH-dependent trend is seen in wild lactic fermentations. Hydrogen ion concentration also has an effect on the incidence of secondary lactic bacterial spoilage (Bousbouras and Kunkee 1971; Davis, et al., 1986, 1988). *Lactobacillus* and *Pediococcus* are generally not reported growing in wines of pH < 3.5.

The concentration of alcohol also affects the growth and survival of lactics in wine. Generally 12 to 14% (vol/vol) is reported as the upper limit of alcohol tolerance for *Leuconostoc* and *Pediococcus*, whereas *Lactobacillus* sp. has been reported to grow in wines at 15% alcohol, and *Lactobacillus trichoides* ("Fresno mold") has been found in wines of 20% alcohol (Vaughn 1955).

The role of sulfur dioxide in the prevention of malolactic fermentation is unclear. Sulfur dioxide exists in a complex equilibrium between its active molecular form and bisulfite and sulfite. Beech et al. (1979) have established that molecular SO_2 levels of 0.8 mg/L are inhibitory to lactic acid bacteria. See Chapter 9 (Table 9.1).

Processing protocol plays a role in predisposing a wine to malolactic fermentation. Prefermentation processing techniques such as cold clarification and fining not only may reduce native populations of lactic bacteria but also may reduce nutrient levels so growth may be impeded. Skin contact time and temperature are known to enhance the growth of lactics, but it is unclear whether this behavior is due to increases in pH and/or to extraction of nutrients.

Certain yeast strains may inhibit the successful growth of lactic bacteria when grown in coculture. This antagonism may result from competition for nutrients and/or from production of soluble antimicrobial agents (King and Beelman 1986). Demand for and accumulation of amino acids by yeast are reported to deplete available pools needed for bacterial growth (Beelman et al. 1982). Certain strains of wine yeast are known to produce sulfur dioxide at levels sufficient to inhibit lactic bacteria (Fornachon 1968). Further, Labatut et al. (1984) report the importance of octanoic and decanoic acid production by yeast in microbial inhibition.

The chemical parameters that affect the activity of lactics include pH, TA, concentration of molecular SO_2 and other inhibitors (e.g., fumaric), alcohol levels, potential nitrogen limitation and sanitation.

ML-INOCULATION

Vintners may intentionally bring about a bacterial fermentation by the addition of high-titer ($>10^6$ cells/mL) inocula of lactic bacteria. Unlike wine yeast addition, however, the selection usually is limited to stains of *Leuconostoc oenos*. Two of the most commonly available strains are *L. oenos* ML-34 and PSU-1. The two strains differ largely in the area of pH tolerance. PSU-1 generally is recommended for use in must/wine in the pH range 3.1 to 3.3, whereas ML-34 is recommended in the pH range 3.4 to 3.6. On rare occasions, *Lactobacillus* sp. has been utilized. It has been the authors' experience, however, that these strains generally are not acceptable.

Over the past several years, a number of commercially available high-titer lyophilizates have been utilized. Their advantage is that the lag time needed to prepare a sufficient volume of active starter is reduced significantly from that needed to bring up cultures stored on laboratory media.

Clementi and Rossi (1984) report the use of dried lactic bacteria in starter preparation. They concluded that cell viability, upon rehydration, was not greatly different (under optimized conditions of dehydration) than that seen in other starters. They recommend storage of dried cultures at 5–12°C until use. In the case of lots stored at 5°C, successful rehydration and subsequent activity was noted after 8 months of storage.

Some wineries with the necessary laboratory support elect to maintain lactics in culture. Upon the transfer of lactic starters to wine, some viability is lost (Beelman, et al. 1982); so prior to use, it is necessary to create a high-titer ($>10^6$ cells/mL) inoculum and expand the final volume of the starter to $>1\%$ of the final volume of the wine. The methodology of propagation and the expansion of lactic starters will have an impact on the final activity of the bacteria upon the addition to wine or must. Although rather elaborate preinoculation growth media have been reported, most techniques call for utilization of grape or apple juice (5–10° B, pH 4.5) supplemented with 0.5% yeast extract. Variations in preinoculation media include culturing in Rogosa broth for 5 to 7 days prior to transfer to the juice. The culture is then expanded to 1% (vol/vol) of the final wine (Beelman et al. 1982).

There is no unanimous opinion as to timing the addition of lactic starters. They may be added along with yeast at the crush/clarification stage, during the course of, or upon completion of, alcoholic fermentation. When sulfur dioxide is added at crush, or shortly thereafter, inoculation at this stage risks inhibition. As fermentation proceeds, the quantity of free molecular sulfur dioxide is reduced, and, depending upon the initial levels of sulfite addition, the probability of successful bacterial conversion is enhanced. When the winemaker wishes to have the malolactic fermentation occur, consideration must be given to initial sulfite addition levels. At levels as low as 50 mg/L, total SO_2 may produce enough sulfite–acetaldehyde addition product to inhibit *Leuconostoc* and *Lactobacillus* sp. (Hood 1984).

The controversy regarding the time of addition of starter cultures centers around adding the culture during the early stages of alcoholic fermentation. The advantages include initiation of bacterial growth before major amounts of alcohol are produced. Further, wine yeast produces nutrients (B-complex vitamins) that are known to be important in the growth and survival of lactics. These substances are depleted as the wine ages.

Much of the concern, among winemakers, over the addition of lactic cultures before the onset of alcoholic fermentation comes from the French literature. In one report, the onset of alcoholic fermentation was delayed by the addition of *Leuconostoc oenos* at a cell density of 10^7 cells/mL (Lafon-Lafourcade et al. 1983). Also, workers reported that during the extended lag phase the unimpeded growth of lactics on the sugars present produced higher than normal levels of acetic and lactic acid (Lucmaret 1981). The American literature, however, has not reported this to be a problem.

Current thought among many winemakers is to use high-titer ($>10^6$ cells/mL) bacterial starter additions during the course of alcoholic fermentation. At this point, potentially inhibitory levels of sulfur dioxide used at crush have been reduced, and growth of yeasts has proceeded to the point where bacteria have little impact on their activity. Further, there is ample nutrient availability in the form of yeast autolysate, which serves as an important source of B-complex vitamins to stimulate bacterial growth and activity (Weiller and Radler 1972). Therefore, addition during active fermentation takes advantage of the presence of soluble metabolites arising from yeast activity as well as the mixing action of the fermentation.

In the case of red wine fermentations, many vintners prefer to add bacterial cultures after pressing and before the fermentation has finished. Some add starters to freshly fermented wines or rely on endogenous populations present in barrels to bring about the conversion of malic acid. Successful completion of the malolactic fermentation at this stage in processing may be difficult to achieve because of alcohol levels and/or nutrient depletion. To overcome potential nutrient depletion, some winemakers may utilize a period of lees contact to make available the necessary nutrients for growth of the bacteria. In addition, sulfur dioxide levels must be kept low to assure bacterial activity.

Once malic acid conversion is completed, winemakers typically add sulfur dioxide. Depending upon its concentration and wine pH, viable cell numbers decrease rapidly. In instances where pH is high and antimicrobal levels of molecular sulfur dioxide are low, secondary growth of other lactic species, namely *Pediococcus* and/or *Lactobacillus*, may occur. The latter are considered to impart spoilage characters to the wine. Further, there is evidence that the growth of these species may inhibit the activity of *Leuconostoc oenos* (Costello et al. 1983).

Wines that have undergone malolactic fermentations may still be susceptible to the growth of other lactics as well as yeast. Hydrogen ion concentration is known to be important in the activity of *L. oenos* and is believed to be even more important in establishing a suitable environment for the growth of spoilage lactics. For example, Vetsch and Mayer (1978) reported secondary growth ($>10^7$ cells/mL) of *Pediococcus cerevisiae* upon completion of malolactic fermentations in a red wine. In this case, the pH of the spoiled wine was 3.9. Lactobacilli of several strains have been implicated in post-malolactic spoilage. *Pediococcus cerevisiae* and *P. pentosaceus* also appear to be important in secondary spoilage, principally in the case of high-pH wines.

In most cases, table wines contain sufficient carbohydrates to serve as energy sources for the growth of lactic bacteria. Most of the carbohydrate components of table wines are utilizable. Lactic acid bacteria have been found to utilize residual sugars in dry wines (Amerine and Kunkee 1968). Melamed (1962) has demonstrated decreased sugar levels in dry wine after bacterial growth, identifying the most commonly used sugars as glucose and arabinose. Glucose levels as low as 0.5 μM/mL have been found sufficient to support the growth of lactic bacteria (Pilone and Kunkee 1972). Thus, serious bacterial decomposition may occur in wines with reducing sugar levels of 0.5% or less.

Pilone and Kunkee (1972) found *L. oenos* able to utilize D-glucose, D-fructose, and ribose as initial sugars. They reported the following sugars as being nonutilizable: raffinose, lactose, maltose, sucrose, galactose, arabinose, and xylose.

Use of Immobilized Lactic Acid Bacteria

Studies have shown the potential for the use of high-titer ($>10^6$ cells/mL) lactic acid bacteria immobilized on different types of substrates including alginates and alginate gels as well as polyacrylamide gels (Spettoli et al. 1982; Rossi and Clementi 1984). These workers report success in bringing about a rapid although incomplete conversion of malic acid. Principal limiting parameters in these studies include pH, alcohol content, and concentration of free sulfur dioxide in the wines. Additionally, the sensory impact on wines treated using this technique has not been reported.

The literature contains at least one report of the unsuccessful use of immobilized malic acid enzyme in bringing about the malic acid conversion (Gestrelius 1982). The worker attributed the failure to instability of the NAD cofactor in wine as well as an inhibitory effect of wine pH on the enzyme complex.

CONTROLLING MICROBIAL GROWTH IN WINE (A SUMMARY)

This section is designed to acquaint the winemaker with techniques that may be utilized to control the growth of undesirable wine microbes. Several parameters act in combination to influence the activity and growth of wine microorganisms, including pH, TA, alcohol levels, temperature, concentration of antimicrobial agents, oxygen levels, carbonation pressure, and phenolic and nitrogen compounds.

Hydrogen Ion Concentration (pH)

As discussed in Chapter 4, pH may be the most important consideration in winemaking. In addition to the direct activity of the hydrogen ion in cell metabolism, it has a major impact on the active concentration of most wine additives active against microbial growth.

The growth of wine yeasts is dramatically impacted by juice and wine pH, their growth rate being approximately twice as fast at pH 4 as at pH 3. However, fermentation rates appear to be similar within this range. The primary role of increased H^+ concentration is enzyme inhibition. This may have application to the winemaker attempting to hold juice for blending later in the year.

The other major role of pH is related to the nature of the antimicrobials used in winemaking. As discussed in Chapter 9, the percentage of molecular SO_2 increases with decreased pH. The same relationship exists for sorbic and benzoic acids (Chapter 11).

Although yeast growth is affected by pH, complete control probably cannot be achieved by manipulating this parameter alone. Acetic acid bacteria generally are very pH-tolerant, requiring low pH levels for optimum growth. These organisms, however, are sensitive to the effects of molecular sulfur dioxide, which is present in proportionately higher concentrations at low pH. The growth of lactic acid bacteria also is influenced by pH. Generally, inhibition begins to occur near pH 3.2, but, in the absence of SO_2, some wine lactics may grow at pH 3.0. However, neither low pH nor high free SO_2 levels can completely guarantee inhibition. Following a malolactic fermentation, increases in pH, coupled with reduction in molecular SO_2, increase the susceptibility of the wine to general biological deterioration.

Alcohol Levels

The alcohol content of a wine plays an important role in limiting the growth of wine microorganisms. Table 12-4 compares the relative alcohol tolerance of major groups of wine microbes.

Temperature and Oxygen

The control of aerobic species (acetics and film yeast) can be effected by limiting the oxygen levels above the wine surface. This is best accomplished by storage in completely filled containers. The properly designed winery should have available a variety of tank volumes to avoid the need for partial fills. When this is not possible, carbon dioxide or nitrogen blanketing may be used to help displace air above the surface of the wine. Because of its limited solubility, nitrogen is the preferred blanketing agent in wine. It should be emphasized that blanketing is not a preferred technique for storage, and only very rigorous blanketing programs can be expected to help control the growth of aerobic species.

Lactic acid bacteria are microaerophilic, requiring very limited amounts of oxygen to support growth. Therefore, controlling oxygen contact is not an adequate means of control of this group.

Wine Additives

Sulfur Dioxide

Sulfur dioxide is an effective inhibitor of microbial growth. The level of molecular free SO_2 is important; generally, 0.8 mg/L of molecular SO_2 is sufficient

Table 12-4. Alcohol tolerance in wine micro-organisms at 20°C.

Group	Max. alcohol level for growth (% v/v)
Wine yeasts	16[1]
Wine spoilage yeasts (non-*Saccharomyces* spp.)	10–13
Acetic acid bacteria	
Acetobacter spp.	10–15[2]
Gluconobacter oxydans	<5
Lactic acid bacteria (except *L. trichoides*)	10–16[3]

[1]Harrison and Graham, 1970
[2]Drysdale and Fleet, 1988
[3]Davis, et al., 1988

to prevent the growth of wine bacteria. Also, levels of bound sulfur dioxide approaching 50 mg/L may be inhibitory toward lactic acid bacteria.

The activity of molecular SO_2 toward the control of microorganisms in wine and juice is dependent upon the following considerations:

The Microbe. Although inhibition may be possible, achieving cell death may require SO_2 levels that are unacceptable. Wine yeasts (*Saccharomyces*) are less sensitive to the effects of molecular SO_2 than are wild or spoilage yeasts. Acetic acid bacteria generally are sensitive to the effects of molecular SO_2. Although bound SO_2 has little inhibitory activity toward yeasts and acetic acid bacteria, it can affect the growth of lactic acid bacteria.

High concentrations of the bisulfite addition product are known to inhibit the malolactic fermentation. Because of this suppression, sulfites used in the prefermentation processing of wines destined to undergo malolactic fermentations must be kept at low concentrations (<50 mg/L).

The Stage of Growth. Compounds such as aldehydes and reducing sugar, with free carbonyl groups that can bind SO_2, serve as major binding substrates that lower the concentration of the active molecular form. During fermentation, yeast produces large amounts of acetaldehyde; and, as discussed in Chapter 9, acetaldehyde very rapidly binds added SO_2. As seen from the dissociation constant ($K_D = 5 \times 10^{-6}$), the reaction of acetaldehyde and SO_2 favors formation of the addition product. This is the principal reason why it is difficult to stop yeast fermentations with the use of SO_2.

Reactivity is slower at lower temperatures, so reducing storage temperatures slows formation of the addition product. Reportedly, the combination of sulfur dioxide use at levels of 200 mg/L and storage at 40°F was effective in storing juice (Splittstoesser, 1981). Successful application of this technique requires the use of clean, sound fruit of low pH as well as lowering initial yeast populations by cold settling or centrifugation.

If the winemaker's goal is to control microbial growth in wine, a single large dose of SO_2 is more effective than several smaller additions. This is particularly important in high-pH wines because of the relatively low percentage of free sulfur dioxide in the active molecular form. Additionally, the formation of bound SO_2 is favored by higher pH.

Sorbic Acid

Added to sweet wines (R.S. > 1.5 g/L) as the salt potassium sorbate, sorbic acid exhibits inhibitory activity toward wine yeasts. Properly used, the compound inhibits the growth of fermentative yeast; but it has little practical inhibitory activity toward acetic and lactic acid bacteria.

Lactic bacteria may utilize sorbic acid in production of the volatile compound described as "geranium tone." Thus, it must be used in combination with SO_2 additions at bottling to prevent further bacterial growth. Depending upon pH, 30 mg/L free SO_2, or more, is needed at bottling to suppress the growth of lactic acid bacteria. In any event, the use of sorbic acid in wines destined for long-term aging is not suggested (see Chapter 11).

Fumaric Acid

At levels of 300 to 400 mg/L (2.5–3.5 pounds/1000 gallons), fumaric acid may be useful in controlling the growth of lactic acid bacteria. However, at least one group reports that levels of 1500 to 2000 mg/L were required to inhibit this growth (Pilone et al., 1974). Fumaric acid may be utilized by yeast; so it must be added only to dry wines *after* the first racking (Ough and Kunkee, 1974).

Although approved for use in the United States, fumaric acid suffers from at least three problems. Initially, it is often difficult to dissolve, and, in some cases, effective concentrations may approach the solubility limits of the acid. Secondly, fumaric acid has its own unique sensory properties, and its contribution to the overall wine profile should be considered. Lastly, the concentration of fumaric acid used appears to be critical to success or failure. When the concentration of the acid is inadequate for bacterial inhibition, it may be utilized, in addition to malic acid, as a carbon source.

As with most other antimicrobial compounds used in wine and juice, the effectiveness of fumaric acid is enhanced by low pH, adequate levels of molecular SO_2, and low bacterial titer. Fumaric acid is a relatively strong organic acid, so its additions may result in drops in pH and corresponding increases in TA (Cofran and Meyer, 1970). Fumaric acid additions of 1 g/L are equivalent to an addition of 1.29 g/L tartaric acid. (The reader is referred to Chapter 4 for regulations governing the use of fumaric acid.)

Carbon Dioxide and Pressure

At atmospheric pressure, carbon dioxide has little effect on yeast growth. However, as pressure increases above one atmosphere, inhibitory effects become apparent. Table 12-5 illustrates the relationships between pressure and yeast growth.

The winemaker can take advantage of CO_2 pressure in the storage of juice for later fermentation or blending purposes. As seen in the table, yeast replication generally is inhibited at pressures exceeding 6 atm. However, fermentative activity may proceed at higher pressures; so it is important that only still (nonfermenting) juice be used. Many wineries use small pressurized vessels held at 70 to 90 psi (4.7–6 atm) with limited amounts of SO_2 (<200 mg/L) for juice storage. The success of the method is predicated upon low initial yeast populations and adequate pressures in addition to low storage temperatures and adequate levels of SO_2.

Although yeast growth is significantly affected by increased pressures, some

Table 12-5. Effect of pressure on yeast cell growth.

Pressure (atm)	Yeast cell titer (cells/mL)
0	104
2	15
3	11
4	6
5	3
6	<1

Source: Schmitthenner (1950).

lactics may grow at pressures exceeding 7 atm; so the technique may not be effective in controlling growth. Because of their requirements for oxygen, the growth of acetics is inhibited in the presence of CO_2.

Nitrogen Availability

The presence of assimilable nitrogen sources has a significant impact on the potential for microbial growth. Except for certain conditions of excessive amelioration and/or clarification, low must nitrogen levels are not limiting factors during the primary yeast fermentations. During primary fermentation, each generation cycle of yeasts utilizes a portion of the available nitrogen content; so nitrogen stores eventually may reach a critical level that will not support complete fermentation. The utilization of available nitrogen has been used as a technique for the production of stable sweet wines in Europe, but such techniques have not proven successful in wines made in the United States. In the case of sparkling wine production, nitrogen additions may be necessary prior to the secondary fermentation. This is particularly true in the case of aged cuvees.

Several post-fermentation processing decisions may affect the potential for bacterial activity. These include extended lees contact (*sur lie*) as well as protein and tartrate stabilization techniques.

Yeast lees provide one of the better-known means of stimulating bacterial activity. Yeast autolysate is an excellent source of exogenous nutrient in the form of vitamins and amino acids required by malolactic bacteria. Therefore, if the vintner wishes to prevent the malolactic fermentation, early and continued racking is needed. Bentonite fining, in white wines, depletes (removes) nitrogen sources important for growth of the organisms. Cold stabilization also is known to impede the activity of lactics, possibly by reducing the levels of critical nutrients and lowering pH (see Chapter 13).

Biological Control

Recent reports of the isolation of bacteriophage specific for *Leuconostoc oenos* may play a key role in the future control of lactic bacteria in wine. In a recent study of Swiss wines that exhibited interrupted malolactic fermentations, workers reported the isolation of four separate lytic phages specific for *L. oenos* (Gnaegi and Sozzi, 1983). Lysogenic phages also have been isolated from lactic bacteria in other countries of Europe and South Africa (Lee, 1978; Nel et al., 1987).

Procedure 12-1a. Isolation and Cultivation of Wine Microorganisms

I. Examination of the Wine

Microbial instabilities in wine may be due to bacterial or yeast growth. However, the analyst should not overlook the potential for haze and/or sediment resulting from nonmicrobial sources. Because these problems are reviewed in Section III of this book, they will not be discussed here except for the suggestion that personnel dealing with instability problems should become aware of the appearance of typical chemical instabilities as well as confirmatory tests for them.

Whether the problem is chemical or biological, a thorough review of processing records for the wine in question often is useful. Also, sensory evaluation of the product may provide some insight into the nature of problems, with the analyst paying particular attention to the nature of precipitates (e.g., crystalline or refractile versus amorphous) as well as turbidity and gas formation. Peculiarities in color, nose, and taste also may be important clues as to the nature of the problem (e.g., geranium tone, hydrogen sulfide, yeastiness).

After forming preliminary opinions, the analyst may wish to order appropriate lab tests, which typically include VA, free and total sulfur dioxide, pH, and possibly malic acid, as well as metals.

If the instability is present as a light haze, it may be necessary to concentrate the suspension so that it can be examined microscopically. This can be accomplished by centrifugation or membrane filtration. In the latter event, the membrane, or pieces of it, may be examined directly under the microscope. In some cases, allowing the bottle to stand upright overnight will yield sufficient sediment for one to collect the necessary amount for examination.

Routinely, isolation (enumeration) techniques employ collection (concentration) of microorganisms on sterile filter membranes. The advantage of this technique is that, because of the membrane porosity and the free flow of nutrient, cells subsequently can be cultured by placing the membrane directly on the appropriate growth media. In the case of prebottling evaluation and samples from the bottling line, potential microbial contamination can be enumerated by filtering a known volume (usually 100 mL) using membranes of appropriate pore diameter (usually 0.45 μm). Membranes for this purpose usually have a grid-marked surface to facilitate counting, and they may be purchased with white or dark backgrounds. The latter are particularly convenient for the visualization of light-colored colonies.

II. Collection of Sample and Isolation of Microbes

(a) Equipment

Autoclavable filter holder and funnel of approx. 300 mL capacity

Vacuum flask and line

Prepackaged sterile Petri plates and 47-mm presterilized membrane filters (0.45 μm)

Pads (media-impregnated)

Forceps

(b) Reagents

Ethanol (70% vol/vol) for sterilizing forceps

Bunsen burner

(c) Procedure

1. Dip forceps in alcohol and flame.
2. Carefully open the filter package and, using sterile technique, place a sterile filter membrane (grid side up) on the sterile filter housing.
3. Connect the vacuum source.
4. In the case of bottling-line samples, aseptically open the bottle. It is recommended that the neck be thoroughly swabbed with alcohol prior to the removal of the cork, which should be removed so as to minimize the intrusion of outside air. Once the cork has been removed, flame the neck area of the bottle, and transfer 100 mL of sample to the funnel.
5. Apply the vacuum.
6. Once the wine has been filtered, allow pressure across the filter to equilibrate, and transfer the membrane to the appropriate growth medium (grid side up!). It is recommended that the filter be placed so that it is in complete contact with the substrate. Voids between the filter and the growth medium prevent the free flow of nutrient through the membrane.
7. For wines in which bacteria and/or yeasts are known to be present, preliminary dilution (or reduction in sample volume) may be required. Owing to the potential for uneven distribution of cells on the membrane surface, one should avoid using initial sample volumes of less than 10 mL. A better distribution of colonies can be achieved by first pouring 20 to 30 mL of sterile water into the funnel and then the measured amount of wine. *Note*: This technique does not create a dilution in the final calculation of cell number.
8. If potassium sorbate or large quantities of other inhibitors have been added to the wine, it may be necessary to follow the wine with sterile water or saline rinses (20–50 mL).
9. Media selection: For analysts not wishing to prepare lab media, pre-

packaged sterile media are commercially available, usually consisting of sterile Petri plates containing media-impregnated absorbant disks (pads). When ready for use, sterile water is added to rehydrate the medium.

For yeast growth, nutrient, grape juice, or wort, agar is routinely used (see Appendix III). Wine yeasts differ in their sensitivity to actidione, and this somewhat diagnostic feature may be evaluated early in the identification procedure by transfering suspect yeast to plates with and without actidione. Use levels of actidione in media vary, but 50 mg/L appears to be the most commonly encountered concentration.

Wine lactics may be cultured on a Rogosa medium. To inhibit yeasts in bacterial isolation, the medium is supplemented with actidione.

In wine samples with probable contamination due to acetics, isolations should be made on calcium-carbonate-supplemented media (see Appendix III). 10 mg/L biphenyl may be added to culture media to inhibit mold growth.

10. Incubate in an inverted position. Both bacterial and yeast isolations may be incubated at room temperature (25°C). Yeast generally develop in 3 days to a week, whereas lactic bacteria take somewhat longer, depending upon the organism. The growth of lactics is somewhat enhanced in a reduced-oxygen atmosphere. Such conditions may be achieved by placing plates in a 1-gallon jar (or dessictor). Position a lighted candle on the top plate, and seal the container. The candle will burn until the majority of the oxygen is depleted.
11. Once colonies have formed, count and express results as cells/mL (CFU/mL), taking into account any preliminary dilution.
12. Microscopically examine the colonies to identify the nature of microorganisms. Examination is facilitated by the use of phase microscopy. Although bright-field scopes are adequate, it generally is necessary to use stains to enhance the contrast between the material to be observed and the background. Such stains or dyes include nigrosin, methylene blue, Gram, Walford's, and Ponceau-S (see Appendix III).

Where bacteria and/or yeasts are noted, make preliminary identification based on cell shape. Prepare appropriate media for isolation and identification. Transfer well-isolated colonies to separate plates. Depending upon the organism (yeast versus bacteria) and the source of isolation (wine suspension versus sediment in bottom of barrel), from one to several subsequent transfers may be necessary in order to obtain a pure culture. Yeasts may be particularly troublesome in this regard because they often secrete a rather extensive capsule when recently isolated from natural sources. The mucilaginous nature of capsules serves to trap bacteria and mold spores that make isolation of the yeast in pure culture difficult. When isolation has been achieved, note the size and shape of the colony, and whether it has a shiny or an opaque surface, as well as any coloration that might be seen.

Once the organism is isolated and in pure culture, it should be stored on appropriate media pending further identification. In the case of aerobes, agar slants work well. Where the organism is thought to be an acetic, the use of calcium-carbonate-supplemented media is useful to neutralize acid as formed. Calcium-carbonate-impregnated slants also should be used if *Brettanomyces* or *Dekkera* is suspected.

If the suspect organism is a lactic bacteria, one should plan to use agar deeps into which the culture is inoculated by the stab technique.

Procedure 12-1b. Identification of Microbial Isolates

This procedure is designed for tentative identification of wine spoilage organisms. The reader is referred to appropriate references (Lodder, 1970; Bergey's *Manual of Determinative Microbiology*, Buchanan and Gibbons 1984) for more definitive testing, if necessary. Table 12-6 will aid in the identification of wine bacteria.

The catalase test is accomplished by placing a drop of 3% (vol/vol) hydrogen peroxide on a microscope slide. Add a loopful of the suspect organism. Bubbling (foam) formation is a positive indication of catalase activity characteristic of aerobic organisms such as *Acetobacter* and *Gluconobacter*.

$$2\,H_2O_2 \xrightarrow{\text{Catalase}} O_2 + 2H_2O$$

I. Wine Lactic Bacteria
   1. Morphology: In the case of wine lactics, typical cell shapes are baciloid (rods), characteristic of the genus *Lactobacillus*, and coccoid, ranging to cocco-bacilloid. The homofermentative genus *Pediococcus* is typically coccoid, whereas its heterofermentative counterpart *Leuconostoc* tends toward cocco-bacilloid.
   2. Generally, the first test of taxonomic value in the identification of bac-

Table 12-6. Physiological characteristics of wine bacteria.

Organism	Gram reaction	Catalase	Oxygen reqs.	Major endprd.	Sporulation
Gluconobacter	neg.	+	Aer.	Acetic	neg.
Acetobacter	neg.	+/−	Aer.	Acetic	neg.
Lactobacillus	pos.	−	Aer./ana.	Lactic	neg.
Leuconostoc	pos.	−	Fac./ana.	Lactic	neg.
Pediococcus	pos.	−	Aer./ana.	Lactic	neg.
Bacillus	pos.	+	Aer.	Several	pos.

teria is the Gram stain reaction. In the case of lactics, the Gram stain is positive (blue). Staining always should be done on young cultures of approximately the same age.

3a. Initial separation of *Pediococcus* from *Leuconostoc* can be accomplished by testing for glucose fermentation. The medium used is Rogosa *except that apple juice is replaced by tomato juice at the same dilution*. This is a test for carbon dioxide formation, so the expected decarboxylation of malic acid present in apple juice would serve only to create positive results!

Inoculate 0.1 mL of actively growing culture into approximately 5 mL of sterile medium culture tubes. Chill the tubes, and overlay them with liquid (hot) sterile vaspar (see Appendix III). For best results, a tight seal must be effected. Incubate tubes at 20°C for 1 week, and examine them for gas formation.

Production of carbon dioxide from glucose in the absence of malic acid is indicative of *Leuconostoc* sp., whereas the absence of gas production suggests *Pediococcus*.

3b. Another diagnostic test separating wine hetero- from homolactic species is formation of mannitol crystals from fructose.

Prepare Apple Rogosa broth as described in Appendix III. Include in the preparation 2% fructose. Inoculate 5 mL of sterile broth with 0.1 mL of active bacterial culture. Incubate the culture at 20°C for 1 week. Transfer the culture to an evaporation dish, and hold it until the liquid phase has evaporated.

The presence of mannitol crystals is indicative of *Leuconostoc*.

4. In the case of some heterofermentative lactics, it may be difficult to distinguish microscopically between elongated cocci and short bacilli. Kunkee (1986) suggests that separation may be made by using ammonia formation from arginine. Heterofermentative lactobacilli are positive, whereas cocci (*Leuconostoc*) are negative.

Prepare arginine medium as described in Appendix III. Inoculate the sterile arginine medium with the suspect heterofermentative lactic, and incubate the sample for 1 week. Transfer culture to a spot plate, and test for ammonia using Nessler's reagent (see Appendix III).

Interpretation: Orange/red coloration is considered positive. No change/pale yellow is negative.

Lab personnel may wish to carry out a positive control using dilute ammonia.

Cell Counting

Monitoring and enumerating yeast and bacterial populations during the winemaking process provide valuable information about changes occurring in the wine. In the case of yeast, total and viable cells numbers as well as the percentage of budding provide vital information to the winemaker in the preparation and maintenance of yeast starters.

Classically, procedures for monitoring the cell number call for collecting representative samples (potentially containing yeast and/or bacteria) and plating on appropriate media that will support growth. In most cases, however, the winemaker requires a quick decision about the presence of contaminants and or physiological status. Thus, the delay associated with colony formation generally is unacceptable. In the case of wine yeast incubated at room temperature, 36 to 48 hours is required. *Brettanomyces, Dekerra* and lactics require several days before growth is seen. Additionally, plate counting presumes that each colony arose from a single cell; but in the case of wine yeasts that tend to aggregate, this may not be a valid assumption.

To shorten the time needed and improve the accuracy of cell counting, a variety of other techniques have been tested, including the use of biochemical methods (e.g., measurement of ATP from living cells) as well as direct count procedures using fluorescence microscopy and dye reduction techniques.

Adenosine triphosphate (ATP) is produced in the course of metabolism by all living cells; and, upon cell death, it rapidly degrades. This characteristic, coupled with the fact that the concentration of ATP in living cells is relatively uniform, makes measurement of ATP an attractive (although indirect) estimate of biomass (and, hopefully, viable cell numbers) per volume of sample. Sample preparation is relatively easy. First it is necessary to inactivate in vivo cell enzyme systems that destroy ATP, a step accomplished by preparing the sample in boiling Tris buffer. The heating step brings about rapid autolysis and liberation of cytoplasmic ATP. The released ATP is measured using the luciferase assay, in which reduced luciferin is oxidized in the presence of oxygen, luciferase enzyme, ATP, and Mg^{2+}. The light emitted by the oxidation is directly proportional to the concentration of ATP in the sample. The components for this assay are commercially available in prepackaged kit form.

Although the technique reportedly is very reproducible (Atlas and Bartha 1981), several potential problems should be noted. Initially, concern lies in relating the concentration of ATP to the actual yeast and/or bacterial biomass. Secondly, ATP is produced concomitantly with metabolic processes, which depend upon the physiological condition of the organism in question. Thus, data collected in the early stages of growth may be significantly different from data collected from samples taken from stuck fermentations.

Several direct-count procedures have been proposed and currently are used in the enumeration of microorganisms. Fluorescence microscopy makes use of the fact that some compounds absorb ultraviolet light and reemit a portion of the absorbed energy as light in the visible spectrum; that is, they fluoresce. Because glass lenses block ultraviolet light, a quartz condenser lens is required to focus radiation from the ultraviolet light source upon the sample. Light in the visible spectrum is reemitted from the sample, so conventional optics are utilized to collect and magnify the sample image. Samples containing microbes are collected on membrane filters and stained with a fluorescent dye such as acridine orange. With this stain, viable yeast/bacteria fluoresce green, whereas dead cells appear orange. However, cells exposed to preservatives such as sorbate, floresce intermediate shades (Meidell 1987) and this makes interpretation difficult. Several other fluorescent viable stains also are available for the purpose of separating living and dead material.

The most commonly employed method of direct counting in the wine industry is dye reduction. The basis for this viable staining technique is that there is a change in the color of a reducible dye (usually methylene blue) when it is used as a hydrogen acceptor in lieu of oxygen. In order to quantitate cell numbers, it is necessary to use a cell counting slide. The hemocytometer counting chamber, used routinely in clinical laboratories for counting blood cells, also can be employed for counting yeast. The hemocytometer (see Fig. 12-2) is specially ruled so that the volume present in the counting area is constant.

In counting wine yeast, the initial problem is the tendency of the yeast to aggregate in groups that not only are difficult to count but also significantly bias the final result. Thus, mixing to break aggregation is critical to good correlation.

Procedure 12-2. Cell Counting

I. Equipment
   Hemocytometer hit
   Pipettes
   Dilution bottles
   Test tubes
   Vortex mixer

II. Reagents (see Appendix II)
   Methylene blue
   Isotonic saline (0.7%)

III. Procedure
   1. Depending upon the stage of yeast growth, prepare the appropriate dilutions from the collected sample in known volumes of isotonic saline. For example, in monitoring starter tanks, initial 10^{-4} to 10^{-5} dilutions may be needed in order to obtain countable slides, whereas post-fermentation counts may be made directly without prior dilution.
   2. Vortex each diluted sample before going to the next dilution in the series.
   3. After vortexing the final dilution in the series, transfer a small volume from the center of the dilution tube (bottle) to the counting slide.
   4. Using the cover slip provided with the kit, carefully position it over the counting grid. Be careful not to create air bubbles in the counting area.
   5. Count all the cells within the bounded area of the grid (400 squares). Tally nonviable and viable as well as percent budding cells present in

DAY ______________

TIME ______________

PERSON COUNTING CELLS ______________________

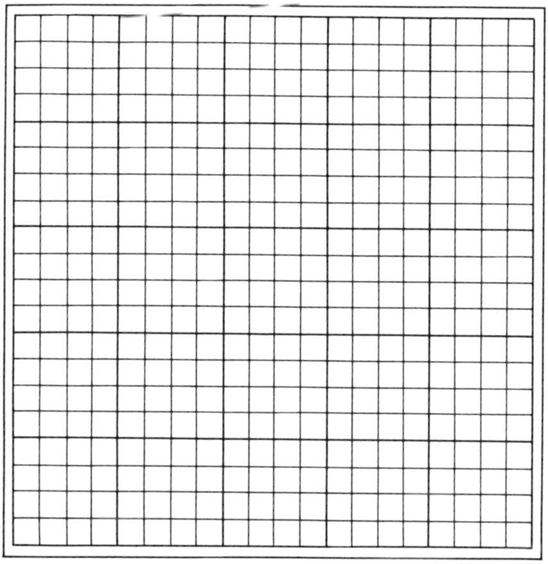

(1) Number yeast side 1 __________

Number yeast side 2 __________

$\bar{x}$ cell number __________

Fig. 12-2. Hemocytometer.

each field. Because the second field represents a duplicate measurement, average the results for each category.

When using methylene blue in the final diluent, one may also score viable/nonviable cells as part of the total cell count. In this case, viable cells with be colorless (i.e., they reduce the blue dye to its colorless of leuco-form). Nonviable ("dead") cells, by comparison, appear blue or have concentrated areas of blue stain within the boundaries of the cell wall (see Supplemental Notes 1 and 2).

6. When one counts all the cells present within the grid network, the expression of results (in cells/mL) may be calculated using the following equation:

$$\text{Cells/mL} = (\text{No. cells counted})\,(10^4) \times \text{Dilution}$$

7. In some instances, one may count only a portion (e.g., one-half) of the entire field. Although *this is not a recommended technique*, these results

also can be calculated by using the above equation. In this case, one need only multiply the result by the appropriate factor (in this case, 2).

IV. Supplemental Notes

1. Although statistically approved procedures exist for counting portions of the grid field, they apply to measurements of blood cells. In cases where uniform distribution of cells is difficult to achieve (e.g., wine yeasts), it is recommended that the entire field of 400 squares be counted. For reproducible results, it is necessary that each laboratory establish whether cells lying on grid boundaries be included in the count or rejected. Depending upon the initial dilution created, a difference of five or more cells may have a significant effect on the final calculation of cell number. Upon inoculation, viable cell numbers should range from 2–5 $\times$ 10^6 cells/mL
2. Methylene blue, itself, eventually will poison living cells. Thus it is imperative, once the dye is mixed with the final dilution, that cell numbers, and especially viability, be determined as soon as possible.
3. When bottling-line samples are sceened microscopically, it is necessary to concentrate the microbes potentially present in the sample. In these cases, known volumes (usually 100 mL or more) are filtered through a membrane filter of sufficient porosity to retain yeast and or bacteria. Portions of the membrane then may be stained and examined directly (see Section III of the book and Appendix III).

Procedure 12-3. Malolactic Fermentation by Paper Chromatography (Semiquantitative Measurement)

The occurrence and/or progress of the malolactic fermentation can be easily monitored by paper chromatographic separation, with the absence of malic acid on the developed chromatograph being a *somewhat reliable indication* of the occurrence of the bacterial fermentation (Ingraham et al. 1960a). In principle, wine acids partition themselves according to their affinity for the mobile solvent and stationary phases. Separation is best when the system is operated in an ascending manner; however, several wineries use descending chromatography successfully—a procedure that involves the purchase of special supplies at significantly greater cost than that of the former method.

A typical paper chromatogram is presented in Fig. 12-3. The solvent used in this procedure has the added advantage of not requiring subsequent development; it already contains an indicator for the acid spots. Thus, after drying, the results can be evaluated directly. The solvent is reusable, but fresh solvent should be prepared weekly. Kunkee (1968) noted that with old solvent there may be excessive trailing of spots, making interpretation of the chromatogram difficult. When not in use, solvent should be stored in a separatory funnel away from exposure to direct sunlight. Storage in a separatory funnel has the advantage that the remaining aqueous phase can be discarded before use.

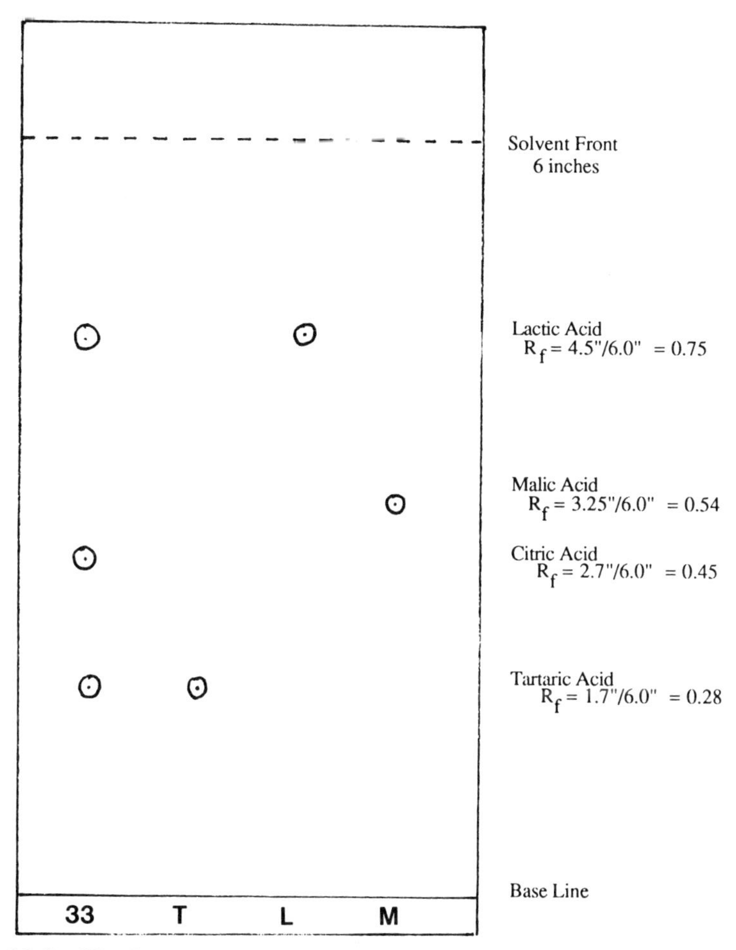

Fig. 12-3. Distribution of wine acids using conventional paper chromatography. Chromatogram shows separation of acids in white wine #33 compared with standards (T) tartaric (L) lactic and (M) malic

The solvent mixture used contains the pH indicator bromcresol green which undergoes color change from yellow to blue in the pH range 3.8–5.4. The presence of an acid thus is indicated as a yellow spot on a blue background. To identify the acid, analysts unfamiliar with the distribution may calculate relative front values (R_f). (The relative front is defined as the ratio of the distance traveled by a spot to the distance traveled by the solvent front.) The R_f values vary with both solvent and solid phases. For the solvent system used in this separation, the appropriate ranges of values are shown in Table 12-7. Hence, the following order of acids would be expected (from solvent front to base line) on a typical chromatogram: succinic and lactic, malic, citric (if present), and tartaric.

Table 12-7. Relative front (R_f) values for wine acids.

Acid	R_f range
Tartaric	0.28–0.30
Citric	0.42–0.45
Malic	0.51–0.56
Lactic	0.69–0.78
Succinic	0.69–0.78

Source: Kunkee (1968)

The analyst should run standard acid(s) as reference(s), thus reducing the possibility of differences in R_f values attributable to experimental conditions. Standard acids are prepared at concentrations of 0.3% (Kunkee 1968).

Depending on the supportive paper used and the ambient temperature, the time required per run will vary. Some workers recommend Whatman No. 1 chromatography paper for routine work. The authors find that the best separation is achieved using Whatman No. 4. The time required for separation, however, is from 2 to 4 hours longer in the latter case, depending upon room temperature.

The presence of a lactic acid spot is not always indicative of a bacterial malolactic fermentation. Being a Krebs cycle acid, malic acid is used to a slight extent by wine yeasts. Although this reaction normally would be expected to occur under aerobic conditions, approximately 10% conversion is reported during winemaking operations (Peynaud 1947). Additionally, the fermentative activities of the yeast *Schizosaccharomyces* may result in complete oxidation of malic acid to ethanol and carbon dioxide. However the normal titer of this yeast is low, and therefore, the probability of malic acid utilization by these organisms in routine fermentations is low.

It has been reported in the literature (Gump et al. 1985) that the paper chromatographic method is not sensitive enough; wine that tests negatively for malic acid has undergone a malolactic fermentation in the bottle. As a result, laboratory workers have begun to use other procedures for malic acid analysis, two of which are included in this book. One, an enzymatic analysis procedure, is included above as Procedure 4-4. The other is the high performance liquid chromatographic method, presented as Procedure 4-3.

I. Equipment
   - Whatman No. 1 chromatography paper
   - Chromatography developing tank
   - Separatory funnel
   - Micropipettes (20 μL) or hematocrit tubes

II. Reagents (see Appendix II)
   - Wine acid standards (0.3%)
   - Chromatography solvent

III. Procedure
1. Taking care to handle chromatography paper only by the edges, cut out a piece of the appropriate size such that it will fit into the developing tank.
2. Using a lead pencil, draw a line parallel to, and approximately 1 inch from, the bottom edge of the paper.
3. Using micropipettes or hematocrit tubes, spot standard acids and wine samples at equal intervals along the base line. The spots should be of as small a diameter as possible (less than 1 cm). It is recommended that respotting at least twice in order to achieve this goal. Spots should be dry prior to respotting. Spots should be at least 2.5 to 3.0 cm apart.
4. Transfer solvent to the developing tank, allowing at least 30 minutes for vapor saturation to occur. A minimum depth of 0.25 inch of solvent is required for adequate development.
5. Immerse the base-line side of the paper in the tank, taking care that solvent moves uniformly up the paper.
6. When the solvent has ascended to near the upper edge of the paper, the chromatogram may be removed and allowed to dry.
7. When it is dry, results may be interpreted by noting the positions of yellow spots (acids) on the blue background (see Fig. 12-3). Identification of various wine acids may be made by comparison to standard acids or by calculation of R_f values.

IV. Supplemental Notes
1. The time required for migration of the solvent front is not critical. To facilitate maximum separation of wine acids, the solvent front should be allowed to move to within close proximity of the top edge. In practice, chromatograms may be successfully interpreted even if the solvent front reaches the top of the paper. However, in such cases R_f values are difficult to determine, and one should rely on standards for comparison.
2. In solvent preparation, formic acid is added to suppress the ionization of acids, which would prevent their separation.
3. Yeast and lactic bacteria may produce lactic acid as a normal product of metabolism. Therefore, the presence of a lactic acid spot on the chromatogram is not necessarily conformation of a malolactic fermentation.
4. Resolution of dried chromatograms can be improved by spraying a solution of dilute ammonia across them.

Procedure 12-4. Malolactic Fermentation: Alternative Paper Chromatographic Technique

This procedure describes a rapid and simple paper chromatographic method for following the progress of a malolactic fermentation during vinification. In contrast to usual techniques using large rectangular shaped filter paper with a correspondingly long run time, the method presented here utilizes triangular shaped filter paper, and shortened run times on the order of 3 hours.

I. Equipment

Chromatographic paper: Whatman No. 1, cut as shown in Fig. 12-4:

(a) In the shape of an isosceles triangle, 5.5 cm × 18 cm.

(b) In the shape of a pentagon with one part triangular (5.5 × 10.0 cm) and the other part rectangular (5.5 × 8.0 cm).

Microsyringe: Hamilton No. 780, 100 μL

Loop: In lieu of a microsyringe, one can use a stainless steel loop (1.40

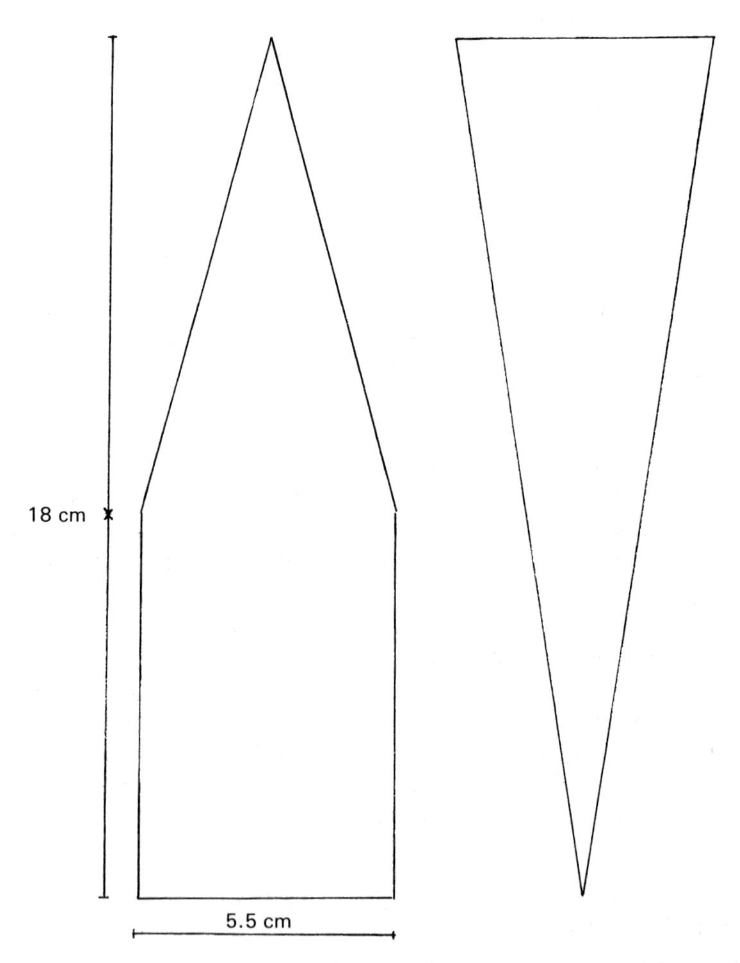

Fig. 12-4. Alternative shapes of paper chromatograms used for analysis of wine acids.

× 1.65 mm) made from 0.5-mm-diameter stainless steel wire. This loop will carry 0.7 μL of solution.

Chromatographic chamber: May be of various sizes and shapes, depending on the number of chromatograms being run simultaneously. The minimum height is 20 cm. The chamber should be filled with solvent to a level of 1 cm.

II. Reagents

Solvent system: *n*-butanol : acetic acid : water (4 : 1 : 5)

Prepare this system by mixing the components and allowing the mixture to separate into two layers; the aqueous (lower) phase is discarded. Add bromophenol blue at a concentration of 1 g/L of butanol. If during use this solvent becomes turbid and a separate aqueous phase is produced, the excess water should be removed.

III. Procedure

1. Each formed chromatographic sheet is used for one sample of wine.
2. Spot 25 to 30 μL of sample 1.5 cm from the vertex of the triangular part, being careful not to touch the edges of the paper. The spot should be 1 cm or less in diameter. It should be deposited in several applications with drying in between each application. The volume of sample applied depends on the state of the malolactic fermentation, usually being 25 to 30 μL.
3. Place the chromatographic paper into the chamber, cover it, and allow the mobile phase to ascend. Under optimum operating conditions the run will take about 3 hours.
4. Remove the chromatogram from the chamber, and dry it in a ventilated place. On humid days it is convenient to store chromatograms between sheets of white paper.
5. Note the presence of organic acids, appearing as yellow spots on a blue background.
6. Compare spot positions to those of standards run on a separate chromatogram.

IV. Supplemtal Notes

1. Qualitative identification of the spots is accomplished by using solutions of standard organic acids. The elution order of the acids is: citric, tartaric, malic, lactic, and succinic.
2. It is best to use control chromatograms run with reference solutions of organic acids at concentrations similar to those found in wine samples. This also aids in spot identification, as R_f values vary with experimental conditions.
3. It is also possible to identify the spots corresponding to the ethyl esters of tartaric and malic acid, if they are in sufficient concentration for detection.
4. The triangular-shaped paper is well suited for small quantities of sample.

The pentagonal-shaped paper causes more lateral expansion of the spots during a chromatographic run; one thus obtains improved resolution, even though large samples are spotted, and some of the spots are large. Lateral expansion, which characterizes the latter type of chromatogram, increases as the width of the paper increases. It allows one to work with large sample volumes, of up to 45 μL, without losing resolution.

5. The solvent may be used a number of times and kept for up to six months (if it remains clear). The running time is approximately 3 hours and produces a solvent front about 15 cm from the origin.
6. The concentrations of the common fixed organic acids in wine may be detected with 30 μL of sample even though the concentration is as low as 0.03 to 0.05 g/L. For malic acid, the absence of a spot from a 40-μL sample indicates a concentration of less than 0.03 g/L. For routine analyses it is best to use between 25 and 30 μL of wine. If the malic acid content is low, it is best to use 40 to 45 μL.

Procedure 12-5. Determination of Biacetyl in Wine

Biacetyl is a by-product of the metabolism of lactic acid bacteria. Its presence as a butter-like aroma in concentrations above ~2 to 4 mg/L (depending on the wine) is an indication that a malolactic fermentation has taken place. In this procedure, biacetyl first is distilled out of a wine sample. The absorbance (520 nm) of a red-colored complex of biacetyl and ferrous ion, formed in the presence of tartrate, is used to quantitate the amount of biacetyl present. The method works well at low concentrations.

I. Equipment

Distillation unit: consisting of a 500-mL 3-neck flask with a distillation column connected to first a connecting adapter, 75° bend, and then a condenser followed by a tube adapter, 105° bend. A second neck of the flask is fitted with a thermometer adapter holding a Pasteur pipette. The position of the pipette is set so that its tip extends to just above the surface of the sample. The third neck of the flask is used for admitting the sample and is otherwise fitted with a stopper.

Heating mantle with rheostat

Steam bath

Visible spectrophotometer with 1-cm cuvettes

CO_2 tank with regulator

Pipettes, 1 mL, 0.0–1.0 mL, 0.0–30.0 mL

Assorted beakers

II. Reagents (see Appendix II)

Biacetyl stock solution (800 mg/L)

Hydroxylamine–acetate solution (hydroxylamine hydrochloride–sodium acetate)

Potassium phosphate–acetone solution

Sodium potassium tartrate (45% wt/vol) in ammonium hydroxide
Ferrous sulfate (3.5% wt/vol) in sulfuric acid (1%)

III. Procedure

(a) Preparation of standards

1. Prepare a biacetyl working standard solution (40 mg/L biacetyl) by diluting 50 mL of biacetyl stock to 1000 mL with distilled water.
2. Prepare biacetyl calibration standards of 5, 10, 20, and 40 mg/L in 100-mL volumetric flasks (see Table 12-8).

(b) Distillation of wine sample to collect biacetyl

3. Transfer 1.5 mL hydroxylamine–acetate reagent and 10 to 15 mL distilled water to a 100-mL graduated beaker.
4. Place the beaker under the delivery tip from the condenser so that the tip is beneath the surface of the liquid in the beaker. Position an ice bath so that the lower part of the beaker is surrounded by ice.
5. Connect a tube from the CO_2 cylinder to the Pasteur pipette in the 3-neck flask. Start the flow of CO_2 through the system at a rate of about one bubble per second, observed in the liquid in the receiver beaker.
6. Add a sample (125 mL) of wine to the distillation flask. Rinse the contents into the flask with distilled water. Place a stopper in the neck of the flask.
7. Begin heating the distillation flask. Once distillation begins, continue it for about 40 minutes or until 60 mL of distillate has been collected in the receiver beaker.
8. Place the receiver beaker on a steam bath and concentrate the volume of distillate to approximately 20 mL. Allow the concentrated distillate to cool.

(c) Spectrophotometric analysis of biacetyl

9. Transfer the concentrated distillate with several small distilled water rinsings into a 50-mL volumetric flask.
10. Transfer 25 mL of each biacetyl calibration standard plus 1.5 mL

Table 12-8. Preparation of biacetyl calibration standards from 40-mg/L working standard.

Volume working standard (mL)	Final diacetyl concentration in 100-mL volumetric flask (mg/L)
12.5	5
25	10
50	20
100	40

hydroxylamine–acetate solution into a separate 50-mL volumetric flask.

11. Prepare a blank by transferring 25 mL of distilled water plus 1.5 mL of hydroxylamine–acetate solution into another 50-mL volumetric flask.
12. To each flask add 4 mL potassium phosphate–acetone solution, mix the solution, and let it sit for 5 minutes.
13. To each flask add 12.4 mL potassium tartrate–ammonium hydroxide solution, 0.8 mL ferrous sulfate (3.5%)–sulfuric acid solution, and enough distilled water to bring the contents to the mark. Mix the solution well.
14. Read the absorbance of each flask at 520 nm using 1-cm cuvettes. Set the instrument to zero (or 100% *T*) with the prepared blank.
15. Prepare a calibration curve from the absorbance readings of the standards. Read the concentration of biacetyl in the wine sample distillate directly from the calibration curve.
16. The concentration of biacetyl in the original wine sample is obtained by multiplying the value obtained from the calibration curve by a factor of 5:

$$\text{Biacetyl in wine} = \text{Value from graph} \times \frac{125}{50} \times \frac{50}{25}$$

IV. Supplemental Notes

1. The first correction factor (125/50) accounts for the fact that 125 mL of wine produced 50 mL of distillate plus reagents. The second correction factor (50/25) accounts for the dilution of standards (plus reagents) into 50-mL volumetric flasks.
2. There is research evidence that some grape varieties produce more biacetyl than others during malolactic fermentations.

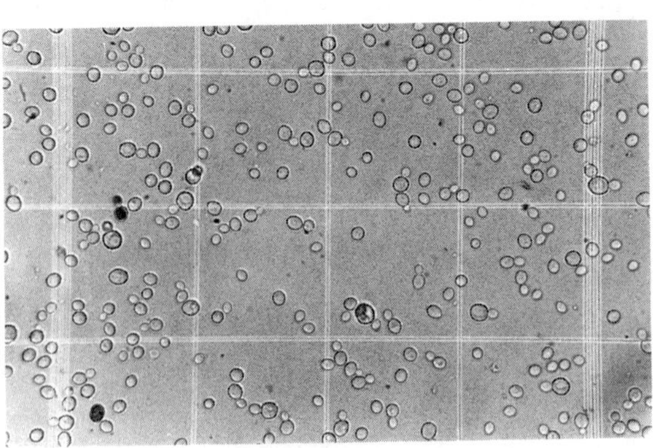

Photomicrograph 12-1. Yeast on cell counting slide. (Courtesy of Dr. Richard Dengré Lallemand, Inc. Montreal, Canada.)

Section III

Chemical Stability

Chemical and physical instabilities in wine frequently lead to the formation of new compounds that are insoluble. Such insoluble compounds may form a finely divided solid phase that remains suspended in solution producing a hazy appearance. These insoluble compounds may also form discrete crystals and precipitate out as a sediment on the cork or on the side/bottom of the bottle. Since a hazy/cloudy appearance or a significant amount of precipitate in a wine is considered to be a defect, these conditions are to be avoided.

The following four chapters consider chemical and physical instabilities caused by the presence of excessive amounts of tartaric acid, salts, the metals, copper and iron, and protein. While tartaric acid is an important component in wine quality, it can form insoluble crystalline precipitates with potassium and calcium naturally present in the wine. To obtain a measure of the stability or potential instability of a wine one can perform the freeze or conductivity tests. One can also directly measure the concentrations of tartaric acid—potassium, and calcium using a variety of chemical and instrumental techniques.

Copper and iron are generally found in very small quantities in wines made in modern wineries. When present these metals may react to form precipitates and fall out of solution as a hazy casse. Since there are treatments that will remove copper and iron from a wine, it may be to the winemaker's advantage to analyze for these metals. Both spectrophotometric and atomic absorption methods are presented in this section.

Proteinaceous compounds are important because they too can form hazy precipitates in wines. There are a number of techniques available for measuring protein stability. Other nitrogenous compounds play important roles in fermentation and processing; it is, again, important to have some measure of the amounts present. Methods are presented for accomplishing this using spectrophotometric and selective ion electrode methods.

Chapter 13

Tartaric Acid and Its Salts

As a principal acid in grapes and wine, tartaric acid plays an important role not only in contributing to overall wine acidity but also in the maintenance of juice and wine buffer systems. As a result of this activity, juice as well as wine is maintained (buffered) at a relatively low pH, which, in turn affects the biological stability and visible color, in addition to the overall acidic taste of the product. The tartaric acid content of grape musts varies from 0.2 to 1%, with 0.1 to 0.6% being found in most grape wines.

Tartaric acid (H_2T) and its salts, potassium bitartrate (KHT) and calcium tartrate (CaT), are normal constituents of wine, and are important from the viewpoint of precipitation and thus wine stability. Potassium bitartrate is believed to be produced after veraison. During ripening, the concentration of undissociated tartaric acid decreases with formation of its mono- and dibasic salts. The tartrate and potassium content of fruit varies not only from region to region, but also with the grape variety and maturity, as well as the soil type and viticultural practices (see Chapter 4).

In grapes and wines, tartaric acid also may be found in its ionized forms: bitartrate and tartrate. The relative distribution of each component as a function of pH is seen in Fig. 13-1. The reaction of bitartrate with potassium yields potassium bitartrate. Although potassium bitartrate is partially soluble in grape juice, the formation of alcohol during fermentation, coupled with lower temperature, decreases its solubility, resulting in a supersaturated solution of KHT that is poised toward precipitation.

Calcium is present in wine at levels of 6 to 165 mg/L (Amerine and Ough 1980) and may complex with tartrate and oxalate anions to form crystalline precipitates. Crawford (1951) reported calcium tartrate precipitate formation in sherry with calcium levels of 50 mg/L. Approximately 30% of the precipitate was present as calcium oxalate. Calcium instabilities usually appear from 4 to 7 months after fermentation.

Several sources have been thought to contribute to increased calcium levels in wine, including the soil in which the grapevine was grown, fermentation or storage in concrete tanks, and the use of calcium-containing fining material and filter pads. Furthermore, where $CaCO_3$ is used in deacidification practices or where "plastering" is employed in adjusting the acidity of shermat material,

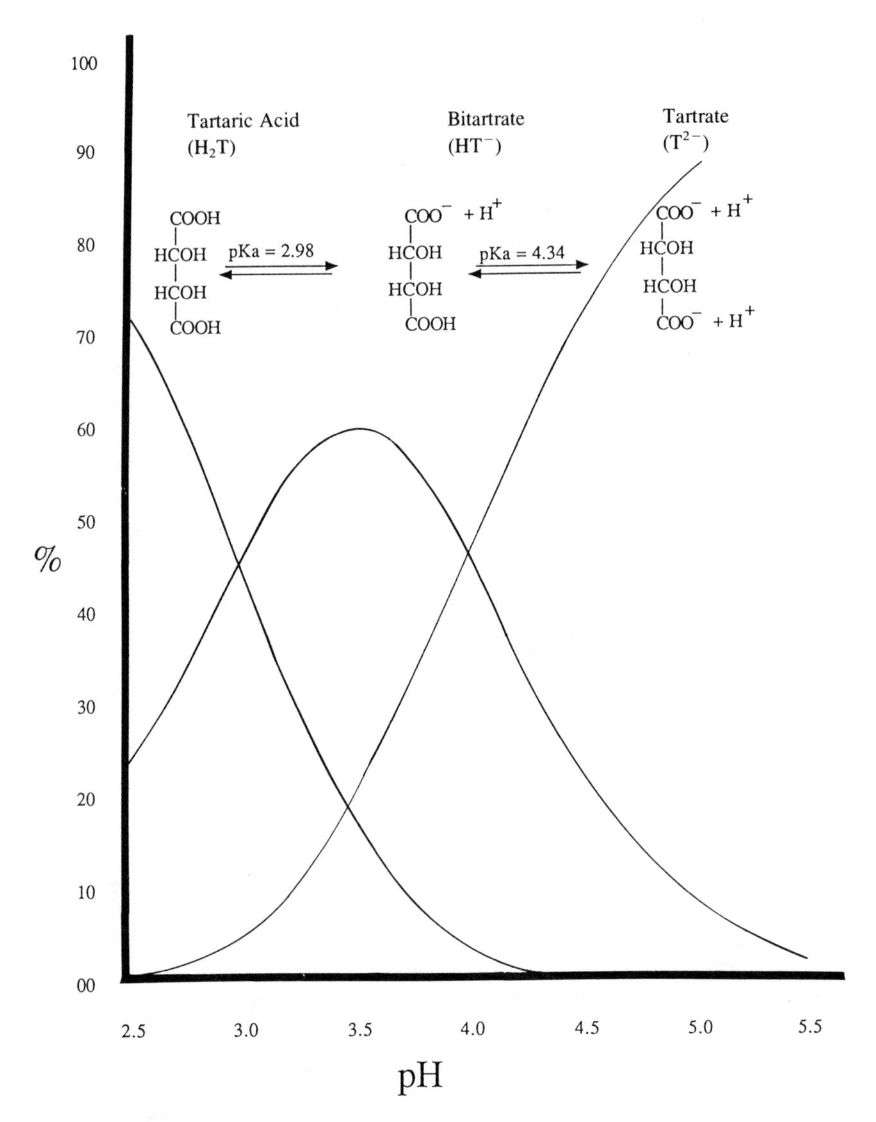

Fig. 13-1. Relative concentration of tartaric acid species in aqueous solution at different pH values.

increased calcium levels may be expected. Largely because of changes in processing technology, calcium tartrate generally is not a problem in the U.S. wine industry today.

All wines differ in their "holding" or retention capacity for tartrate salts in solution. If that capacity is exceeded, precipitation occurs, resulting in the formation of "tartrate casse." In wine, tartrate salt solubility is largely dependent upon alcohol content, pH, and temperature, as well as interactive effects of the solution matrix created by the various cations (particularly potassium and calcium) and anions.

As seen in Fig. 13-1, the percentage of tartrate present as HT^- is greatest at pH 3.7; and, if other factors permit, precipitation will be maximal at this point. Thus any wine treatment causing changes in pH, such as blending, the occurrence of a malolactic fermentation, and so on, may affect subsequent bitartrate precipitation. Therefore, the winemaker must be concerned with identifying the potential for precipitation and then correcting the problem. The following section covers the commonly used techniques for predicting bitartrate tartrate instabilities in wine.

BITARTRATE/TARTRATE STABILITY

Bitartrate/tartrate stability in wine is important to the winemaker because of the potential for precipitation of insoluble salts of potassium bitartrate and/or calcium tartrate once the wine is bottled. As the presence of these crystalline deposits in bottled wines is considered by many consumers to represent a defect, winemakers generally strive to reduce the potential for secondary precipitation.

Crystallization depends upon: (1) the concentration of the salt and other components that may be involved in the crystallization equilibrium; (2) the presence of nuclei of that salt, on which crystalline growth may occur; and (3) the presence of complexing factors that may impede crystal growth. In general, a certain level of supersaturation is necessary for adequate nucleation. Once nucleation has occurred, the further growth of crystals on those nuclei results in precipitation.

During alcoholic fermentation, potassium bitartrate, present in the must, becomes increasingly insoluble, and the resulting young wines may become supersaturated with respect to KHT. As the degree of supersaturation is affected by temperature, winemakers frequently enhance the potential for supersaturation and precipitation of salts by lowering the temperature of wine in storage.

Potassium bitartrate stability traditionally has been achieved by chilling or ion exchange (Chapter 18) or by combinations of both techniques. In conventional cold stabilization (chill-proofing), wines are chilled to a selected low temperature in order to decrease potassium bitartrate solubility and, optimally, bring about precipitation. Perin (1977) has calculated the optimum temperature needed for bitartrate stabilization using the following empirical relationship:

$$\text{Temperature } (-\,^\circ\text{C}) = \frac{(\text{alcohol } \%)}{2} - 1 \qquad (13\text{-}1)$$

Potassium bitartrate precipitation occurs in two stages. During the initial induction stage, the concentration of KHT nuclei increases because of chilling. This is followed by the crystallization stage, where crystal growth and development occur. The precipitation rate for potassium bitartrate at low temperatures is more rapid in table than in dessert wines. Further, precipitation from white wines is faster than from reds (Marsh and Guymon 1959). During conventional chill-proofing, precipitation usually is most rapid during the first 12 days. After the initial period, KHT precipitation decreases considerably, a reduction due to decreased levels of KHT saturation in solution. Temperature fluctuations during cold stabilization may have a significant effect in reducing precipitation rates because of the effect of this variable on the speed of nucleation needed for crystal growth. Without crystal nuclei formation, crystal growth and subsequent precipitation cannot occur. Simply opening the cellar doors in the winter, although seemingly cost-effective, may not be ideal for KHT precipitation. Further, oxygen is more soluble in wine at low temperature; and because of this potential for increased absorption of oxygen in wines held at low temperatures, alternatives to convention cold stabilization have been sought. (Problems related to oxidation are discussed in Chapter 8.)

Mixtures of calcium carbonate and calcium tartrate have been studied for their ability to increase the rate at which tartrate stability is achieved in wines. Upon addition of this mixture to a wine, the added calcium carbonate reacts with the tartaric acid in the wine to form insoluble calcium tartrate. The precipitation is aided by the seeding effects of the calcium tartrate crystals also present. In laboratory trials the tartaric acid concentration can be reduced to a low-enough level that cold stability is achieved (Clark et al. 1988). One potential advantage of this technique is that these materials function at room temperature.

Complexing Factors

Complexing factors can greatly affect potassium bitartrate formation and precipitation. As a result, wine—especially red wine—may retain the supersaturated condition longer than a corresponding alcohol–water solution. As seen in Fig. 13-2, it is thought that wine can support a supersaturated solution of potassium bitartrate because portions of the tartrate, bitartrate and potassium ions are complexed with wine constituents and thus resist precipitation.

Metals, sulfates, proteins, gums, polyphenols, and so on, can form complexes with free tartaric acid and potassium, thus inhibiting the formation of potassium bitartrate (Pilone and Berg 1965). The complexes are formed mainly between polyphenols and tartaric acid in red wines and between proteins and tartaric acid in whites.

In a study of white Bordeaux wines, Peynaud et al. (1964) found that sulfate was the most important factor in stability after potassium and tartrate. This

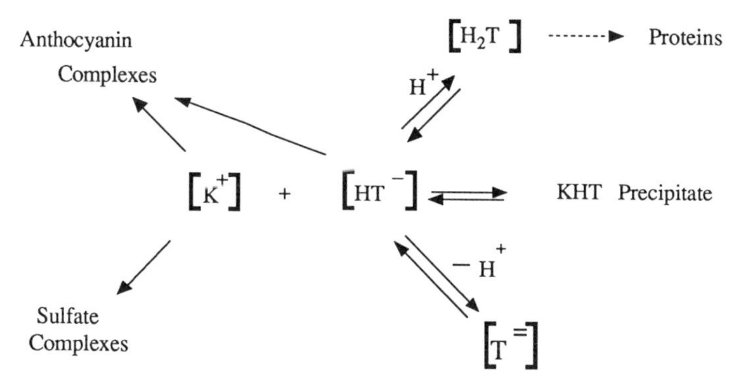

Fig. 13-2. Interaction equilibria between potassium bitartrate and "complexing factors" present in wine.

influence would appear to be due to complex formation between sulfate and potassium (Chlebek and Lister 1966). Almost one-half of the sulfate in white wines and 100% of the sulfate in red wines is thought to complex with potassium ion to form K_2SO_4 or KSO_4^- (Bertrand et al. 1978). The extent to which tartrate complexes form and their relative ability to impede precipitation depend upon the wine in question.

Red wine pigments often are involved in complexes formed with tartaric acid (Balakian and Berg 1968). As the wine oxidizes and pigment polymerization occurs, the holding capacity for tartaric acid diminishes resulting in delayed precipitation of tartaric acid. Additionally, the cultivar, growing conditions, and the growing region may well affect the holding capacity.

Pilone and Berg (1965) and Balakian and Berg (1968) have suggested the importance of colloids in bitartrate precipitation equilibria. It is thought that pectins and other polysaccharides such as glucans, produced by the mold *Botrytis cinerea*, may inhibit bitrartrate crystallization, perhaps because of adsorption of these compounds onto the surface of the growing crystal and prevention of further growth. However, in a study of German white wines, Neradt (1977) could find no inhibition of crystallization by either gelatin or gum arabic. Thus, each wine, because of its unique composition, will achieve unique solubility and equilibria under imposed temperature conditions.

Occasionally, winemakers choose to add complexing agents or inhibitors to prevent potassium bitartrate formation. In theory, the correct inhibitor, at the proper concentration, could reduce the need for cold stabilization. Unfortunately, none of the wine additives approved for such purposes is completely satisfactory.

Perhaps the best-known inhibitor of potassium bitartrate crystal formation is metatartaric acid, the hemipolylactide of tartaric acid, a compound produced by heating tartaric acid at 170°C for 120 hours. Although it has been approved as a wine additive in certain countries, metatartaric acid currently is not permitted in the United States.

It has been reported that 50 to 100 mg/L of metatartaric acid protects young wines from bitartrate precipitation even when stored at low temperatures for several months (Peynaud and Guimberteau 1961). The mechanism of inhibition appears to be an interference with the formation and growth of potassium bitartrate crystals, due to the coating of growing bitartrate crystals by metatartaric acid (Peynaud 1984).

After its addition to wine, metatartaric acid is slowly hydrolyzed to tartaric acid with a corresponding loss of activity. Its period of effectiveness is a function of the wine storage temperature. Peynaud (1984) reports that wines stored at 0°C were stable for several years, whereas metatartaric acid disappeared after two months in wines stored at 25°C. Hence, it would appear that metatartaric acid has its greatest application in wines that are expected to be consumed rather early.

Carboxyl methyl cellulose is another known inhibitor of potassium bitartrate precipitation (Cantarelli 1963). Purified apple pectin and tannin also inhibit crystal formation. Tannin addition at levels of 1 g/L strongly inhibits precipitation of tartrates (Wucherpfennig and Ratzka 1967).

Rather than adding inhibitors, a more common approach is to reduce or eliminate those complexing factors present in the wine that prevent or slow bitartrate precipitation. Thus, there is a close relationship between wine fining and potassium bitartrate stabilization. For example, it is well known that condensed polyphenols interfere with tartrate precipitation; so perhaps removal of a portion of these polyphenols by selective addition of fining agents prior to cold stabilization would enhance subsequent potassium bitartrate precipitation (Zoecklein, 1988).

Cold stabilization procedures (chill-proofing) result in precipitation of both potassium bitartrate and wine proteins. In white wines, the proteins may actively affect the tartrate holding capacity, inhibiting precipitation (Pilone and Berg 1965). White wines contain relatively large insoluble proteins; and as white wine phenols oxidize and polymerization occurs, there soon is binding and coprecipitation with the wine proteins. This precipitation affects tartrate holding equilibria and thus wine stability. However, white wines usually are too deficient in phenols for them to initiate protein precipitation. Bentonite fining decreases the tartrate holding capacity of wines by reducing levels of both wine proteins and phenolics (Berg and Akiyoshi 1971). Additionally, if the wine pH is below 3.65, chill-proofing may cause a downward shift in pH, which also enhances protein precipitation. Because of this reduction in pH and precipitation of proteins during cold stabilization, some winemakers elect to bentonite-fine during or following bitartrate stabilization. Bentonite fining during cold stabilization allows potassium bitartrate crystals to help compact bentonite lees. The addition of 2.5 or more pounds of bentonite per 1000 gallons has been found to reduce concentration product (CP) values of dry white wines from 15 to 18% and CP values of dry red wines from 25 to 32% (Berg et al. 1968) (see p. 298). Bentonite fining in conjunction with cold stabilization was covered extensively in a paper by Gorinstein et al. (1984).

Reducing sugars are known to impact tartrate stability. Berg (1960) reported 20% higher CP values in sweet sherries (6° B) when compared with sherries at 1° B. Further, final tartrate deposition took 23% longer in the higher-sugar group. A similar trend was demonstrated by Berg and Akiyoshi (1971) in examination of wines from California.

Bitartrate Stabilization and Changes in TA and pH

Winemakers should expect reductions in both pH and titratable acidity using conventional cold stabilization and seeding in wines with initial pH values below 3.65, because of the generation of one free proton per molecule of KHT precipitated. Using these techniques, one might expect pH levels to drop by as much as 0.2 pH unit with a corresponding decrease in titratable acidity of up to 2 g/L. By comparison, potassium bitartrate precipitation in wines with initial pH values above 3.65 results in higher pH and a corresponding decrease in titratable acidity, as a result of the removal of one proton per tartrate anion precipitated. The above values represent ranges seen in practice and may vary considerably. With contact seeding and potentially with conventional cold stabilization, some winemakers report reduction in extract content after treatment (see Chapter 3). This may present a problem in the case of low-extract wines.

Definition of Stability

Opinions among winemakers differ with regard to the definition of cold stabilization, as well as cellaring techniques used to achieve cold stability; and there is also variation in laboratory measurements used to predict this instability. Table 13-1 summarizes responses in a survey of California wineries, highlighting the multitude of procedures used to determine cold stability (Cooke and Berg 1984). The fact that there is no industry standard for the determination of cold stability is readily apparent from this survey. This is not surprising in view of the fact that stability is a relative term that can and may be defined differently by each winery.

METHODOLOGY FOR ESTIMATING COLD STABILITY

Three methods are presented for determining potassium bitartrate stability: the freeze or slush test, measurement of conductivity, and concentration product (CP) values. For details regarding each technique, consult the procedures section of this chapter.

Table 13.1. Techniques used for determination of cold stability among several California wineries.

Winery	Test
A	Hold sample at 40°F for two weeks. Examine for crystalline deposits.
B	Two weeks at 32–34°F. Examine at 7 and 14 days.
C	Two weeks at 34°F. Examine while sample is still cold.
D	Hold sample for 4 hours at 12°F. Examine.
E	Hold sample for 30 days at 30°F. Run C.P.
F	Hold sample for 72–96 hours at 40°F. Examine. Alternatively, freeze sample for 12 hours. Thaw sample and examine at room temperature.
G	Freeze sample for 12 hours and/or hold for 72 hours at 28°F. Examine while sample is cold and at room temperature. Stability determination is based upon the amount of precipitate formed.
H	Hold sample 3–4 weeks at 35°F. Examine while sample is cold and at room temperature.
I	Hold sample for 96 hours at 18°F. Examine. Allow chilled sample to sit at room temperature for 24 hours. Examine. Precipitate formed at 18°F should redissolve upon standing for 24 hours at room temperature.
J	Filter sample. Hold for 24 hours at 25°F. Examine.
K	Filter sample thru 0.65 μm membrane. Freeze for 24 hours. Thaw and examine at room temperature.
L	Filter sample through 0.45 μm membrane. Run tests on two samples: (1) Freeze sample for 12 hours. Thaw, examine after 12 hours at room temp. Record results after each treatment. (2) Hold sample at 38–42°F for 72 hours. Bring to room temperature and hold for 12 hours. Examine after each treatment. Run C.P.
M	Filter sample through 0.45 μm membrane. (1) Hold sample at 10°F for 16 hours. Thaw, examine at room temperature. (2) Hold second sample at 15°F for 16 hours. Examine at room temperature.
N	Pad filter sample. Membrane filter if needed. (1) Hold one aliquot at 26°F for 48 hours. Hold second aliquot at 100°F for 48 hours. Examine both aliquots. (2) Hold both aliquots at 28°F for 48 hours. Examine both aliquots. Hold both above aliquots and additional 48 hours at room temperature and examine.
O	Filter sample through 0.65 μm membrane. Seed with KHT crystals. Hold at 25°F for 36 hours with mixing. Measure changes in total tartrates.
P	Not determined. Wine considered stable after prolonged aging in cool cellar.
Q	Not done.
R	Concentration Product (C.P.).

Cooke, G. M. and Berg, H. W. 1984. A re-examination of varietal table wine processing practices in California. II. Clarification, stablization, aging, and bottling. *Am. J. Enol. and Vitic. 35(3)*: 137–42.

The Freeze Test

This procedure relies on the formation of potassium bitartrate crystals as a result of holding wine samples at reduced temperatures for specified time periods. As seen in Table 13-1, a wide variety of procedures are employed for this analysis. Often one of the samples is frozen and then thawed to determine the development of bitartrate crystals and whether the crystals formed subsequently resolubilize. The absence of crystal formation and resolubilization of those that do form is taken to indicate that the wine in question is stable with respect to potassium bitartrate precipitation. As ice formation increases, there is an increase in the relative concentrations of all species in the sample, including alcohol, thus enhancing nucleation and crystallization. In fact it is difficult to accurately relate crystal formation, in this concentrated wine sample, with potassium bitartrate instability.

In common practice, preliminary laboratory treatment of samples may call for some sort of filtration. However, such filtration also removes crystal nuclei that may affect test results, and therefore one should not filter. If one has filtered the wine sample, the freeze test is, in reality, a measurement of the crystallization rate (Boulton 1983). Despite deficiencies in the test, as seen in Table 13-1, it is probably the most widely used technique for attempting to predict cold instability.

The Conductivity Test

As noted above, the monitoring of crystal formation in a wine sample held at low temperature (slush/freeze test) is essentially a measure of the precipitation rate—that is, the formation of nuclei (nucleation), secondary crystal growth, and subsequent precipitation. Unless one provides seed crystals, precipitation over the relatively short period of the test is, in fact, a measure of the wine's ability to form nuclei and precipitate. And, what one often sees from such results is variation in this relative ability. For example, two wines may be equally unstable, but after filtration and storage at low temperature, one precipitates bitartrate and is considered "unstable" whereas the second does not form nucleation sites at the same rate and is deemed "stable" (see discussion in Procedure 13-1). Thus, the freeze or slush test is of limited value in the prediction of bitartrate stability in wine.

To improve the accuracy of the determination of potassium bitartrate stability, the sample may be seeded with finely ground potassium bitartrate powder. The oversaturated portion of tartaric acid and potassium present in the wine deposits on the added seed.

Estimation of KHT stability then is based upon electrical conductivity of the juice or wine solution. To perform this determination, one saturates a sample with finely ground potassium bitartrate seed powder at a defined and carefully controlled low temperature. As K^+ is the major conducting species in wine, bitartrate precipitation can be measured by changes in conductance between the beginning and the end of the test. Changes of less than 5% in conductance during the test period generally are interpreted to mean that the wine in question is stable. However, stable results are valid only *at (or above) the temperature at which the test was run*. The wine may not be stable with respect to bitartrate precipitation at lower temperatures. Thus the test should be carried out at the lowest temperature that the wine is expected to encounter after bottling. In practice, most winemakers select 0°C as the stabilization temperature for whites and 5°C for reds.

Thus, the conductivity test provides a final stable conductivity value that is unique for the wine tested. Complexing or "fouling" factors that may be present and possibly affect potassium bitartrate crystal formation are taken into account.

The Concentration Product

Prediction of precipitation based upon the solubility of products is widely known. In the case of wine, Berg and Keefer (1958) attempted to quantify solubility limits of potassium bitartrate in model solutions. Later, Berg and Keefer (1959) and DeSoto and Yamada (1963) established similar solubility limits for calcium tartrate.

In their review of stability evaluation (Table 13-1), Cooke and Berg (1984) found only two out of 18 California wineries using concentration product values to establish stability in wines. This is in part due to the difficulty in measuring the necessary parameters without having available suitable analytical equipment.

By measuring the concentration of potassium, calcium, total tartrates (as bitartrate or tartrate), pH, and alcohol, one can, in theory, calculate the solubility product for potassium bitartrate:

$$CP = (K^+ \text{ moles/L})\,(\text{Total tartrate moles/L})\,(\%HT^-) \qquad (13\text{-}2)$$

The values for % bitartrate and/or tartrate at measured pH and alcohol levels are taken from Tables 13-3 and 13-4.

As an example, consider the typical data for a white wine presented in Table 13-2. Carrying out the calculations (referring to Table 13-3):

$$CP = \frac{1100 \times 10^{-3}}{39.1} \times \frac{1600 \times 10^{-3}}{150} \times 0.661$$

$$CP = 19.8 \times 10^{-5}$$

Table 13-2. Analytical data from white wine 87-013a.

Analysis	Result
K^+	1100 mg/L
Ca^{++}	76 mg/L
Tartrate	1600 mg/L
pH	3.74
Alcohol	12% (v/v)

From Table 13-5, the minimum CP for a dry white wine is 16.5×10^{-5} at 0°C. Because the calculated CP, in this example, exceeds values considered "safe," we conclude that the wine will throw tartrates.

The original work on tartrate solubility products was done in model systems. These data were then extrapolated to the wine system. In this case, however, the theory is at variance with what is observed in practice. In addition to the inherent problem of being unable to quantify the contribution of complexing agents, the methodology of component analyses presents other problems.

Potassium

The levels of potassium in grape musts range from 600 mg/L to over 2500 mg/L in certain red varieties. During veraison, potassium from the soil is translocated into the fruit where it forms soluble potassium bitartrate. Factors affecting potassium uptake are reported to include soil type, cultivar, rootstock, and so on (see Chapter 4), as well as cultural practices. Although potassium bitartrate is often soluble in unfermented juice, the presence of ethanol during and after fermentation greatly decreases its solubility, resulting in a young wine that is supersaturated with respect to the salt and poised toward precipitation. Potassium content may be measured by a number of analytical techniques—atomic absorption analysis (Procedure 13-5), flame photometry (Procedure 13-6), and potentiometry (using the potassium ion selective electrode—Procedure 13-8) being the most common ones. The shelf life of the potassium ion selective electrodes is somewhat limited; otherwise the cost of the equipment required (a good pH or selective-ion meter) is less than for the other two techniques.

Calcium

The calcium content in American wines ranges from 6 to 165 mg/L (Amerine and Ough 1980). As is the case with potassium, calcium forms insoluble salts with tartrates present in wine. Unlike potassium salts, however, calcium tartrate precipitation occurs more slowly, occasionally requiring months for complete crystallization. As a result, there is a potential for its precipitation in bottled wine.

Table 13-3. Changes in percent bitartrate (HT) in total tartrates with changes in pH and alcohol (% v/v) at 20°C.

alcohol (%) 10		11		12		13		14		16		17		18		19		20		21	
pH	%HT	pH	%HT	pH	%HT	pH	%HT	pH	%HT	pH	%HT	pH	%HT	pH	%HT	pH	%HT	pH	%HT	pH	%HT
2.81	37.7	2.83	38.0	2.84	38.3	2.85	38.6	2.87	38.9	2.89	39.5	2.91	39.8	2.92	40.0	2.93	40.3	2.95	40.6	2.96	40.9
2.91	42.9	2.93	43.2	2.94	43.5	2.95	43.8	2.97	44.1	2.99	44.7	3.01	45.0	3.02	45.3	3.03	45.6	3.05	45.9	3.06	46.2
3.01	48.1	3.03	48.4	3.04	48.7	3.05	49.0	3.07	49.3	3.09	49.9	3.11	50.3	3.12	50.5	3.13	50.8	3.15	51.1	3.16	51.4
3.11	53.0	3.13	53.3	3.14	53.6	3.15	53.9	3.17	54.2	3.19	54.8	3.21	55.1	3.22	55.4	3.23	55.7	3.25	56.0	3.26	56.3
3.21	57.4	3.23	57.7	3.24	58.0	3.25	58.3	3.27	58.6	3.29	59.1	3.31	59.4	3.32	59.7	3.33	60.0	3.35	60.3	3.36	60.6
3.31	61.1	3.33	61.4	3.34	61.7	3.35	61.9	3.37	62.2	3.39	62.8	3.41	63.1	3.42	63.4	3.43	63.6	3.45	63.9	3.46	64.2
3.41	63.9	3.43	64.2	3.44	64.4	3.45	64.7	3.47	65.0	3.49	65.5	3.51	65.8	3.52	66.1	3.53	66.4	3.55	66.6	3.56	66.9
3.51	65.6	3.53	65.9	3.54	66.2	3.55	66.4	3.57	66.7	3.59	67.2	3.61	67.5	3.62	67.8	3.63	68.0	3.65	68.3	3.66	68.6
3.61	66.2	3.63	66.5	3.64	66.7	3.65	67.0	3.67	67.3	3.69	67.8	3.71	68.1	3.72	68.3	3.73	68.6	3.75	68.9	3.76	69.1
3.71	65.6	3.73	65.9	3.74	66.1	3.75	66.4	3.77	66.7	3.79	67.2	3.81	67.5	3.82	67.8	3.83	68.0	3.85	68.3	3.86	68.6
3.81	63.9	3.83	64.1	3.84	64.4	3.85	64.7	3.87	65.0	3.89	65.5	3.91	65.8	3.92	66.1	3.93	66.4	3.95	66.6	3.96	66.9
3.91	61.1	3.93	61.4	3.94	61.7	3.95	61.9	3.97	62.2	3.99	62.8	4.01	63.1	4.02	63.4	4.03	63.6	4.05	63.9	4.06	64.2
4.01	57.4	4.03	57.7	4.04	58.0	4.05	58.3	4.07	58.6	4.09	59.1	4.11	59.4	4.12	59.7	4.13	60.0	4.15	60.3	4.16	60.6
4.11	53.0	4.13	53.3	4.14	53.6	4.15	53.9	4.17	54.2	4.19	54.8	4.21	55.1	4.22	55.4	4.23	55.7	4.25	66.0	4.26	56.3
4.21	48.1	4.23	48.4	4.24	48.7	4.25	49.0	4.27	49.3	4.29	49.9	4.31	50.2	4.32	50.5	4.33	50.8	4.35	61.1	4.36	51.4

Berg, H. W., and R. M. Keefer. Analytical determination of tartrate stability in Wine. I. Potassium bitartrate. *American Journal of Enology* 9: 180–83. (1958)

Table 13-4. Changes in percent tartrate ($T^=$) in total tartrates with changes in pH and alcohol (% v/v) at 20°C.

Percent alcohol by volume																					
10		11		12		13		14		16		17		18		19		20		21	
pH	% $T^=$	pH	% $T^=$	pH	% $T^=$	pH	% $T^=$	pH	% $T^=$	pH	% $T^=$	pH	% $T^=$	pH	% $T^=$	pH	% $T^=$	pH	% $T^=$	pH	% $T^=$
2.81	1.5	2.83	1.5	2.84	1.5	2.85	1.5	2.87	1.5	2.89	1.5	2.91	1.5	2.92	1.5	2.93	1.5	2.95	1.5	2.96	1.5
2.91	2.2	2.93	2.2	2.94	2.2	2.95	2.2	2.97	2.1	2.99	2.1	3.01	2.1	3.02	2.1	3.03	2.1	3.05	2.1	3.06	2.1
3.01	3.1	3.03	3.1	3.04	3.0	3.05	3.0	3.07	3.0	3.09	3.0	3.11	3.0	3.12	2.9	3.13	2.9	3.15	2.9	3.16	2.9
3.11	4.3	3.13	4.3	3.14	4.2	3.15	4.2	3.17	4.2	3.19	4.1	3.21	4.1	3.22	4.1	3.23	4.0	3.25	4.0	3.26	4.0
3.21	5.8	3.23	5.8	3.24	5.8	3.25	5.7	3.27	5.7	3.29	5.6	3.31	5.6	3.32	5.5	3.33	5.5	3.35	5.4	3.36	5.4
3.31	7.8	3.33	7.8	3.34	7.7	3.35	7.7	3.37	7.6	3.39	7.5	3.41	7.4	3.42	7.4	3.43	7.3	3.45	7.2	3.46	7.2
3.41	10.3	3.43	10.2	3.44	10.1	3.45	10.1	3.47	10.0	3.49	9.8	3.51	9.7	3.52	9.7	3.53	9.6	3.55	9.5	3.56	9.4
3.51	13.3	3.53	13.2	3.54	13.1	3.55	13.0	3.57	12.9	3.59	12.7	3.61	12.6	3.62	12.5	3.63	12.4	3.65	12.3	3.66	12.2
3.61	16.9	3.63	16.8	3.64	16.6	3.65	16.5	3.67	16.4	3.69	16.1	3.71	16.0	3.72	15.8	3.73	15.7	3.75	15.6	3.76	15.4
3.71	21.1	3.73	20.9	3.74	20.8	3.75	20.6	3.77	20.4	3.79	20.1	3.81	19.9	3.82	19.8	3.83	19.6	3.85	19.4	3.86	19.3
3.81	25.9	3.83	25.7	3.84	25.5	3.85	25.3	3.87	25.1	3.89	24.7	3.91	24.5	3.92	24.3	3.93	24.1	3.95	23.9	3.96	23.7
3.91	31.1	3.93	30.9	3.94	30.7	3.95	30.4	3.97	30.2	3.99	29.8	4.01	29.5	4.02	29.3	4.03	29.1	4.05	28.8	4.06	28.6
4.01	36.8	4.03	36.5	4.04	36.3	4.05	36.0	4.07	35.8	4.09	35.3	4.11	35.0	4.12	34.8	4.13	34.5	4.15	34.3	4.16	34.0
4.11	42.8	4.13	42.5	4.14	42.2	4.15	41.9	4.17	41.7	4.19	41.1	4.21	40.8	4.22	40.6	4.23	40.3	4.25	40.0	4.26	39.8
4.21	48.9	4.23	48.6	4.24	48.3	4.25	48.0	4.27	47.7	4.29	47.1	4.31	46.9	4.32	46.6	4.33	46.3	4.35	46.0	4.36	45.7

Berg, H. W., and R. M. Keefer. Analytical determination of tartrate stability in wine. II. Calcium tartrate. *American Journal of Enology 10*: 105–109. (1959)

Table 13-5. Suggested maximum CP levels for potassium acid tartrate and calcium tartrate for California wines.

	Potassium Acid Tartrate ($\times 10^{-5}$)		Calcium Tartrate ($\times 10^{-8}$)	
Wine type	Highest level found	Suggested safe level	Highest level found	Suggested safe level
White table	18.5	16.5	230	200
Red table	34.7	30.0	590	400
Pale dry sherry	14.6	10.0	170	90
Sherry	15.0	17.1	202	125
Cream sherry	15.0	10.0	171	120
Muscatel	18.0	17.5	310	250
White port	11.0	10.5	175	155
Port	23.3	20.0	410	275
Dry vermouth	7.0	6.0	69.5	50

Yamada, H. and R. T. DeSoto, Relationship of solubility products to long range tartrate stability, *American Journal of Enology* 14:43–51 (1963).

Tartrate

The tartaric acid content of grape must ranges from 2.0 to 10 g/L (Dunsford 1979). Depending upon pH, the ratios of $H_2T/HT^-/T^=$ can vary greatly and thus significantly influence the potential for precipitation of insoluble salts.

Ethanol Concentration

Potassium bitartrate becomes increasingly insoluble during fermentation and in the wine, owing to the lower dielectric constant for ethanol–water mixtures (as compared with water). Because of this physical property of solutions, the ethanol content is a necessary chemical parameter in any determination of tartrate stability in wine using CP relationships.

pH

As seen in Fig. 13.1, the distribution of tartrate/bitartrate species in the total population of tartrate species is pH-dependent. The percentage of tartrate as either bitartrate or tartrate was calculated by Berg and Keefer (1958, 1959) as a function of pH and ethanol concentration in model solutions. The resultant value must be used in the calculation of concentration products.

Determination of CP for a wine yields a final numerical quantity related to the solubility product for the particular wine. By comparison to the solubility values provided in tables, the winemaker may calculate proportions of blend components that will yield a theoretically stable final product. As with any blending operation, the winemaker should not assume that blending two or more stable wines will result in a stable final blend. Changes in alcohol, pH, tartrate and cation concentrations, as well as undefined contributions of complexing agents, may contribute to potential instability.

Procedure 13-1. The Freeze Test for Wine Stability

The freeze or slush test involves freezing, or chilling, a wine sample at a defined low temperature for a prescribed period of time. Upon thawing, the sample is examined for the presence of crystals. If they are present, the wine is judged to be "unstable"; conversely, the absence of crystals is taken to mean stability. In spite of its limits, the technique continues to be a widely utilized measure of stability.

It has been common practice to prefilter the sample, to remove noncrystalline amorphous material, but this step has been called into question. Unless one provides seed crystals, precipitation over the short time period of the test is, in fact, a measure of the ability to form nuclei and precipitate. As Clark et al. (1985) point out, two wine samples may be equally unstable, but upon filtration and storage at low temperature one precipitates tartrates and is considered "unstable," whereas the second does not form nucleation sites at the same rate and is considered "stable."

In addition to conventional freeze tests, many wineries routinely evaluate stability at refrigeration temperatures. In this case, however, the instability, seen as a haze, is due to protein–tannate complexes that become insoluble at lower temperatures.

I. Equipment
   Refrigerator or freezer for constant-temperature storage

II. Procedure
   1. Place a clarified wine or juice sample in a 4-oz vial.
   2. Place the sample in the freezer for a predefined period of time, determined by experience (see Table 13-1 for examples of the time). Retain a similarly treated "control" at room temperature.
   3. Evaluate the sample, upon thawing, to see whether it has precipitated crystals or not. Note the crystal volume, being sure to distinguish between crystals and precipitated amorphous material (see Supplemental Note 4). The sample is reevaluated at room temperature to note the extent of crystal resolubilization.
   4. If all the crystalline precipitates that were formed redissolve upon warming of the sample to room temperature, some winemakers consider the original wine to be stable. However, subsequent exposure to low temperatures may bring about recrystallization.

III. Supplemental Notes
   1. Should one use membrane filtration to clarify the sample, the foci needed for crystallization are removed. In this instance the procedure

probably is not a measure of bitartrate stability, but rather the capacity of the wine to form nuclei and their subsequent rate of precipitation.

2. The above procedure is subjective. Various wineries utilize different times for the test and different temperatures, as well as different interpretations of stability based on the amount of crystal formation.
3. In addition to bitartrate crystals, other solid materials may fall out of solution during the freeze test. These materials consist of proteins, phenols, pectins, and colloidal materials, and may be indicators of wine instability.
4. In order to distinguish between bitartrate crystals and other solid materials, one should use a high-intensity light source held at right angles to the line of vision. Check for refraction of light from the bitartrate crystals. (See Section IV remedial actions for further identification tests.)

Procedure 13-2. The Conductivity Test for Bitartrate Stability

This determination employs saturation of a wine sample with potassium bitartrate powder (seed crystals) at a defined and controlled temperature. Potassium ion is the major conducting species in wine; so bitartrate precipitation can be measured by changes in conductance from the beginning to the end of the test. Changes of less than 5% in conductance during the test interval are considered an indication that the wine in question is stable. However, stable results are valid only at, or above, the temperature at which the test was run; the wine may be unstable at lower temperatures (see text, above).

I. Equipment

Conductivity meter: Unit should be capable of measuring in the range of 100–1000 micro-Siemens. A Siemens unit is 0.94073 International Ohm. Meters also are available that combine conductivity, pH, and mV capabilities.

Controlled-temperature water bath, +/−1.0°C

Magnetic stirrer

Thermometer

II. Reagents

Potassium bitartrate powder (40–70 μm). See Supplemental Note 1 for further details.

III. Procedure

1. Using a 100 mL volume, equilibrate juice or wine to the desired stability temperature before proceeding. Measure the conductivity of the original sample.
2. Add 1.0 g of potassium bitartrate powder. Place the conductivity probe in the sample.
3. While constantly mixing the sample, take readings (conductivity and

temperature) at 2-minute intervals. Temperature fluctuations should not exceed ±1 °C over the test period. Continue mixing at the defined temperature until conductance readings stabilize—usually 20–30 minutes or until several successive readings indicating no change in conductance.

IV. Interpretation

The final conductivity value corresponds to that of the stable wine. It can be used for a comparison with conductivity values of samples taken from the winery during full-scale stabilization treatment to determine if and when stability has been reached. The difference between the conductivity value before the addition of powdered potassium bitartrate and the final value is a measure of potential KHT instability.

1. Conductance reading changes of more than 5% between the beginning (unseeded sample) and the end of the test (stable conductivity reading) suggest that the wine is unstable with respect to bitartrate precipitation.
2. Conductance reading differences of less than 5% are taken to mean that the sample is stable. (Some procedures use a value of 3%.)

V. Supplemental Notes

1. The bitartrate seed used in laboratory evaluations should be the same as that used in cellar operations. Optimally, the crystal size should be 40 to 70 μm.
2. Conductivity values are affected by temperature. Therefore, the sample temperature must be constant throughout the test. As the temperature rises, conductivity readings will increase because of temperature effects.

Procedure 13-3. Tartaric Acid Analysis by Metavanadate (Carbon Decolorization)

Several methods of tartaric acid determination are in use in winery laboratories. The metavanadate method involves a colorimetric reaction between sodium (or ammonium) metavanadate and tartaric acid in acetic acid solution. This methodology frequently requires initial treatment of wine samples and standards with activated charcoal for decolorizing, prior to reaction of the extract with metavanadate.

Several studies have shown that variations in this first step lead to significant differences in the final measurement of tartrate (Clark et al. 1985). Carbon removes different amounts of tartaric acid from standards than from wine samples. Thus, the metavanadate procedure (with carbon pretreatment) generally will give high results.

Hill and Caputi (1970) reported that the presence of sugar and lactic acid also interferes with the analysis. These workers utilized a preliminary separation of the tartaric acid on acetate-charged resin prior to analysis (see Procedure 13-4). Clark's remedy for the problem with carbon was to use a high performance liquid chromatographic technique (see Procedure 4-3), the one used for organic acids in general. By avoiding the use of carbon, both Caputi's and Clark's methods for tartrate analysis should produce accurate results. However, both the methodology for determination of tartrates and technical expertise may vary from one laboratory to the next; so analytical results may vary significantly.

I. Equipment
 125-mL Erlenmeyer flasks
 100-mL volumetric flasks
 Electric hot plate
 Volumetric pipettes (2, 4, 25 mL)
 Graduated cylinder (100 mL)
 Spectrophotometer (visible)
 Matched cuvettes
 Whatman #1 filter paper
 Glass funnels
 Wash bottles

II. Reagents
 Tartaric acid stock standard solution (10 g/L)
 1 *N* HCl
 Concentrated acetic acid
 Sodium metavanadate (5%)
 Decolorizing carbon
 Boiling chips

III. Procedure
 (a) Preparation of standard curve
 1. Using 125-mL Erlenmeyer flasks, prepare a series of standards according to the instructions in Table 13-6.
 2. With these 20-mL standards, proceed with steps 4 through 11 below. Plot the concentration (g H_2T/L) vs. absorbance at 520 nm.
 (b) Wine samples
 3. To a clean 125-mL Erlenmeyer flask, volumetrically transfer 20 mL of wine.

Table 13-6. Preparation of standard solutions of tartaric acid from 10 g/L stock.

Standard solution (mL)	Volume H_2O (mL)	Final concentration H_2T (g/L)
2	18	1.00
4	16	2.00
6	14	3.00
8	12	4.00
0 (blank)	20	0.00

4. Add 2 mL 1 *N* HCl, 30 mL distilled water, 0.50 g (accurately weighed) decolorizing carbon, and several boiling chips.
5. Boil the solution on a hot plate 2 to 3 minutes; then cool it to room temperature.
6. Filter the solution into a 100-mL volumetric flask using Whatman #1 filter paper.
7. Examine it for clarity. If necessary, refilter the solution. Take care to thoroughly rinse the Erlenmeyer flask and filter paper with distilled water. Multiple rinses using small amounts of water are recommended.
8. Bring the solution to a 100-mL volume with distilled water.
9. Transfer 25 mL decolorized solution to another 100-mL volumetric flask. Add 2 mL concentrated acetic acid and 10 mL filtered metavanadate solution, and fill the flask to the mark with distilled water.
10. Allow 30 minutes for color development. Record absorbance on the spectrometer at 520 nm.
11. With reference to a standard curve, determine the tartaric acid content of the sample.

IV. Supplemental Notes

1. As noted, the above procedure suffers from interferences due to the use of carbon in the decolorization step as well as the presence of sugar and lactic acid.
2. In step 7, the rinsing operation must be thorough and uniform for the samples.
3. All glassware must be thoroughly clean prior to the beginning of work.
4. Sodium metavanadate is only partially soluble, and its concentration will vary with mixing and storage time. Solutions should be filtered before each run, and a new standard curve should be prepared daily.
5. Keep developed solution away from direct sunlight.
6. For wines that are being stabilized at low temperatures, it is essential that the following routine be observed in regard to the collection and analysis of H_2T: While the wine is under refrigeration, the sample is collected at either middle or top "draws" (avoid collection at the bottom valve!), into a cool sample bottle. It is then rushed to the laboratory and filtered while cold prior to analysis. Allowing the sample to warm prior to filtration may result in resolubilization of some bitartrate crystals.

Procedure 13-4. Alternative Procedure for Tartaric Acid Analysis (Ion Exchange Sample Preparation)

The following procedure of Hill and Caputi (1970) employs a preliminary separation of wine interferences by anion exchange. The sample is placed on an acetate-charged resin and eluted with sulfate ion. The tartaric acid–containing fraction, now isolated from potential interferences, is analyzed in a manner similar to the preceding method.

I. Equipment
   Spectrophotometer (visible) and matched cuvettes
   Chromatography column (500 × 12 mm) with sintered glass plate and needle control valve
   Strong anion exchange resin (Duolite A101D)
   50-mL volumetric flasks
   Volumetric pipettes

II. Reagents (See Appendix II)
   Tartaric acid standard solution (10 g/L)
   Acetic acid (glacial)
   Acetic acid solutions (30% vol/vol and 0.5% vol/vol)
   Sodium sulfate (0.5 *M*)
   Sodium metavanadate (3% wt/vol)

III. Procedure
   (a) Separation
   1. Pack 20 mL anion exchange resin into the column.
   2. Using 20-bed volumes of 30% acetic acid, charge the resin.
   3. Follow this step with a wash of 30 mL acetic acid (0.5%) and 50 mL distilled water. Steps 2 and 3 (and all following steps) should be carried out at a flow rate of 2.5 to 5.0 mL/min.
   4. Add 10 mL wine and 20 mL 0.5% acetic acid sequentially to the column, followed by 50 mL distilled water.
   5. Add 0.5 *M* sodium sulfate to elute acids. Eluates are collected in graduate cylinders. The tartaric acid fraction elutes from the 46th through the 75th mL.

   (b) Analysis
   6. Transfer 25 mL of the tartaric acid fraction to a 50-mL volumetric flask.
   7. Add 2 mL metavanadate and 0.5 mL glacial acetic acid. Bring the solution to volume with distilled water.
   8. Immediately prepare a reagent blank, according to steps 6 and 7, using 25 mL 0.5 *M* sodium sulfate.
   9. Thoroughly mix each solution and hold at room temperature for 80 minutes.
   10. Measure the absorbance of sample(s) at 480 nm vs. the blank and compare the result to the standard curve.

   (c) Standard curve preparation
   11. Dissolve 2.50 g tartaric acid in a 100-mL volumetric flask and bring the solution to volume with 0.5 *M* sodium sulfate.
   12. Using 50-mL volumetric flasks, prepare the series shown in Table

Table 13-7. Preparation of standard solutions of tartaric acid from stock (25 g/L H_2T).

Stock solution (mL)	Final concentration H_2T (g/L) (in 50-mL volume)
2	1.00
4	2.00
6	3.00
8	4.00
0	0.00

13-7. Each secondary standard is brought to volume using 0.5 *M* sodium sulfate (Table 13-7).

13. Using 25-mL aliquots of these standards, proceed according to steps 6 through 10 above.
14. Plot concentration (g/L tartaric acid) vs. absorbance at 480/nm.

IV. Supplemental Notes

1. The above analysis has the added benefit that, under the conditions of separation, any lactic acid present in the sample is eluted in the first 30 mL.
2. Once developed, color is reported to be stable for 1 hour (Hill and Caputi 1970).
3. Hill and Caputi did not specify the resin particle size. However, for best results, the smallest available particle size should be used.
4. Because resins and packing methods differ, it is advisable to test the ion exchange column by running a standard tartaric acid solution. Ten-milliliter fractions should be collected preceding and following the tartaric acid fraction.

Procedure 13-5. Potassium Analysis by Atomic Absorption

Analyses of potassium in wine frequently are carried out using flame spectroscopic methods (below) and ion selective electrode techniques (Procedure 13-8). Both flame emission and flame absorption techniques can be used in the wine laboratory. Emission techniques (Procedure 13-6) measure the radiation (either ultraviolet or visible) emitted by atomic species excited by the flame during their return to ground state. Absorption techniques measure the radiation (either UV or visible) absorbed from an external light source by ground-state atomic species in the flame. Sample elements of interest in liquid form are aspirated into the flame, desolvated, and converted into a cloud of ground-state atoms. These atoms are available to participate in the atomic absorption or emission process. The energy absorbed or emitted at specific wavelengths is monitored using a typical monochrometer and photo-detector.

In emission techniques the intensity of the emitted light is measured in terms of transmittance ($\%T$), whereas in absorption techniques the intensity of light from an external source that is absorbed by the sample is measured in terms of absorbance (A).

I. Equipment
Atomic absorption spectrometer
Potassium (or Na/K) hollow cathode lamp

II. Reagents (See Appendix II)
1000 mg/L stock potassium solution

III. Procedure

1. Prepare a 100 mg/L working standard solution from the potassium stock by pipetting 10 mL of stock into a 100-mL volumetric flask and diluting the contents to the mark with deionized water.
2. Prepare a series of calibration standards of 1, 2, 5, and 10 mg/L potassium by pipetting 1, 2, 5, and 10 mL of the working standard into four 100-mL volumetric flasks. Dilute to volume with deionized water.
3. Install the potassium hollow cathode lamp in the atomic absorption instrument. Turn on and set the lamp power supply to the milliamp (ma) value listed in the operating instructions. Set the wavelength and slit width to the appropriate values.
4. Adjust acetylene and air pressures to the values stated in the instrument's operating instructions. Likewise adjust fuel and support gas (oxidant) settings. Align the burner head in the optical path, and ignite the flame. Adjust the flame to lean conditions (small blue flame cone), and while aspirating deionized water, adjust the lamp emission signal to 100$\%T$. While aspirating one of the standard potassium solutions, adjust the burner height to maximize the measured absorbance reading.
5. Prepare a standard curve by aspirating each standard in turn from the lowest to the highest concentration and recording the respective indicated absorbance values. Plot concentration vs. absorbance readings. (Alternatively, if the instrument has a "concentration" mode, one may calibrate the instrument to read directly in concentration units.)
6. Dilute 1 mL of wine sample to 100 mL with deionized water. aspirate the sample into the burner and read absorbance (or concentration). If the value obtained for the wine is not within the range of standard values, prepare a second sample at a more appropriate dilution level.
7. Read the potassium concentration from the calibration curve. Multiply by 100 (or other dilution value, if used) to obtain the potassium concentration in the original wine sample.

IV. Supplemental Note

1. Samples with high concentrations of sodium, such as may arise from cation exchange treatment, may produce high values for the potassium

concentration, an effect due to the small but significant amount of ionization of sodium atoms in the air–acetylene flame. One way to eliminate or swamp out this interference is to make up all standards and samples so that they contain from 2000 to 5000 mg/L of added sodium. In this case, the added sodium acts as an ionization buffer.

Procedure 13-6. Potassium Analysis by Flame Emission

Flame photometric emission techniques have traditionally been used in the wine laboratory. In this method the heat of the flame excites the potassium atoms, which then emit radiation at their characteristic wavelength. Flame emission techniques generally are being replaced by atomic absorption and plasma emission techniques.

I. Equipment
   Flame photometer (or atomic absorption spectrometer)
II. Reagents
   1000 mg/L stock potassium solution
III. Procedure
   1. From a stock potassium solution (1000 mg/L), volumetrically transfer 10 mL into a 100-mL volumetric flask, and bring the solution to volume with deionized, distilled water. The concentration of K^+ in this working solution is 100 mg/L.
   2. A series of calibration standards is then prepared according to Table 13-8.
   3. A blank, consisting of deionized, distilled water, is used to zero the instrument to 100%*T*.
   4. Consult the operator's manual for the instrument setup and calibration.
   5. Using scrupulously clean 10-mL beakers, introduce each standard solution to the burner. Allow 15–30 sec for the reading to stabilize, and record %*T*.
   6. Prepare a standard curve by plotting %*T* against the appropriate concentration.

Table 13-8. Preparation of calibration standards from 100-mg/L potassium working standard.

Volume working standard (mL)	Final concentration K^+ (mg/L) in 100-mL volume
1	1.0
2	2.0
5	5.0
10	10.0

7. Dilute the wine sample 100× with deionized distilled water, and introduce it to the flame.
8. Using the standard curve, determine the sample's concentration. Remember, the final concentration of the sample must reflect its dilution!

IV. Supplemental Notes

1. Because solution components are present at levels of less than 10 mg/L, the analyst's best technique is required. Scrupulously clean glassware is an absolute necessity, as is proper pipetting procedure.
2. In quantitative flame photometry, one must be careful to minimize fluctuations during analysis. Assuming a properly operating instrument, two techniques commonly are employed to reduce the possibility of analytical interference:
   a. Frequently, quantitative flame emission methods call for use of internal standards of known concentration to compensate for uncontrollable variables. These internal standards are added to standard solutions as well as unknowns. Ideally, such internal standards should have characteristics similar to those of the element analyzed. Most important, the excitation line(s) should be in the same region as the element of interest so that flame temperature fluctuations affect both similarly. In K^+ and Na^+ analyses, lithium frequently is used as the internal standard.
   b. To stabilize aspiration rates of standard solutions and unknowns, procedures may call for the inclusion of detergents (sold under several proprietary names) at levels of less than 0.1%.
3. On a daily basis, many workers prefer not to prepare a standard curve. Instead, the concentration of an unknown may be related to that of a single standard via the following relationship:

$$\text{Conc unknown} = \frac{\%T \text{ unknown}}{\%T \text{ standard}} \times \text{Conc standard (mg/L)}$$

   This procedure usually is limited to instruments that employ filters rather than more sophisticated methods of wavelength approximations.
4. Because the emission lines of Ca^{2+} are close to those of Na^+, significant interference may be expected in flame photometric analyses of the former unless a monochromator is used for wavelength selection. By comparison, the emission lines of Na^+ and K^+ are sufficiently far apart that interferences are not commonly encountered even with the use of a relatively crude filter.
5. Relative accuracy levels of flame photometric analyses are on the order of ±2 to 5%, according to Fritz and Schenk (1974). These authors

point out that this level of accuracy generally is sufficient when the element is present in small quantities.

6. Laboratory data sometimes are expressed in units of milliequivalents / liter (meq / L). To convert results expressed in mg / L to meq / L, it is necessary to divide the former by the appropriate atomic weight (39 in the case of K^+).

Procedure 13-7. Potassium Analysis by Ion Selective Electrode

An alternative technique for potassium analysis involves the use of potentiometry and an ion selective electrode. The potassium electrode is constructed so that it responds preferentially to potassium ion. It requires a high-quality expanded-scale pH meter (mV scale capable of reading to the nearest 0.1 mV) or a select-ion meter, and a reference electrode designed for use with selective-ion electrodes.

I. Equipment
   Expanded-scale pH / mV meter or specific-ion meter
   Potassium selective-ion electrode
   Single-junction reference electrode
   Magnetic stirrer and stir-bar

II. Reagents (See Appendix II)
   1000 mg / L potassium stock solution
   Ionic strength adjuster (ISA)
   Reference electrode filling solution

III. Procedure
   1. Prepare working standards of 100 mg / L and 10 mg / L from the potassium stock solution. Prepare the 100-mg / L standard by pipetting 10 mL stock solution and 2 mL ISA solution into a 100-mL volumetric flask. Dilute to the mark with distilled/deionized water. Prepare the 10-mg / L standard in similar fashion, using the 100-mg / L standard.
   2. Place electrodes in the 10-mg / L standard solution. Stir it, and allow the meter reading to stabilize. Record the final mV reading.
   3. Rinse the electrodes, dry them, and place them in the 100-mg / L standard solution. Stir it, and allow the meter reading to stabilize. Record the final mV reading.
   4. Prepare a calibration curve by graphing millivolt readings vs. log concentration. This may also be accomplished by using semilog paper (graph concentration values on the log axis).
   5. Dilute 10 mL wine (or juice) to 100 mL in a volumetric flask. Transfer the diluted sample into a 150-mL beaker, add 2 mL ISA, and stir the solution. Place clean, dry electrodes in the sample, allow the meter to stabilize, and record the mV reading. Refer to the calibration curve, and read the potassium concentration from the graph.

6. Multiply the value obtained in step 5 by a factor of 10 (dilution factor) to obtain the potassium concentration of the original sample.

IV. Supplemental Notes
1. The electrodes should be recalibrated after every few hours of use.
2. Most electrode operating manuals also contain instructions for electrode use with other ion meters, as well as for known addition techniques.

The standard or known addition technique frequently is used in order for the standards to match the matrix of the wine samples being analyzed. The electrode potential for the wine sample, with appropriate dilution buffer added, is measured first. A known amount of standard potassium solution is added to this sample, and a second potential reading is taken. The difference between the two potential readings and the known concentration of the potassium standard utilized are used to calculate the potassium concentration in the original wine sample.

Procedure 13-8. Calcium Analysis by Atomic Absorption

Analyses of calcium in wine often are carried out using spectroscopic techniques. Both flame absorption and plasma emission techniques can be used in the wine laboratory. Absorption techniques measure the radiation (either ultraviolet or visible) absorbed from an external light source by ground-state atomic species in the flame. As in the potassium analysis (Procedure 13-5), sample elements in liquid form are aspirated into the flame, desolvated, and converted into a cloud of ground-state atoms, which then are available to participate in the atomic absorption process. Again, the energy absorbed at specific wavelengths is monitored using a typical monochrometer and photo-detector. The intensity of light from an external source that is absorbed by the sample is measured in terms of absorbance (A).

I. Equipment
Atomic absorption spectrometer
Calcium hollow cathode lamp

II. Reagents (See Appendix II)
1000 mg/L stock calcium solution
Lanthanum oxide/HCl solution

III. Procedure
1. Prepare a 100-mg/L working calcium standard by diluting 10 mL of calcium stock solution to 100 mL in a volumetric flask.
2. Prepare 1, 2, 5, and 10 mg/L calibration standards in 100-mL volumetric flasks as follows (Table 13-9). Dilute to final volume with a solution of 1% (W/V) lanthanum oxide in 1% (vol/vol) HCl.
3. Install the calcium lamp in the atomic absorption spectrometer. Set the appropriate wavelength, slit width, and power supply current as listed in the instrument's operating instructions.

Table 13-9. Preparation of calibration standards from 100 mg/L working calcium standard.

Volume of working standard (mg/L)	Final Ca^{++} concentration (mg/L) in 100-mL volume*
1	1.0
2	2.0
5	5.0
10	10.0

*Dilute to volume in a solution of 1% (w/v) La_2O_3/HCl.

4. Following the method outlined in Procedure 13-5, light and adjust the burner flame, optimize analytical conditions, and prepare the calibration curve.
5. Dilute 1 mL wine to 10 mL with the lanthanum oxide/HCl solution. Aspirate the sample in the flame, and read the absorbance value (or concentration). If necessary, prepare a second sample at a more appropriate dilution level.
6. Read the calcium concentration from the calibration curve and multiply by 10 (or other dilution factor, if used) to obtain the calcium concentration in the original wine sample.

IV. Supplemental Notes

1. Phosphorus in wine will react with calcium in the air–acetylene flame, resulting in a depression of the measured calcium absorbance. This interference is overcome by adding lanthanum at a level of about 10,000 mg/L to all standards and samples. Phosphorus preferentially reacts with lanthanum so that the former cannot bind with calcium in the flame.
2. One can avoid the phosphorus interference by using the hotter nitrous oxide–acetylene flame. In this instance, lanthanum is not needed as an additive. At higher flame temperatures, an ionization interference may occur, due to a certain amount of ionization of sodium and potassium atoms in the nitrous oxide–acetylene flame. To eliminate the effects of this interference, all standards and samples should be prepared to contain 2000 to 5000 mg/L potassium.
3. Analyses conducted at the low mg/L concentration level require careful attention to cleanliness and careful pipetting and dilution of both standards and samples.

Chapter 14

Copper

Copper, in trace amounts, is an important inorganic catalyst in the normal metabolic activities of yeasts and other microbial organisms. The copper content of typical musts and wines made in the United States ranges from less than 0.1 to 0.30 mg/L. At higher levels, the metal plays an important role in catalyzing the chemical oxidation of wine phenols. In this regard, copper and copper complexes are more active than iron and its complexes. Aside from chemical and biochemical involvements, copper at concentrations exceeding 1 mg/L may be sensorily detectable (Amerine et al. 1972). Furthermore, copper, along with cadmium and mercury, ranks among the most toxic of heavy metals; and at levels exceeding 9 mg/L it becomes a metabolic toxin that inhibits or delays alcoholic fermentation (Suomalainen and Oura 1971). Fortunately, the latter situation is found infrequently in modern-day winery operations.

SOURCES OF COPPER IN WINE

Measurable copper levels in the finished wine may be attributed to three sources: (1) winery equipment, (2) vineyard sprays, and (3) additions of copper salts ($CuSO_4 \cdot 5H_2O$) in winery operations. Abnormal copper accumulations usually are due to the contact of must and wine with copper-containing alloys such as brass during processing. Thoukis and Amerine (1956) reported a 40 to 89% reduction in the levels of the copper present in must upon the completion of alcoholic fermentation. These workers subsequently demonstrated that the loss could be attributed to absorption onto the yeast cell. Hsia et al. (1975) studied the effect of juice solids prior to fermentation on the copper and iron content in the resultant wines, and reported that wines produced from juices with 10 to 38% suspended solids had higher than average iron contents (10 to 20 mg/L) but were low in copper. These workers concluded that the suspended solids were rich in sulfates that subsequently were reduced to sulfide, precipitating copper as copper sulfide.

Copper Instability

Instability, manifested initially as a white haze and later as a reddish-brown amorphous precipitate, may develop upon the storage of a wine with copper contamination in excess of 0.2 mg/L (Amerine 1965). The precipitate formed, termed "casse," develops only under strongly reducing conditions such as those found in bottled wine. As evidence, reoxidation by exposure to air or hydrogen peroxide may cause the precipitate to disappear. Furthermore, casse formation is contingent upon low levels or absence of iron in the wine.

Ribereau-Gayon (1933) first developed the theory for copper instability in wine, proposing that the cloud existed largely as cupric (copper II) sulfide in combination with wine colloids. Rentschler and Tanner (1951b), on the other hand, reported that the sediment from centrifuged wine with copper instability was high in cuprous (copper I) sulfide. Other workers have reported turbidity due to copper–protein complexes and amino acid–copper complexes (Kean 1954). Subsequently the precipitate has been reported to be high in protein nitrogen and rather low in sulfur-containing compounds (Kean and Marsh 1956a,b). Thus casse formation resulting from Cu^{2+} reduction is not entirely the result of copper and sulfur interactions as once thought, but is instead a mixture of copper compounds.

Protein levels in white wine may act as limiting factors in cloud formation. Thus, removal or reduction in the level of protein by bentonite fining may help to prevent casse formation. Heat and light are known to accelerate casse formation. Peterson et al. (1958) found that under light conditions the copper complex formed is due to sulfite reduction and subsequent precipitation as copper (II) sulfide, whereas under dark conditions protein denaturation results from sulfite interaction. The copper–protein complex yields sulfate upon oxidation.

Table 14-1 summarizes the conditions that favor or hinder copper casse formation. Several compounds have been studied in regard to removal or reduction of the levels of copper and iron in wine. The most important of these agents are "Cufex" and "Metafine." (For further consideration of this topic, the reader is referred to Chapters 17 & 19.) Screening procedures for determining the nature of wine casses are given in Section IV-remedial actions.

Procedure 14-1. Copper Analysis by visible spectrophotometry

Copper will react with diethyldithiocarbamate to form a colored complex measured spectrophotometrically at 450 nm. Extraction of this complex into an amyl acetate–methanol solvent mixture isolates it from other colored com-

Table 14-1. Factors favoring and inhibiting copper casse formation in wines.

Conditions necessary for copper casse formation	Preventive measures
Strongly reducing conditions (as seen in bottled wine)	Maintain copper levels at less than 0.3 mg/L
Iron absent or present in very low concentrations	Cold-stabilize and bentonite-fine to remove protein in white wines
Light and/or heat, which may hasten formation	Limit SO_2 additions

pounds in the wine sample. Although this is a straightforward and sensitive procedure, amyl acetate is very volatile; so it is good practice to conduct the extractions in a hood.

I. Equipment

Because of the small amounts of copper normally present in samples and the chance for laboratory contamination, it is strongly recommended that glassware be set aside exclusively for copper determinations.

Pyrex test tubes (of more than 25 mL capacity)
Spectrometer (Bausch and Lomb Spectronic 20 or equivalent)
Matched curvettes
Volumetric pipettes (1.0, 2.0 mL)
Volumetric flasks (100 mL)
Whatman #40 filter paper
Glass funnels
Watch glasses (2.5–3 inches in diameter)

II. Reagents (see Appendix II)

Stock copper solution (1000 mg/L)
1% Sodium diethyldithiocarbamate
5 *N* Ammonium hydroxide
HCl–citric acid solution
Amyl acetate–methanol solvent (2 : 1)
Distilled water

III. Procedure

1. Prepare a 100-mg/L working copper standard by diluting 10 mL of the copper stock solution to 100 mL in a volumetric flask.
2. Prepare 0.5, 1.0, and 2.0-mg/L calibration standards in 100-mL volumetric flasks according to Table 14-2.
3. Volumetrically transfer 10-mL portions of the calibration standards into separate test tubes. Blank preparation consists of 10 mL of deionized distilled water.
4. To each tube, add with thorough mixing:
   a. 1 mL HCl–citric acid solution.

Table 14-2. Preparation of copper calibration standards from 100-mg/L working standard solution.

Volume of working standard solution (mL)	Final Cu^{2+} concentration in 100 mL volume (mg/L)
0.5	0.50
1.0	1.0
2.0	2.0

b. 1 mL 5 *N* NH_4OH.
c. 1 mL sodium diethyldithiocarbamate
Mix the contents well, and allow tubes to stand for approximately 1 minute.

5. Add 15 mL amyl acetate–methanol solvent to each, and mix each tube's contents thoroughly.
6. Allow phase separation to occur.
7. With a 10-mL graduated pipette, draw off the upper organic phase, and filter it through Whatman #40 filter paper directly into a cuvette. Cover the funnel with a watch glass during filtration.
8. Set the spectrophotometer at 450 nm. Using the distilled water reagent blank (following steps 3–6 above), determine the absorbance or transmittance of standards.
9. Plot the measured or calculated absorbance against the appropriate concentration (mg/L) of the copper standard.
10. Using 10 mL of the wine sample, proceed as in the preparation of standards, comparing absorbance measurements to the standard curve.

IV. Supplemental Notes

1. All glassware must be exceptionally clean. The laboratory water and reagents should be checked for copper contamination. Failures in using the above procedure generally can be attributed to contamination from these sources. It is recommended that laboratory personnel maintain separate glassware exclusively for copper analyses. Sloppy technique and/or reagent contamination may be detected as color development in the blank.
2. Post-filtration sample opacity is attributed to incomplete separation of water into the aqueous phase. Such samples may be "cleared" by the addition of anhydrous sodium sulfate powder.
3. In the Marsh procedure (presented above), HCl–citric acid is used to chelate any iron in solution that would interfere with color development.
4. To avoid interference with wine pigments, the copper–carbamate complex is extracted with amyl acetate–methanol solution. Methanol acts to prevent or reduce the tendency for emulsion formation between wine and amyl acetate.
5. Standard solutions should be made fresh each time a standard curve is prepared. Metal ions, in low concentrations, are absorbed by glass upon storage, producing anomalous results.
6. The concentration of standards should be such that the absorbance falls within the range 0.12 to 1.0 (10 to 75%*T*). In this range, the probable error is on the order of 0.5%*T*, which corresponds to concentration errors of ±2%. Outside the recommended range, readings become increasingly inaccurate (Fritz and Schenk 1974).

Procedure 14-2. Copper Analysis by Atomic Absorption

Atomic absorption (AA) is used in many laboratories as a rapid alternative to wet chemical procedures. Caputi and Ueda (1967) were among the first to report its value in heavy metals analysis in wine.

In principle, AA measures the absorption of ultraviolet or visible light by neutral (ground-state) atoms present in the gaseous phase. It is similar to solution spectroscopy except that the absorption lines for atoms are very narrow compared to the relatively broader bands characteristic of absorption spectra of molecules in solution. As a result, in atomic absorption the light source is an interchangeable hollow cathode lamp that emits radiation of the same wavelength as the element to be measured. Because absorption of light is restricted to those species absorbing at the same wavelength as the source, the method is specific and relatively free of interferences. Its accuracy is reported at $\pm 2\%$ (Fritz and Schenk 1974), with sensitivity in the range of 1 mg/L.

The principal drawback is cost. In addition to the initial high cost of the instrument, compressors, and exhaust venting, special source lamps must be maintained for each element analyzed. At present, some 68 elements may be determined using atomic absorption spectrophotometry.

I. Equipment
   Atomic absorption spectrophometer equipped with a copper hollow cathode lamp (or a multielement lamp containing copper as one of its constituents).
   Volumetric flasks (100 mL)
   Volumetric pipettes
II. Reagents (see Appendix II)
   Ethanol (200° proof)
   Glucose
   Copper stock solution (100 mg/L)
   Deionized, distilled water
III. Procedure
   1. Prepare a 10.0 mg/L working copper standard by diluting 10 mL of the copper stock solution to 100 mL in a volumetric flask.
   2. Prepare 0.2 to 1.0 mg/L copper calibration standards in 100-mL volumetric flasks according to Table 14-3. Final dilution is made with 12% (vol/vol) ethanol.
   3. Consult the operator's manual for instrument setup and calibration.
   4. Aspirate the blank (12% vol/vol ethanol) and calibration standards through the "sample inlet," recording absorbance after readings stabilize. Between samples, aspirate a sufficient volume of solvent to flush the system.
   5. Aspirate filtered wine samples, recording absorbance readings.
   6. Plot the absorbance of standards vs. their respective concentrations.

Table 14-3. Preparation of copper calibration standards from 10.0 mg/L working standard.

Volume of working standard (mL)	Final Cu^{2+} concentration in 100 mL volume* (mg/L)
2	0.2
4	0.4
5	0.5
8	0.8
10	1.0

*Dilution is made with 12% (v/v) ethanol.

Compare absorbance of unknown(s) to the standard curve, expressing concentration in mg/L.

IV. Supplemental Note

1. Caputi and Ueda (1967) call for making up standard solutions in 12% (vol/vol) ethanol and 5% (wt/vol) glucose for sweet wines. Values are reported to be within ±10% of the actual copper concentration. The authors have found, however, that in table wine analysis, 12% ethanol alone generally is sufficient.

Chapter 15

Iron and Phosphorus

Yeasts require trace levels of various mineral compounds including iron. At low concentrations, iron plays an important role in metabolism as an enzyme activator, stabilizer, and functional component of proteins. At higher-than-trace levels, iron has other roles: altering redox systems of the wine in favor of oxidation, affecting sensory characteristics, and participating in the formation of complexes with tannins and phosphates resulting in instabilities. This complex formation, also termed ''casse,'' is seen initially as a milky white cloud and later as a precipitate. Two iron-containing casses may form in wines: ''white'' (ferric phosphate) and ''blue'' (ferric tannate casse). The former represents the most commonly encountered iron-related casse.

A recent survey of commercial table wines showed phosphorus levels ranging from 135 to about 200 mg/L (Chow and Gump 1987). Thus there appears to be sufficient phosphorus in wines to promote casse formation, and iron is the limiting factor in this regard.

Compared to white casse, the blue form is not commonly observed. Resulting from tannin–iron complexation, blue casse, when observed, generally is found in white wines after tannic acid additions.

FERRIC PHOSPHATE CASSE

White casse formation is dependent upon several parameters: (1) iron content, (2) pH, (3) redox potential, (4) phosphate content, and (5) the nature and concentration of the predominating acid. Normally, the levels of iron are limited in grapes, even when they are grown on high-iron soils. If no contamination occurs, a typical must has from 1 to 5 mg/L iron (Amerine and Ough 1974). Dupuy et al. (1955) studied sources of iron in new wines, and could find no relationship between grape variety and iron content, although wines produced from different varieties grown on the same soil varied considerably. The most important source of iron in wine is contact with iron-containing alloys during processing. Modern-day winery operations have largely eliminated this problem by use of stainless steel processing equipment.

Iron instability, as ferric phosphate casse, is reported to occur only within the pH range 2.9 to 3.6. Thus, if iron is present at critical levels, winery practice (i.e., blending) that alters the pH so that it falls into this range may affect the potential for casse formation.

Although iron is important in metabolic activities at trace levels, an iron content exceeding 20 mg/L may inhibit fermentation. Yeasts may serve to remove a portion of the must iron content. Thoukis and Amerine (1956) reported that 45 to 70% of the iron in must was removed during fermentation by incorporation into the yeast or adsorption to the cell wall and membrane. In regards to this observation the yeast cell surface has a net negative change that could rapidly and reversibly bind with exogenous divalent cations such as iron.

Under normal winery conditions, most of the iron is present in the ferrous, or iron II, state. The ratio of Fe^{3+} to Fe^{2+} depends upon the oxidation state of the wine, with the ferrous form predominating when oxygen levels are low. If oxidative conditions occur, Fe^{2+} is converted to Fe^{3+} via a simple electron transfer reaction:

$$Fe^{+2} \longrightarrow Fe^{+3} + e^{-} \qquad (15\text{-}1)$$

Subsequent reaction of Fe^{3+} with the phosphates normally present in the wine may yield ferric phosphate casse:

$$Fe^{+3} + PO_4^{-3} \longrightarrow FePO_4\ (s) \qquad (15\text{-}2)$$

Iron may form complexes with several organic acids in wine that render it inactive in subsequent reaction with phosphate. Citric acid has the greatest affinity for iron. When the iron present in a wine approaches 5 mg/L, some winemakers may elect to add citric acid at levels of 1 to 2 pounds per 1000 gallons. At these levels, citric acid acts as a chelating agent, preventing iron from reacting with phosphates. At levels greater than 5 mg/L, special fining techniques may be required to remove the metal. (These techniques are discussed in greater depth in Chapter 19 procedure 19.1.)

Table 15-1 summarizes the factors that favor or hinder ferric phosphate casse formation. Screening procedures for determining the nature of wine casses are given in section IV-Remedial Actions.

Table 15.1. Factors affecting ferric phosphate casse formation in wine.

Conditions necessary for casse formation	Conditions impeding or inhibiting casse formation
Redox potential that favors the presence of Fe III over Fe II	Iron levels of less than 5 mg/L
pH 2.9–3.6	Clarification with bentonite and cold stabilization
Iron concentrations in excess of 7–10 mg/L	Citric acid additions at 1–2 lb/1000 gal

Procedure 15-1. Iron Analysis by Visual Spectrophotometry

This analysis as presented depends on quantitative reaction of the thiocyanate anion (SCN^-) with Fe^{3+}, and subsequent spectrophotometric measurement of the color formed. The colorimetric reaction depends upon the concentration of iron present, the concentration of thiocyanate, and the acidity of the solution. The method permits determination of both Fe^{2+} and Fe^{3+} via hydrogen-peroxide-induced oxidation from the iron II to the iron III state.

I. Equipment
   Pyrex test tubes (of more than 25 mL capacity)
   Volumetric pipettes (1.0, 2.0, 5.0, 10.0 mL)
   Volumetric flasks (100 mL)
   Whatman #41 filter paper
   Spectrophotometer (visible spectrum)
   Matched cuvettes
   Glass funnels

II. Reagents (see Appendix II)
   Iron standard solution (1000 mg/L)
   Amyl acetate–methanol solvent (2:1)
   8% Potassium thiocyanate
   5% HCl
   30% Hydrogen peroxide diluted to 3% immediately prior to use

III. Procedure
   1. Prepare a 100 mg/L working standard by diluting 10 mL of the stock solution (1000 mg/L) to 100 mL in a volumetric flask.
   2. Prepare 1.0, 2.0, 5.0, and 10.0 mg/L iron calibration standards in 100-mL volumetric flasks according to Table 15-2.
   3. Volumetrically transfer 2.0 mL from each standard solution into separate Pyrex test tubes. The blank consists of 2.0 mL of deionized distilled water.
   4. To each tube, add with thorough mixing:
      a. 1 mL of 5% HCl.
      b. 2 drops of H_2O_2.
      c. 1 mL of 8% KSCN. (*Caution: Do not pipette by mouth*!)

Table 15-2. Preparation of iron calibration standards from working standard (100 mg/L).

Volume of 100 mg/L standard (mL) solution	Final Fe^{3+} concentration (mg/L) in 100 mL volume
1.0	1.0
2.0	2.0
5.0	5.0
10.0	10.0

5. Allow each to stand 1 minute.
6. Add 15 mL amyl acetate–methanol solvent to each, and mix with mild agitation.
7. Allow phases to separate, and withdraw the upper organic phase with a 10-mL transfer pipette.
8. Filter the organic phase through Whatman No. 41 filter paper directly into a cuvette.
9. Determine absorbance or transmittance at 520 nm using the reagent blank to zero the spectrophotometer. Plot absorbance vs. concentration (mg/L).
10. Using 2.0 mL of wine, proceed from step 4 above, comparing the absorbance with the standard curve.

IV. Supplemental Notes

1. Post-filtration cloudiness may be cleared by the addition of anhydrous sodium sulfate (see Proc. 14: Supplemental Note 2 Chapter 14).
2. The colorimetric analysis, as presented, utilizes thiocyanate to bind with Fe^{3+}. The intensity of the color developed is dependent upon iron content, concentration of thiocyanate, and pH of the medium. The latter variable is adjusted by the addition of HCl.
3. The addition of H_2O_2 effects oxidation of iron present in the sample, from iron (II) to the iron (III) state.
4. The amyl acetate–methanol solvent is employed to extract the colored complex $Fe(CNS)_6^{3-}$ from the aqueous phase, thus preventing interferences from wine pigments. The inclusion of methanol in the solvent mixture reduces the tendency for amyl acetate and wine to form an emulsion. In some instances, red wine pigments may partition into the solvent layer, thereby producing erroneously high results. One may compensate for the increased absorbance due to pigmentation by running a wine sample without the inclusion of thiocyanate. Subsequent subtraction of the absorbance of this blank sample from that with iron development yields the true absorbance for comparison to the standard curve.
5. It should be noted that if the blank shows significant color development, the reagents, water, and glassware may be contaminated. In such cases, it is necessary to prepare a new blank.
6. As with copper analysis, exceptionally clean glassware is essential to success. It is highly recommended that laboratory personnel maintain separate glassware for each of these analyses, apart from the routine laboratory glassware.

Procedure 15-2. Iron Analysis by Atomic Absorption

As with copper, atomic absorption is finding increased application in the analysis of iron in wine. For a brief discussion of the subject, the reader is referred to the relevant portions of Chapter 14 (Procedure 14-2).

I. Equipment
   Atomic absorption spectrophotometer with iron hollow cathode
   Lamp
   Volumetric flasks
   Volumetric pipettes
II. Reagents (See Appendix II)
   Ethanol (200° proof)
   Glucose
   Iron stock solution (1000 mg/L)
   Deionized, distilled water
III. Procedure
   1. Prepare a working iron standard, 100 mg/L, by diluting 10 mL of iron stock to 100 mL in a volumetric flask.
   2. Prepare 2.0 to 10.0 mg/L iron calibration standards in 100-mL volumetric flasks following the dilution scheme in Table 15-3. Final dilution is made with 12% (vol/vol) ethanol.
   3. Instrument calibration: Consult the operator's manual for instrument setup and calibration.
   4. Run a blank consisting of 12% (vol/vol) ethanol in distilled water to set zero absorbance.
      Beginning with the standard solution of the lowest concentration, aspirate solution through the sample inlet. Record absorbance (or read direct by the concentration). Between samples, aspirate a sufficient volume of solvent to flush the system.
   5. Aspirate filtered wine samples, recording absorbance (or concentration).
   6. Plot absorbance of standards vs. their respective concentrations. Compare absorbance of unknown(s) to the standard curve, expressing the concentration in mg/L.

Table 15-3. Preparation of iron calibration standards from 100-mg/L working standard.

Volume working standard (mL)	Final iron concentration (mg/L) in 100 ml volume*
2.0	2.0
4.0	4.0
6.0	6.0
8.0	8.0
10.0	10.0

*Dilution is made with 12% (v/v) ethanol.

Procedure 15-3. Phosphorus Analysis by Atomic Absorption

This method involves an indirect determination of phosphorus (Chow and Gump 1987). Flame absorption spectrometry is not sensitive enough to directly measure the phosphorus content in wine, but phosphorus will form a compound with molybdenum (molybdophosphoric acid), and molybdenum equivalent to the phosporus can be determined with good sensitivity. Acidified ammonium molybdate reagent forms molybdophosphoric acid with the phosphorus in wine. This acid is soluble in ether and is thereby removed from other potentially interfering compounds in wine. In the presence of an alkaline buffer, molybdophosphoric acid loses a proton, and the molybdophosphate ion transfers out of the ether layer and into the aqueous layer.

I. Equipment
   Atomic absorption spectrophotomer with molybdenum hollow cathode lamp
   Volumetric flasks (100 mL)
   Volumetric pipettes (1, 5, 10 mL)

II. Reagents (See Appendix II)
   Phosphorous standard solution (1000 mg/L)
   Ammonium molbydate reagent (10% wt/vol)
   Basic buffer

III. Procedure
   (a) Instrument setup
   1. Set up the atomic absorption instrument with an air–acetylene flame (strongly reducing), a molybdenum hollow cathode lamp (resonance line at 313.3 nm), and a 0.2-nm slit width.

   (b) Preparation of wine samples
   1. Pipette 10 mL wine into a 125-mL separatory funnel.
   2. Add 1 mL 1 + 1 HCl, and adjust the volume to about 50 mL.
   3. Add about 4 mL molybdate solution, and allow the mixture to stand for 10 minutes.
   4. Add an additional 5 mL HCl, and allow the mixture to stand for another 5 minutes.
   5. Add 45 mL diethyl ether, and shake the funnel vigorously for 4 to 5 minutes.
   6. Allow the phases to separate, and discard the aqueous (lower phase). Rinse the tip of the separatory funnel with distilled water.
   7. Wash the ether phase with 10 mL 1 + 9 HCl. Discard the aqueous acid layer (contains excess molybdenum reagent), again rinsing the tip of the separatory funnel with distilled water.
   8. Add 30 mL buffer to the separatory funnel, shake it for 30 sec, and collect the aqueous (lower) layer in a 50-mL volumetric flask. Repeat this process with an addition 15 mL buffer.

Table 15-4. Preparation of phosphorus calibration standards from 1000 mg / L phosphorus stock.

Volume phosphorous stock (mL)	Final phosphorous concentration (mg/L) in 100 mL volume
5.0	50.0
10.0	100
25.0	250
50.0	500
No dilution	1000

9. The combined aqueous layers are diluted to the 50-mL mark with distilled water and mixed.

(c) Preparation of phosphorous standards

10. Prepare phosphorous calibration standards from the 1000-mg/L stock solution in 100-mL volumetric flasks as indicated in Table 15-4.
11. Extract the phosphorus standards, following the same procedure as presented for wine samples.
12. Aspirate phosphorus standards into the AA instrument. Construct a calibration curve from the absorbance values of the standards.
13. Aspirate wine samples into AA instrument. Read the concentration of phosphorus in the extracted wine sample directly from the calibration curve prepared in step 12.

IV. Supplemental Note

1. This method is also successful using a nitrous oxide–acetylene flame. Because this is a hotter flame than air–acetylene, the method is more sensitive, and the standards and wine samples are diluted 1 + 9 prior to the extraction step.

Chapter 16

Nitrogenous Compounds

The nitrogenous components of must and wine play important roles in fermentation, as well as subsequent clarification, and potential microbial instability. Additionally, they may directly or indirectly affect the development of wine aroma and bouquet. The various nitrogenous constituents of wine include proteins, polypeptides, peptones, amino acids, amides, and ammonia.

Proteinaceous compounds are of major concern to the winemaker because of their tendency to impede clarification and stability in white table wine. Protein comprises from about 5 to 10% of the total nitrogen content of must (Moretti and Berg 1965). In wine, the levels are higher, near 38%.

Polypeptides exist as protein fragments composed of amino acids linked via peptide bonds. Although of lower molecular weight than proteins, polypeptides constitute a significant proportion of the total nitrogen content in wine. Ribereau-Gayon and Peynaud (1958) reported that polypeptides comprise between 60 and 90% of the total. Depending upon processing techniques (such as thermal treatment of musts), the levels of polypeptides may be higher.

Ammonia (present at the pH of juice as NH_4^+) serves as the primary form of available nitrogen for yeast metabolism. As grapes mature, there is a decrease in ammonia nitrogen with increases in protein and peptide nitrogen (Peynaud and Maurie 1953). This may help to explain why wines produced from overripe fruit sometimes ferment slowly.

Amino acids, which serve as the building blocks of polypeptides and protein, occur chemically as molecules with both amino and carboxylic acid groups (see Fig. 16-1). Most of the 20 commonly occurring amino acids are found in must and wine. Although the amount of each varies with grape variety as well as with cultivation and processing techniques, proline generally is found in the highest concentration in must and wine. In must, arginine also is present in relatively high concentration.

A variety of amines are found in wines, generally in low concentrations. Biogenic amines, such as histamine and tyramine, result from bacterially mediated decarboxylation of the corresponding amino acids. Both *Lactobacillus* sp. and *Leuconostoc gracile* have been reported to produce histamine (Lafon-Lafourcade 1976).

Nitrates (NO_3^-) and nitrites (NO_2^-) are present in wine at low levels, usually

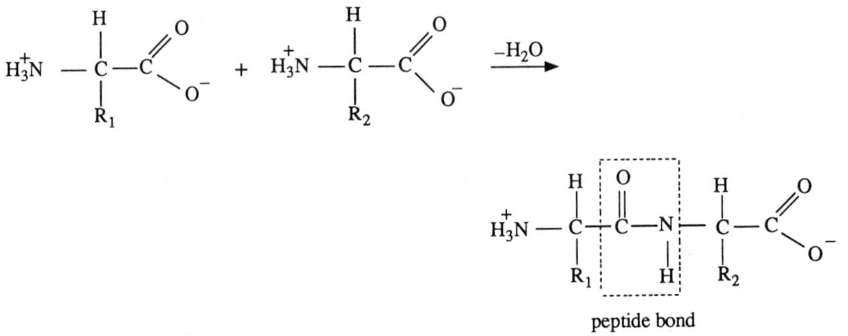

$$H_3\overset{+}{N}-\underset{R_1}{\overset{H}{C}}-C\begin{smallmatrix}\nearrow O\\ \searrow O^-\end{smallmatrix} + H_3\overset{+}{N}-\underset{R_2}{\overset{H}{C}}-C\begin{smallmatrix}\nearrow O\\ \searrow O^-\end{smallmatrix} \xrightarrow{-H_2O} H_3\overset{+}{N}-\underset{R_1}{\overset{H}{C}}-\overset{O}{\overset{\|}{C}}-\underset{H}{N}-\underset{R_2}{\overset{H}{C}}-C\begin{smallmatrix}\nearrow O\\ \searrow O^-\end{smallmatrix}$$

peptide bond

Fig. 16-1. Structure of typical amino acids showing peptide bond formation (leading to protein).

less than 0.3% of total nitrogen. In European wines, Amerine and Ough (1980) have reported values for nitrate levels ranging from less than 7 mg/L, in the case of certain German wines, to an average level of 1.65 mg/L in Italian white wines. In the latter, nitrite levels were reported as low as 30 μg/L.

Nitrate levels in table wine may not be of great importance, but concern is mounting about the eventual role(s) of nitrate in groundwater contamination resulting from the discharge of winery and distillery wastes. In view of the potential problem of regulation of winery discharge, Procedure 16-6, a method for nitrate determination, has been included in this chapter.

CHANGES IN NITROGEN LEVELS

Vineyard management techniques, climate, maturity, grape variety, and method of vinification may all affect the nitrogen content of a finished wine. The amount and form of soil nitrogen can have a notable effect on the total nitrogen content of the grape. Also, the protein content of the grape increases during maturation, generally being higher in warmer regions. Lower crop levels have been associated with higher protein and total nitrogen levels (Ough and Anelli 1979).

Newly fermented wines have lower total nitrogen levels than do the musts from which they were produced. Gorinstein et al (1984) report that from 30 to 46% of the total nitrogen is assimilated by yeast during fermentation, the majority being rapidly utilized after the onset of full fermentation. Each generation cycle of yeast reduces the nitrogen content, a reduction that has been suggested as a value in sweet wine stabilization (Schanderl 1959).

Ammonium nitrogen serves as the primary available form for normal yeast metabolism, and, as such, its concentration is rapidly reduced as fermentation begins. The amino acid content of must and wine varies considerably, with factors that influence concentration including grape variety, growing region, temperature during the growing season, and processing protocol. Although yeasts vary in their amino acid requirements, the amino acid content of must serves as an important nitrogen source.

Quantitatively, proline is usually present in must and wine in highest concentrations. Ough (1968) reports levels of 742 mg/L in grapes and 869 mg/L in wine. Unlike most amino acids in must, proline is not utilized by yeast under fermentative conditions; and, during the course of growth, yeasts may excrete proline into the fermentation. The initial unavailability of proline stems from inhibition of the transport enzyme, proline permease, by ammonium ion present in the must. Once the concentration of ammonium ion is low enough to permit permease activity, the fermentation has reached a point where the obligate oxygen requirements of the catabolic enzyme proline oxidase cannot be met.

Ingledew et al. (1987) point to the potential utilization of proline as a nitrogen source by wine yeast. They suggest that, in addition to conventional nitrogen supplementation, proline utilization significantly increases available nitrogen for yeast growth. These workers conclude that oxygen (air) incorporation into yeast starters and during fermentation may diminish the potential for sticking. Under such conditions, sufficient oxygen may be incorporated to stimulate proline oxidase enzyme and yet not oxidize the wine. (The issue of oxygen incorporation is considered in Chapter 8.)

The development of yeast strains with a strong demand for a growth-limiting amino acid may provide a practical future method of stabilizing sweet wines. Conversely, the selection of strains that cannot synthesize particular amino acids may be advantageous in specific applications.

Under certain conditions, grapes may be low in assimilable nitrogen. Factors that may contribute to this problem include overly mature fruit, or fruit that has been subject to attack by certain yeasts and molds. In each case, nitrogen levels may become depleted to the point where yeast metabolism and fermentation are affected. Ammonium salts additions generally increase the fermentation rate as well as the yeast titer. In France, Peynaud (1984) has suggested addition of ammonium nitrate when endogenous levels fall below 25 mg/L. In the case of fermentations of high-sugar musts, nitrogen supplementation may improve vigor, resulting in improved fermentation rates and higher final alcohol levels. Currently, in the United States diammonium phosphate may be added at levels of up to 1.7 pounds/1000 gallons in the case of table wines. Several workers have pointed out that nitrogen supplementation at these levels results in questionable increases in fermentative activity (Ingledew and Kunkee 1985; Ingledew et al. 1987; Wahlstrom and Fugelsang 1988).

Sparkling wine cuvees are often rather low in available nitrogen. Thus, cuvees may benefit from nitrogen additions prior to secondary fermentations. Legal addition levels of diammonium phosphate in these cases are up to 8 pounds/1000 gallons. Fruit musts (other than grape) are often deficient in nitrogen and require the addition of supplemental nitrogen. As of this writing, the federal government is about to approve raising the addition level of nitrogen salts to 8 pounds/1000 gallons for all wine types (Gahagan 1989).

Because of the potential for nitrogen depletion, protein stabilization of sparkling wine cuvees using bentonite should be done with caution. Bentonite nonselectively removes proteins, peptides, and amino acids, and such a reduction may adversely affect the subsequent fermentation rate. It has been demonstrated that there may be as much as 50% reduction in total nitrogen content, including a major reduction in amino acid content, with the use of bentonite (Ferenczi 1966). Additionally, bentonite may reduce or remove flavor and aroma components (see Chapter 17).

The practice of settling and racking white grape juice prior to fermentation has been found to reduce the total nitrogen content from 10 to 15% (Koch 1963). Somers and Ziemelis (1973a) compared the effects of using bentonite fining during and after fermentation on total nitrogen as well as nitrogen fractions in wines. Post-fermentation bentonite additions resulted in more effective removal of residual wine proteins, with nearly equal amounts of protein and nonprotein nitrogen removed at each level of bentonite used. In Europe, the practice of bentonite addition during rather than after fermentation is believed to result in more even fermentation rates. The addition of bentonite at this stage in processing may reduce or eliminate the need for post-fermentation additions except in those cases where must protein content may be unusually high (Ferenczi 1966). Additional advantages of the use of bentonite during fermentation include reduction in subsequent wine lees volume as well as possible sensory benefits (see Chapter 17). Somers and Ziemelis (1973a) also found that the use of bentonite during fermentation resulted in a greater net reduction of nonprotein nitrogen (approximately twofold) when compared with losses of protein nitrogen at each addition level to wine. Vos and Gray (1979), however, report increased levels of hydrogen sulfide during fermentation in contact with bentonite. They believed this resulted from reduction in the free amino nitrogen (FAN) content of fermenting juice. Reductions in ammonium and FAN nitrogen and subsequent increases in hydrogen sulfide levels due to fermentation in conjunction with bentonite have not been observed in California—perhaps because of the addition of nitrogen-containing fermentation adjuncts. (For more detail on the interaction of FAN and hydrogen sulfide production, refer to Chapter 10.)

Aside from the use of fining agents in prefermentation clarification, the total nitrogen content of a must may decrease for several reasons. One is the effect of alcohol as a denaturant and its role in causing flocculation and precipitation of proteinaceous compounds. As evidence, following wine spirits additions (WSA) wines may precipitate large quantities of proteinaceous lees. However, alcohol levels of 10 to 12% are seldom sufficient to cause significant protein precipitation. Also important is the interaction between phenolic compounds and protein; phenol complexation in red and white wines removes/reduces the concentration of some proteins in solution. In white wines, the relatively lower phenol levels usually do not remove enough protein, so instability may become a problem.

During aging, there is an increase in total nitrogen prior to the first racking (Gorinstein et al 1984). Resulting from yeast autolysate, the majority (87%) of this increase is attributed to amine nitrogen, which reaches maximum levels after about two months of storage on the lees. The balance is in the form of amide nitrogen and protein. It is believed that proteins from yeast autolysate do not contribute to protein instability (Boulton 1980e).

The Burgundian practice of *sur lie* or extended lees contact (including stirring the lees) is a frequently utilized stylistic tool in the production of certain table wines. During autolysis, cellular proteolytic enzymes bring about hydrolysis of cytoplasmic proteins, the products of which are released into the wine. In view of the variety of compounds released in the autolysate, it should be expected that the sensory properties of the wine would be affected; and it is generally accepted that storage of wine in contact with lees results in increased complexity. As lees contact time increases, winemakers note changes in fruity aromas to more muted "vinous" aroma. On the palate, *sur lie* wines often show increases in middle body when compared with conventionally produced lots. Winemakers using the extended-lees protocol report increased aging potential, perhaps due to increases in the pool of available oxidizable substrate. In the case of Methode Champenoise, the autolysate imparts unique aromas and flavors.

Increasing the contact time of a wine with the lees can increase its susceptibility to biological activity. As nitrogenous components, primarily in the form of amino acids, are liberated into the product, it may become an excellent medium for microbial growth, especially for lactic acid bacteria. Thus if the winemaker wishes to induce a malolactic fermentation, additional lees contact time may be employed (see Chapter 12). However, breakdown of sulfur-containing amino acids (methionine and cysteine) may result in the production of hydrogen sulfide (see Chapter 10).

EFFECT OF PROTEIN ON WINE STABILITY

Protein instability is a problem mainly in white wines. It is rarely encountered in red wines owing to their relatively high levels of flavonoid phenols, particularly tannins, which complex with and precipitate the protein. The total protein content of wines varies from 10 to 250 mg/L (Boulton 1980e). A wine's total protein content is not a good index of stability, and thus it cannot be used accurately to predict protein instability. Protein clouding is due not only to the precipitation of thermally labile proteins; it also results from formation of insoluble protein–tannin complexes. The grape is the predominant source of protein in wine. According to Somers and Ziemelis (1973b), about half of the total wine protein is bound to grape phenols; and they believe this portion to be responsible for protein haze formation. Yeast cells may excrete small amounts of protein during fermentation (Bayly and Berg 1967), but much larger amounts end up in the wine upon completion of fermentation as a result of autolysis.

Proteins originating from yeast autolysate are not thought to be involved instability.

Although numerous reports on the subject of protein instability appear in the literature, this complex issue is yet to be fully resolved. At least eight protein fractions have been reported in wine, ranging in molecular weight from 11,000 to 28,000 (Boulton 1980e). The presence and levels of any fraction depend on grape variety, growing region, seasonal variations, and processing protocol, such as skin contact, enzyme utilization, and so on. Proteins also may serve as nuclei around which soluble iron, copper, and other heavy metals may deposit (see Chapters 14 and 15).

THE NATURE OF WINE PROTEINS

The solubility of wine protein depends primarily on temperature, alcohol level, ionic strength, and pH. Changes in any parameter may affect the potential for protein precipitation.

The pH of wine is very close to the isoelectric point for many unstable wine proteins. At the isoelectric or isoionic point (pI), protein molecules have an equal number of positive and negative charges; and wine proteins are least soluble at their isoelectric points. If the wine pH is above the isoelectric point of the protein fraction in question, the net charge on the fraction will be negative, and the protein will bind electrostatically with positively charged fining agents. Conversely, if the wine pH is lower than the isoelectric point for the fraction, the net charge on the fraction will be positive. This relationship is seen in Fig. 16-2. The greater the difference is between the wine pH and the isoelectric point of the protein fraction, the greater the net charge on the protein and the greater its binding affinity toward oppositely charged fining agents.

Using Table 16-1, one can compare the effect of additions of bentonite (principally a negatively charged fining agent) to a wine of pH 3.2. This pH is below the isoelectric point of all the protein fractions of White Riesling, but it is below the isoelectric point of only about 67% of the protein fractions of Malvasia. Therefore, for White Riesling at pH 3.2, 100% of the protein fractions are positively charged and accessible to the negatively charged bentonite; whereas,

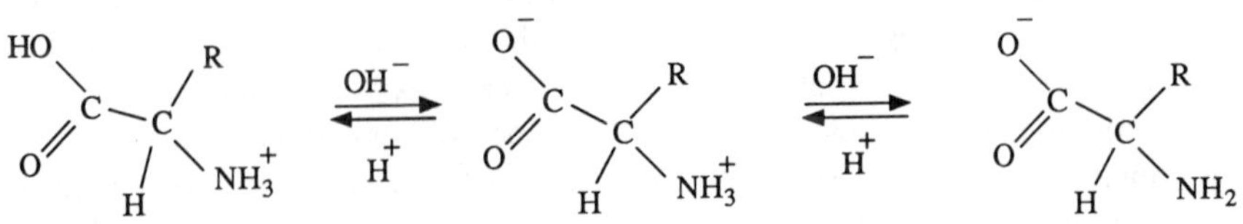

$$\mathrm{HO{-}C(=O){-}CH(R){-}NH_3^+} \underset{H^+}{\overset{OH^-}{\rightleftharpoons}} \mathrm{{}^-O{-}C(=O){-}CH(R){-}NH_3^+} \underset{H^+}{\overset{OH^-}{\rightleftharpoons}} \mathrm{{}^-O{-}C(=O){-}CH(R){-}NH_2}$$

Fig. 16-2. Effect of additions of acid or base on proteins at their isoelectric points.

Table 16-1 Isoelectric points and percentage composition of protein in Malvasia Istriana and White Riesling.[a]

Variety	pH	Total Protein (%)
Malvasia Istriana	2.5	18
	2.8	11
	3.1	4
	4.6	30
	6.5	13
	7.1	5
	8.3	9
	8.7	10
White Riesling	3.6	19
	3.9	53
	6.7	17
	7.1	11

[a] From Anelli (1977)

in the case of Malvasia, only 33% of the protein fractions are positively charged and accessible to the bentonite. The charge characteristics of various protein fractions partly explain why, in some wines, protein stability can be achieved only with excessive amounts of bentonite. In these cases, the necessary use level may "strip" the wine of much of its character.

Hsu and Heatherbell (1987) determined the effect of bentonite additions on protein removal and subsequent protein stability. Bentonite removed intermediate molecular weight (32,000–45,000) fractions (pI 5.8–8.0) first. To achieve stability as measured by the heat test (80°C for 6 hours followed by 4°C for 12 hours), the workers found it necessary to remove lower molecular weight (10,000–32,000) fractions with pI of 4.1 to 5.8. In this group, the glycoproteins are the major fraction. Thus, heat-labile proteins appear to be comprised mainly of glycoproteins of molecular weight less than 30,000. (For additional discussion of the role of bentonite, the reader is referred to Chapter 17.)

PROCESSING CONSIDERATIONS AND PROTEIN STABILITY

The treatment of juice and fermenting and fermented wines with bentonite to adsorb proteins is a general method of obtaining protein stability. However, it has disadvantages, including the formation of a relatively large volume of lees, as well as a possibly detrimental impact on wine flavor. Further, for grape varieties such as Sauvignon Blanc and Gewürztraminer, stabilization could require bentonite additions that, if carried out, would exceed governmentally approved levels (>8 pounds/1000 gallons). Thus, alternative methods of protein stabilization have been investigated, including the use of immobilized tannin (Weetall et al. 1984) and the utilization of proteases (Heatherbell et al. 1984).

Ultrafiltration has been evaluated as a means of achieving protein stabilization in white wines (Heatherbell et al. 1984; Broome et al. 1985). Ultrafiltration is a tangential-flow membrane filtration process for separating molecules based upon size; ultrafiltration membranes are selectively permeable, allowing smaller molecules to pass through while retaining larger molecules. Thus, proteins and oxidized and polymerized phenols may be removed using this technique. Ultrafiltration offers several significant advantages over bentonite for obtaining protein stability, including recovery of wine that would be lost in the lees, as well as a reduction in the "stripping effect" on wine flavor that is often associated with the use of bentonite.

Both heat and wine spirits additions may initiate denaturation and subsequent flocculation and precipitation of wine protein and/or protein complexes. Skin contact (in white grapes), especially at elevated pomace temperatures, also can increase protein extraction.

When protein instability is known to present a problem and bentonite fining is necessary, it is essential that laboratory fining trials simulate as closely as possible the conditions under which the product will be used in the cellar. For example, the same bentonite (from the same lot) should be used for both cellar and laboratory activities. The bentonite should be hydrated using the same methodology in both laboratory trials and cellar operations. For example, Waring-type blenders used for preparation of laboratory slurries exert a shear force that cannot be duplicated in the cellar.

Another less obvious difference between laboratory and cellar treatments with bentonite involves the contact time. Bentonite reacts almost immediately in binding proteins, but proteins are bound electrostatically and, in time, will begin to sluff off the bentonite platelet. Laboratory fining trials that have demonstrated protein stability should be duplicated by the same wine–bentonite contact time in the cellar (Lee 1985). (For more detail regarding the use of bentonite, the reader is referred to Chapter 17.)

Winery operations may play a significant role in protein stability. Thus, bitartrate stabilization, a malolactic fermentation, as well as blending and acidification, may render a previously stable wine unstable with respect to protein (as a result of pH shifts). It is essential that protein stability be determined after all cellar operations are complete and prior to bottling.

DETERMINATION OF PROTEIN STABILITY

The formation of a protein haze in bottled wine is always a concern for the winemaker. There is no uniform industry standard for determining the potential for protein instability in wine.

Table 16-2 summarizes a study by Cooke and Berg (1984) in which they reviewed methods for the laboratory evaluation of wine protein instabilities in several contributing California wineries.

Table 16-2 Methodology of heat stability evaluation.

Winery	Protocol
1	Filter sample, hold at 120°F for 24 hr visually inspect.
2	Filter 0.65 μm membrane. 120°F for 48 hr. Compare to control held at room temp.
3	120°F for 48 hr. Evaluate visually under high intensity light.
4	120°F for 96 hr. Hold at room temp. for 24 hr. Inspect.
5	140°F., for 24 hr. Visually inspect.
6	Filter, 140°F., for 24 hrs. Visually inspect.
7	100°F for 48 hr. 40°F for 48 hr. Evaluate sample when cold and at room temp.
8	140°F for 48 hr. Hold at room temp. for 24 hr. Visually inspect.
9	120°F. for 24 hr. 140°F. for 24 hr. 40°F for 24 hr. Examine.
10	120°F. for 24 hr. Evaluate. 40°F for 2 weeks. Examine.
11	Pad filter sample. 140°F for 48 hr. 35°F for 24 hr. Evaluate.
12	Membrane filter (0.45 μm) hot sample. Hold at 145°F for 16 hr. Monitor changes daily.
13	Pad and membrane filter sample. Trial 1. "Sample A." 26°F. for 48 hr. "Sample B." 100°F. for 48 hr. Inspect each. Trial 2. "Sample C" 26°F. for 96 hr. "Sample D" 100°F. for 48 hr. followed by 26°F for 48 hr. Inspect each. Trial 3. Hold smaples from Trial 2 at room temp. for 48 hr. before examination.
14	Membrane filter sample (0.45 μm). Trial 1. 130°F for 72 hr. 40°F for 12 hr. Hold at room temp. 12 hr. Evaluate progress daily. Trial 2. Boil sample 7 min. Cool overnight. 130°F for 72 hr. 40°F for 12 hr. Hold at room temp. for 12 hr. Evaluate progress daily.
15	Membrane filter (0.65 μm) sample. 176°F for 6 hr. 40°F for 12 hr. Monitor samples before and after treatment with nepholometer.
16	Trichloroacetic Acid (TCA)
17	No evaluation made.

Source: Cooke, C.M. and Berg, H.W. 1984. "A Reexamination of Varietal Table Wine Processing Practices in California. II. Clarification, Stabilization, Aging, and Bottling." *American J. Enol. and Vitic.* Vol. *35*: (3): 13T-142.

Heat Stability Testing

As seen in Table 16-2, most techniques involve some exposure of the wine to elevated temperatures for varying periods of time. Precipitation of a colloid such as a protein is affected not only by the exposure temperature but also by the duration of heating. Virtually all wine protein may be precipitated by heat, so wines exhibit varying degrees of heat stability. Pocock and Rankine (1973) evaluated sample treatment temperatures and time over the temperature range 50 to 90°C. With respect to the temperature of treatment, they reported precipitation of approximately 40% of wine protein when a sample was held at 40°C for 24 hours. By comparison, holding at 60°C for the same time period precipitated 95 to 100% of the protein. They also demonstrated that the time necessary for haze formation decreased with increasing temperature. Based upon their work, they recommended that 80°C (6 hours) be used in heat stability evaluation. (Holding at lower temperatures, even for 24 hours, did not yield maximum haze formation in the wines tested.) Increasing the temperature to 90°C and decreasing the exposure time to less than 6 hours increased apparent haze formation even in samples where little protein was present. Troost (1961) reported temperature ranging from 30°C (86°F) to 60°C (140°F) for time intervals ranging from 15 minutes to 28 hours. Ribereau-Gayon and Peynaud (1961) considered wines heated to 80°C (176°F) for 10 minutes to be stable if no haze developed upon cooling.

Some workers recommend chilling wine samples following heat treatment. Visible haze formation in these cases is slightly more than that seen in samples without subsequent cooling. Berg and Akiyoshi (1961) recommended holding the sample at 49°C (120°F) for four days followed by cooling to −5°C (23°F) for 24 hours. Upon warming of the wine to room temperature, haze and/or precipitate formation is evaluated.

Chemical Precipitation Tests

In addition to a wide array of laboratory methods involving heating, a number of chemical methods have been employed to predict stability of wine proteins. These include precipitation using ethanol, ammonium sulfate, trichloroacetic acid, phosphomolybdic acid, phosphotungstic acid, and tannic acid.

The Bentotest

This test, developed by Jakob (1962), uses a solution of phosphomolybdic acid prepared in hydrochloric acid to denature and precipitate wine proteins. Precipitation of wine protein occurs by neutralization and aggregation as the result of cross links formed with the molybdenum ion. Haze formation is proportional to the amount of protein present in the wine and may be used to determine bentonite addition levels.

Rankine and Pocock (1971) demonstrated that the Bentotest is more sensitive than the heat test technique using 70°C and 15 minutes of exposure. Of the 125 samples tested, 90 gave the same results for both tests, 27 were positive for the Bentotest only, and 8 were positive only with exposure to heat.

The Trichloracetic Acid (TCA) Test

The TCA test (Berg and Akiyoshi 1961) involves the use of 1 mL of reagent added to 10 mL of filtered wine. The solution is then heated in boiling water for 2 minutes, after which it is cooled to room temperature. The presence of haze is indicative of heat-labile protein. The methodology for this procedure is presented in Procedure 16-2.

Berg and Akiyoshi reported that this technique correlated well with the heat test protocol where samples are held at 49°C (120°F) for four days with subsequent holding at 23°F (−5°C) for two days and at room temperature for one day. In the study mentioned above, the Bentotest was more sensitive than either the TCA test or storage at elevated temperatures.

Saturated Ammonium Sulfate Test

Koch (1963) recommended a testing procedure involving the addition of a 5% solution of saturated ammonium sulfate to wine samples. Upon addition, the sample was heated to 45°C (113°F) for 9 hours and subsequently cooled to −6°C (21°F) for 15 minutes. The presence of a precipitate indicates protein instability.

DETERMINATION OF TOTAL PROTEIN AND NITROGEN-CONTAINING COMPOUNDS

As can be seen from earlier discussion, there is no uniform industry standard for determining the potential for protein instability in wine. The most common procedure calls for subjecting wine samples either to heat or to a chemical oxidant, such as trichloroacetic acid, and subsequent examination for haze development. An alternative method would be to analyze directly for protein content. Unfortunately, the analysis for total protein does not correlate well with observed protein instability in wine. Protein cannot be accurately estimated from a total nitrogen determination (Kjeldahl); the values obtained will be erroneously high. It can be precipitated from a juice or wine sample using trichloroacetic acid and then successfully determined by the Kjeldahl or another protein-specific method.

An alternative to the Kjeldahl method for total protein analysis is to utilize a commercial protein assay kit. One example of these kits, the Bio-Rad Protein Assay kit, utilizes a protein-dye (Coomasie Blue) binding reaction that shifts the absorbance maximum of the dye from 465 nm to 595 nm. The dye binding reaction appears to be insensitive to the potential interferences found in wine and juice samples. Also, Tyson et al. (1981) have developed an HPLC method for determining soluble protein in wine and must.

Procedure 16-1. Evaluation of Protein Stability by Exposure to Heat and Visual Examination

A clarified sample of wine is subjected to an elevated temperature (49°C) for 24 hours. At the end of this time period, the sample is visually inspected for haze formation relative to a control (unheated) sample of the same wine. As seen in Table 16.2, there are several variations on this procedure in terms of both temperature and time. Additionally, some wineries subject the wine to low-temperature storage as well.

I. Equipment
   Two 4-oz clear glass screw-cap or stoppered bottles
   Incubator or water bath at 49°C (120°F)
   0.45-μm membrane and filter housing
   High-intensity light source
II. Procedures
   1. Membrane-filter a sufficient quantity of wine to fill two 4-oz sample bottles.
   2. Fill the same bottles, labeling the first as "room temperature" and the second as "one day at 49°C."
   3. Examine each sample under high-intensity light, and record your impressions of initial clarity.
   4. Place the "one day at 49°C" sample in the incubator or water bath, noting the temperature.
   5. At 24 hours, examine each sample carefully.
   6. Clouding and/or precipitate formation in the heated sample vs. a clear control sample is indicative of protein instability.
III. Supplemental Notes
   1. Prior to evaluation of results, the wine should be brought to room temperature or chilled (see p. 338). Interpretation depends upon the winemaker's experience. Although the absence of haze is judged to indicate stability, a faint haze may be considered acceptable based upon prior experience with the test and wine type. As noted in the text, temperature and time regimes may be adjusted to suit individual needs.

Procedure 16-2. Evaluation of Protein Stability by the Trichloracetic Acid Test with Nephelometric Evaluation

This procedure is adapted from Berg and Akiyoshi (1961). Passage of a light ray through a turbid medium results in scattering and apparent energy loss in the incident beam. In fact, energy is not lost but undergoes directional changes as a result of scattering. This scattering effect may be measured at any angle relative to the plane of incident light.

The degree of scattering depends primarily upon particulate number, size, and shape. These parameters, themselves, depend upon several variables, including temperature, pH, concentration of reagents, and mixing procedures. Because of inherent difficulties, nephelometric procedures are largely empirical.

I. Equipment
   Boiling water bath
   Pyrex test tubes (20-mL capacity)
   High-intensity light source

Pipettes (1 mL)
Turbidimeter (nephelometer)

II. Reagents (See Appendix II)
Trichloroacetic acid (55%)

III. Procedure
1. Fill two test tubes with 10 mL each of filtered wine sample.
2. Examine both sample tubes for clarity under a high-intensity lamp.
3. To one sample add 1 mL of 55% trichloroacetic acid and transfer the sample tube to a boiling water bath for 2 minutes.
4. At the end of the reaction period, remove the sample tube and visually compare its clarity with that of the control (untreated) sample. Haze in the heated sample is indicative of protein instability.
5. After its removal from boiling water bath, hold the treated sample for 15 minutes for the reaction to be completed.
6. Consult the operator's manual for turbidimeter (nephelometer) setup and operation.
7. Determine "nephlos units" of sample(s).

IV. Supplemental Notes
1. Rankine and Pocock (1971) suggest using a nephlos value (< 19) as the upper limit for protein stability in table wines.
2. Some wineries prefer to compare the absorbance of heated vs. control samples, using a spectrophotometer at a wavelength around 430 nm. Others simply evaluate relative clarity differences visually between control and treated wines using a high-intensity light source.

Procedure 16-3. Ammonia (Ammonium Ion) by Ion Selective Electrode

The ammonia selective ion electrode has been successfully utilized for analysis of ammonia (NH_3) and ammonium ion (NH_4^+) concentrations in must and wine (McWilliams and Ough 1974). The electrode is constructed with an interior pH-sensing combination electrode separated from the outside by an ammonia-permeable membrane. When the electrode is placed in a sample, ammonia diffuses across this membrane and into the electrode body. The ammonia then reacts with the filling solution of the electrode, producing a shift in pH that is sensed (and measured) by the internal electrode. In an alkaline sample, ammonium ion is coverted to ammonia and measured in similar fashion.

I. Equipment
Expanded-scale pH/mV meter or specific ion meter
Ammonia selective ion electrode
Magnetic stirrer and stir-bar

II. Reagents (See Appendix II)
Ammonia stock solution, 1000 mg/L
Sodium hydroxide, 10 *M*
Distilled water

III. Procedure

1. Prepare calibration standards of 1.0, 10.0, and 100.0 mg/L from the ammonia stock solution according to Table 16-3. Bring solutions to final volume with distilled water.
2. Transfer 100 mL of the 1.0-mg/L standard to a 150-mL beaker. Place the electrode in the standard, add 1 mL NaOH, and begin stirring. Allow the mV reading to stabilize, and record it.
3. Repeat the process with 10.0- and 100.0-mg/L standards. Between readings, rinse the electrode and blot it dry.
4. Construct a calibration curve of mV readings vs. log concentration. This can best be done using semilog graph paper.
5. Treat the wine sample as in step 2. Record the reading, and determine the ammonia concentration by reference to the calibration curve.
6. If the reading is off the scale, measure the ammonia concentration in a diluted sample as follows: Place 20 mL sample in beaker, add 80 mL distilled water, and proceed as instructed.
7. Multiply the answer obtained from the calibration curve by 5 to account for the sample dilution.

IV. Supplemental Notes

1. The calibration curve should be checked periodically by rerunning one or more of the standards. Experience will determine how often one must do this.
2. Most electrode manuals will contain instructions for use with different ion meters, as well as for known addition techniques. These meters and techniques should produce answers equivalent to those obtained using the above procedure.

Procedure 16-4. Determination of Alpha Amino Nitrogen (Arginine) in Must

Seven major alpha amino acids are found in musts and wine: arginine, proline, alanine, threonine, serine, glutamic, and aspartic acids. Quantitatively, this group represents the major amino acid constituents of must and wine; so its measurement provides an indication of the potential for problem fermentations as well as the need for nitrogen supplementation. Unfortunately, published analytical results appear to be inconsistent, reflecting problems with the method and interferences associated with it.

Table 16-3. Preparation of ammonia calibration standards from 1000 mg/L stock solution.

Volume of stock (mL)	Final NH_3 concentration (mg/L) in 1-L volume
1.0	1.0
10.0	10.0
100.0	100.0

Following removal of proteins and large peptides from a must sample, 2,4,6-trinitrobenzene sulfonic acid (TNBS) will react with the primary amino groups of amino acids to form trinitrophenylated amino complexes absorbing at 420 nm (Crowell et al. 1985). The reaction initially is carried out in pH 9.5 borate buffer; the pH then is lowered to 7 with phosphate to avoid interference from hydroxyl ion. Sulfite ion concentrations of between 0.5 and 2 n*M* are optimum for full color development.

I. Equipment
   Spectrometer capable of measurements at 420 nm
   Vortex mixer
   Various pipettes
   Centrifuge

II. Reagents (See Appendix II)
   Sodium sulfite/sodium dihydrogen phosphate solution
   Sodium borate, 0.1 *M* in 0.1 *N* NaOH
   2,4,6-Trinitrobenzene sulfonic acid (TNBS)
   Arginine · HCl (200 μ*M* or 2.8 mg/L), primary amine stock standard
   Trichloroacetic acid (60% wt/vol)

III. Procedure
   1. Prepare must samples by centrifugation at 2000 rpm for 20 minutes. Filter the supernatant through Whatman No. 1 filter paper and then through a 0.45-μm membrane filter. Add 2 mL of 60% (wt/vol) trichloroacetic acid (TCA) to 1 mL sample, mix the solution thoroughly, and refilter it through a 0.45-μm membrane filter. Dilue 1.5 mL of the clarified must sample to 50 mL with distilled water.
   2. Dilute 0.25, 0.50, 0.75, and 1.00 mL of the 200 μ*M* stock arginine standard to 1.00 mL with distilled water. These calibration standard solutions are equivalent to 0.7, 1.4, 2.1, and 2.8 mg primary amino nitrogen/L.
   3. To 0.5 mL of diluted sample or working standard in a small glass test tube add 0.5 mL borate buffer. Mix the sample, add 20 μL TNBS, and mix it again. Set the test tube aside for exactly 5 minutes. After 5 minutes, add 2.0 mL of the sodium sulfate/sodium dihydrogen phosphate solution to stop the reaction.
   4. Measure the absorbance of the orange-colored solution within 10 minutes at 420 nm, against a reagent blank prepared by substituting 0.5 mL water for the sample.
   5. Calculate the concentration of primary amino nitrogen using the following equation:

$$\text{Primary amino nitrogen (mg/L)} = \frac{A(420\text{ nm})}{\text{Slope}} \times \text{Dilution factor} \quad (16\text{-}1)$$

   where the dilution factor = 100.

IV. Supplemental Notes

1. The results obtained from this test are expressed as arginine equivalents. Alanine, glycine, serine, and threonine give comparable responses to TNBS.
2. Proline does not react with TNBS, as it lacks a primary amino group. Although proline is the major amino acid present in wine grapes, it usually is not utilized by yeast.
3. Although this method has been tested on wines, Crowell et al (1985) have found that it gives falsely high results and thus is of limited utility.

Procedure 16-5. Determination of Proline in Juice and Wine

Of the various amino acids present in grape musts and wines, proline is present in the greatest concentration. Although its average level in grapes is reported to be 742 mg/L (Amerine and Ough 1980), its concentrations are known to vary with variety as well as growing season. Cabernet Sauvignon and Chardonnay musts typically have higher proline levels than other varieties (Ough and Stashak 1974). Higher levels of the amino acid also are found in fruit produced in cooler growing seasons, compared to fruit harvested in warmer years (Winkler et al. 1974).

Under the conditions normally followed in wine production, this amino acid probably is not utilized by yeast, for two reasons. First, the two enzymes involved in the uptake and utilization (proline permease and proline oxidase) are repressed by the presence of ammonium ion in must. The permease is irreversibly inactivated by ammonium ion. Second, oxidase enzymes require oxygen for activity. Ingledew et al. (1987) reported that wine yeasts may utilize proline as a nitrogen source when oxygen is present. They suggested that in addition to nitrogen supplementation, incorporation of air into yeast starters may prove valuable in reducing the frequency of stuck fermentations. This view was also presented by Fugelsang (1987).

Proline may be determined via its reaction with ninhydrin in the presence of formic acid, yielding a colored product that absorbs visible radiation at 517 nm (Amerine and Ough 1980).

I. Equipment
Spectrometer capable of reading at 517 nm
130 × 15 mm screw-cap test tubes
100-mL volumetric flasks
Boiling water bath
Various pipettes from 0.25 to 10 mL

II. Reagents (See Appendix II)
Stock proline standard (575 mg/L)
Methyl cellosolve–3% ninhydrin solution
Isopropanol : water, 1 : 1 mixture
Formic acid

III. Procedures
1. Pipette 0, 1, 2, 3, 5, 7, and 10 mL of the proline standard into 100-mL volumetric flasks. Dilute to the mark with distilled water. This will produce proline working standards of 0, 5.75, 11.50, 17.25, 28.75, 40.25, and 57.50 mg/L.
2. Pipette 1 mL of wine or grape juice into a 50-mL volumetric flask, and dilute to the mark with distilled water.
3. Transfer 0.5 mL of each proline working standard and wine or juice sample into 130 × 15 mm screw-cap test tubes.
4. Add 0.25 mL formic acid and 1 mL of the ninhydrin–methyl cellosolve solution to each tube, mix the contents well, cap the test tubes, and place the tubes in the boiling water bath for exactly 15 minutes.
5. After 15 minutes, remove the test tubes, and place them in a 20°C water bath to cool. Add 5 mL of the isopropanol–water mixture to each tube, mix the solutions, and transfer the contents to appropriate cuvettes.
6. Read the absorbance of each standard and sample at 517 nm against the water blank carried through the procedure. Plot absorbance vs. concentration of the standards, and read the concentrations of the samples directly from the plot.
7. Calculate the proline concentration in the original wine or juice sample by multiplying the value taken from the calibration plot times the dilution factor (50 mL/1 mL if sample was diluted according to step 2).

IV. Supplementary Notes
1. If the sample of wine or grape juice has an absorbance value that exceeds that of the most concentrated standard, dilute the sample 1 + 1 with the isopropanol–water mixture and repeat the determination.
2. Loss of color for the solutions is reported to be about 2% per hour. For greatest accuracy, one should run the calibration curve with each batch of samples.

Procedure 16-6. Nitrate Ion by Ion Selective Electrode

Nitrate ion (NO_3^-) concentrations in water, waste streams, and so on, can be measured with the nitrate ion selective electrode. The electrode is constructed with an interior silver/silver chloride reference element separated from the outside by an ion selective membrane. The membrane consists of an organic liquid ion exchanger immobilized in a polyvinyl chloride matrix material. When the electrode is placed in a sample, nitrate is sensed preferentially at the ion exchange membrane. A double junction reference electrode is used in conjunction with the nitrate electrode. To measure nitrate levels, standards and samples should be maintained at the same ionic strength. An ionic strength adjustment buffer is prepared and used for this purpose.

I. Equipment
Expanded-scale pH/mV meter or specific ion meter

Table 16-4. Preparation of nitrate calibration standards from 1000 mg / L working standard.

Volume of working standard (mL)	Final NO_3 concentration (mg/L) in 1-L volume
1.0	1.0
10	10.0
100	100

Nitrate selective ion electrode
Double junction reference electrode
Magnetic stirrer and stir-bar

II. Reagents (See Appendix II)
Sodium nitrate ($NaNO_3$) stock solution (10,000 mg/L as nitrate)
Ammonium sulfate, $(NH_4)_2SO_4$, ionic strength adjustment buffer (2.0 *M*)
Distilled water

III. Procedure

1. Prepare a working standard of 1000 mg/L nitrate from the nitrate stock. Pipette 100 ml of stock solution into a 1-L volumetric flask, and dilute to the mark with distilled water.
2. Prepare calibration standards of 1.0, 10, and 100 mg/L from the nitrate working standard solution (see Table 16-4). Bring the solutions to final volume with distilled water.
3. Transfer 50 mL of the 1.0-mg/L standard to a 150-mL beaker. Add 50 mL ionic strength adjustment buffer, place the electrodes (nitrate ISE and double-junction reference) in the standard, and begin gentle stirring. Allow the mV reading to stabilize, and record it.
4. Repeat the process with 10.0, 100.0, and 1000.0 mg/L standards. Between readings place the meter in the "standby" position, rinse the electrodes with a small amount of the next standard, and blot them dry.
5. Construct a calibration curve of mV readings vs. log concentration. This can best be done using four-cycle semilog graph paper with the concentration (mg/L) on the log axis.
6. Treat the sample as in step 3. Rinse the electrodes with a small amount of sample, and blot them dry before placing them in the beaker. Record the reading, and determine nitrate concentration by reference to the calibration curve.
7. If the reading is off the scale, dilute the sample prior to measuring its nitrate concentration. Be sure to multiply the answer obtained from the calibration curve by the appropriate dilution factor.

IV. Supplemental Notes

1. Please see Supplemental Notes for Procedure 16-3.

Section IV

Remedial Actions

Occasionally, laboratory personnel must evaluate wines containing sediment and haze that are not biological in origin. Based on its visual appearance, Quinsland (1978) categorizes such sediment into three classes: crystalline, fibrous, and/or amorphous.

The following paragraphs will acquaint the laboratory analyst with several rapid screening procedures that may serve as a basis for the more exhaustive testing that should follow.

PRELIMINARY SAMPLE PREPARATION

Generally, it is recommended that sediment and/or haze be concentrated somewhat prior to identification. In the case of sediment, this may be accomplished by pipetting or decanting the wine off the sediment and collecting the latter in a convenient container. With haze, however, it may be necessary to centrifuge the sample for 5 to 10 minutes or membrane filter the sample in order to concentrate the suspension. Once it has been concentrated, Tanner and Vetsch (1956) recommend washing the collected sediment in 5 mL of 95% alcohol. This is most conveniently done by resuspending the collected sediment in the alcohol and recentrifuging.

CRYSTALLINE DEPOSITS

Potassium bitartrate and calcium tartrate instabilities are discussed in Chapters 13 and 18. For details regarding the formation of bitartrate/tartrate precipitates as well as their analytical determination, the reader is referred to these chapters. Additionally, filter acid (diatomaceous earth) may occasionally be seen (see photomicrograph IV-5), p. 353.

I. Equipment

  Bright field microscope and several slides and cover slips

  Membrane filters and appropriate housing. Most laboratory filtration units utilize 47-mm membranes. For general purposes, 1–5-μm cellulose acetate filters are useful.

  Clinical centrifuge

  Magnesium oxide rods

  Cobalt filter

II. Reagents (See Appendix II)
   H_2SO_4 (1 + 3)
   Sodium metavanadate solution (3%)
III. Procedure
   1. Using the filter apparatus and appropriate membrane, collect a portion of wine sample containing the suspect sediment.
   2. Rinse sediment/crystals with a small volume of distilled water, and apply vacuum to remove water.
   3. Transfer the membrane to a watch glass. Place a drop of dilute sulfuric acid on the precipitate.
   4. Add a drop of metavanadate solution.
IV. Interpretation
   1. Tartrate present in sediment turns yellow-orange in color.
V. Alternative Test for Potassium
   1. "Load" a magnesium oxide ("magnesia") rod with sediment using the following technique:
      a. Using the flame from a Bunsen burner, heat the end of the rod and insert it into the collected sediment.
      b. Reheat the rod, and again insert it into the sediment.
      c. Repeat (a) and (b) if necessary.
   2. When the sediment is concentrated into the magnesia rod, hold the loaded end in the outer portion of the flame.
   3. Using a cobalt filter disk, the presence of potassium is indicated by a rose-color flame.
VI. Microscopic Examination.
   1. See photomicrographs IV.I and IV.2.

CRYSTAL-LIKE SEDIMENTS

Cork dust appears "crystal-like" microscopically, and it may be confused with bitartrate unless further examination is performed. The recommended procedure is to use a stain that reacts with lignin, a structural macromolecule in cork.

I. Equipment
   The equipment required is the same as presented in the preceding method.
II. Reagents (See Appendix II)
   Phloroglucinol stain (made fresh daily)
III. Procedure
   1. Collect a portion of wine containing the debris by filtering it through an 8-μm membrane filter. (Membrane filters of 8 μm pore size may be ordered through supply houses carrying the Nucleopore product line.)
   2. Wet the filter and sediment with phloroglucinol stain. Hold the stain in contact with sediment for 5 minutes.

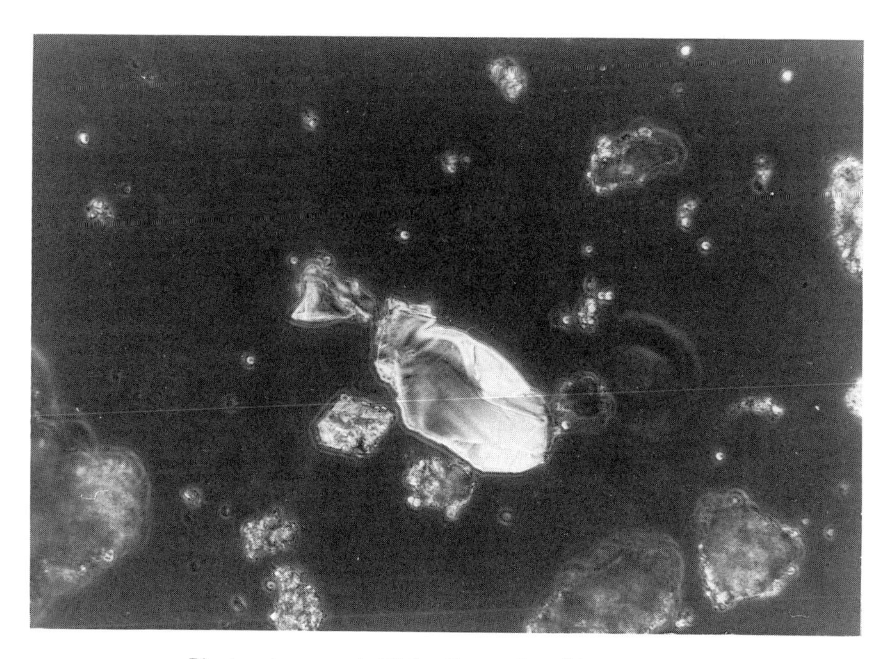

Photomicrograph IV-1. Potassium bitartrate.*

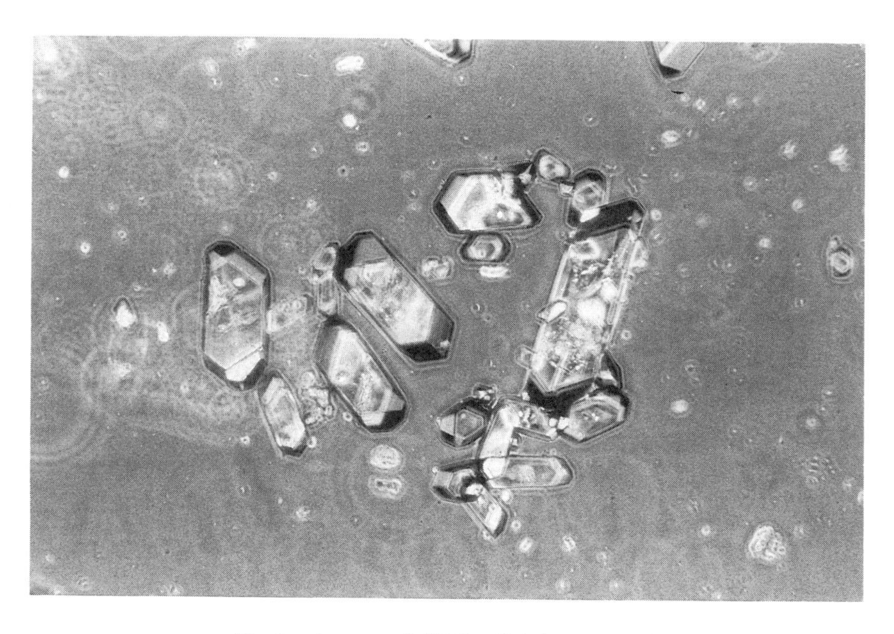

Photomicrograph IV-2. Calcium tartrate.

*Photomicrographs IV-1–IV-5 courtesy of Ms. Wendy Single-Davidson, CSU, Fresno.

3. Apply vacuum to the filter to remove the stain. Rinse the filter with distilled water.

IV. Interpretation

1. Examine the sediment microscopically. Cork debris appears as red crystal-like aggregations of cells (see photomicrograph IV.3).
2. Case lint also stains red using this technique. However, it is fibrous in appearance (see photomicrograph IV.4).

FIBROUS (CELLULOSIC) MATERIALS

Fibrous materials found in finished wines usually are cellulosic in nature, originating from the filter pad matrix or from case lint that may get into bottles prior to filling.

I. Equipment
As above.

II. Reagents (See Appendix II)
Phloroglucinol stain
Cellulose stain

III. Procedure

1. Collect by membrane filtration a portion of the wine containing the sediment onto two separate membranes.
2. Stain the first membrane using the phloroglucinol technique presented

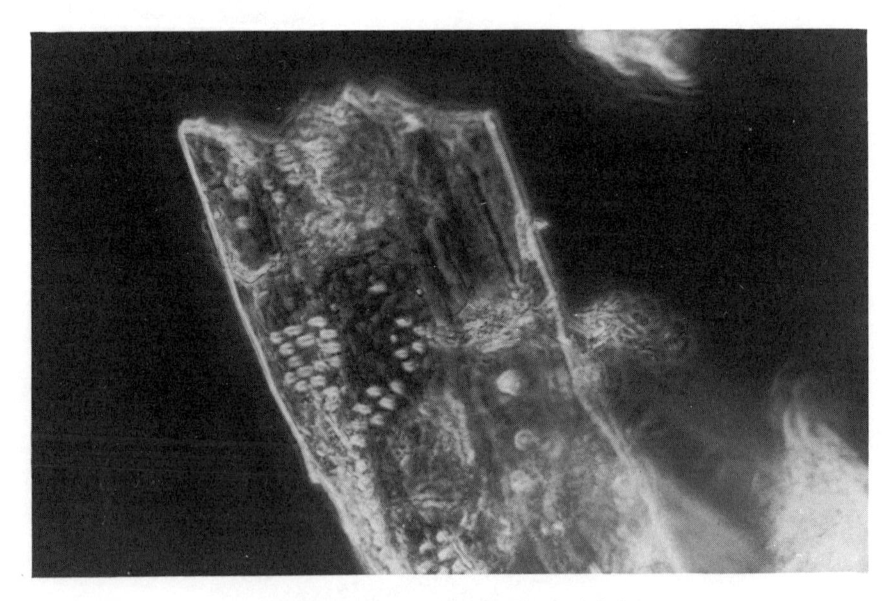

Photomicrograph IV-3. Cork debris.

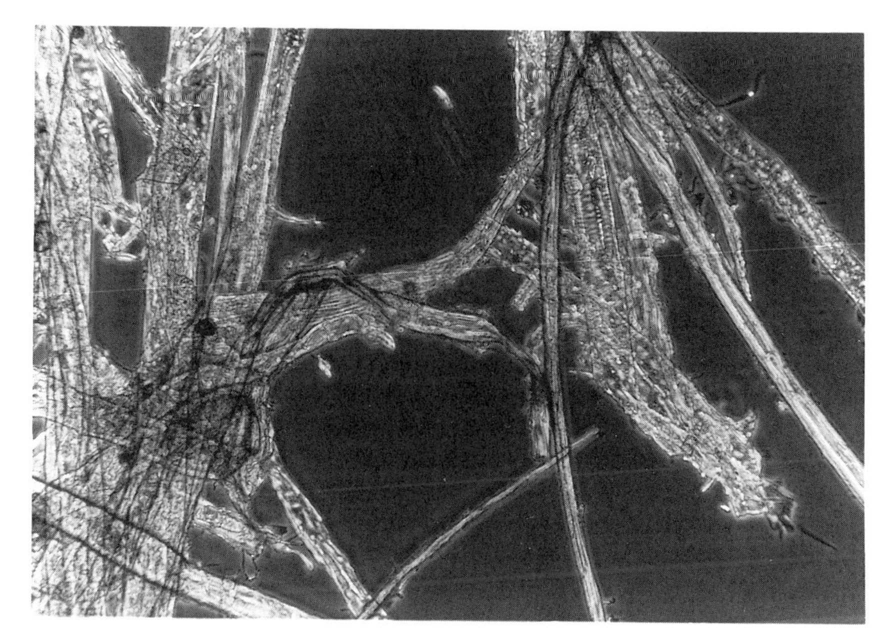

Photomicrograph IV-4. Cellulosic debris.

above. *Interpretation:* Case lint (containing lignin) appears as red fibrous material.

3. Treat the second membrane with cellulose stain.
   a. Flood the filter and particulates with stain. Hold 5 minutes.
   b. Filter to remove the stain. Rinse with several milliliters of distilled water. *Interpretation:* Microscopically, cellulosic material not containing lignin appears light blue. (*Note:* Best results are obtained when the preparation is examined fresh. Color intensity diminishes after 30 minutes.)

IV. Supplemental Note
   1. Asbestos fibers are not stained using either of these techniques.

AMORPHOUS MATERIALS

Particulates with no defined shape, include protein and phenolics (and complexes of the two), as well as paraffin used to coat corks.

I. Equipment
   Polycarbonate membrane filters (47 mm) and housing

II. Reagents (See Appendix II)
    Amido black 10-B protein stain
    Eosin Y protein stain
    Folin Ciocalteu reagent
    H_2SO_4
    0.5% (wt/vol) $K_4Fe(CN)_6 \cdot 3H_2O$

III. Procedure for Protein
1. Filter sample containing sediment through an 8-μm polycarbonate membrane. Cellulose acetate membranes are unstable in the presence of protein stain.
2. Wet the membrane with stain, and hold 10 minutes.
3. Remove the stain by applying vacuum. Rinse with methanol–acetic acid solvent until the filter is white. *Interpretation:* Proteinaceous materials will stain blue-black microscopically.
4. With the above procedure, Eosin Y stain may be used instead of amido black 10-B.

IV. Procedure for Phenolic Precipitates
1. Collect sediment by filtering sample through an 8-μm membrane filter.
2. Rinse with several milliliters distilled water and, with a spatula, transfer sediment onto a watch glass.
3. Using previously diluted Folin-Ciocalteu reagent, add a drop to sediment. *Interpretation:* Phenolic complexes dissolve to yield a slate gray to blue turbid solution.

V. Alternative Procedure
1. Collect a portion of sediment in another test tube. Add 1 mL of concentrated H_2SO_4 to the sample, and gently heat it. *Interpretation:*
   a. Phenolics (pigments and tannins) present in the sample turn dark red.
   b. Carbonization is suggestive of protein.

PECTIN INSTABILITIES

1. To a 25-mL aliquot of juice or wine containing unidentified haze, add 50 mL of 95% ethanol or isopropanol.
2. The formation of a gel after several minutes is indicative of pectin.

METAL INSTABILITIES

On occasion, metal instabilities in the form of iron and/or copper complexes (casse) may be seen in bottled wines. Several screening tests have been proposed that provide insight into the nature of the instabilities (Tanner and Vetsch 1956; Amerine et al. 1980).

Preliminary acidification of the suspect sample using 10% HCl is useful for separation of metal-containing complexes from complexes of protein and phenolics.

I. Procedure

1. Collect approximately 20 mL of the suspect wine. Add 3 to 5 mL of HCl (10%), and note whether the haze dissipates or remains. If the haze solubilizes, proceed using the diagnostic scheme presented below. If the haze remains, the instability is probably due to protein or complexes of protein, protein–phenolics, or phenolics–phenolics (e.g., pigment–tannin). A method for the presumptive identification of proteins and phenolics is presented above.

(a) Flame test for copper and organics

2. Collect 15 to 20 mL of the suspect wine.
3. Add 5 drops of H_2O_2
4. If the haze dissipates, Cu^{2+} is suspected. If the haze remains, see options below under (b).
5. If the haze can be concentrated by centrifugation or if sediment is present, collect a sufficient amount on a stainless steel laboratory spatula. *Slowly* dry the sediment over a Bunsen burner, and, when it is completely dry, attempt to ignite it by more intensive exposure to the flame. If the haze consists primarily of complexes of copper and organics, the sediment is partially burnt. However, copper sulfide and ferric phosphate casse will not burn.

(b) Color formation tests for copper and iron

1. Collect 20 mL of the turbid wine in two test tubes.
2. To Tube 1 add several 5 mL of postassium ferrocyanide (0.5%). *Interpretation:* Formation of red coloration is a positive presumptive test for copper and its complexes.
3. To Tube 2 add 5 mL of potassium ferrocyanide (0.5%) and 5 mL of HCl (10%). *Interpretation:* Formation of blue coloration is a positive presumptive test for iron.

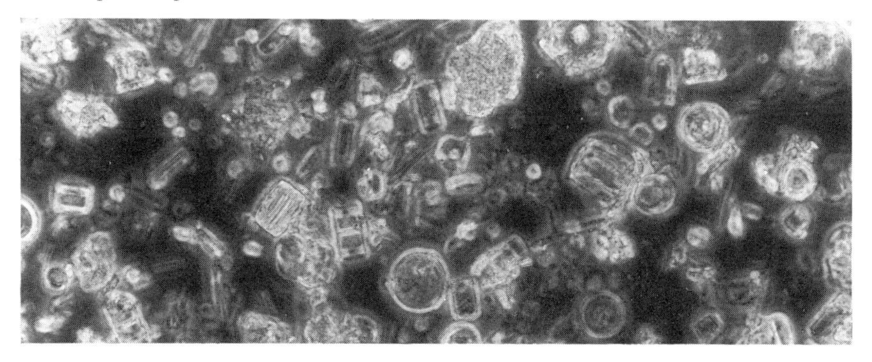

Photomicrograph IV-5. Diatomaceous earth.

Chapter 17

Fining and Fining Agents

GENERAL PRINCIPLES OF FINING

Fining is the addition of a reactive or adsorptive substance to remove or reduce the concentration of one or more undesirable constituents. As such, fining agents are added for the purposes of clarity, color, flavor and/or stability modification. These agents can be grouped according to their general nature:

1. Earths: bentonite, kaolin
2. Proteins: gelatin, isinglass, casein, albumen
3. Polysaccharides: agars
4. Carbons
5. Synthetic polymers: PVPP, nylon
6. Silicon dioxide (kieselsols)
7. Others, including metal chelators, enzymes, and so on

The mechanisms of action for fining agents may be electrical (charge) interaction, bond formation, and/or absorption and adsorption. In the case of electrical interaction, those particles of opposite charge to the fining agent are induced to coalesce with the fining agent, forming larger particles. Owing to its greater density, the growing complex eventually settles from solution.

Turbidity in juice or wine may be due to the presence of particulate matter in the form of grape tissue, yeast and bacteria, or colloidally suspended particles derived from the grape, or may be the result of changes occurring during aging or storage. These particles may be present in the form of proteins, pectins and gums, metallocolloids (formed by flocculation of insoluble oxidized salts), and degradation products of polyphenols. Enhanced filterability, due to absorption/adsorption of colloidal and suspended material by the fining agent complex, is a frequent benefit of using fining agents.

The commonly utilized fining agents are compared in Table 17-1 with respect to their activity (and effectiveness) as well as the potential for problems. The table compares fining agents in order, from most to least effective, for several typical fining goals.

Table 17-1 ignores method of preparation, relative addition levels, temper-

Table 17-1. Comparison of selected fining agents with respect to desired effects and potential problems listed in the order of most effective to least effective.

Color reduction	Tannin reduction	Volume of lees formed
carbon	gelatin	bentonite[1]
gelatin	albumen	gelatin
casein	isinglass	casein
albumen	casein	albumen
isinglass	bentonite	isinglass
bentonite	carbon	ferrocyanide*
ferrocyanide*	ferrocyanide*	Cufex
Cufex	Cufex	carbon
Clarity and stability	**Potential for overfining**	**Quality impairment**
bentonite	gelatin	carbon
ferrocyanide[2]	albumen	bentonite
Cufex	isinglass	casein
carbon	casein	gelatin
isinglass	ferrocyanide*	albumen
casein	Cufex	isinglass
gelatin	—	ferrocyanide*
albumen	—	Cufex

[1]Usually twice the lees of other agents.
[2]Not permitted in United States.
* Source: Berg (1981).

ature of addition, and age of the wine. For example, with age monomeric (positively-charged) charged anthocyanins polymerize, yielding colloidal macromolecules that may eventually precipitate. The use of bentonite (net negative charge) on a young wine may result in color loss due to the charge interaction of the monomeric pigments with bentonite. With age and pigment polymerization, the phenolic charge is lost, and bentonite has less of effect on color. Conversely, utilization of a protein fining agent, such as gelatin, (net positive charge) on a young red wine has a much less dramatic effect on color removal than its addition to an older red wine (Bergeret 1963).

The effectiveness of fining agents depends upon the agent, the method of preparation and addition, the quantity employed, wine pH, metal content, temperature, and age as well as previous treatments. Fining is a surface reaction, so the method of hydration and addition is important.

To duplicate laboratory trials in the cellar, the same lot of fining agent must be prepared for each purpose and used in the same way. Carefully controlled laboratory fining trials must be performed prior to cellar addition. In evaluating fining trials, the winemaker must note and record how each fining addition alters clarity, lees production, and compaction, stability, color, body (front, middle and finish), astringency, bitterness, and the nose characteristics in general. Other considerations include changes in fruit, finish, aging potential, and overall wine palatability. The reader is referred to Table 17-2 for guidelines in laboratory fining operations. A discussion of commonly employed fining agents is presented in the following sections.

It is essential that fining agents be added properly in the cellar if plant trials are to duplicate laboratory results. Common methods of adding the fining agent include making the addition:

(a) Uniformly and slowly through a "Y" on the suction side of a pump while transferring or mixing.
(b) Uniformly and slowly through a proportioning "in line" pump.
(c) Uniformly and slowly through a "T" into a Guth-type tank mixer.
(d) Slowly, in slurry form, to a barrel, using a dowel stirred in a figure-8 motion or a barrel mixer through the bung hole.

BENTONITE

American bentonite is a volcanic claylike material that exists as a complex hydrated aluminum silicate with exchangeable cationic components ($Al_2O_3 \cdot 4SiO_2 - H_2O$). Known as Montmorillonite clay after the French town where

Table 17-2. Laboratory fining trial equivalents to pounds per 1000 gallons.

Level of treatment per 1000 gal (pounds)	Fining Trial Treatment Level Using 700 mL of Juice or Wine (soln. vols. in mL)						
	Tannic acid soln. (1%)	Gelatin soln. (1%)	Sparkolloid stand. soln.* (1.4%)	Metafine or Cufex stand Soln.*	Bentonite (5%)	Casein (2%)	Any solid (g)
1/4	2.1	2.1	0.90	0.17	0.40	1.1	0.021
1/2	4.2	4.2	1.8	0.35	0.80	2.1	0.042
3/4	6.3	6.3	2.6	0.52	1.3	3.2	0.063
1.0	8.4	8.4	3.5	0.70	1.7	4.2	0.084
1.5	12.6	12.6	5.25	1.0	2.5	6.3	0.13
2.0	16.8	16.8	7.0	1.4	3.4	8.4	0.17
3.0	25.2	25.2	10.5	2.1	5.0	12.6	0.25
4.0	33.6	33.6	14.0	2.8	6.7	16.8	0.34
5.0	42.0	42.0	17.5	3.5	8.4	21.0	0.42
6.0	50.4	50.4	21.0	4.2	10.0	25.2	0.50
7.0	58.8	58.8	24.5	4.9	11.8	29.4	0.59
8.0	67.2	67.2	28.0	5.6	13.4	33.6	0.67
9.0	75.6	75.6	31.5	6.3	15.1	37.8	0.75

*See manufacturer's instruction or Appendix II for preparation of standard solutions.

bentonite was first mined, the material is mined in Wyoming and also called "Wyoming clay." Calcium, sodium, and magnesium forms of bentonite are available, the most commonly used form in the United States being sodium bentonite, largely because of its increased protein binding ability when compared to other forms.

Bentonite finds its principal application in the removal of proteins from white wine and juice. Additionally, it is active in reducing enzymes such as polyphenol oxidases, which may catalyze oxidation and browning in juice. The mechanism of removal is adsorptive interaction between the flat negatively charged surfaces of the bentonite platelets and the positively charged proteins (see Fig. 17-1). Because the platelet edges are positively charged, some limited binding of negatively charged proteins may occur. Wines of lower pH (greater net positive charge) require less bentonite for stabilization than those of higher pH.

Reduction in levels of unstable protein usually is attempted using bentonite. Because of variations in the fundamental chemistry of wine proteins, the types of bentonites used, and addition and mixing protocols, the entire question of protein stabilization is often confusing. Protein instability in juice and white table wine is a major production problem. Unless proteins are reduced to levels at which precipitation will not occur, sediment formation is likely to occur later in the bottled wine. Despite an abundance of literature dealing with protein instability, definitive levels at which instability will occur have not been reported. Levels of protein nitrogen reported in wine range from 10 to 275 mg/L (Boulton 1980e). Variables affecting the concentration of protein in wine include grape variety, growing season, maturity, condition of fruit at harvest, pH, and processing protocol. With regard to variety, some cultivars present more problems than do others. For example, Sauvignon blanc, Gewurztraminer, and muscats often have large concentrations of unstable proteins and thus may require extensive treatments to achieve stability. Wine proteins are a mixture of those derived from the grape as well as those from yeast autolysate. Protein from yeast autolysate has not been reported to be involved in white wine instability. Reports indicate that about half of the total wine protein is bound to flavonoids originating in the grape. It is this unstable complex that is thought to be responsible for protein haze (Somers and Ziemelis 1973b).

Wine proteins may be characterized by molecular weight as well as electronic charge. Up to eight separate fractions have been reported, ranging in molecular weight from 11,000 to 28,000 (Boulton 1980e). Based upon the distribution of charged groups on the protein, the point of maximum instability and, hence, tendency toward precipitation is reached when positive and negative charges are equal. The pH at which charge is distributed equally is referred to as the isoelectric point for the particular protein (see Ch. 16).

The greater the difference between wine pH and the isoelectric point of the protein fraction, the greater the net charge on that fraction and thus the greater the potential for reaction with fining materials of opposite charge. Hence, correcting protein instability would appear to be most difficult to correct when the wine pH is near the isoelectric point of the various unstable fractions.

Bentonite may indirectly adsorb some phenolics via binding with still-reactive protein that has complexed with the phenolics. However, as seen in Table 17-1, its activity toward phenolics is relatively low.

Bentonite is known to affect red wine color directly by binding with positively charged anthocyanins. Reaction with bentonite may result in up to 15% removal of color (Kroll 1963). This removal or reduction in red wine color by bentonite is dependent upon the age of the wine. For example, bentonite may remove more color in younger wines than in their older counterparts, largely because of the greater action of bentonite on colloidally colored material found in younger wines (Bergeret 1963). In this regard, the addition of bentonite to red wines at levels of 1/2 to 1 pound per 1000 gallons also is reported to improve membrane filtration, presumably because of a reduction of colloidally suspended particles.

Bentonite fining is known to indirectly prevent or impede formation of copper and possibly iron casse in wines where metal levels may be a problem. In the case of copper casse, this effect is probably due to removal or a reduction in the levels of proteins and peptides known to be involved in the formation of haze and precipitate.

Preparation of Bentonite

Preparation of bentonite for addition to wine greatly impacts is activity toward protein. In solution, bentonite swells to many times its dehydrated dimensions, its activity in solution being much like that of a multi-plateleted, long-chained, linear, negatively charged molecule (Singleton 1967). During the hydration phase, charged platelets repel each other and begin to separate. Water molecules partially neutralize and separate the exposed surfaces, thereby exposing a very large matrix of reactive surface. Each silicaceous platelet has calculated dimensions of 1 nm × 500 nm. Flat surfaces are negatively charged, whereas the edges carry positive charges. When fully hydrated, 1 g of sodium bentonite has an extremely large adsorption surface. As seen in Fig. 17-1, when properly dispersed, bentonite exists as a network resembling a "house of cards." The presence of water molecules within the network prevents flocculation and precipitation.

The methodology of bentonite preparation largely determines its reactivity toward proteins when added to wine. Most bentonites should be slowly hydrated in hot water to avoid clumping. Typically, the bentonite-to-water ratio in working slurrys is on the order of 5 to 6% (wt/vol). In any case, the total quantity of water must not exceed 1% of the wine volume treated. One recommended hydration procedure calls for mixing bentonite at 96 g/L into warm

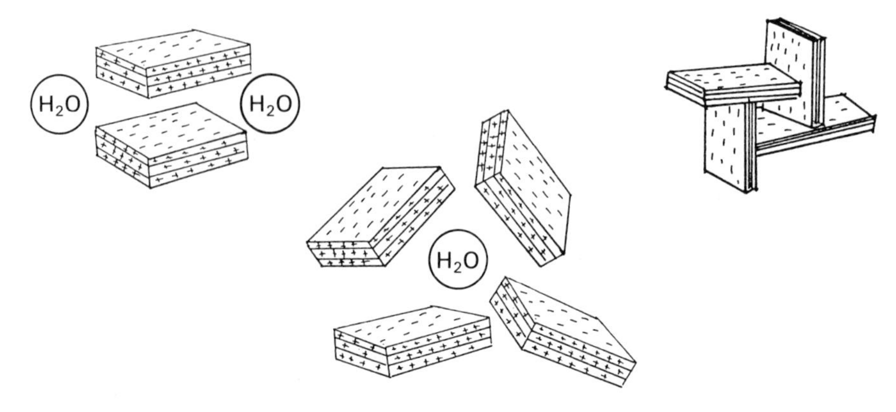

Fig. 17-1. Fully hydrated bentonite molecule depicting "house of cards" configuration.

(46°C) water with continued agitation until a smooth slurry is achieved. Heating allows the platelets to separate fully; when it is properly carried out, the slurry resembles a gel. The winemaker should refer to necessary technical literature when preparing bentonite.

The water used in the hydration phase should have a low mineral content. Dissolved metal cations present in slurry water will preferentially replace sodium ions clustered on the sodium bentonite clay surface and thereby detrimentally affect the hydration, viscosity, and binding capacity of the bentonite (American Colloid Co.).

Bentonites are mined in several areas of the world, typically differing in level of purity, particle size, adsorption, and swelling capacity. The type and source of the bentonite used may affect protein removal, with differences in protein removal generally due to variations in the swelling capacity and cation exchange capacity of the bentonite in question. In the refining process bentonites from several sources are stockpiled and tested for purity, adsorption, as well as lead and iron content. They then are dried, ground, and packaged. Because of these variations, there may be differences between lots; so it is imperative that lab trials be conducted using the same lot as will be used in the cellar treatment.

Utilization of Sodium and Calcium Bentonite

The sodium form of bentonite finds widest application in the United States, largely as a result of its superior swelling capacity when compared with calcium bentonite. By comparison, calcium bentonite platelets tend not to separate as well as the sodium form; so calcium bentonite suffers from a reduced exposed surface area for protein binding. Further, calcium bentonite tends to settle from solution at a slower rate than does the sodium form. However, lees compaction is better in calcium-bentonite-treated wines compared to those treated with the sodium form (Ferenczi 1966). Calcium bentonite finds its widest acceptance in Europe where sodium levels are restricted, and as a riddling aid in Methode Champenoise production. Expected sodium pickup from sodium bentonite may range from 1.7 to 3.5 g/100 g bentonite (Amerine and Joslyn 1970).

The most commonly reported problem with sodium bentonite is excessive lees production and the rather loose compaction of those lees. Typically, bentonite lees volumes range from 5 to 10% of the total volume of juice or wine treated. Several techniques may be employed to deal with and minimize these problems.

1. Properly hydrated, bentonite requires only minutes to react with and precipitate protein. Seventy five percent of the total protein so removed is bound in the first minute after contact (McLaren et al. 1958). Because adsorbed protein may "slough off" bentonite platelets upon standing, prolonged contact may result in less efficient removal of proteins. Therefore, the winemaker might consider using in-line centrifugation or depth filtration to remove bentonite rather than traditional gravity clarification.
2. Another method of reducing lees volume is to hydrate bentonite in the wine to be treated rather than in water. Although this may significantly reduce its binding capacity due to premature flocculation, the final lees volume should be about half that obtained by conventional hydration protocols.
3. An additional way to enhance bentonite lees compaction is to bentonite-fine during conventional cold stabilization. Potassium bitartrate crystal formation significantly compacts and reduces the lees volume. Cold stabilization and protein fining may be effectively linked, especially if the wine pH is below the pK_1 value for tartaric acid (3.65). At pH values of less than 3.65, formation of potassium bitartrate releases a free hydrogen ion into solution and thereby maintains or causes decreases in pH. Because lower pH favors increased positive charges on proteins, the activity of bentonite also is enhanced. Additionally, cold stabilization, itself, may reduce the concentration of proteins in wine.
4. Counterfining of bentonite with Sparkolloid, kieselsol, gelatin, and so on, is often a means of aiding bentonite precipitation and compaction.

An additional fining procedure which may be helpful in preserving varietal aroma and aroma intensity is to ferment in contact with bentonite vs. using bentonite in the wine for protein stabilization. Such a practice avoids or minimizes the need for subsequent bentonite addition into wine.

Fermentation in contact with bentonite has several advantages. Possible sensory benefits may result due to the fact that only juice components are adsorbed onto bentonite not fermentation or barrel aging constituents. Fermentation lees have a lower monetary value than does finished wine lees. Thus protein stabilization or partial stabilization during fermentation may be an important economic consideration. The procedure for fermentation of white juice in contact with bentonite is as follows:

1. Settle juice to remove non-soluble solids. This may be done with refrigeration and/or the use of fining agents (A high solids level could foul the bentonite utilized during fermentation and reduce overall efficiency). Add the desired quantity of bentonite in line while racking into the fermentor.
2. Make any yeast nutrient, sugar or acid addition need to the juice.
3. Add yeast innoculum on to juice surface. The bentonite may bind with the yeast, pull the yeast to the bottom of the fermentor and thus delay the fermentation rate. For this reason, mixing is avoided.

Yeast nutrient addition is a preferable step in fermentations occurring in the presence of bentonite. Bentonite may deplete assimilable nitrogen content of the must due to electrostatic binding and adsorption. This may result in fermentation sticking and/or H_2S production. The addition of an exogenous source of nitrogen eliminates these potential problems (See Ch. 10).

A determination of the quantity of bentonite to add to the juice to attain a protein stable wine is done empirically or analytically. Many winemakers fermenting in contact with bentonite simply add several pounds per 1000 gallons of juice. Methods for predicting specific bentonite levels needed in the juice for subsequent wine stabilization are available (see Ch. 16).

The use of bentonite to remove or reduce unstable proteins in wine is well documented. Kean and Marsh (1956b) report reductions in wine protein levels of 40 to 90 mg/L from initial values. However, bentonite is not specific in its reaction with wine components; significant changes may be seen in the sensory properties of the product after treatment. Bentonite additions, especially those exceeding 4 pounds per 1000 gallons, may strip wine flavor, body, and, in the case of young red wines, significant amounts of color. Further, bentonite may impart an earthy character to the wine.

Bentonite fining may remove peptides and some amino acids, potentially affecting the rate and completion of fermentation. Bubble retention and quality are, in part, related to the residual concentration of protein and peptides; so sparkling wines made from heavily bentonited cuvees may produce a finished product lacking in these areas. Some winemakers facing large bentonite additions may elect to utilize multiple additions rather than a single large dose. This approach may be successful in reducing the overall bentonite requirement, especially if the wine in question is low in suspended solids.

The difficulties encountered with the use of bentonite to obtain protein stability in wine have led to examination of other techniques and materials, including fermentation in contact with bentonite, and ultrafiltration.

POLYSACCHARIDES

The polysaccharides most commonly used for fining are the alginates. Alginic acid or algin is extracted from marine brown algae where it serves as the structural polymer comprising the cell wall. Alginic acid exists as a high molecular weight, long-chained polymeric salt of β-1,4-D-manuronic acid and L-guluronic acid.

The extent of polymerization is an important consideration in fining. In wine, low-viscosity, short-chained polymers find the greatest application (Ribereau-Gayon et al. 1972). Alginic acids are positively charged and usually are bound to some inert carrier such as diatomaceous earth to facilitate settling. Although their reactivity is somewhat unpredictable, clarification is best accomplished if the juice or wine pH is less than 3.5. Clarification may be accelerated with small additions of counterfining agents such as gelatin or bentonite. Recognizing this, some proprietary products include 5 to 10% (wt/wt) gelatin.

Sparkolloid and Klear-mor are two commercially available polysaccharides. In solution, both are present as positively charged alginates on a diatomaceous earth carrier. Sparkolloid is sold as a cream-colored powder that, when added to water, forms a viscous colloidal solution. Sparkolloid and Klear-mor find their principal application in enhancing clarity and filterability. As these compounds have little absorptive capability, they do not affect color, odor, or flavor. Both fining agents usually are prepared by hydration in hot water (1 pound/1–2 gallons water at 180°F) and are added to juice or wine while still hot. Their addition levels generally parallel those of bentonite, but as with any other fining agent, laboratory trials should precede actual addition.

Compared to bentonite, these fining agents have the advantage that they produce relatively compact lees, thus facilitating subsequent processing. They occasionally have been used as "top dressing" agents following bentonite fining to aid in lees compaction and after treatment for metal instabilities with Cufex or Metafine.

CARBONS

Activated carbon adsorbents have long been used as a means of rapidly modifying the sensory character of juices, wines, and spirits. As such, activated carbon is a relatively nonspecific adsorptive agent that tends to bind with weakly polar molecules, especially those containing benzene rings or their derivatives. The phenolic compounds in wine exist principally as ring structures (or their derivatives), so they are effectively removed by carbon addition. Owing to the nature of the carbon particle surface, smaller phenolics are bound preferentially. Singleton (1967) pointed out that the active adsorptive surface of most carbons is confined to micropores, so that compounds larger than flavonoid dimers are excluded. Carbon-catalyzed oxidation of phenols to quinones has been reported

(Singleton and Draper 1962); however, quinones are strongly adsorbed and removed from solution.

The activated carbons used in the wine industry are usually of two types: (1) decolorizing carbon, used primarily for removal of undesirable brown coloration from white juice or wines; and (2) deodorizing carbon, for removal of undesirable odors and flavors. Decolorizing carbon can effectively "strip" color from a white juice or wine and, at excessive levels, impart a carbon taste to the product. In addition, wines treated with decolorizing carbon will, at times, undergo significant browning from the above-mentioned carbon-catalyzed oxidation. Like all additions of this type, careful laboratory trials should precede additions to the wine. In the case of decolorizing carbon, it is often desirable to artificially age the treated sample in the laboratory and measure its subsequent browning rate against an untreated, heated control. Such aging may be accomplished by holding the samples in an incubator at 120°F for 1 to 3 days. Recommended use levels for decolorizing carbon range from 0.25 to 5 pounds/1000 gallons.

Deodorizing carbon is often added to wine after the addition of wine spirits, as well as to the spirits themselves. In limited amounts, it can be effective in masking the hot or harsh character of newly fortified wines. Much of the vinous character may be removed by charcoal addition, so care must be taken to establish acceptable fining levels by preliminary laboratory trials. Use levels for deodorizing carbon seldom exceed 4 pounds/1000 gallons.

Activated carbons contain a great deal of air (oxygen) within the carbon particles, so their early and thorough removal from the wine may aid in reducing potential carbon-catalyzed oxidation. In addition to oxidation of wine phenols, activated carbons may induce oxidation of alcohol, resulting in increased levels of acetaldehyde. Because carbon particles tend to remain suspended in solution, it may be necessary to follow carbon fining with a bentonite addition to facilitate settling. The amounts of bentonite required will, of course, depend upon the degree to which the haze is a problem. Some winemakers add carbon with the diatomaceous earth body feed during filteration. Such "in-line" additions, and removal, minimize the carbon–wine contact time and reduce the possibility of carbon-induced oxidation. Singleton and Draper (1962) reported that the oxidative properties of carbon in wine were reduced by adding ascorbic acid, and they concluded that such additions did significantly diminish the activity of decolorizing charcoal. In contrast, others have found that lowering the pH and elevating the temperature increased the adsorptive activity of the carbon (Amerine et al. 1980). The addition of carbon to juice rather than to wine helps to minimize carbon induced oxidation.

KIESELSOL (COLLOIDAL SILICON DIOXIDE)

Kieselsol, which is a generic name for aqueous suspensions of silicon dioxide, was used first in Germany as a substitute for tannic acid in gelatin fining. Tannins, like (−)kieselsol, electrostatically bind with positively charged proteins (such as gelatin) and initiate flocculation and settling. In the replacement of tannin with kieselsol, the same reaction occurs. Kieselsols have the ability to both electrostatically bind and adsorb compounds onto their particle surfaces. To meet the variety of needs in the wine and juice industries, several kieselsol products are available. The characteristics of these products depend upon several factors, the most important of which is the method of production (Hahn and Possman 1977), which controls particle size, the shape and nature of the particle surface (perhaps the most important parameter), particle size distribution within the suspension, and particle charge density. The charge density depends on the number of hydroxy groups on the surface of the kieselsol particle that are available for binding, which is determined, in part, by the pH of the kieselsol suspension.

As mentioned, the most frequent utilization of kieselsol is in clarification, as a replacement for tannic acid during protein fining. The effectiveness of the kieselsol–gelatin complex depends on the above-mentioned parameters and upon the "bloom number" of the gelatin used. For kieselsol finings, a low bloom number usually is required (between 70 and 130). Using a higher-bloom gelatin in conjunction with kieselsol may result in unreacted gelatin being left in solution.

Some of the advantages of kieselsol over tannin counterfining are the following:

1. The volume of lees produced by kieselsol–gelatin is less than that produced by tannin–gelatin.
2. Flocculation and precipitation of the kieselsol–gelatin complex proceed at a faster rate than that of the tannin–gelatin.
3. Clarity is often superior with the use of kieselsol.
4. The action of gelatin appears to be more "gentle" with kieselsol–gelatin than with tannin–gelatin. That is, stripping of the wine character is noticeably reduced with the use of kieselsol.

In finings where proteinaceous agents (e.g., gelatin) are used to reduce phenolic levels in wine, kieselsol plays a role not only in precipitating protein but also in helping to bring down protein–tannate complexes. Hence, kieselsol plays the secondary role, in this case, of enhancement of clarification, as well as compaction of lees.

Practical Considerations Regarding Kieselsol

The winemaker should weigh several considerations in using kieselsols:

1. Because of the wide variety of silica products available, winemakers should thoroughly review the manufacturer's recommendations prior to use. As with any fining agent, laboratory trials must precede cellar additions.

2. Most kieselsol products presently on the market have a limited shelf life (less than two years). Care should be taken to avoid freezing the product, as that can detrimentally affect particle suspension.
3. Generally, gelatin or other protein fining agents are employed in conjunction with kieselsol at a level of 0.1 to 0.2 "times" the quantity of kieselsol. The order of addition may be important; it is usually recommended that gelatin be added first, followed by the kieselsol.
4. With few exceptions, kieselsol utilization should not exceed 10 pounds/1000 gallons. The legal limit for kieselsol products is 20 pounds/1000 gallons (of a 30% colloidal slurry). Government regulations state that silicon dioxide must be completely removed by filtration.
5. Some manufacturers recommend that their kieselsol products be added prior to the protein fining agents if the goal is phenol reduction. In such cases, care must be taken to avoid overfining.

PROTEIN FINING AGENTS

Protein and protein-like fining agents have selected affinity for wine polyphenols. The mechanism of reaction is hydrogen bonding between the phenolic hydroxyl and peptide bonds of the protein component (Fig. 17-2).

Compared with other types of chemical bonding, hydrogen bonds are rather weak; so the capacity of a protein fining agent is, in part, a function of the number of potential hydrogen bonding sites per unit weight. Thus, the techniques used for hydration or swelling, as well as addition and mixing, are of importance in achieving the desired effect.

The selectivity of protein fining agents is based partly on bond strength (number of potential bonds formed) between the fining agent and wine phenols. The combination of phenol and fining agent that produces the strongest total

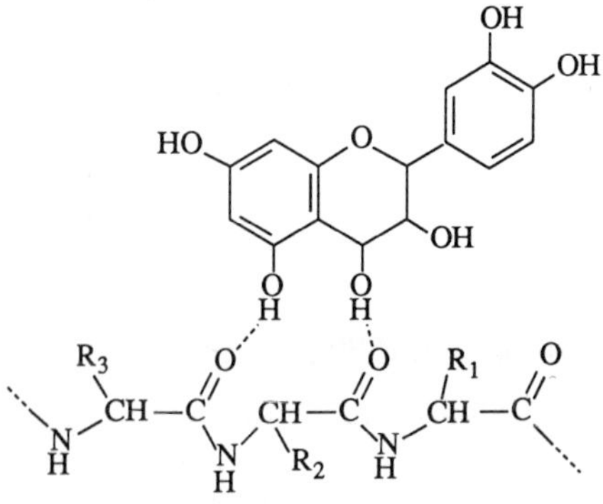

Fig. 17-2. Hydrogen Bond interaction between phenolic compound and protein.

hydrogen bonding usually occurs preferentially; so larger phenolics with more available hydroxyls and thus more potential hydrogen bonding sites are adsorbed preferentially.

As wines age, some of the monomeric phenols polymerize to form larger molecules. Rossi and Singleton (1966) have shown that proteinaceous fining agents such as gelatin, casein, and isinglass preferentially remove condensed tannins of molecular weight over 5000. Therefore, the strongest complexes usually are formed between dimeric and larger phenols, whereas only weak bond formation occurs with monomeric species.

As polymeric phenols are largely responsible for the perception of astringency and monomeric forms for bitterness, the use of protein fining agents may, by selective removal of the polymeric group, unmask a wine's bitterness.

Another consideration in the selection of this type of fining agent is its solubility and flexibility. To be effective, the fining agent must align itself with phenolic hydroxyls. Insoluble agents (such as PVPP) generally have a much lower capacity for phenol removal than soluble species because of their smaller number of available binding sites for bond formation. In addition, overfining occasionally is a problem with protein fining agents, its results being principally body stripping. Juices and young wines are much more forgiving of the action of protein fining agents than are aged wines.

Gelatin

Gelatin is prepared from collagen, the major structural protein in skin and bones. Hydrolysis of the multistranded polypeptide in solutions of acid and base causes strand separation, with resultant production of gelatin. The peptide chain in gelatin ranges in size from 15,000 to over 140,000 molecular weight, and contains high levels of the amino acids glycine, proline, and hydroxyproline, compared to most proteins (Singleton 1967).

The isoelectric point of gelatin is pH 4.7. Therefore, it occurs in wine as a positively charged entity capable of reaction with negatively charged species such as tannins via hydrogen bond formation. It finds principal application in clarification as well as in modification of overly astringent wines. It also is employed to reduce harshness (astringency) and improve clarity in juice prior to fermentation.

In terms of phenol adsorption capacity, gelatin preferentially binds with larger molecules that have more phenolic groupings and more extensive hydrogen bonding than smaller molecules (Singleton 1967). Thus the fining agent has a less dramatic effect on color and tannin reduction in younger wines than in older products, as the latter generally have a greater percentage of larger polymeric phenolic compounds that readily bind with gelatin. As a generality, gelatin additions may result in color shifts in red wines from tawny (brown) to a more ruby red, perceived visually as a shift in hue (the ratio of absorbances at 420 nm and 520 nm). (A more detailed discussion of wine color is presented in Chapter 7.)

Gelatin occasionally is employed to help reduce the phenol level and brown color in juice prior to fermentation. Such applications usually occur in conjunction with kieselsol fining. Cofining with kieselsol and occationally bentonite may aid in rapid clarification.

In any fining operation, only good-quality gelatin, free of undesirable flavors and odors, should be used. Commercial gelatin, which is available in several forms and grades, usually is rated according to purity as well as bloom. Bloom refers to the ability of the gelatin to absorb large quantities of water, usually six to ten times its weight. Determination of bloom number is done by allowing a 6.66% solution of gelatin to age for 18 hours at 10°C. The weight (in grams) required to force a 0.5-inch stamp into the gel to a depth of 4 mm determines the bloom number (Hahn and Possman 1977); the higher the bloom rating is, the greater the adsorbing capability. Gelatin recommended for wine treatment ranges from 80 to 200 bloom.

As the number of potential bonding sites determines its effectiveness, the size of the gelatin molecule is also an important consideration. The use of lower molecular weight gelatin is believed to reduce the rate of precipitation but enhances clarification and lees compaction.

The quantities of gelatin needed to achieve the desired level of clarification may reduce wine astringency to undesirably low levels. This can be a problem especially for white wines, most of which have such a low precipitable phenolic content that an exogenous source of tannic acid or kieselsol may be desirable for reaction with the excess gelatin remaining in solution. If tannic acid additions are made they occur 24 hours prior to gelatin fining of white wines. The ratio of tannin to gelatin is usually 1:1 (wt/wt), but will vary depending on the individual wine.

Most winemakers prefer to counterfine with kieselsol rather than tannin. The replacement of tannin with kieselsol is reported to moderate the activity of gelatin on wine flavor, and it reduces the amount of gelatin needed and the volume of lees produced. The rate and quality of clarification also are improved (Hahn and Possman 1977). For use in conjunction with kieselsol, a gelatin that is soluble at elevated temperatures (e.g., 70–130 bloom) is recommended (Mobay 1976). Gelatin fining occasionally is employed in conjunction with bentonite additions to remove residual haze. In this case, the negatively charged planar surface of the bentonite platelets permits reaction with positively charged gelatin, and the two substances precipitate from solution. This technique also may help in compaction of troublesome bentonite lees. Overfining with gelatin, as with most protein fining agents, may render the wine hazy and unstable with respect to heat-labile proteins. Additionally, where copper levels exceed "safe limits," the presence of residual gelatin may increase the possibility of casse formation.

Most commonly available as a powder, gelatin also may be purchased as a liquid concentrate of 30 to 46% gelatin, or in sheets. Liquid gelatins are produced by hydrolysis in order to lower the molecular weight and prevent gelatinization at high concentration. Liquid gelatin concentrates typically are stabilized with benzoates and/or sulfur dioxide.

Prior to use, dry gelatin must be hydrated in warm water (1 pound gelatin/2 gallons water at 112°F) and added to juice or wine by the methods outlined earlier in this chapter. Prolonged or excessive heating of gelatin solution may result in denaturation of the protein and reduced activity. Thus hydration procedures must be standardized between laboratory and the cellar. Because of potential biological deterioration, gelatin solutions should not be stored over extended periods. Addition levels in wine generally range upward from 1/16 pound/1000 gallons. In red wines, levels often range from 0.4 to 0.8 pound/1000 gallons. Much larger doses are used in juice fining, especially press juice. Heavy press juice may require upwards of 4 pounds/1000 gallons to reduce astringency and oxidized color. Counterfining with kieselsol or bentonite is recommended (Zoecklein 1984). Gorinstein et al (1984) confirmed that low temperatures aid in gelatin precipitation.

Casein

As the principal protein in milk, casein occurs in solution as a positively charged macromolecule with a molecular weight of approximately 375,000. It is presently available to wineries in two forms: (1) purified milk casein, which is insoluble in acid but soluble in alkali solution; and (2) sodium or potassium caseinate, which is soluble in water. Some proprietary caseinates contain potassium bicarbonate, which enhances the solubility of the casein and its salts in water. By comparison to caseinate, which is simply hydrated with water prior to use, milk casein first must be dissolved in water at a pH greater than 8.0. Regardless of the particular casein or casein preparation, it should be hydrated in water, never juice or wine. The agent must be fully hydrated prior to use.

Upon its addition to wine, a drop in pH causes casein to flocculate, with the resulting precipitate adsorbing and mechanically removing suspended material as it settles from solution. This property also minimizes the potential for the presence of the residual protein in the wine.

In general, casein is used in white wines to remove off background flavors, reduce oxidized color, and, on occasion, aid in clarification.

Casein has been used to impede or prevent pinking in susceptible wines such as Pinot blanc. It may be employed as a substitute for carbon in color modification in juice and white wines, and it frequently is used in reducing or removing the dark color and cooked flavor character from baked sherries. Although less effective than carbon, casein does not catalyze the oxidative deterioration associated with the use of carbon in wine.

In cases where metal instability is a problem, casein has been reported to be effective in reducing the concentration of both copper and iron. Reductions of up to 45% of copper and 60% of ferric ion from wines or musts have been reported.

The use of casein for clarification may not yield the desired results. However, when used for this purpose, it usually is prepared in a more concentrated form. The use of working solutions prepared at concentrations of at least 25 g/L appears to improve clarification.

As is the case in gelatin fining, casein finings of white wine often are preceded by tannic acid or kieselsol additions. In this case, however, the ratio of tannic acid to casein is on the order of 0.5 : 1 (wt/wt). Such additions usually are made approximately 24 hours before planned casein addition and should be preceded by careful laboratory fining trials.

It is essential that only the purest available grade of casein be used in order not to impart off flavors and characters to the wine. Addition levels to wines usually range from 1/8 to 2 pounds/1000 gallons, depending upon the purpose and the particular form of casein used. Laboratory trials should be performed by pipetting casein to the bottom of the trial vessel and mixing, to avoid clot suspension.

Egg Albumen

Egg albumen is a relatively common fining agent for red wines. Albumens are found in egg whites, blood and milk. Fresh egg whites contain approximately 12.5% (wt/wt) protein, which corresponds to approximately 3 to 4 g of active product per white (Peynaud 1984). The principal proteins in egg whites are albumen and globular proteins. Albumens are soluble in water, whereas globular fractions are not; the latter dissolve in neutral dilute salt solutions. Egg whites contain 7 to 8 g/L salts; and if the whites are diluted with water rather than wine prior to the addition to wine, the salt concentration is lowered so that the globulin fraction becomes partially insoluble, producing a turbid solution. In order to solubilize the entire protein content and, hence, maximize fining efficiency, a pinch of potassium chloride often is added to the water albumen slurry. Sodium chloride addition is not permitted in the United States.

The Federal Register gives the following guidelines for the use of albumen: A working solution is prepared by dissolving 1 oz (28.35 g) of potassium chloride and 2 pounds (907 g) of egg white in one gallon (3.8 L) of water. Use levels of this solution may not exceed 1.5 gallons/1000 gallons of wine.

In preparation for fining, fresh or frozen whites should be gently whipped with a whisk prior to addition to the main volume of wine. Excessive mixing, although effective in reducing the troublesome gelatinous character of the white, yields a foam, some of which will remain on the wine surface. As compared with frozen whites, fresh egg whites appear to have a larger phenol adsorption capability, which may amount to as much as a twofold difference. Egg whites generally are added to barrels and mixed either with a dowel through the bung hole rotated in a figure-8 motion, or by the use of specially designed barrel mixers.

Egg whites seldom are used in white wine fining because of the need for counterfining agents. Compared to gelatin, egg white appears to remove fewer phenols and less of the fruit character (see Table 17.1). The precautions noted for casein (above) regarding potential instabilities also apply to egg white fining. Addition levels range up to eight whites per 225-L barrel.

Isinglass

Produced from sturgeon collagen, isinglass is a positively charged protein fining agent. Used at levels of 1/8 to 1/3 pound/1000 gallons, isinglass finds its principal application in white wine fining. Isinglass is used to clean up the aroma and modify the finish without significantly affecting tannin levels (see Table 17-1). Additions after aging and immediately prior to bottling are common to round the palate, possibly freshen the nose, and improve clarity. Isinglass like gelatin and bentonite sometimes is used as a riddling aid in Methode Champenoise.

Isinglass is available in two forms: (1) sheet isinglass, which must be hydrated and rinsed repeatedly to remove undesirable odors; and (2) flocculated isinglass, which, because of its purity, usually needs only to be hydrated. Hydration should be carried out in cold water (<60°F). If prepared in hot water, isinglass undergoes partial hydrolysis, with formation of smaller molecules. The change in molecular weight from 140,000 to lower molecular weight species (15,000–58,000) results in differences in its fining characteristics (Rankine 1984). The product is more gelatin-like in its activity.

A typical hydration protocol for flocked isinglass is that of Rankine (1984). To a 250-L container, add 60 L of water at 10°C, 500 g citric acid, and 140 g potassium metabisulfite. Mix these ingredients, and slowly add 1 kg of powdered isinglass. Continue mixing until the mixture is uniform. Following dispersion, the isinglass is allowed to stand overnight to ensure complete hydration. The temperature during the hydration phase should not exceed 15°C. The next day the solution is brought to 200 L final volume. The final concentrations are: isinglass, 5 g/L; citric acid, 2.5 g/L; and sulfur dioxide, 350 mg/L. Prior to addition to the wine, the mixture should be mixed thoroughly. It should be noted that some flocked isinglass already contains citric acid and potassium metabisulfite.

Isinglass finds its principal application in white wine fining where the product is used to bring out or unmask the fruit character without significant changes in tannin levels. It is less active toward condensed tannins than either gelatin or casein (Rankine 1984). As condensed phenolics are the principal substances responsible for astringency, isinglass has a less dramatic effect on the reduction of both wine astringency and body than do most other protein fining agents. It has the added benefit of not requiring extensive counterfining as compared with other protein fining agents. Ribereau-Gayon (1972) reports that most wines contain sufficient quantities of the endogenous tannin needed to glean the residual protein present after fining.

Isinglass may be used at several stages in the winemaking process. Many vintners fine with the agent after aging and before bottling to produce a brilliantly clear white wine without the stripping effect seen with other protein fining agents.

Isinglass has several advantages over gelatin in the fining of white wines. The agent is active at lower concentrations than gelatin, producing enhanced clarification and a more brilliant wine. Clarification is much less temperature-dependent using isinglass than with gelatin, which shows enhanced properties at low temperature (Ribereau-Gayon et al. 1972).

Isinglass does have several significant drawbacks, however. The low density of the flakes formed after its addition to the wine may result in rather voluminous lees formation (>2%). Additionally, the particulates tend to hang on the sides of barrels and casks. This problem may be corrected by using counterfining agents such as bentonite. Depending upon the age and source of the isinglass, extensive preliminary treatment may be required to clean up the fishy odors noted in the working slurry. For this reason, only fresh isinglass of the highest quality should be used.

Polyvinylpolypyrolidone (PVPP)

PVPP is a synthetic, high molecular weight fining agent composed of cross-linked monomers of polyvinylpyrolidone (PVP) (see Fig. 17-3). It is a protein-like fining agent with affinity for low molecular weight polyphenols. The mechanism of its action is hydrogen bond formation between carbonyl groups on the

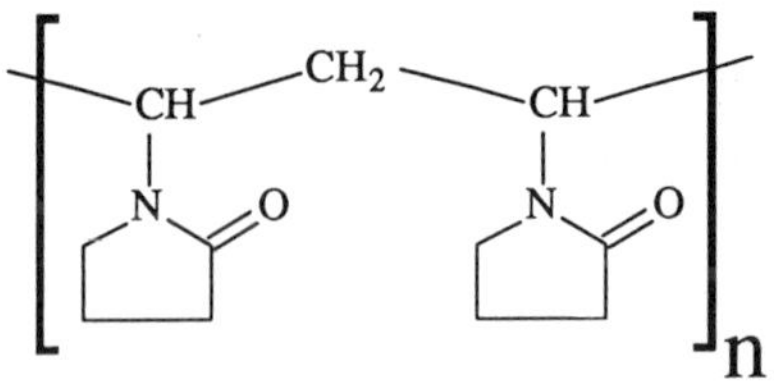

CH — CH_2 — CH
N O N O
n

Fig. 17-3. Polymer of PVPP.

polyamide (PVPP) and the phenolic hydrogens. Unlike the soluble protein fining agents, which may selectively remove larger polyphenols by virtue of their ability to conform with the molecule and interact with many hydroxyl groups, insoluble PVP is able to contact relatively few of the reactive groups needed for bond formation. Therefore, PVPP finds its major application in binding with and removing smaller phenolic species such as catechins and hydroxy-cinnamates that conform to the PVPP molecule.

As a selective phenol adsorbent, this proprietary compound is available in several forms. Polyclar 10 is chemically equivalent to the more commonly used insoluble form, Polyclar AT. Their difference lies in particle size, that of Polyclar 10 being much smaller than that of its counterpart. Because of its smaller particle size and greater surface area, the adsorptive capacity of Polyclar 10 is superior to that of Polyclar AT. However, for routine winery application the benefits of Polyclar 10 must be considered against the relative ease of clarification and filtration seen in Polyclar AT.

The activity of Polyclar AT is specific for low molecular weight phenols such as catechins (GAF 1975). The latter compounds are precursors to browning in white wines and browning and bitterness in red; so Polyclar AT may be effective in "toning down" bitterness or the potential for this problem in wines. Ough (1960) found that Polyclar AT removed more tannin and anthocyanin than did gelatin—behavior that represents a potential problem in use of this agent in red wine fining.

In addition to the properties described above, Polyclar AT finds application in the removal of browning or pinking precursors in juice and young white wines. When employed for browning removal, it often may be used in conjunction with decolorizing carbon. In cofining with carbon, Polyclar AT can aid precipitation. The recommended use levels for Polyclar AT range from 1 to 6 pounds/1000 gallons. PVPP has the ability to strip wine complexities; so it is desirable to fine juice or young wines prior to the development of the aged bouquet. Federal regulations require the fining agent to be removed from the wine by filtration prior to bottling.

TANNIN

Tannin or tannic acid (as oak bark tannin) occasionally is used as a fining agent. In solution, tannin is negatively charged. Historically, it has been used most frequently in conjunction with gelatin fining to enhance clarification of the latter. However, tannin may find application in increasing the astringency in wines deficient in grape tannins. Tannic acid additions to wines are usually at levels of less than 0.25 pounds/1000 gallons. In the laboratory, tannic acid usually is prepared as a 1% (wt/vol) solution in 70% ethanol.

ENZYMES

Pectolytic Enzymes (Pectinases)

Grapes contain a number of insoluble solids and colloids that are responsible for difficulties in pressing, clarification, and filtration of juices and wines. The polysaccharide, pectin, and pectic acid, the product of its hydrolysis, are important members of this group. Pectins are present in plants as structural components of the cell wall, which are functionally analogous to animal collagen. Pectins occur as highly methylated polymers of galacturonic acid linked via α-1,4-glycosidic bonds.

Pectolytic enzymes are active in hydrolyzing pectin and pectic acid at specific locations along the polymer. Three such enzymes are identified by their specific reaction site:

1. Polymethygalacturonase (PMG), which attacks the α-1,4-glycosidic linkages between adjacent methylated galacturonic acid groups.
2. Polygalacturonase (PG), which causes hydrolysis of 1,4-glycosidic bonds but without the methylation required for PMG. Usually two groups of PG-enzymes are active: Endo-PG enzymes lyse bonds within the polymer, whereas Exo-PG counterparts attack 1,4-linkages at the terminal ends of the polymer.
3. Pectin methylesterase (PME), which hydrolyzes the methyl ester of galacturonic acid.

The various sites of enzyme activity on the pectin polymer are presented in Fig. 17.4.

Grape enzyme systems bring about limited hydrolysis of pectin polymers. However, because native enzyme systems are relatively poor in Endo-PG activity, only terminal groups are lysed, so there is little overall breakdown of

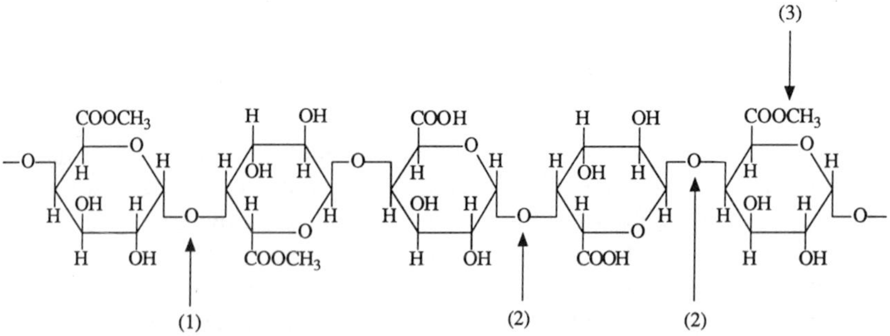

Fig. 17-4. Portion of a pectin polymer showing sites of pectinase activity. (1) site of PMG activity, (2) PG-activity and (3) PME-activity.

the polymer chain. From a processing point of view, the problem is manifested as difficulties in pressing, juice clarification, filtration, and so on. For these and other reasons, some winemakers add commercially prepared enzymes that are high in Endo-PG activity. Pectolytic enzymes find particular application with grapes of high pectin content such as varieties of *Vitis labrusca*. During fermentation, 30 to 90% of grape pectic compounds are removed via enzyme activity as well as by ethanol production (Amerine et al. 1972).

The presence of undegraded pectin in juice and wine can be readily detected using an alcohol precipitation test. One part juice or wine is added to two parts 95% ethanol or isopropanol. The sample is mixed and examined after several minutes for the presence of a flocculant precipitate or jelly like curdle. A clear solution or one developing a slight haze indicates adequate depectinization.

Pectinolytic enzymes are available in different forms and grades; one should consult the manufacturer's recommendations for the appropriate use levels. In general, the use of enzymes increases juice yield, clarity, red wine color, and filterability.

Glucanases

The growth of *Botrytis cinerea* on grapes can have a significant effect on the polysaccharide content of the resultant juice and wine. For example, the pectin content often is reduced as a result of the production of polygalacturonases by the fungus. However, winemakers often experience difficulty in clarifying juices and wines produced from grapes degraded by *B. cinerea*. Ribereau-Gayon (1988) reports a 900,000 PM glucan believed to be responsible for clarification difficulties in wines produced from *B. cinerea*-infected grapes. The structure of this chain is composed principally of β-1–3 linkages with 1–6 β-linked branch chains. Several commercial β-1–3 glucanases exist that can hydrolyze the glucan and thus significantly aid in clarification and filtration.

FINING AND WINE STABILITY

Wine fining can have a significant effect on wine stability by removing complexing factors. These mainly are complexes formed between polyphenols and tartaric acid in red wines and proteins and tartaric acid in whites (see Chapters 13 and 18), so there is an intimate relationship between wine fining and potassium bitartrate and protein stabilization. For example, it is known that condensed polyphenols interfere with potassium bitartrate precipitation (Balakian and Berg 1968), which suggests that the use of fining agents, particularly protein fining agents, can have an effect on potassium bitartrate stability.

If one compares the fining abilities of clear versus unclear wines, there is often a striking difference. This difference is due to the reduction of protective colloids. It is these protective colloids that make the fining of young wines and wine produced from botrytized grapes very difficult. Gelatin is particularly sensitive to this phenomenon, albumen less sensitive, and casein and isinglass are least affected by the protective colloids in wines (Ribereau-Gayon et al. 1972). For this reason, if initial fining does not give sufficient clarification, a second one following racking often will lead to acceptable results.

The effects of protein fining agents can be unpredictable, so carefully controlled laboratory fining trials must be performed and evaluated prior to cellar treatment. It is generally accepted that juice prior to fermentation and young wines are much more "forgiving" than older wines of the action of protein fining agents.

Due to their phenol binding abilities, protein fining agents can have a dramatic effect on color, texture, body (front, middle, and finish), astringency, bitterness, nose characteristics in general, fruit, finish, aging potential, stability, lees production, lees compaction, and overall palatability. It is vital that careful fining trials be performed and evaluated prior to the use of these agents in the cellar. The same lot of fining material prepared exactly the same must be used in both laboratory and cellar addition. Protein fining agents must be of the highest purity, properly hydrated, and properly added in the cellar to be effective.

SUMMARY OF IMPORTANT CONSIDERATIONS IN FINING

1. In any fining operation, one should strive to use the smallest quantity of fining agent necessary to achieve the desired result. Only fining agents of the highest purity, free from undesirable odors and flavors, should be used.
2. Whenever possible, one should limit the contact time between fining agent(s) and juice or wine to just that necessary for complete reaction and formation of lees. In many cases, centrifugation may be used effectively to shorten exposure time.
3. It is essential that careful laboratory trials be carried out prior to cellar treatment. Additionally, laboratory fining agents should be prepared by the method intended for cellar treatment. For example, the sheer force exerted by laboratory blenders cannot be duplicated in the cellar.
4. Trials should consider not only the levels of agent needed to achieve the desired effect, but also changes in the sensory profile of the wine.
5. To ensure maximum utilization and reactivity of a fining agent, it is essential that thorough mixing of the material and wine be achieved. Techniques commonly used for mixing were reviewed earlier in this chapter. See page 357.
6. Wines to be clarified by use of fining agents should be low in dissolved CO_2. The evolution of gas from the solution will serve to maintain particulates in solution and impede settling.

7. The effectiveness of a fining agent is dependent upon the agent, method of preparation and addition, levels of addition, pH, metal content, temperature, presence of CO_2, and prior wine treatments.
8. Often a wine with a lower pH will require less fining agent for clarification than the corresponding wine of higher pH, because of the greater number of positively charged species in solution.
9. A high metal content in either the fining slurry, juice or the wine can adversely affect flocculation and could reduce the activity of the agent. Although the action of some agents such as gelatin reportedly depends on the presence of metal ions (iron), many winemakers choose to use deionized water for hydration.
10. Because temperature plays an important role in the reaction time, it is essential that laboratory fining trials be carried out at the same temperature at which the wine will be treated.

Chapter 18

Correction of Tartrate Instabilities

The formation of crystalline deposits of potassium bitartrate and occasionally calcium tartrate, although a normal phenomenon in aging wine, generally is unacceptable to consumers.

Potassium bitartrate stability traditionally has been accomplished by chilling, ion exchange, or both. Also, a number of other techniques have been attempted: filtration (Scott et al. 1981), electrodialysis (Postel and Prasch 1977), reverse osmosis (Wucherpfennig 1978), crystal flow (Riese 1980), contact seeding, and so on. Of these processes, filtration, crystal flow, and contact seeding are finding greatest commercial application (Ewart 1984).

In conventional cold stabilization (chill-proofing) procedures, wines are chilled to a temperature designed to decrease potassium bitartrate (KHT) solubility to a level that optimally, will cause precipitation of the salt. Important variables affecting precipitation of potassium bitartrate during chilling include: (1) the concentration of reactants, specifically tartaric acid; (2) availability of nuclei for crystal growth; and (3) the solubility of the potassium bitartrate formed. Perin (1977) proposes the following empirical relationship for determination of the temperature needed for bitartrate precipitation:

$$\text{Temperature } (-°\text{C}) = \frac{(\%\text{ ethanol})}{2} - 1 \qquad (18\text{-}1)$$

Conventional cold stabilization of table wines involves reducing the temperature of the wine for a period that may last for several weeks or more. Dessert wines with higher alcohol and sugar levels require lower temperatures and longer holding times than are needed for table wines.

Due to the energy costs of holding large volumes of wine at low temperature for long periods, alternatives have been developed to accelerate the cold stabilization process. Seeding techniques are used to reduce the time and improve efficiency. In their review of cold stabilization techniques, Cooke and Berg (1984) reported that approximately half of California wineries surveyed had used or were currently using potassium bitartrate seeding. In principle, the addition of an excess of finely powdered potassium bitartrate creates a supersaturated solution. In such a system, the driving force is toward rapid precipitation on already formed crystals. Thus, the technique eliminates the limiting step in crystal formation, namely, nucleation.

Several processing considerations that are important in achieving stability using bitartrate seeding techniques are outlined in the following section.

QUANTITY OF KHT AND CRYSTAL SIZE

Because of the involvement of complexing agents in precipitation equilibria, the amount of KHT seed required will depend somewhat on the wine in question. The weight of KHT seed used always must produce a supersaturated solution. Table 18-1 compares changes in tartaric acid, potassium ion, and final CP values in wine treated with different levels of seed. As seen from the table, tartaric acid, potassium, and concentration product values decrease with increased levels of the potassium bitartrate seed. Rhein and Neradt (1979) reported optimal levels of seed to be 4 g/L. In practice, stability may be achieved by using lower seed concentrations. However, the winemaker is cautioned to evaluate carefully any changes from recommended seed levels before treatment. This may be accomplished by conducting laboratory trials as outlined in Procedure 13-2.

The seed particle size should range from 30 to 150 μm (Guimberteau et al. 1981). Rhein and Neradt (1979) report that the maximum reactive surface area is achieved at a particle size of 40 μm. Because the reaction rate depends on available surface area, the use of larger seed particles will slow the rate of crystal growth. Hence, from the winemaker's point of view, the processing time is extended.

Agitation

Cold stabilization using seeding should optimally be conducted in a small tank (<2000 gallons) where mixing can be controlled. As crystal growth is dependent upon the available interactive surface area, agitation is essential, and effective seeding requires intimate contact between the wine and KHT seed added. Table 18-2 compares a mixed and a static system with regard to final stability.

Table 18-1. Influence of potassium bitartrate seed (40 μm) level on wine components at 0°C.

Seed addition level (g/L)	Tartaric acid (g/L)	$[K^+]$	CP ($\times 10^{-5}$)
Control	1.58	920	15.1
1	1.11	808	9.3
2	1.03	794	8.5
4	0.93	765	7.6
8	0.78	754	6.2

SOURCE: Blouin et al. (1982).

Table 18-2. Effect of agitation on final tartrate, K^+, and CP values in white wine seeded with KHT at 4 g/L (0°C).

Wine	Tartaric acid (g/L)	K^+ (mg/L)	CP ($\times 10^{-5}$)
Control	1.58	920	15.1
Static system	1.38	870	12.5
Agitated system	1.17	805	9.8

SOURCE: Blouin et al. (1982).

The data of this table show that proper mixing is critical in producing a stable wine with this method.

Time and Temperature

When bitartrate seeding is employed, the enormous surface area presented by the addition of powdered KHT reduces or eliminates the energy-consuming nuclei-induction phase and permits immediate crystal growth. The procedure can be carried out at any temperature. However, when the seeding process is used, the treatment temperature should be identical to the desired stability temperature. For example, many wine producers seed white wines at 0°C and reds at +5°C. If the stabilization procedure is correctly performed, wines held at or above these temperatures should be stable with respect to potassium bitartrate precipitation.

During the first hour of contact seeding, there is a rapid reduction in tartaric acid, potassium, and concentration product values. This reduction slows after the first hour and then levels off, in most wines, at the end of three hours (Blouin et al. 1982). For security, it is desirable to have a minimum contact time of four hours (Guimberteau et al. 1981). It has been suggested that the use of 40 μm KHT seed allows stabilization in 90 minutes (Neradt 1977). Reduction of the level of seed used prolongs the stabilization period. Also, filtration of the wine following contact seeding is essential; to prevent resolubilization of potassium bitartrate crystals, this step should be performed at the same temperature as the stabilization procedure.

Table 18-3 compares several wine parameters after cold stabilization chilling, contact seeding, and ion exchange.

Ribereau-Gayon and Sudraud (1981) reported a comparison of potassium bitartrate stabilization techniques in 16 white wines and 11 reds. Each lot was prefiltered and evaluated after 14 and 21 days of storage at −4°C, and the results were compared with the contact procedure. Concentration product values were determined for each wine after respective treatments. Eighty percent of the wines treated by contact seeding had lower final concentration product values than did their conventionally cold-stabilized counterparts. In all cases, stability

Table 18.3. Sparkling wine cuvee prior to and after KHT stabilization by several methods

Stabilization Method	Alcohol (% vol)	Sugar-Free Extract (g/L)	pH	Total Acid (g/L)	HT (g/L)	K^+ (mg/L)
Untreated	9.37	21.26	3.31	7.55	2.50	720
Chilling	9.43	21.74	3.23	7.25	2.50	715
Contact	9.46	20.34	3.20	7.15	1.95	565
Ion Exchange	9.35	20.74	3.26	7.45	2.40	360

SOURCE: Rhein and Neradt (1979).

values for contact-seeded wines were at least equivalent to conventionally cold-stabilized wines.

Potassium bitartrate crystals can be reused after removal from the treated wine. After repeated use, the crystals increase in size, thus decreasing the surface area available for growth (Rhein & Neradt 1979). Regrinding of the crystals eventually is necessary to optimize performance, although small wineries may not find it economically feasible to regrind the crystals. At any rate, the winemaker must be aware of the potential for microbiological contamination due to the reuse of tartrate crystals.

Ion Exchange

The basic principles of ion exchange techniques are discussed in Procedure 13.1 and Appendix I. In wine, the practice of ion exchange most generally has been applied to effecting bitartrate stability and in pH and titratable acidity adjustment.

BITARTRATE STABILIZATION AND TA, pH, AND EXTRACT VALUES

Titratable acidity, pH, and extract values vary, depending upon the technique used for bitartrate stabilization. Winemakers should expect reductions in both pH and titratable acidity using conventional cold stabilization and seeding in wines with initial pH values of <3.65, as a result of the generation of one free proton per molecule of KHT precipitated. Using these techniques, one might expect pH levels to drop by as much as 0.2 pH unit with a corresponding decrease in titratable acidity of up to 2 g/L. By comparison, potassium bitartrate precipitation in wines with initial pH values of >3.65 results in increased pH levels and corresponding decreases in titratable acidity. The above values represent ranges seen in practice and may vary. With contact seeding and potentially with conventional cold stabilization, winemakers report reduction in extract content after treatment (see Chapter 3). This may represent a problem in some low-extract wines.

EVALUATION OF WINE STABILITY

In forming potassium bitartrate, 1.0 g/L of tartaric acid combines with 0.26 g/L of potassium, producing 1.26 g/L of the KHT salt. In principle, comparison of tartaric acid or potassium levels before and after seeding should provide an indication of wine stability. Wines have been reported to be stable if the difference in tartaric acid levels before and after stabilization was less than 200 mg/L (SWK 1978). In the case of K^+, changes of less than 40 mg/L are considered as stable.

The evaluation of wine stability also may be achieved by measurement of conductance before and after mixing of the wine sample with potassium bitartrate seed crystals. Changes in conductance of less than 5% before and after the addition of seed indicate a stable wine. This technique has the advantage of requiring relatively inexpensive analytical equipment (conductivity meter and probe) when compared to measurement of ions by atomic absorption spectrophotometry. Further, minimal reagent preparation is required. It should be noted, however, that during the course of reaction the temperature must be carefully maintained. For more details regarding the test and its interpretation, the reader is referred to Procedure 13-2.

The use of concentration products (CP) (see Chapter 13) for estimation of stability is another approach. In this case, samples are compared before and after treatment with regard to the components of reaction; namely, K^+, Ca^{++}, total tartrates (as HT^-, $T^=$), alcohol, and pH. Results are compared to stability values considered as "safe" for each particular wine type. (Table 13-5). Wines are considered stable if the calculated value is less than published values. Because this technique produces a final solubility value, the winemaker has the option of preparing blends of stable wines with others where the concentration product is known. As with any blending operation, the winemaker should not assume that blending two or more stable wines will result in a stable final blend. Changes in alcohol, pH, tartrate, and cation concentrations, as well as undefined contributions of complexing agents, may contribute to potential instability of some other type (e.g., protein haze) and, in such cases, testing of laboratory blends should be performed.

Procedure 18-1. Cation and Anion Exchange

Ion exchange may be used to bring about potassium bitartrate stabilization, but this generally is not considered a practice for premium wine production. Ion exchange occasionally is used in conjunction with refrigeration to avoid deterioration in wine palatability.

The precipitation of potassium bitartrate from solution is dependent upon the concentrations of K^+ and bitartrate HT^- present in solution. The relationship is defined in the following equation:

$$K_{sp} = [K^+][HT^-]$$

K_{sp} = solubility product constant

$[K^+]$ = concentration of K^+

$[HT^-]$ = concentration of HT^- (18-2)

It is apparent that increases in either or both of the right-hand members in the above equation will result in increases in the solubility coefficient and, hence, increases in the likelihood of precipitation. This generalized equation is dependent upon the two-step dissociation of tartaric acid in solution:

$$H_2T \overset{K_1}{\rightleftharpoons} H^+ + HT^- \quad pK = 3.14 \qquad (18\text{-}3a)$$

$$HT^- \overset{K_2}{\rightleftharpoons} H^+ + T^{-2} \quad pK = 4.32 \qquad (18\text{-}3b)$$

The distribution of species diagram for the above equilibria is presented in Fig. 13-1.

With ion exchange the cation concentration of a wine can be reduced by exchanging it with either H^+ or Na^+. Reduction in the concentration of the cation component (K^+) results directly in a decrease in the solubility coefficient and hence a decreased likelihood of salt precipitation at low temperatures. Unlike the case of chill-proofing, however, the concentration of tartaric acid (tartrates) does not change. Hence, there are not the often troublesome changes in titratable acidity and pH that may affect stability with storage of the wine at low temperatures.

I. Equipment
   Glass ion exchange column
   pH meter
II. Reagents (See Appendix II)
   Cation exchange resin (Duolite C-20 or equivalent)
   Anion exchange resin (Duolite A-7 or equivalent)
   2 *N* HCl

1.5 *N* NaOH
10% wt/vol NaCl
Deionized water
0.1 *N* silver nitrate

III. Procedure

(a) Column preparation

1. Load column with *presoaked* resin slurry.
2. Pass one bed-volume of deionized water through the column.
3. Backwash with sufficient water pressure that the resin bed is expanded by approximately 50%. Allow the resin to reclassify.
4. Flow two bed-volumes of deionized water through the column. The volume of water, at this point, should not fall below the top of the resin bed.

(b) Cation exchange

(1) Sodium cycle

5. After the cation exchange resin has been placed in the column, backwashed, settled, and drained (steps 1–4 above), flow two bed-volumes of 2 *N* HCl through the resin at a flow rate of four bed-volumes/hour.
6. Wash the acid from the column with five bed-volumes of deionized water.
7. Regeneration: Flow two bed-volumes of 10% NaCl through the column at four bed-volumes/hour.
8. Wash the column with five bed-volumes of deionized water.
9. Add wine, adjusting the flow rate to five to six bed-volumes/hour. Continue to operate the exchanger until the column is exhausted as determined by analysis of K^+ in the exchanged product. When K^+ levels approach those found in the untreated wine, the column is exhausted.
10. Drain the column completely, and follow with five bed-volumes of water. Rinse with three bed-volumes of citric acid–SO_2 solution (equivalent to 8 pounds citric acid + 2 pounds potassium metabisulfite per 1000 gallons).
11. Recharge the column as described above. Every fourth time the column is recharged, begin at step 1 with the inclusion of 2 *N* HCl. The latter helps keep the determinate sites free of debris.

(2) Stability evaluation

12. Compare the K^+ and total tartrates content of the base wine used in the exchange operation with that of the product. Calculate CP values for both. Consult Chapter 13 for interpretation of the results.
13. Determine the percentages of base wine and exchanged wine needed to prepare a stable blend.
14. Make the above blend, filter it, and store duplicate samples under

defined conditions of low temperature. Make observations, and record results.

(3) Cation exchange hydrogen cycle

15. To operate the column in the hydrogen cycle, follow steps 5 and 6 above. Regeneration with NaCl is omitted.
16. Exhaustion is monitored by comparing pH increases in the exchanged product with those observed in the product early in the run.

(c) Anion exchange

(1) Use of anion exchange column with wine

17. Using Duolite A7 anion exchange resin, proceed according to steps 1 through 4 above.
18. Regeneration: Flow two bed-volumes of 1.5 *N* NaOH through the resin.
19. Rinse with five bed-volumes of deionized distilled water. The final pH should be approximately 7.0.
20. Introduce the wine. The degree to which the final objective is achieved is dependent upon flow rate (generally, four to five bed volumes/hour). Exhaustion is monitored by TA titration (in the case of tartrate reduction), and by color ($A_{420} + A_{520}$) (in the case of red wine and rosés), as well as by decreases in phenolic concentrations.
21. Depending upon the wine and the objective of the exchange, it is necessary to ''wash'' the resin once every three or four runs with 2 *N* HCl.

(2) Evaluation

22. Anion exchange finds its principal application(s) in the reduction of tartrate levels, color (in red and rosé-style wines).
23. Anion exchange may effect significant decreases in the concentration of SO_2. Exchanged wines should be monitored for SO_2 losses and adjusted accordingly.
24. Anion exchange is effective in the removal of aromatic components in the wine, and may be effectively employed when the objective is the production of neutral products (white ports and mutes).
25. Column efficiency, as measured by tartrate reduction and time of operation, may be greater for white wines than for reds and rosés, because of the presence of precipitable pigments and phenolics in the latter.

IV. Supplemental Notes

Included in the following comments are several in-plant considerations that may also apply to laboratory-scale operations.

1. Blends made with ion-exchanged wines should be filtered prior to eval-

uation. In the exchange process, some particulate matter may slough from the determinate sites on the resin and be found in the product.

2. Only filtered wine should be run through the column. Large quantities of particulate matter in the wine will block determinate sites and, hence, the exchange capacity. Wines containing fining agents should not be exchanged.
3. Whenever possible, wines should be bentonite-fined before ion exchange, thus reducing the potential for premature exhaustion of resin-reactive sites. Difficulties in fining, particularly with bentonite, have been reported after ion exchange.
4. Some workers have reported difficulty in using Cufex (Metafin) on ion-exchanged wines. Therefore, it is important to guard against secondary metal contamination in the exchanged product. Concrete tanks, which may leach Ca^{2+} into the wine, should be avoided in the storage of ion-exchanged wines.
5. The percentage of ion-exchanged wine used in a sparkling wine cuvee should be limited. Ion exchange removes trace elements that are necessary for the secondary fermentation. Where such exchanged products are used, many winemakers recommend the addition of yeast nutrient.
6. Ion exchange resins expand upon hydration, a phenomenon that should be taken into account in both plant and laboratory work. Manufacturers describe different procedures for storage of an idle column, so one should consult the appropriate user's guide for the resin in question. Generally, cation resins should be stored in SO_2–citric acid solutions to prevent microbial growth within the column. When idle, the column storage water should be monitored routinely for possible decomposition.
7. In the lab, water used in rinsing the column should be ion-free. This requirement may not be feasible in plant operations, where softened water frequently is used.
8. BATF Section 27CFR240.1051 Subpart ZZ, "Materials Authorized for Treatment of Wine," places the following limitations on the use of ion-exchange technology:

   Anion, cation, and nonionic resins, except those anionic resins in the mineral acid state, may be used in batch or continuous column processes as total or partial treatment of wine, provided that after complete treatment:

   a. the basic character of the wine has not been altered.
   b. the color of the wine has not been reduced to less than that normally contained in such wine.
   c. inorganic anions in the wine have not been increased by more than 10 mg/L.

d. the metallic cation concentration in the wine has not been reduced to less than 300 mg/L.
e. the natural or fixed acid in grape wine has not been reduced to less than 4 ppt (0.4 g/100 mL) for red wines, 3 ppt (0.3 g/100 mL) for white wines, or 2.5 ppt (0.25 g/100 mL) for all other grape wines. The natural or fixed acid content in wine other than grape wine, may not be reduced below 4 ppt.
f. the pH of the wine has not been reduced below pH 3 nor increased above pH 4.5.
g. the resins used have not imparted to the wine any material or character (incidental to the resin treatment) which may be prohibited under any other section of the regulations.

Conditioning and/or regenerating agents consisting of water, fruit acids common to the wine being treated, and inorganic acids, salts, and/or bases may be employed, provided the conditioned or regenerated resin is rinsed with water until the resin and container are essentially free from unreacted (excess) conditioning or regenerating agent prior to introduction of the wine. Tartaric acid may not be used in treating wines other than grape.

9. It is generally accepted that the quality of wines stabilized by ion exchange is less than that of wines stabilized by refrigeration. However, the large difference in cost may warrant the use of ion exchange for production of standard wines.

Procedure 18-2. Analysis of Sodium Ion by Atomic Absorption

Sodium (Na^+) in non-ion-exchanged wines is present at relatively low concentrations: 10 to 172 mg/L. Its level in wine is attributed, in part, to the proximity of grapes to saline water or soils. Although generally not important in wines, Na^+ becomes a concern in certain ion exchange applications, as discussed above.

I. Equipment
   Atomic absorption spectrometer with sodium (or Na/K) hollow cathode lamp
II. Reagents (See Appendix II)
   1000 mg/L stock sodium chloride solution
   Distilled water
III. Procedure
   1. Prepare a 100 mg/L (ppm) working sodium standard by diluting 10 mL of the sodium stock solution to 100 mL in a volumetric flask.
   2. Prepare calibration standards of 1, 2, 5, and 10 mg/L from the working standard, as shown in Table 18-4. All dilutions are made in distilled water.

Table 18-4. Preparation of calibration standards from 100 mg / L sodium working standard.

Volume of stock standard (mL)	Final concentration (mg / L) in 100-mL volume
1.0	1.0
2.0	2.0
5.0	5.0
10.0	10.0

3. Install the sodium lamp in the AA instrument. Set the appropriate wavelength, slit width, and power supply current as listed in the instrument's operating instructions.
4. Ignite the burner and adjust the flame, optimize the analytical conditions, and prepare a calibration curve by aspiration of the standards prepared above.
5. Dilute 1 mL wine to 50 mL with deionized water. Aspirate sample into the flame, and read the absorbance (or concentration) value. If necessary, prepare a second sample at a more appropriate dilution.
6. Read the sodium concentration from the calibration curve and multiply by 50 (or other dilution factor, if used) to obtain the sodium concentration in the original wine sample.

IV. Supplemental Notes

1. Samples with high potassium concentrations may produce high values for the sodium concentration. This effect is caused by the small but significant amount of ionization of potassium atoms in the air–acetylene flame. One way to eliminate or swamp out this interference is to make up all standards and samples so that they contain about 2000 to 5000 mg / L of added potassium (which acts as an ionization buffer).
2. Analyses conducted at the low mg / L concentration level require careful attention to cleanliness, and careful pipetting and diluting of the standards and the samples.

Procedure 18-3. Analysis of Sodium Ion by Flame Emission

Sodium in wine lends itself to analysis by traditional flame photometric (atomic emission) methods. The emission analysis can be accomplished using an atomic absorption spectrometer, but without the sodium hollow cathode lamp. As in other emission techniques, the sodium signal is read on the percent transmittance scale (more sodium, more emission, and thus more apparent transmittance).

I. Equipment

Flame photometer (or atomic absorption spectrometer)

II. Reagents

1000 mg/L stock sodium chloride solution

III. Procedure (See Appendix II)

1. Using a dilution series identical to that prepared for Na^+ analysis by atomic absorption (see Table 18-4), prepare a standard curve of sodium concentration vs. $\%T$.
2. Dilute a wine sample 50× with deionized, distilled water, and aspirate it into the flame.
3. Using the standard curve, determine the concentration of Na^+ in the sample. Remember to multiply this value by the appropriate dilution factor.

IV. Supplemental Note

1. Sodium analyses by flame emission are quite similar in type to those for potassium. Refer to Procedure 13-6 for relevant comments.

Chapter 19

Removal of Copper and Iron—The Hubach Analysis

BLUE FINING

As discussed in Chapters 14 and 15, the presence of copper and iron in wine may catalyze premature oxidation and, hence, deterioration of the wine. The levels at which the respective metals may cause problems vary with wine type and chemistry. It is generally accepted that copper levels in excess of 0.3 mg/L and iron levels greater than 5 to 7 mg/L may result in instabilities. Metal contamination usually originates from contact of must and/or wine with iron- or copper-containing alloys as well as the use of Cu(II) sulfate in removal of H_2S (see Chapter 10).

Historically, German winemakers found that excess iron could be removed from wine by the addition of ferrocyanide. Also known as *Blauschonung* or "blue fining," the practice has been used in Europe under very strict governmental regulation. Because the major decomposition product of the reaction is cyanide, the practice is not permitted in the United States. However, use of ferrocyanide-containing compounds (Cufex and Metafine) is permitted. The use of these compounds in wine requires subsequent examination by means of the Hubach test. Wines with residual cyanide in excess of 1 mg/L are unacceptable under federal standards.

The reaction of ferrocyanide with iron (Fe^{3+}) is slow, requiring up to seven days (Castino, 1965). Some workers recommend the use of ascorbic acid, at 50 mg/L, to reduce the iron present to the Fe^{2+} state, thus increasing the rate and completeness of the reaction for precautions, see Ch. 8. Properly used, ferrocyanide removes not only the excess iron but also most of the copper without adversely affecting the flavor or the bouquet of the wine.

The proprietary compounds Cufex and Metafine react preferentially with copper (subsequently with iron), usually settling out of solution rapidly and producing very little sediment. Flavor and bouquet are not damaged. Both preparations are sold in the form of a cream or as solutions.

Iron and copper analyses provide an indication of the amount of fining agent required. Generally, one can expect a reduction in copper levels of approximately 1 mg/L for each 1 pound/1000 gallons equivalent of Cufex or Metafine used. It is recommended that complete laboratory fining trials be run prior to addition to the total volume of wine. For further information on metals, refer to Chapters 14, 15, and Section IV—Remedial Actions.

All wine additives are strictly controlled by federal and state governments. For further information regarding permissible additives, use levels, legal documentation, and so on, consult Part 240 of Title 27 of the Federal Regulations, and appropriate state administrative codes.

REACTION OF METALS AND BLUE FINING AGENTS

At excess levels in wine, copper and iron may affect stability by forming a precipitate and haze or by altering the sensory character of the wine. Historically, the most widely used procedure for removal of these metals employed the addition of unbuffered potassium ferrocyanide, a practice commonly referred to as blue fining. The method takes advantage of the fact that cyanide ion forms very stable complex ions with transition metal ions:

$$Cu^{+2} + 3CN^{-} \longrightarrow Cu(CN)_3^{-} \qquad (19\text{-}1a)$$

$$Fe^{+3} + 6CN^{-} \longrightarrow Fe(CN)_6^{-3} \qquad (19\text{-}1b)$$

$$Fe^{+2} + 6CN^{-} \longrightarrow Fe(CN)_6^{-4} \qquad (19\text{-}1c)$$

$$3\,Fe(CN)_6^{-4} + 4Fe^{+3} \longrightarrow \underset{\text{(precipitate)}}{Fe_4[Fe(CN)_6]_3} \qquad (19\text{-}1d)$$

However, excess ferrocyanide remaining in solution may, as a result of degradation, produce traces of free hydrocyanic or "Prussic" acid (HCN):

$$Fe(CN)_6^{-4} + H_2O \longrightarrow Fe(CN)_5^{-3} + OH^{-} + HCN \qquad (19\text{-}2)$$

As previously stated, the practice of blue fining is not permitted in the United States, but the proprietary compounds Cufex and Metafine are allowed. The active component in both compounds is potassium ferrocyanide, so their addition essentially constitutes blue fining. Federal and California regulations specify that no soluble residue in excess of 1 mg/L shall remain in the finished wine. Thus it is necessary for the winemaker to analyze for residual cyanide in the final product prior to bottling and sale.

DETERMINATION OF RESIDUAL CYANIDE

The current procedure for determination of residual cyanide in wine is Hubach's modification of an earlier analysis developed by Gettler and Goldbaum (1947) for the medical field. The procedure involves reduced-pressure distillation of a dealcoholized aliquot of wine through specially prepared filter paper (described in Procedure 19-2). The reaction surface consists of an alkaline ferrous sulfate–impregnated paper interpositioned between the vacuum adaptor and a cold water condenser. (Flange *A* in Fig. 19-1). A positive reaction is demonstrated by

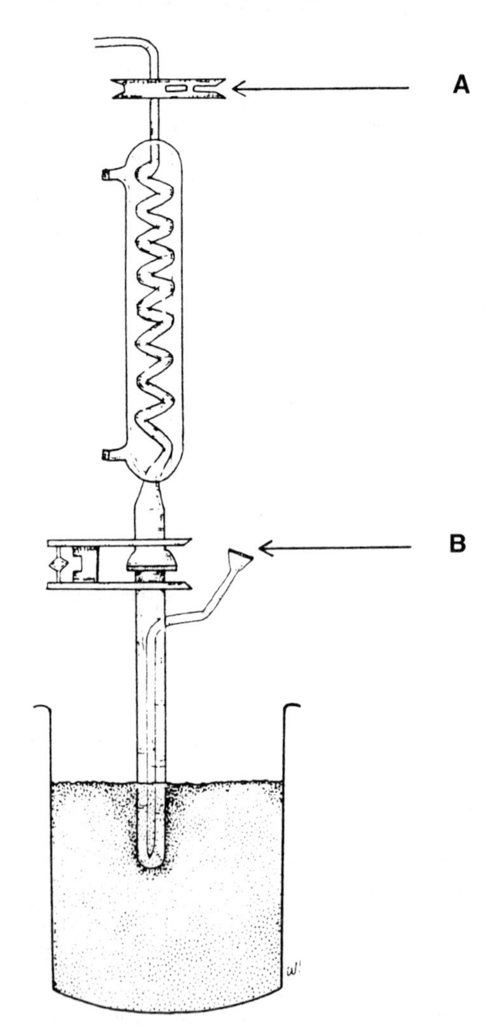

Fig. 19-1. Hubach apparatus. (A) Flange for holding Hubach paper. (B) Receiving port for addition of samples and standards.

formation of a characteristic blue spot corresponding to Prussian blue. The reaction between ferrocyanide and iron(III) is specific with no known interferences:

$$\underset{\text{in wine}}{K_4Fe(CN)_6} + \underset{\text{on paper}}{Fe^{+3}} \longrightarrow \underset{\text{Prussian Blue}}{KFe[Fe(CN)_6]} + 3K^{+} \qquad (19\text{-}3)$$

The equilibrium in the above reaction is displaced to the right-hand side. Analysis calls for reaction of a neutralized, dealcoholized wine sample in the presence of cuprous chloride catalyst and 1 + 1 H_2SO_4. Subsequent reduced-pressure distillation is carried out at 80 to 90°C. In the presence of the cuprous chloride catalyst and sulfuric acid, hydrocyanic acid is liberated quantitatively from any remaining ferrocyanide. The HCN thus formed then reacts with iron on the test papers, forming Prussian blue according to the reaction presented above (Equation 19-3). Upon completion of the distillation step, the Hubach paper is acid-treated to dissolve iron hydroxides, which tend to mask the development of Prussian blue (Bonastre, 1959):

$$[Fe(CN)_6]_4 + 4Fe(OH)_3 + 6H_2SO_4 \longrightarrow Fe_4[Fe(CN)_6] \qquad (19\text{-}4)$$

The intensity of color formation is proportional to the concentration of HCN and may be determined by comparison to the color development of standards. The detection limit of the Hubach test is 1 mg/L (Roberts, 1988).

Procedure 19-1. Laboratory Fining Trials Using Cufex and Metafine

This procedure is intended to familiarize laboratory personnel with a technique for establishing the amount of fining agent required to achieve desired reductions in copper and iron levels in wine. In practice, Cufex and Metafine are equally effective.

I. Equipment
   Laboratory benchtop mixer
   250-mL Erlenmeyer flasks or 100-mL graduated cylinders
II. Reagents
   Cufex (available from Cellulo-Finer Filter Products, Fresno, California)
   Metafine fining agent (available from Scott Labs, Petaluma, California)
III. Procedure
   1. Determine the concentration of iron and copper in the wine sample (refer to Chapters 14 and 15).
   2. Prepare a working solution of Cufex fining agent (a paste) by diluting 50 g of stock in 350 mL of deionized water.
   3. Blend the working solution in the mixer to uniform consistency.
   4. Transfer 100-mL aliquots of wine into six 250-mL Erlenmeyer flasks.

5. Determine the total amounts of copper and iron to be removed.
6. Cufex removes copper first and then iron. Properly prepared, 1 pound Cufex/1000 gallons will remove 1 mg/L of metal.
7. Using the working slurry prepared in steps 2 and 3, (0.1 mL of the slurry is equivalent to a cellar addition of 1 pound Cufex/1000 gallons of wine), determine the total amount of metals to be removed. As an example, the following data were collected for white wine 88-0098:
   Copper: 2.0 mg/L
   Iron: 10.0 mg/L

   The winemaker would like to reduce the copper to 0 mg/L and the iron to 5 mg/L. Thus, a total of 7 mg/L of metal (both copper and iron) need to be removed. Using the relationship presented in step 7 above, prepare a series of additions that bracket the desired removal level.
8. Add the calculated amount of Cufex slurry to each wine sample, mix, and set it aside overnight.
9. After 8 to 12 hours, decant the wine from the sediment and filter the supernatant through a 0.45-μm membrane.
10. Rerun metal analyses on lab trials to verify the predicted results (see Table 19-1).
11. Run the Hubach analysis. Residual HCN must be less than 1.0 mg/L.
12. Add the calculated amount of working solution to the total volume of wine to be treated. Mix the treated wine for 1 hour (or longer in the case of larger lots).

IV. Supplemental Notes
   1. To aid in clarification, the winemaker may elect to use "top dressing" agents after the mixing period. The selection of clarifying agents may include bentonite, gelatin, and/or kieselsol, either singly or in combination.

Table 19-1. Typical lab setup and expected results for 100-mL wine samples.

Sample	Vol. working sol (mL)	Final [Cu] (mg/L)	Final [Fe] (mg/L)
Flask 1	0.3 (=3#/M)	0	9
Flask 2	0.5 (=5#/M)	0	7
Flask 3	0.7 (=7#/M)	0	5
Flask 4	0.9 (=9#/M)	0	3

#/M = pounds/1000 gal

Procedure 19-2. Hubach Analysis

Potassium ferrocyanide, in the presence of acid, decomposes to form volatile hydrogen cyanide (HCN), which can be distilled from solution and allowed to pass through a filter paper impregnated with iron(II). The HCN will react with the iron, producing Prussian blue, with the intensity of the blue spot formed directly related to the amount of residual cyanide present in the sample (Hubach, 1948).

I. Equipment
   Hubach apparatus (see Fig. 19-1)
   1-L Pyrex beaker
   Hot plate
   Thermometer, range 0–110°C
   5-inch evaporating dish
   Vacuum source

II. Reagents (see Appendix II)
   Potassium ferrocyanide ($K_4Fe(CN)_6 \cdot 3H_2O$ stock solution (100 mg/L as cyanide)
   1 + 3 HCl
   1 + 1 H_2SO_4
   6 *N* NaOH
   Cuprous chloride catalyst (see Appendix II)
   Hubach test papers (see Appendix II). (If the analyst does not wish to prepare papers, they may be purchased from Scott Laboratories, Petaluma, California.)

III. Procedure
   (a) Standard preparation
      1. Prepare 1.0- and 10.0-mg/L cyanide working standards from the 100-mg/L stock solution (see Table 19-2).
      2. Thoroughly clean the Hubach apparatus, using distilled water, to remove traces of residual cyanide. Allow it to air-dry.
      3. Preheat the water in the beaker to 80 to 90°C. The water depth should be above the wine level in the aeration tube, as indicated in Fig. 19-1. *Note*: To ensure complete hydrolysis of ferrocyanide, it is essential that the temperature of the water bath remain between 80 and 90°C during the distillation step.
      4. Immerse the apparatus in the water bath, using utility clamps for support.
      5. Turn on the condenser water.

Table 19-2. Preparation of standard solutions from stock (cyanide = 100 mg/L) solution.

Volume of stock (mL)	Final concentration (mg/L) in 100-mL volume
10	10
1.0	1.0

6. Transfer 20 mL of standard solution to receiving port *B*.
7. Add 1 mL of CuCl, being sure to mix the suspension thoroughly before use. Add 1 mL 1 + 1 H_2SO_4.
8. With a pair of forceps, insert Hubach paper at the union of the aeration tube and the condenser (*A*). Moisten the paper with one or two drops of distilled water.
9. Connect the aeration tube and the condenser, using a Thomas clamp, and turn on the vacuum slowly so as to draw air through the system at a rate just short of forming a steady bubble stream.
10. Aspirate the sample for 10 minutes.
11. Remove the paper from the unit using forceps, and place it in an evaporating dish containing 1 + 3 HCl.
12. Allow the paper to soak until it is white.
13. A blue stain indicates a positive test for cyanide. Rerun the test using a distilled water blank. When the unit is clean (as demonstrated by the absence of a blue spot in the water blank distillation), continue with the next standard.

(b) Sample preparation and analysis

14. Prior to each analysis, a 20-mL distilled water blank should be run to demonstrate that the apparatus is free of residual cyanide.
15. Wine samples to be run are first neutralized with 1 + 1 NaOH to prevent excessive foaming and then dealcoholized by evaporation over a steam bath until they occupy approximately a 5-mL volume. *Note*: Samples must be brought back to volume with distilled water before distillation.
16. Substituting the dealcoholized wine sample for the standard, continue according to steps 6 through 13 above.

IV. Supplemental Notes

1. It has been reported by the BATF San Francisco Laboratory (Roberts, 1988) that the aeration oxidation glassware (of Procedure 9-3) can be adapted to perform Hubach analyses. Interested laboratories should contact BATF personnel for details regarding the methodology.
2. It has been the authors' observation that the Hubach test paper must be wetted with a drop of distilled water prior to its placement in the Hubach apparatus.
3. The production of HCN from potassium ferrocyanide is accelerated in acid solution (1 + 1 H_2SO_4).
4. Cuprous chloride acts as a catalyst in the formation of HCN and, as such, is continuously regenerated.
5. The 1 + 3 HCl solution is added to dissolve iron oxides that otherwise would mask formation of the Prussian blue spot on the test paper.
6. To ensure the integrity of the unit and the test paper, it is essential that standards be run prior to each day's work. After standard testing, the

unit should be thoroughly cleaned, and a blank, consisting of reagents and distilled water, run to make certain that the apparatus is free of residual HCN. Thorough cleaning between trials in any day's operation is also essential.

7. Hubach test papers must be kept in a dessicator and are viable only for a limited period of time.
8. Failure to produce a definite blue stain on Hubach test papers *with standards* may result from:
   a. Improperly prepared test papers.
   b. Improper addition of reagents.
   c. Incorrect water bath temperature. The test sample must be kept between 80 and 90°C to ensure complete hydrolysis of the ferricyanide.
   d. Condenser water that is too warm.
   e. An insufficient aeration period.
   f. Leakage around flanges.

Appendixes

Appendix I

Chromatographic Techniques

Chromatography is an analytical technique used to separate molecules that differ from one another in some way. It is a very powerful tool in its modern variations, capable of performing separations of molecules that differ only slightly. The separations are achieved because different molecules interact differently with their environment, because of differences in size, functional group, geometry, charge and charge distribution, and solubility. With the proper chromatographic technique, each of these types of interaction can act as a ''handle'' by which one can separate molecules.

Chromatography, then, is a technique for separating various solute molecules or compounds in a mixture. The separation occurs because these molecules will distribute themselves to different extents in two immiscible phases that move with respect to one another. The main considerations here are the concepts of differential distribution and of moving phases. Mechanically, in order to have phases moving with respect to one another, it is necessary to fix one phase (immobile or stationary) and let the other move (mobile or carrier). The fixed phase can be a solid material with some surface activity, or a liquid, which is either spread in a thin film or chemically bonded onto an inert solid material. The mobile phase must be fluid, and thus is either a liquid or a gas.

A chromatographic system consists of a packed bed of some solid (or liquid-coated solid) material with a fluid phase percolating through it. Solutes placed in this system are distributed between these two phases and move when they are in the mobile phase. To the extent that these solutes remain on/in the stationary phase (and are stationary), they are caused to move through the overall chromatographic system at different rates, which effect a separation.

Partition. A number of different mechanisms are used in chromatographic systems to cause separations, one of the most common being that of partition. This is the process that occurs in a separatory funnel during liquid–liquid extraction. A solute is distributed or dissolved in the two immiscible phases according to its solubility in each of the phases. Once equilibrium is achieved in a separatory funnel, any possible separation of solutes has occurred. In contrast, a chromatographic system continuously moves a solute to a fresh stationary phase as equilibrium steps occur continuously. Instead of just one partition step, countless partitions can take place.

The ratio of the solubility of the solute in the stationary phase to the solubility of the solute in the mobile phase is called its *partition coefficient*. A high coefficient indicates that the solute has a high affinity for the stationary phase. Two different solutes, with different partition coefficients, will move through the chromatographic system independently, at their own rates, so that a separation takes place. In a separatory funnel two solutes will completely separate only if one partition coefficient is very large and the other very small. However, with the large number of partition steps that occur in a chromatographic system, small differences in the partition coefficients of two solutes may be enough to effect a separation.

Adsorption. In this common chromatographic mode, solute molecules distribute themselves between an active solid surface and a mobile phase (liquid or gas). Solutes are attracted to the surface because of dipole–dipole interactions, H-bonding, van der Waals forces, and so on. An *absorption coefficient* can be constructed from the ratio of the concentration of solute on the active surface to the concentration of solute in the mobile phase. Again, the larger the coefficient, the more the solute is retarded relative to the mobile phase. Solutes are separated when they have sufficiently different adsorption coefficients.

Ion Exchange. This chromatographic mechanism is used to obtain separations of mixtures of ions. In this mode the stationary phase (the ion exchanger) is synthesized with charged groups incorporated into it. Counter-ions (ions of the opposite charge) are present as well, to maintain electrical neutrality. These counter-ions are exchangeable; that is, they may be exchanged for another ion of the same charge. A *cation exchange* material is one in which the counter-ion is a cation that may exchange with different cations in the mobile phase. Similarly, an *anion exchange* material has an anion counter-ion.

Different cations in solution have exchange coefficients that are measures of their ability to exchange with the cation present on the particular cation exchange material being used for a chromatographic procedure. The same is true for different anions in solution and a particular anion exchange material.

Ion exchange materials are packed into columns through which a mobile phase percolates. A sample, containing various ions, is placed at the head of the column. If the column is packed with an anion exchange material, then anions in the sample will exchange with the anion counter-ions to the extent determined by their exchange coefficients. Some ions will move faster through the column than others, and separation of these ions will be achieved. An analogous scheme holds for cations, with the column packed with a cation exchange material.

Size Exclusion. In this commonly used chromatographic mechanism, the stationary phase consists of a porous solid, with the size of the pores controlled. Molecules that are too large to fit into the pores are excluded from the internal volume of the stationary phase. These molecules are carried through the chromatographic system with the moving phase, in a total volume of mobile phase equal to the volume of the packed column minus the volume of the solid stationary phase. This volume, external to the solid particles, is called the column void volume; it is the volume between the packed particles.

Smaller solute molecules may partially fit into the pores of the stationary phase. Because they "see" a larger column volume, they require more mobile phase to move them through the chromatographic system. Molecules that are small enough to completely fit into all the pores will require the largest volume of mobile phase to be carried through and out of the column. Thus we have a separation based on size (and thus approximately based on the molecular weight) of the solutes present in the sample.

Common Types of Chromatography. The common forms are:

Paper chromatography (PC)
Thin-layer chromatography (TLC)
Gas chromatography (GC)
High performance liquid chromatography (HPLC)

These techniques are listed in order of the complexity of the mechanical hardware required to perform them. That is, HPLC is far more complex with respect to equipment than is paper chromatography.

Paper chromatography is performed using the partition mechanism. In this case the chromatographic "column" is a sheet of filter paper placed inside a jar or chamber. Water adsorbed onto the cellulosic framework of the paper acts as the stationary phase. The mobile phase consists of solvents that have some limited solubility in water. A sample (which, for example, could be a mixture of organic acids) is placed at one end of the paper in a process called spotting, and the paper is dipped into the mobile phase in the chromatographic chamber. One has to be careful that the mobile phase level is below all sample spots so that the spots are not just washed off the paper. In practice the mobile phase and water usually are combined and together put into the chromatographic chamber. The phases climb the paper by capillary action, the water adsorbs onto the paper's cellulose structure, and the organic solvent(s) component flows up past the adsorbed water. The solute molecules partition themselves between the stationary water (polar phase) and the mobile organic solvents (nonpolar phase), and are carried upward as a function of their individual partition coefficients. The larger their partition coefficients, the less distance the spots move.

To prevent the solvents from evaporating off the surface of the paper (especially at the solvent front), the entire system is enclosed in a chamber with a lid. When evaporation occurs, the solvent front is skewed and obscured. The chamber may be as simple as a jar or beaker with a lid (plastic wrap), or it may be a more elaborate commercial glass chamber.

The chromatogram (record of what happened in the chromatographic process) is allowed to develop until the solvent front has climbed a sufficient distance to allow the various solute components to separate. In this, and in the thin-layer technique, the solutes (spots) are not eluted or removed from the chromatographic system; so care must be exercised to stop the development before the solvent front reaches the top of the paper. For qualitative analysis purposes one needs to be able to measure the distance that the solvent front has traveled from the place where the spots were originally put on the paper (called the origin).

Because solutes are not eluted from the end of the chromatographic system in this technique, one cannot measure the retention time or the mobile phase retention volume of the solutes. Instead, one measures the distance the spot travels. To calculate spot travel, one measures the distance that the center of the spot has moved from the origin relative to the distance that the solvent front has moved from the origin:

$$R_f = \frac{\text{Distance spot travels}}{\text{Distance solvent front travels}}$$

One can find R_f values in the literature, and as long as one's chromatographic system exactly duplicates the one described, similar R_f values should be obtained. Because chromatographic systems are difficult to reproduce exactly, R_f values most likely will approximate those in the literature. One generally runs mixtures of standards along with unknowns so that direct comparisons of unknowns and standards may be made on the same paper chromatogram.

Most chromatographed materials are not naturally colored, so one must generate a color for a spot by chemical means. There are many references in the literature to color-generating reagents for different chemical systems. These materials are sprayed onto the finished chromatogram (after drying) and the positions of the resulting colored spots noted. Semiquantitation sometimes can be accomplished by comparing spot size and color density with one or more standards chromatographed at the same time.

Thin-layer chromatography is quite similar to paper chromatography. The sorbent material is spread as a thin layer on some support material such as glass, aluminum, or plastic, producing what are called thin-layer plates. Typical sorbents used to make these plates are silica gel, alumina, and crystalline cellulose. The plates are spotted in the same way as above, except that the spots generally are kept as small as possible (a few millimeters in diameter). If the plates are dried, and the mobile phase consists of anhydrous organic solvent(s), then separation is achieved by the adsorption mechanism. If water is included in the mobile phase mixture of solvents, then the partition mechanism is involved in achieving separation. Thin-layer plates are placed in chambers, as described above, and the developed spots visualized by use of chemical sprays.

As the particle size of the sorbents used is quite small, good chromatographic efficiencies are achieved with this technique. Thin-layer plates may be purchased, or they may be made (coated with sorbent) in the laboratory. Manufactured plates tend to be more uniform in performance, and also more efficient, than "home-made" plates. Traditional glass-backed plates are most commonly used, but plastic-backed plates with small-volume solvent chambers are growing in popularity.

In terms of hardware, *gas chromatography* and *high performance liquid chromatography* (GC and HPLC, respectively) techniques are considerably more sophisticated than those of paper chromatography and thin-layer chromatography. The heart of both GC and HPLC systems is the column, packed with a stationary phase through which the mobile phase percolates. In order to get mobile phase flow through a somewhat impermeable bed of fine particles, a pressurized mobile phase source is required. Sample placement at the inlet of the chromatographic column requires special injection devices. Finally, because in GC and HPLC techniques sample components are eluted off the column, some type of detector or sensor is required to "see" these components.

The mobile phase is moved through these pressurized systems with a pump. In the case of GC, with a gas as the mobile phase, the pump is very simple, a cylinder of compressed carrier gas. Gases can be compressed to hundreds of atmospheres. Upon expansion through a two-stage regulator, gases can provide adequate controlled flow rates through most GC columns. This very simple pump system is found on all gas chromatographs. HPLC uses liquid mobile phases. Liquids are essentially noncompressible, so that a mechanical pump is required to provide adequate flow rates of the mobile phase through the column. Such pumps range from fairly simple syringe pumps to sophisticated units with dual pumping heads driven by eccentric cams that provide pulseless flows of mobile phase liquids. These pumps tend to be rather expensive and contribute to the relatively high cost of HPLC.

Sample placement in GC is accomplished by use of a syringe that pierces a synthetic rubber septum and places the sample into a heated injection port. Small syringes that contain 1 to 10 μL of sample are used. Because the internal pressures involved in GC are not very high (3–6 atm), a rubber septum closure will not leak (until it has been pierced by too many injections, and then it is changed). With the higher pressures employed in HPLC to move relatively viscous liquids through a bed of fine particles, high pressure injection valves are required. Such a valve has several ports, allowing the sample to be placed into the valve without stopping the carrier flow. The valve is then activated and the sample injected; that is, the sample is shunted into the flow path of the mobile phase. Sample valves occasionally are used in GC for the injection of gaseous samples. These valves are more expensive than the relatively simple septum-type injection systems.

Both GC and HPLC columns contain packed beds of particles making up the stationary phase. Because gases are fairly nonviscous fluids, and the stationary phase particles are not too fine, GC column lengths on the order of 2 to 3 meters (packed columns) or 12–25 meters (capillary columns) are common. These columns are coiled in order to fit into the instrument's column oven, and they are heated in order to keep all the solutes in the gas phase and to control solute solubility in the stationary phase. By controlling solubility, one also controls the elution time (retention time) of the solutes. As sample mixtures contain components having widely different boiling points, the column temperature may have to be increased during the chromatographic run in order to speed up the elution of high-boiling compounds. This technique is called programmed temperature gas chromatography (PTGC).

The HPLC mobile phases are relatively viscous fluids (liquids). In addition, stationary phase particles are quite small in size (3–10 μm) to achieve chromatographic efficiency and good separations. As a result, HPLC columns tend to be short (10–30 cm) and straight. Column ovens are used with certain types of stationary phases; they are set to a specific temperature and kept there during the run (isothermal operation).

GC Detectors. The detectors used in GC and HPLC sense the presence of sample components as they elute from the exit end of the column. Since GC columns elute solutes as gases and HPLC elutes them as liquids or solutions, the detectors used in the two techniques are of different types and employ different principles for sensing the presence of the solutes. The common detectors used in gas chromatography are: the thermal conductivity detector, TCD; the flame ionization detector, FID; the electron capture detector, ECD; the flame photometric detector, FPD; and the mass selective detector, MSD. Very briefly, these detectors function as follows:

TCD. Carrier gas passing over a continuously heated filament conducts heat away from the filament so that it maintains a constant temperature and therefore a constant resistance. When a solute/carrier mixture passes over this filament, less heat is conducted away, with a resultant change in temperature and resistance. The filament is included in an electrical circuit that senses changes in resistance. A change in resistance produced by the presence of the solute is registered as a peak on the recorder/signal processor attached to the chromatograph. This detector will sense virtually everything and is referred to as a universal detector.

FID. Solutes eluting from the end of the column are burned in an air–hydrogen flame. During the combustion process, ions are formed and are attracted to a charged collector electrode positioned above the flame tip. The collected charges constitute a small current, which is amplified by a sensitive amplifier (an electrometer). The electrometer output provides a signal that drives the chart recorder or integrator attached to the chromatograph. This detector senses organic compounds and other combustible (oxidizable) materials, but not inorganic gases such as CO_2, N_2, O_2, and so on, or water.

ECD. A radioactive foil (^{63}Ni) produces a stream of beta particles (nuclear decay particles with the mass and charge of an electron). These particles are collected at an electrode, producing a standing current. Solute molecules with a high electron affinity (e.g., those containing halogen, oxygen, or nitrogen atoms or unsaturated bonds) can absorb some of these electrons, causing a decrease in this current. This decrease can be amplified by an electrometer and drive a chart recorder or integrator. The EC detector is a selective detector; it senses certain compounds preferentially to others.

FPD. A microburner system such as that described above burns the solutes as they elute from the column. Certain elements in these solutes become excited and emit radiation with characteristic wavelengths. This radiation is carried by a fiberoptic cable to a filter photometer and photo-detector, where it produces the chromatographic signal. Flame photometric detectors are tuned to be selective to a specific element such as phosphorus, sulfur, or nitrogen, and will sense those elution peaks containing the specific element.

MSD. This is essentially a mass spectrometer that uses a gas chromatograph as a sample inlet. In the mass spectrometer, high-energy electrons smash into the solute molecules, creating charged molecular fragments. These charged fragments are passed through a mass filter and are collected in order of increasing mass. The resulting mass spectrum provides explicit qualitative information about the chemical identity of the solute. This detector is universal and provides excellent qualitative information. Unfortunately, it is rather expensive.

HPLC Detectors. Common detectors used in HPLC are the refractive index, UV, and conductivity detectors.

Refractive Index Detector. This detector utilizes the principles of refractometry to sense solutes eluting from the column. The mobile phase is selected to have a refractive index value different from those of the solutes. When solutes pass through the detector cell, the change in refractive index is measured and produces the signal that drives the chart recorder/integrator. This is a universal detector, capable of sensing all typical solutes. It is very sensitive to changes in temperature and mobile phase composition or flow rate.

UV Detector. This detector is essentially a small filter or grating spectrophotometer with a micro-cuvette through which the solutes flow. Simpler versions use a mercury lamp as the energy source and operate at a fixed wavelength of 254 nm. More sophisticated versions use both deuterium and tungsten lamps to permit operation at any wavelength in the UV–VIS spectrum. Whereas the latter detector is fairly universal (even alkanes absorb in the low UV—195 nm), the simpler version is limited to compounds that have some measured absorbance at 254 nm.

Conductivity Detector. This detector is used in ion exchange chromatography to sense the presence of ionic solutes. The mobile phase is selected for its minimal electrical conductivity. Ionic solutes cause a measurable increase in the conductivity of the solute/mobile phase solution passing through the detector, generating a signal that drives the recorder/integrator. Inorganic anions or cations may be separated by "ion chromatography" and detected with good sensitivity with this device. This detector has some of the same limits as the refractive index detector, although in some of the newer instruments gradient (variable composition as the run progresses) mobile phases can be used.

Appendix II

Reagent Preparation for Laboratory Procedures

Conversions between Temperature Scales

Temperatures can be converted from one scale to the other through the simple relationships:

$$°F = (°C + 40) \times 9/5 - 40$$
$$°C = (°F + 40) \times 5/9 - 40$$

Conversions Between mg / L and mg / Kg (ppm) Concentration Units

Concentrations of low-level constituents are usually calculated in mg/L units. For solutions with densities (specific gravities) less than 1.00, mg/L units are not equal to parts-per-million (technically, mg/Kg) units. One can convert mg/L units to mg/Kg (ppm) units through the following relationships:

$$\text{mg/Kg} = \text{mg/L} \times [1/\text{Density (g/mL)}]$$
$$\text{mg/Kg} = \text{mg/L} \times [1/\text{Specific Gravity}]$$

STANDARD ACID–BASE SOLUTIONS

Standard Sodium Hydroxide Solutions

Preparation of Stock solution (1 + 1): In a 1-L Erlenmeyer flask, carefully mix one part CP grade NaOH and one part distilled water. Upon dissolution and after the reaction has cooled, transfer the mixture to a polyethylene container for storage. After Na_2CO_3 precipitation is complete (several days), the following solutions may be prepared by dilution of the filtered stock solution (Table II-1).

Standardization: Because of the indeterminative effect of carbonate, the above solutions represents only an approximate concentration. It is necessary to standardize them relative to a primary standard acid such as potassium hydrogen phthalate (KHP), which can be purchased in prestandardized liquid

Table II-1. Dilution procedures for preparation of working solutions from stock sodium hydroxide.

Desired Approximate Normality	Milliliters of 1 + 1 Stock Solution per 1.0 L
0.01	0.54
0.02	1.08
0.10	5.40
0.50	27.00
1.00	54.00

form or as a powder of defined purity. In either case an accurately determined quantity is titrated with base to the phenolphthalein end point. If prestandardized liquid KHP is used, the normality of base may be calculated according to the following relationship:

$$\text{Normality of base} = \frac{(\text{Normality KHP})\ (\text{Volume KHP used})}{(\text{Volume of base used})}$$

In the case of powdered KHP, the following relationship applies:

$$\text{Normality of base} = \frac{(\text{grams of KHP used})\ (1000)}{(\text{Volume of base used})\ (204.229)}$$

Standard Hydrochloric Acid Solutions

The concentration of stock HCl ranges from 35 to 37%. The dilutions from stock required to achieve the approximate concentrations desired are presented in Table II-2.

Standardization: For routine laboratory uses, anhydrous sodium carbonate

Table II-2. Dilution procedures for preparation of working solutions from stock hydrochloric acid.

Desired Approximate Normality	Milliliters of Stock H_2SO_4 per 1.0 L
0.01	0.89
0.02	1.78
0.10	8.90
0.50	44.50
1.00	89.00
2.00	178.00

(ACS grade) may be used for standardizing HCl solutions. Dissolve 1 to 3 g (accurately weighed) of ACS-grade anhydrous sodium carbonate (Na_2CO_3) in 40 mL distilled water. Titrate with HCl using four drops of methyl orange indicator until the solution begins to change color slightly. At this point, transfer it to an electric burner and boil it gently for 2 minutes. Cool the solution, and titrate it until it is the color of the reference solution. The reference solution consists of 80 mL boiled distilled water and four drops of methyl orange indicator. The normality of HCl may be calculated using the following relationship:

$$N \text{ HCl} = \frac{(\text{g Na}_2\text{CO}_3)(1000)}{(\text{mL HCl})(52.994)}$$

Standard Sulfuric Acid Solutions

Table II-3 presents the volume of stock sulfuric acid that must be used per 1.0 L of final solution.

Standardization. Accurately weigh and dissolve 3 to 5 g of sodium borate ($Na_2B_4O_7 \cdot 10H_2O$) in approximately 40 mL of boiled distilled water. Stopper the solution, and allow for clarification. Add five drops methyl red indicator and titrate to the equivalence point. Calculate the normality of the acid according to the relationship:

$$\text{Normality} = \frac{(\text{g Na}_2\text{B}_4\text{O}_7 \cdot 10\text{H}_2\text{O})(1000)}{(\text{mL of acid})(190.69)}$$

Note: Acid solutions also can be standardized against previously standardized sodium hydroxide solutions. Using a volumetric pipette, carefully transfer an aliquot of standard NaOH to an Erlenmeyer flask. Add several drops of indicator (phenolphthalein or methyl red), and titrate to an end point. Calculate the normality of the acid using the following relationship:

$$N_{\text{acid}} = \frac{V_{\text{OH}} N_{\text{OH}}}{V_{\text{Acid}}}$$

Table II-3. Dilution procedures for preparation of working solutions from stock sulfuric acid.

Desired Approximate Normality	Milliliters of Stock H_2SO_4 per 1.0 L
0.01	0.28
0.02	0.57
0.10	2.84
0.50	14.18
1.00	28.35

REAGENT PREPARATION FOR SPECIFIC LABORATORY PROCEDURES

Procedure 1-1. Juice Preparation for Aroma Evaluation

No reagent preparation is required for this procedure.

Procedure 1-2. Mold (Botrytis) Estimation via Laccase Assay

(a) 0.1% Syringaldazine: Weigh 0.1 g syringaldazine and add it to 100 mL absolute ethanol. Place the container in a sonic bath to completely dissolve the solid. Store it in a capped bottle.
(b) Sodium acetate (0.1 *M*): Dissolve 0.82 g reagent grade sodium acetate in distilled water. Bring to volume in a 100-mL volumetric flask.
(c) Activated PVPP: Activate commercial PVPP at 100°C in 6 *M* HCl for 1 hr. Follow by thorough rinsing with distilled water until neutral.

Procedure 1-3. Total Soluble Solids Determination by Hydrometry

No reagent preparation is required for this procedure.

Procedure 1-4. Total Soluble Solids Determination by Refractometry

No reagent preparation is required for this procedure.

Procedure 1-5. HPLC Analysis is Glycerol, Acetic Acid, and Ethanol in Grape Juice

(a) Glycerol–acetic acid–ethanol standard solution: In approximately 800 mL of distilled water, dissolve 500 mg glycerol, 200 mg acetic acid, 1.0 mL ethanol, 115 g glucose, and 115 g fructose. Mix the solution thoroughly. Adjust it to standard temperature, and bring it to final volume of 1 L with distilled water.
(b) Mobile phase: Prepare this reagent from chromatography grade water by adjusting it to pH 2.7 with sulfuric acid (ACS grade).

Procedure 1-6. Monoterpene Analysis

(a) Linalool stock solution (1 mg/mL): Accurately weigh approximately 50 mg of reagent grade linalool into a 50-mL volumetric flask. Add 10 mL absolute ethanol to the flask, and swirl to dissolve the reagent. Bring solution to the mark with distilled water.
(b) Linalool standard solution (0.1 mg/mL): As needed, dilute linalool stock 1 + 9 with distilled water.

(c) Vanillin-sulfuric acid reagent (2% wt/wt): Weigh 2 g reagent grade vanillin, and dissolve it in concentrated sulfuric acid. Store the solution in a brown glass bottle at 0 to 4°C.
Note: This reagent is corrosive. Its addition to aqueous solutions generates considerable heat. Safety glasses and gloves are recommended when one is using this reagent!

(d) Sodium hydroxide (20% wt/vol): Add 20 g reagent grade sodium hydroxide pellets to a beaker. Dissolve the reagent in approximately 80 mL distilled water, and cool. When it is cool, bring the solution volume to 100 mL. Store it in a plastic bottle.

(e) Phosphoric acid (50% vol/vol): Add 50 mL reagent grade phosphoric acid to approximately 50 mL distilled water to make 100 mL of solution. Store it in a glass bottle.

Procedure 2-1. Alcohol Determination by Ebulliometry

(a) Sodium hydroxide (1%) cleaning solution: Dissolve 10 g sodium hydroxide in 990 mL tap water. Identify as "Ebulliometer Cleaning Solution."

Procedure 2-2. Alcohol Determination by Distillation and Hydrometric Analysis

(a) Sodium hydroxide: (2 *N*): There is no need to standardize this reagent. Dissolve 82 g sodium hydroxide in approximately 800 mL distilled water, and bring it to a 1-L final volume at room temperature.

Procedure 2-3. Alcohol Determination by Dichromate Oxidation

(a) Ferrous ammonium sulfate: In a 2-L volumetric flask, dissolve 270 g $FeSO_4(NH_4)_2SO_4 \cdot 6H_2O$ in approximately 1500 mL distilled water. Carefully add 50 mL concentrated H_2SO_4, and bring the solution to volume with distilled water at a defined temperature.

(b) Potassium dichromate solution: In a 2-L volumetric flask, dissolve 67.536 g $K_2Cr_2O_7$ in approximately 1000 mL distilled water. Carefully add 650 mL concentrated H_2SO_4, cool, and bring the solution to volume with distilled water at a defined temperature.

(c) 1,10-Phenanthroline ferrous sulfate indicator: In a 500-mL volumetric flask, dissolve 3.48 g ferrous sulfate ($FeSO_4 \cdot 7H_2O$) in approximately 250 mL distilled water. Add 7.43 g *o*-phenanthroline, and dilute the solution to volume with distilled water. Transfer it to convenient-sized dropper bottles.

Procedure 2-4. Alternative Procedure of Dichromate Analysis

Prepare reagents as for the acid dichromate procedure presented above. The preparation of a standard alcohol series is presented in the body of the procedure.

Procedure 2-5. Gas Chromatographic Analysis of Ethanol

(a) Internal standard solution of 2-propanol (0.2% vol/vol): Dilute 2.0 mL reagent grade 2-propanol to 1000 mL with distilled water. Mix the solution well to assure homogeneity. Each standard or sample analyzed will require approximately 100 mL of this solution.
(b) Ethanol–distilled water standard solution (10%): Dilute 10.0 mL absolute ethanol to 100 mL in a volumetric flask at a defined temperature (20°C) with distilled water. Standards can be similarly prepared to cover other anticipated sample concentration ranges (dessert wines, coolers, etc).

Procedure 2-6. Gas Chromatographic Analysis of Fusel Oils

(a) Stock fusel oil solution: In a 100-mL volumetric flask, transfer 2 mL each of ethyl acetate, acetaldehyde, and methanol. To this mixture add 1 mL each of *n*-propyl alcohol, *n*-butyl alcohol, isobutyl alcohol, isoamyl alcohol, and 2-methyl-1-butanol (active amyl alcohol). Bring the solution to volume using a solution of water–ethanol (1 : 1).
(b) Ethanol–water (50% vol/vol): Dilute 500 mL absolute ethanol to 1 L with distilled water.

Procedure 3-1. Extract Determination by Specific Gravity

No reagent preparation is required for this procedure.

Procedure 3-2. Extract Determination Using Brix Hydrometer

No reagent preparation is required for this procedure.

Procedure 4-1. Hydrogen Ion Concentration (pH)

(a) Buffer (pH 3.55): Add excess potassium acid tartrate to 500 mL distilled water (approximately 5 g/500 mL). Mix this on a magnetic stirring table for 5 minutes. Allow undissolved crystals to settle and decant the solution, filtering it as necessary. At 25°C, the pH of this solution is 3.55.
(b) Buffers (pH 7.00 and 4.00): Use commercially available buffer or tablets, according to label instructions. Prepared buffers with indicators for detection of dilution or breakdown are commercially available through chemical supply houses. Depending upon individual laboratory needs, these buffers may be superior to lab-prepared products.
(c) KCl internal filling solution: Purchase this from chemical supply house.
(d) Cleaning solution (75% methanol): Dilute 75 mL absolute methanol to 100 mL with distilled water. Keep the container tightly capped.

Procedure 4-2. Titratable Acidity

(a) Standard NaOH: See "Standard Acid–Base Solutions" section of this appendix for preparation of standard base or normality of 0.10 or less.
(b) 1% Phenolphthalein indicator: Dissolve 1 g indicator in approximately 70 mL 95% ethyl alcohol. Add sufficient dilute NaOH ($<0.1\ N$) to achieve a very light pink solution. Dilute it to 100 mL with distilled water.
(c) Standard buffers: See above (Procedure 4-1 reagent preparation).

Procedure 4-3. HPLC Analysis of Organic Acids

(a) Citric acid internal standard solution: In a 100-mL volumetric flask, dissolve 2.0 g citric acid in distilled water. Bring the solution to volume at a defined temperature with distilled water.
(b) Formic acid internal standard solution: In a 100-mL volumetric flask, dilute 2.0 g stock formic acid to volume using distilled water.
(c) Sulfuric acid mobile phase ($0.01\ N$): Dilute 0.28 mL concentrated sulfuric acid to 1 L with distilled water.
(d) H_2SO_4 (1 + 4): Place 80 mL distilled water in a beaker, and carefully add 20 mL sulfuric acid. Mix it well. *Warning:* Considerable heat will be generated during this dilution.

Procedure 4-4. Enzymatic Analysis of L-Malic Acid

See product literature for preparation of reagents.

Procedure 5-1. Volatile Acid Determination by Steam Distillation (by Cash Still)

(a) Preparations of standard base and phenolphthalein indicator already have been discussed (under "Standard Acid–Base Solutions," above, and in reagent preparation for Procedure 4-2).
(b) For preparation of the reagents required for SO_2 correction, the reader is referred to Procedure 9-1.

Procedure 5-2. Acetic Acid Determination by Enzymatic Assay

See the product literature for reagent makeup.

Procedure 5-3. Gas Chromatographic Analysis of Acetic Acid

(a) Acetic acid standard stock (100 g/L): Dilute 10 mL glacial acetic acid to 100 mL in a volumetric flask with distilled water.

(b) *n*-Pentyl alcohol internal standard solution: Dilute 1.3 g *n*-pentyl alcohol to 1 L in a volumetric flask with distilled water.
(c) Ethanol (10% vol/vol): Dilute 100 mL absolute ethanol to 1 L with distilled water.

Procedure 6-1. Rebelein Method for Reducing Sugars

(a) Alkali Rochelle salt solution: *Component A:* In approximately 400 mL distilled water, dissolve 250 g sodium potassium tartrate. *Component B:* In approximately 400 mL distilled water, dissolve 80 g sodium hydroxide. Carefully, combine Component A and Component B in a 1-L volumetric flask, and bring the solution to volume with distilled water.
(b) Copper sulfate solution: In a 1-L volumetric flask, dissolve at 41.92 g copper sulfate ($CuSO_4 \cdot 5H_2O$) in approximately 600 mL distilled water. Add 10 mL 1 *N* H_2SO_4, mix, and bring the solution to volume with distilled water.
(c) Potassium iodide solution: In a 1-L volumetric flask, mix 100 mL 1 *N* sodium hydroxide and 300 g potassium iodide. Add approximately 300 mL distilled water. Upon complete dissolution, bring the solution to volume with distilled water.
(d) Sodium thiosulfate: To a 1-L volumetric flask, add 50 mL 1 *N* NaOH and 13.777 g sodium thiosulfate ($Na_2S_2O_3 \cdot 5H_2O$). Add approximately 200 mL distilled water, and dissolve the solid. Upon complete dissolution, bring the solution to volume with distilled water.
(e) Starch solution: *Component A:* In 500 mL boiling water, dissolve 10 g soluble starch. *Component B:* In 500 mL distilled water, dissolve 20 g potassium iodide and add 10 mL 1 *N* NaOH. Carefully combine Components A and B.
(f) Sulfuric acid (16% vol/vol): *Carefully*, and with mixing, add 175 mL 95% sulfuric acid to 825 mL distilled water.
(g) Sulfuric acid (1 *N*): See Table II-2 for instructions.

Procedure 6-2. Reducing Sugar by Modified Lane-Eynon Procedure

(a) Fehling's A solution: In a 2-L volumetric flask, dissolve 138.556 g cupric sulfate ($CuSO_4 \cdot 5H_2O$) in approximately 1800 mL distilled water. Bring the solution to volume at a defined temperature. Fehling's A solution should be refrigerated when not in use.
(b) Fehling's B solution: In a 2-L flask, dissolve 692 g sodium potassium tartrate and 200 g sodium hydroxide in enough distilled water to create a solution. Bring it to volume at a defined temperature with distilled water. One need not use an analytical balance for preparation of this reagent.
(c) Dextrose solution (0.5% wt/vol): In a 2-L flask, dissolve 10.000 g anhydrous dextrose in distilled water. Bring the solution to volume at a defined

temperature. Store it in a refrigerator. Discard solutions that appear hazy or have sediment. Alternately, add 2 g sodium benzoate and 1 g citric acid before bringing the solution to volume.

(d) Methylene blue indicator solution: Dissolve 1 g methylene blue in 100 mL distilled water. Store it in a dropper bottle.

Procedure 6-3. Enzymatic Analysis for Reducing Sugars

See the product literature for reagent preparation.

Procedure 6-4. Invert Sugar Analysis

(a) Ammonium hydroxide solution (1 + 1.5): Dilute 400 mL stock NH_4OH to 1 L with distilled water.

(b) HCl solution (1 + 1): Carefully mix 100 mL stock HCl with 100 mL distilled water.

Procedure 6-5. HPLC Analysis of Carbohydrates

(a) Standard solution (for use in juice): In an appropriate volume of distilled water, dissolve 10 g glucose, 10 g fructose, and 0.5 g sucrose. Bring the solution to 1 L with distilled water.

(b) Standard solution (for use in wine): In an appropriate volume of distilled water, dissolve 0.1 g glucose, 0.1 g fructose, and appropriate amounts of other sugars of interest. Bring the solution to 1 L with distilled water.

(c) Mobile phase (0.01 N H_2SO_4): See above, under "Standard Acid–Base Solutions," for instructions.

Procedure 7-1. Phenolics by Folin-Ciocalteu

(a) Phenol standard (5000 mg/L gallic acid): Prepare phenol stock solution by dissolving 500 mg of gallic acid in distilled water and bringing the solution to 100 mL final volume at a defined temperature.

(b) Folin-Ciocalteu reagent: Transfer 700 mL distilled water to a 2-L round-bottom boiling flask. Add the following: (1) 100 g sodium tungstate and (2) 25 g sodium molybdate. When dissolution is effected, add: (3) 50 mL stock phosphoric acid and (4) 100 mL concentrated hydrochloric acid. Add several glass boiling beads, connect the reflux condenser with cooling water, and reflux the solution for 10 hours. After 10 hours of refluxing, cool the flask's contents and rinse the reflux column into the boiling flask with several milliliters of distilled water. Add 150 g lithium sulfate (monohydrate) and several drops of bromine. Under a fume exhaust hood, boil the flask's contents for 15 minutes. Cool the reagent, volumetrically bring it to a 1-L final volume with distilled water, and filter it. The reagent thus prepared should be stored in an amber bottle. *Note:* The final color of the reagent should be a definite yellow. There should be no indication of the earlier green coloration. To reoxidize older solutions of Folin-Ciocalteu reagent that may have blue-orange tinges, add several drops of bromine and reboil them.

(c) Sodium carbonate solution: Into approximately 700 mL distilled water, add 200 g anhydrous sodium carbonate. Bring the mixture to volume (1 L) and boil it until the solid is completely dissolved. When dissolution is complete, cool, add 2 to 3 g additional sodium carbonate, and hold the solution for 24 hours. Filter it prior to use.

Procedure 7-2. Polyphenol Index Determination (Permanganate Index)

(a) Potassium permanganate (0.01 *N*): In a 1-L volumetric flask, dissolve 316 mg $KMnO_4$ in approximately 500 mL distilled water. Bring the solution to volume with distilled water.

(b) Indigo carmine solution: Transfer 150 mg indigo carmine to approximately 500 mL distilled water in a 1-L volumetric flask. Add 50 mL dilute sulfuric acid solution (see "c" below). Bring to volume with distilled water.

(c) Dilute sulfuric acid (1 + 2): To two volumes of water, add one volume of stock sulfuric acid.

Procedure 7-3. HPLC Determination of Pigments

(a) Mobile Phase I (acetic acid–water 15:85): Dilute 150 mL glacial acetic acid to 1 L with distilled water. Sparge with an inert gas (helium) before use.

(b) Mobile phase II (water–acetic acid–methanol 65:15:20): Dilute 150 mL glacial acetic acid plus 200 mL methanol to 1 L with distilled water. Sparge with an inert gas (helium) before use.

(c) Anthocyanidin standards (1 mg/L): Prepare a stock solution (1 mg/mL) by dissolving 100 mg of the appropriate standard anthocyanidin in 100 mL distilled water. Dilute 1 mL of stock to 1 L to prepare working standards (1 mg/L).

Procedure 7-4. Wine Color Specification by Hue

Concentrated sulfuric acid
Distilled water
These reagents are used for dilution of red wine samples.

Procedure 7-5. The Ten-Ordinate Method for Color Specification

Concentrated sulfuric acid
Distilled water
These reagents used for dilution of red wine samples

Procedure 7-6. Spectral Evaluation of Juice and Wine

(a) Sodium metabisulfite solution [20% (wt/vol)]: Dissolve 2 g sodium metabisulfite in 10 mL distilled water.
(b) aqueous acetaldehyde solution [10% (wt/vol)]: Dilute 1.26 mL freshly distilled acetaldehyde to 10 mL with cold distilled water. Store mixture in a refrigerator.
(c) hydrocloric acid (1 *M*): See instructions at beginning of section for preparation of dilute acids.

Procedure 8-1. Carbon Dioxide Determination by Titrimetric (Enzymatic) Analysis

(a) Carbonic anhydrase (0.1 mg/mL): Dissolve 10 mg carbonic anhydrase enzyme in 100 mL distilled water. Store the solution under refrigeration until use.
(b) NaOH (50% wt/wt): Carefully dissolve 50 g reagent grade sodium hydroxide in 50 mL distilled water. Allow the solution to cool, and store it in a plastic bottle.
(c) Sulfuric acid (≤ 0.100 *N*): See Table II-3 for preparation.

Procedure 8-2. Determination of Oxygen Using a Dissolved Oxygen Meter

(a) Half-saturated KCl: Prepare ~ 100 mL saturated KCl solution. Pour off the supernatant liquid, and dilute the solution to 200 mL with distilled water. Add four drops Kodak Photo Flo to this solution.

Procedure 8-3. Enzymatic Analysis of Acetaldehyde

See the procedure accompanying the enzyme kit for reagent preparation.

Procedure 9-1. Sulfur Dioxide by the Ripper Method

(a) Stock iodine solution: Dissolve 12.9 g iodine and 25 g potassium iodide in approximately 100 mL distilled water. When dissolution is complete, transfer the solution to a 1-L volumetric flask, and bring it to volume with distilled water at a defined temperature.

(b) Working iodine solutions: To prepare iodine solution of the approximate normality desired, dilute required volumes of stock with distilled water to a final volume of 1 L, according to Table II-4. *Standardization of iodine solutions:* Because these dilutions yield only approximate concentrations, it is necessary to standardize working solutions against the primary standard sodium thiosulfate, which may be purchased in liquid, prestandardized form. Analysts wishing to prepare their own primary standard may employ the following procedure:

1. Accurately weigh approximately 300 mg solid reagent grade sodium thiosulfate, and transfer it to a 100-mL volumetric flask.
2. Dissolve the reagent in approximately 50 mL distilled water. Add a pinch of sodium bicarbonate, and dilute the solution to volume with distilled water
3. The normality (molarity) of this solution may be calculated using the following formula:

$$N = \frac{\text{Wt } Na_2S_2O_3(\text{mg}) \times \% \text{ Purity}}{158.1 \text{ mg/mMol} \times 100 \text{ (mL)}}$$

(b) Sodium hydroxide (1 N): Because this reagent is used in a hydrolysis step, there is no need for standardization. The reagent is prepared by dissolving one equivalent weight (40 g) of sodium hydroxide in 1 L distilled water.

(c) Sulfuric acid (1 + 3): This reagent also need not be standardized. For preparation, carefully dilute one volume of concentrated acid into three volumes of distilled water.

(d) Starch indicator (1%): Mix 10 g soluble starch and 1 L distilled water. Heat the solution to incipient boiling. Cool, and store it in a refrigerator. *Note:* Discard any starch indicator that appears turbid.

Table II-4. Dilution procedures for preparation of working solutions from stock iodine.

Approximate Noramlity Desired	Dilution
0.01	100 mL stock + 900 mL distilled water
0.02	200 mL stock + 800 mL distilled water

Procedure 9-2. Alternative Iodate Reagent Procedure

(a) Potassium iodate/iodide standard solution (0.02 N): In a 1-L volumetric flask, dissolve 713.3 mg dried reagent grade potassium iodate in approximately 800 mL distilled water. Add 50 mL 1 + 3 sulfuric acid with mixing, and dilute the solution to volume with distilled water.

(b) Potassium iodide–starch solution: In a 1-L volumetric flask, dissolve 10 g potassium iodide and 2.5 g soluble starch in approximately 800 mL distilled water. Bring the solution to volume with distilled water.

Procedure 9-3. Aeration Oxidation Procedure for Sulfur Dioxide

(a) Sodium hydroxide (0.01 *N*): See above, "Standard Acid–Base Solutions" section, for preparation of this standardized reagent.
(b) Hydrogen peroxide (0.3% vol/vol): Immediately prior to use, volumetrically transfer 5.0 mL stock H_2O_2 (30% wt/wt) to approximately 400 mL distilled water in a 500-mL volumetric flask. Bring the solution to volume using distilled water at a defined temperature.
(c) Phosphoric acid (1 + 3): Using *o*-phosphoric acid (85% stock) carefully prepare a 1 + 3 solution with distilled water.
(d) Indicator solution: Dissolve 50 mg methylene blue and 100 mg methyl red in approximately 90 mL 50% ethanol–water solvent. When dissolution is complete, dilute the solution to a final volume of 100 mL using the solvent.

Procedure 9-4. Modified Monier-Williams Procedure for Sulfur Dioxide

See the text for preparation of the reagents.

Procedure 10-1. Sensory Determination of Hydrogen Sulfide and Mercaptans

(a) Copper (II) sulfate (1% wt/vol): In approximately 90 mL of distilled water, dissolve 1 g $CuSO_4 \cdot 5H_2O$. Bring the solution to a final volume of 100 mL using distilled water.
(b) Cadmium sulfate (1% wt/vol): In approximately 90 mL distilled water, dissolve 1 g $CdSO_4 \cdot 8H_2O$. Bring the solution to a final volume of 100 mL using distilled water.
(c) Ascorbic acid (10% wt/vol): In approximately 90 mL distilled water, dissolve 10 g ascorbic acid. Bring the solution to a final volume of 100 mL using distilled water.

Procedure 11-1. Colorimetric Analysis for Sorbic Acid

(a) Sorbic acid stock (100 mg/L): Weigh 134 mg potassium sorbate, and transfer it to a 1-L volumetric flask. Dissolve the reagent, and bring the solution to volume with distilled water. This solution is equivalent to 100 mg/L sorbic acid.
(b) Sulfuric acid (0.3 *N*): Dilute 15 mL 2 *N* H_2SO_4 with distilled water to a final volume of 100 mL.

(c) Potassium dichromate solution: Dissolve 147 mg $K_2Cr_2O_7$ in distilled water, bringing the final volume to 100 mL.
(d) Thiobarbituric acid (0.5% wt/vol): In a 50-mL volumetric flask, dissolve 250 mg thiobarbituric acid in 5 mL of 0.5 *N* NaOH. Warm the flask's contents under a stream of hot water to speed dissolution. When dissolution is complete, add 20 mL distilled water adjusted with 3 mL 1 *N* HCl. Dilute the solution to volume with distilled water. *Note:* This solution must be prepared daily.

Procedure 11-2. Ultraviolet Analysis for Sorbic Acid

(a) Sorbic acid stock (100 mg/L): See instructions for Procedure 11-1 reagents.
(b) HCl (0.1 *N*): See above, under "Standard Acid–Base Solutions."

Procedure 11-3. Alternative Extraction Procedure for Sorbic Acid

(a) Sorbic acid stock (500 mg/L): Weigh 670 mg potassium sorbate, and dissolve it in 12% (vol/vol) ethanol reagent to a final volume of 1 L in a volumetric flask. This solution is equivalent to 500 mg/L sorbic acid.
(b) Isooctane: Used without further preparation.
(c) Phosphoric acid: Used without further preparation.
(d) Ethanol (12% vol/vol): Dilute 120 mL absolute ethanol to a final volume of 1 L with distilled water. In cases where denatured or less than absolute ethanol is to be used, the preparer of this solvent should account for the differences by increasing the ethanol volume proportionately.

Procedure 11-4. Spectrophotometric Determination of Benzoic Acid

(a) Benzoic acid stock (1000 mg/L): Weigh 1.00 g reagent grade benzoic acid, dissolve it in diethyl ether, and bring the solution to volume in a 1-L volumetric flask. Keep it tightly closed during storage.
(b) Saturated NaCl solution: Add enough sodium chloride to 1 L distilled water in a beaker to saturate the solution. Decant the saturated solution, and store it in a bottle.
(c) HCl (1 + 999): Dilute 1 mL concentrated HCl to 1 L with distilled water.
(d) Diethyl ether: Used without further preparation.

Procedure 11-5. HPLC Determination of Benzoic Acid and Sorbic Acid

(a) Benzoic acid stock (100 mg/L): Weigh 118 mg reagent grade sodium benzoate, and bring it to 1 L in a volumetric flask with the mobile phase.
(b) Sorbic acid stock (100 mg/L): Weigh 134 mg potassium sorbate, and bring it to 1 L in a volumetric flask with the mobile phase.

(c) Mobile phase (85% 0.01 *N* H_2SO_4: 15% acetonitrile): Prepare ~0.01 *N* H_2SO_4 by adding 2.8 mL concentrated acid to 10 L chromatography grade distilled water. To the 10 L dilute acid, add 1.76 L chromatography grade acetonitrile, and mix the solution. Store this reagent in sealed containers.

Procedure 11-6. Analysis of Dimethyldicarbonate by Gas Chromatography

(a) Carbon disulfide: In a separatory funnel wash 200 mL carbon disulfide with 20 mL fuming nitric acid. Draw off the excess acid, and wash the organic layer with 20-mL portions of distilled water until neutral.
(b) Ethyl methyl carbonate (EMC) standard (500 mg/L): Place 50 mg EMC in a 100-mL volumetric flask. Bring the solution to volume with 95% vol/vol ethanol.
(c) Diethyl carbonate (DEC) internal standard (500 mg/L): Place 50 mg DEC in a 100-mL volumetric flask. Bring the solution to volume with 95% vol/vol ethanol.

Procedure 12-1a. Isolation and Cultivation of Wine Microorganisms

See Appendix III for preparation of media and reagents used in this procedure.

Procedure 12-1b. Identification of Microbial Isolates

See Appendix III for preparation of media and reagents used in this procedure.

Procedure 12-2. Cell Counting

See Appendix III for preparation of reagents used in this procedure.

Procedure 12-3. Malolactic Fermentation by Paper Chromatography (Semiquantitative Measurement)

(a) Mobile phase: In a separatory funnel mix 100 mL *n*-butanol, 100 mL distilled water, 10.7 mL stock formic acid, and 15 mL indicator solution consisting of 1 g water-soluble bromcresol green in 100 mL water. Mix the reagents thoroughly, and allow the phases to separate. Discard the lower (aqueous) phase.
(b) Wine acid standard (0.3% wt/vol). Dissolve 0.3 g of each of the acids of interest in 100 mL distilled water. A drop or two of 6 *N* NaOH may aid in dissolving the acids. Store in a refrigerator.

Procedure 12-4. Malolactic Fermentation (Alternative Paper Chromatographic Technique)

(a) Mobile phase (*n*-butanol : acetic acid : water—4 : 1 : 5): Mix appropriate volumes of the components in a separatory funnel. Allow the layers to separate, and discard the lower (aqueous) layer. Add bromcresol blue indicator at a concentration of 1 g/L of butanol used.
(b) Prepare wine acid standards as above in Procedure 12-3.

Procedure 12-5. Determination of Biacetyl in Wine

(a) Biacetyl stock solution (800 mg/L): Dissolve 0.800 g fresh or redistilled diacetyl in 100 mL alcohol. Transfer the solution to a 1-L volumetric flask, and dilute to volume with distilled water.
(b) Hydroxylamine–acetate reagent:
   1. Hydroxylamine hydrochloride: Dissolve 43.75 g reagent grade hydroxylamine hydrochloride in approximately 500 mL distilled water. Transfer the solution to a 1-L volumetric flask, and dilute it to the mark with distilled water. Prepare this reagent fresh every two weeks.
   2. Sodium acetate: Dissolve 52.75 g anhydrous sodium sulfate in approximately 200 mL distilled water. Transfer the solution to a 250-mL volumetric flask, and dilute it to the mark with distilled water. Prepare this reagent fresh every two weeks.
   3. Mixed reagent: Mix 200 mL hydroxylamine hydrochloride solution with 50 mL sodium acetate solution. Filter the mixture through Whatman No. 1 paper. Prepare this reagent fresh every two weeks.
(c) Potassium phosphate–acetone reagent: In a 1-L volumetric flask, dissolve 144 g potassium hydrogen phosphate ($K_2HPO_4 \cdot 3H_2O$) in approximately 500 mL distilled water. Add 200 mL reagent grade acetone, and dilute the solution to the mark with distilled water. Mix well, and filter the solution through Whatman No. 1 paper. Prepare this reagent fresh every two weeks.
(d) Sodium potassium tartrate in ammonium hydroxide (45% wt/vol): Dissolve 900 g Rochelle salt (sodium potassium tartrate) in approximately 1200 mL distilled water with gentle warming. Cool the solution to room temperature. Add 387 mL concentrated ammonium hydroxide, and mix the components. Dilute to 2000 mL with distilled water, and filter the solution through Whatman No. 1 paper. Prepare this reagent fresh every two weeks.
(e) Ferrous sulfate (3.5% wt/vol) in sulfuric acid (1%): Add 1.45 mL concentrated sulfuric acid to ~200 mL distilled water. Dissolve 8.75 g ferrous sulfate in this solution, mix it, and dilute to 250 mL with distilled water. Filter the solution through Whatman No. 1 paper. Prepare this solution fresh daily.

Procedure 13-1. The Freeze Test for Wine Stability

No reagent preparation is required for this procedure.

Procedure 13-2. The Conductivity Test for Bitartrate Stability

Potassium bitartrate powder (40–70 μm). See Supplemental note 1 in the procedure for preparation:

Procedure 13-3. Tartaric Acid Analysis by Metavanadate (Carbon Decolorization)

(a) Tartaric acid stock solution (10 g/L): Accurately weight 1 g reagent grade tartaric acid and dissolve in 100 mL distilled water in a volumetric flask.
(b) Metavanadate solution (5%): In a 100-mL volumetric flask, dissolve 5 g $NaVO_3$ in hot (less than 70°C) distilled water. Cool the solution and bring to volume at a defined temperature. *Note*: Sodium metavanadate does not completely dissolve and must be filtered through Whatman No. 2 paper prior to use.
(c) HCl (1 *N*): See above, under "Standard Acid–Base Solutions."

Procedure 13-4. Alternative Procedure for Tartaric Acid Analysis (Ion Exchange Sample Preparation)

(a) Tartaric acid stock solution (10 g/L): prepare as directed above in Procedure 13-3.
(b) Acetic acid (30% vol/vol): To a 1-L flask containing approximately 300 mL distilled water, add 300 mL glacial acetic acid. Bring the solution to volume with distilled water.
(c) Acetic acid (0.5% vol/vol): Proceed as above, except dilute 5 mL of glacial acetic acid to 1 L final volume using distilled water.
(d) Sodium sulfate (0.5 *M*): Dissolve 71 g Na_2SO_4 in approximately 500 mL distilled water. Bring the solution to a final volume of 1 L when dissolution is complete.
(e) Sodium metavanadate (3%): Prepare as for Procedure 13-3, except use 3 g $NaVO_3$/100 mL distilled water.

Procedure 13-5. Potassium Analysis by Atomic Absorption

(a) Potassium standard stock solution (1000 mg/L): In a 1-L volumetric flask, dissolve 1.907 g dried reagent grade KCl in approximately 800 mL distilled water, and dilute to volume. Alternatively, premade stock potassium solutions are commercially available and may be conveniently utilized.

Procedure 13-6. Potassium Analysis by Flame Emission

(a) Potassium standard stock solution (1000 mg/L): Prepare as for Procedure 13-5.

Procedure 13-7. Potassium Analysis by Ion Selective Electrode

(a) Potassium standard stock solution (1000 mg/L): Prepare as for Procedure 13-5.
(b) Ionic strength adjustor (ISA): In a 100-mL volumetric flask, dissolve 35.1 g reagent quality NaCl in distilled water, and bring the solution to volume.
(c) Reference electrode filling solution: Transfer 2 mL ISA to a 100-mL flask, and dilute the solution to 100 mL with distilled water. Add silver nitrate (1 *M*) dropwise until a cloud persists.

Procedure 13-8: Calcium Analysis by Atomic Absorption

(a) Standard calcium solution (1000 mg/L): Dissolve 2.500 g predried $CaCO_3$ in approximately 100 mL deionized distilled water. Add 5 to 7 mL stock HCl, and warm the solution on a hot plate to drive off CO_2. *Quantitatively* transfer the solution to a 1-L volumetric flask, and bring it to volume with deionized distilled water at a defined temperature. *Note:* The distilled water used in the makeup of working and calibration standards should be boiled (and cooled to a defined temperature) prior to use to eliminate entrapped CO_2. Commercially available calcium standards may be conveniently substituted for the above reagent preparation.
(b) Lanthanum oxide/HCl: Dissolve 10 g La_2O_3 in 10 mL concentrated HCl. When dissolution is complete, dilute the solution to 100 mL with distilled water.

Procedure 14-1. Copper Analysis by Visible Spectrophotometry

(a) Standard copper solution (1000 mg/L): In a 1-L volumetric flask dissolve 1.000 g copper metal (available as either foil, powder, shot, or turnings) in several milliliters of dilute nitric acid. When dissolution is complete, bring the solution to volume with distilled water at a defined temperature. *Note:* The copper used should be designated as "suitable for standardization" or be of accurately defined purity. Alternatively, many laboratories use copper salts of defined purity as well as commercially available standard solutions of copper.
(b) Ammonium hydroxide (5 *N*): Dilute 666 mL stock NH_4OH to a 2-L final volume with distilled water.
(c) HCl–citric acid solution: In a 1-L volumetric flask contain ~500 mL dis-

tilled water, dissolve 94 g reagent grade citric acid and 62 mL concentrated HCl. Bring the solution to volume with distilled water.

(d) Amyl acetate–methanol solvent (2 : 1): In a convenient-size flask, mix two volumes of amyl acetate with one volume of methanol. *Note:* The fumes are very volatile. Hold the solvent in a ventilated area (under fume hood) until needed.

(e) Sodium diethyldithiocarbamate (1%): Transfer 1 g sodium diethyldithiocarbamate to a 100-mL volumetric flask, dissolve it, and bring the solution to volume with distilled water. This solution should be prepared daily.

Procedure 14-2. Copper Analysis by Atomic Absorption.

(a) Stock copper solution (100 mg/L): See preparation of standard copper solution for Procedure 14-1.

Procedure 15-1. Iron Analysis by Visible Spectrophotometry

(a) Standard iron solution (1000 mg/L): In a 1-L volumetric flask, dissolve 1.000 g iron metal (available as wire or powder of defined purity) in several milliliters of concentrated HCl. When dissolution is effected, bring the solution to volume with distilled water at a defined temperature. Alternatively, many laboratories use iron-containing salts (e.g., ferrous ammonium sulfate) or commercially available standards.

(b) Potassium thiocyanate (8% wt/vol): In a 100-mL volumetric flask, dissolve 8 g KSCN in 95 mL distilled water. Bring the solution to volume when dissolution is complete.

(c) Hydrochloric acid (5% vol/vol): Dilute 12-mL concentrated HCl to 100 mL with distilled water.

(d) Amyl acetate–methanol solution (2 : 1): Immediately prior to use, mix two volumes of amyl acetate with one volume of methanol. *Note:* The fumes are very volatile. Store the solution in a ventilated area (in fume hood) until it is used.

(e) Hydrogen peroxide (3% vol/vol): Dilute 10 mL stock 30% H_2O_2 to 100 mL with distilled water immediately prior to use.

Procedure 15-2. Iron Analysis by Atomic Absorption

(a) Stock iron solution (1000 mg/L): See preparation of iron solution for Procedure 15-1. Alternatively, prestandardized solutions are commercially available.

Procedure 15-3. Phosphorus Analysis by Atomic Absorption

(a) Standard phosphorus solution (1000 mg/L): In a 1-L volumetric flask, dissolve 4.4 g dried (2 hours at 110°C) reagent grade potassium dihydrogen phosphate in distilled water. Bring the solution to volume with distilled water, and store it in amber-colored glassware to prevent deterioration. Prepare the reagent fresh weekly.
(b) Ammonium molybdate reagent (10% wt/vol): In a 250-mL volumetric flask, dissolve 25 g reagent grade ammonium molybdate in distilled water. Bring the solution to volume, and store in amber-colored glassware. Prepare the reagent fresh weekly.
(c) Basic buffer: In 500 mL distilled water, dissolve 53.3 g ammonium chloride. Add 70 mL ammonium hydroxide, and bring the solution to 1 L with distilled water.

Procedure 16-1. Evaluation of Protein Stability by Exposure to Heat and Visual Examination

No reagent preparation is required for this procedure.

Procedure 16-2. Evaluation of Protein-Stability by the Trichloroacetic Acid Test with Nephelometric Evaluation

(a) Trichloroacetic acid (55% wt/vol): Dissolve 55 g trichloroacetic acid in distilled water, and bring the final volume to 100 mL.

Procedure 16-3. Ammonia (Ammonium Ion) by Ion Selective Electrode

(a) Ammonia stock solution (1000 mg/L): In a 1-L volumetric flask, dissolve 3.141 g reagent grade ammonium chloride (NH_4Cl) in distilled water. Mix the solution and bring it to volume with distilled water. The concentration of this solution is $5.9 \times 10^{-2} M$ expressed as NH_3.
(b) Sodium hydroxide (10 *M*): In a 1-L volumetric flask dissolve 400 g NaOH in approximately 500 mL distilled water. Mix the solution, and bring it to volume with distilled water.

Procedure 16-4. Determination of Alpha Amino Nitrogen (Argenine) in Must

(a) Primary amine standard (L-arginine monohydrochloride): In a 1-L volumetric flask, dissolve 2.11 g amino acid in 500 mL distilled water, and bring the solution to volume. Prepare a 200-μM dilution by diluting 1 mL of stock to a final volume of 50 mL using distilled water.

(b) Sodium sulfite (0.1 *M*): In a 100-mL volumetric flask, dissolve 1.26 g sodium sulfite in distilled water, and bring the solution to volume.
(c) Sodium dihydrogen phosphate (0.1 *M*): In a 100-mL volumetric flask, dissolve 1.38 g reagent in distilled water, and bring the solution to volume.
(d) Sodium sulfite/sodium dihydrogen phosphate solution: Dilute 1.5 mL of 0.1 *M* sodium sulfite with 98.5 mL sodium hydrogen phosphate solution. This solution should be prepared daily.
(e) Sodium borate solution (0.1 *M*). In a 100-mL volumetric flask, dissolve 2.01 g sodium borate in 0.1 *N* NaOH. Bring the solution to volume using 0.1 *N* NaOH.
(f) 2,4,6-Trinitrobenzene sulfonic acid (TNBS) (also known as picrylsulfonic acid): In a glass test tube, dissolve 645 mg TNBS in 2.0 mL distilled water.
(g) Trichloroacetic acid (TCA) 60% w/v.

Procedure 16-5. Determination of Proline in Juice and Wine

(a) Stock proline solution (575 mg/L): In a 100-mL volumetric flask, dissolve 57.5 mg proline in distilled water, and bring the solution to volume.
(b) Methyl cellosolve (2-methoxyethanol)-ninhydrin solution: Dissolve 3 g of ninhydrin in 100 mL methyl cellosolve.

Procedure 16-6. Nitrate Ion by Ion Selective Electrode

(a) Sodium nitrate (10,000 mg/L NO_3): In a 100-mL volumetric flask, dissolve 1.38 g predried reagent grade sodium nitrate. Bring the solution to volume with distilled water.
(b) Ammonium sulfate ionic strength adjustment buffer (2.0 *M*): Dissolve 26.41 g ammonium sulfate $(NH_4)_2SO_4$ in approximately 75 mL distilled water, and bring the solution to a 100-mL final volume.

Section IV. Remedial Actions: Reagents

(a) Metavanadate solution: Prepare as for Procedure 13-4.
(b) Phloroglucinol stain: Mix 2 g phloroglucinol in 100 mL 10% HCl. This is a near-saturated mixture, and the supernatant must be decanted from any crystals that do not dissolve. The solution should be made fresh each time it is used. Therefore, unless many samples are to be run, the analyst may wish to reduce proportionately the volume made at one time.
(c) Cellulose stain: In 100 mL distilled water, dissolve 200 g zinc chloride. Add 20 mL iodine solution. The iodine solution for cellulose stain is made by dissolving 10 g KI and 4 g I_2 in distilled water, and bringing the solution to a 100-mL final volume with distilled water.
(d) Amido black stain: Dissolve 2 g amido black 10-B in 100 mL methanol:acetic acid solvent. The solvent is prepared by mixing 90 mL methanol and 10 mL acetic acid.

(e) Eosin Y Protein stain (available commercially prepared).
(f) Folin-Ciocalteu reagent: Prepare as for Procedure 7-1.
(g) Potassium ferrocyanide (0.5%): Dissolve 0.5 g $K_4Fe(CN)_6 \cdot 3H_2O$ in 95 mL distilled water. When it is completely dissolved, bring the solution to a 100-mL final volume.
(h) Sulfuric acid (1 + 3): Carefully add 2 mL concentrated sulfuric acid to 6 mL water.

Chapter 17. Fining and Fining Agents: Reagents

The following stock fining agents are presented for use in conjunction with Table 17-2. Consult supplier's instructions for preparation of these materials.

(a) Stock tannic acid solution (1%): Slowly dissolve 1 g tannic acid in 100 mL ethyl alcohol–water solution. The latter is prepared by mixing equal volumes of water and fortifying alcohol at greater than 180 proof.
(b) Stock gelatin solution (1%): Dissolve 1 g gelatin (100–200 bloom) in 100 mL (112°F) water. If the agent is to be stored for any period of time, add 0.5 g citric acid and 0.1 g sodium benzoate.
(c) Stock Sparkolloid (hot mix) solution: Transfer 2.4 g Sparkolloid with thorough mixing to approximately 580 mL of boiling water. When a uniform consistency has been achieved, bring the solution to 1-L volume with hot water. Maintain the mixture at near boiling until its addition to wine (Scott Laboratory Product Notes).
(d) Stock Cufex and Metafine: Transfer 50 g reagent to a blender. Add 350 mL of iron- and copper-free distilled water, and mix the solution until it is uniform in consistency.
(e) Stock bentonite solution (5%): Add 5 g bentonite to 100 mL hot (>180°F) water. Mix the reagent until a uniform suspension is obtained.
(f) Stock casein solution (2%):
   1. Sodium caseinate: Dissolve 2 g caseinate in 100 mL water, bringing the reagent to volume when solution is effected.
   2. Milk casein: Dissolve 2 g powder in 98 mL water. *Note:* The pH of the water used must be adjusted to 11.0 using NH_4OH.

Procedure 18-1. Cation and Anion Exchange

(a) Hydrochloric acid (2 *N*): See above, under "Standard Acid–Base Solutions."
(b) Sodium chloride (10% wt/vol): Dissolve 200 g of NaCl in approximately 1500 mL water and bring the solution to a 2-L volume.
(c) Sodium hydroxide (1.5 *N*): Dissolve 120 g NaOH in approximately 1500 mL water, bringing the solution to a 2-L final volume.
(d) Silver nitrate (0.1 *N*): Dissolve 1.69 g $AgNO_3$ in distilled water, bringing the final volume to 100 mL at defined temperature.

Procedure 18-2. Analysis of Sodium Ion by Atomic Absorption

(a) Standard sodium solution (as NaCl): Dissolve 2.542 g predried NaCl in deionized, distilled water. Bring the solution to a 1-L final volume at a defined temperature. This solution is equivalent to a Na concentration of 1000 mg/L. Alternatively, prestandardized sodium standards are commercially available.

Procedure 18-3. Analysis of Sodium Ion by Flame Emission

Use standard sodium solution as prepared for Procedure 18-2.

Procedure 19-1. Laboratory Fining Trials using Cufex or Metafine

If prepared slurry is purchased, no reagent preparation is required for this procedure. If purchased as paste, see 17(d) above.

Procedure 19-2. Hubach Analysis

(a) Potassium-ferrocyanide stock solution (100 mg/L as cyanide): Dissolve 270.6 mg $K_4Fe(CN)_6 \cdot 3H_2O$ in one liter of distilled water in a volumetric flask. Mix Well.
(b) Sulfuric acid solution (1 + 1): To one volume of distilled water, carefully add one volume of concentrated sulfuric acid. Cool the solution to room temperature before use.
(c) Hydrochloric acid solution (1 + 3): To three volumes of distilled water, add one volume of concentrated HCl. Cool the solution to room temperature before use.
(d) Sodium hydroxide solution (6 *N*): Dissolve 200 g reagent grade NaOH in approximately 800 mL distilled water. When the solution has cooled to room temperature, bring it to a 1-L volume with distilled water.
(e) Cuprous chloride: Because CuCl acts as a catalyst and is regenerated during the distillation, the quantity used is not crucial. Dissolve 1.5 g reagent grade CuCl in 1 *N* H_2SO_4, decanting the blue supernatant. The process is repeated two times, or until only a light blue color remains in the acid solution. Store cuprous chloride in an amber bottle, taking care to avoid exposure to air.
(f) Hubach test papers: Test papers are most conveniently purchased at nominal cost through Scott Laboratories (Petaluma, California). However, analysts wishing to prepare their own may use the following guide:
  1. Ferrous sulfate solution: Dissolve 5 g ferrous sulfate ($FeSO_4 \cdot 7H_2O$) in 50 mL distilled water, adding a drop or two of concentrated sulfuric acid to remove turbidity. This solution should be prepared shortly before use. Immediately before use, add one drop of concentrated sulfuric acid.

2. Alcoholic sodium hydroxide: From stock 50% (wt/vol) NaOH, transfer 11 g solution to a 100-mL volumetric flask, bringing the solution to volume with 95% ethanol.

Using a Whatman No. 50 filter paper of convenient size, immerse the paper in the ferrous sulfate solution for 5 minutes. Remove the paper, and let it air-dry; the drying step should be carried out at room temperature. When the paper is completely dry, immerse it in the alcoholic sodium hydroxide solution; allow 5 minutes for soaking. Remove the paper, and let it air-dry; again, this step must be carried out at room temperature. When completely dry, the paper should be light green to tan in color. At this point, it may be cut into appropriate-size sections for analysis. It is recommended that test papers be stored in a desiccator jar away from moisture and light.

Appendix III

Laboratory Media and Stains

Both the isolation and identification of microorganisms and the routine maintenance of yeast and bacterial cultures for winemaking purposes require preparation of appropriate media to support the nutritional needs of the strains of interest.

Besides providing for the maintenance of cultures, selective utilization of media serves as a valuable aid to the identification of troublesome microbes that may be found in wines. Among strains of ascomycetous yeasts, this may include growing isolates on selective substrates that promote sporulation. In most cases, speciation requires establishing nitrogen and carbon utilization profiles, which is accomplished by growing the isolate on carbohydrate- and nitrogen-free media supplemented with the appropriate substrates. Media also may be prepared to include dyes or other components that react with metabolites characteristic of certain strains, so that color development in and around colonies provides a means of rapid screening. In wine microbiology labs, calcium carbonate–ethanol plates may be considered examples of such specialized media. In this case, clearing around colonies suggests the presence of acid-producing bacteria or yeasts. In both yeast and bacterial identification, fermentation broths are used routinely for identification of fermentative utilization of sugar.

Laboratory growth and differentiation media are made (or commercially available) as either liquid "broths" or solid substrates. Where a solidified medium is required, it is achieved by the inclusion of 1.5 to 2% agar in the preparation.

The following section includes examples of media (both solid and liquid) that are useful to the wine microbiologist in the maintenance of cultures as well as diagnostic work.

It should be pointed out that one may obtain many of the media needed for identification from chemical supply houses. In some cases, special orders can be accepted when the product is not in general use.

GENERAL GROWTH AND ISOLATION MEDIA (YEAST AND BACTERIA)

Several media are used commonly for isolation, differentiation, and cultivation of wine microorganisms. Most are commercially available in dehydrated form. In some instances, modification of standard media (biphenyl for mold and actidione for yeast inhibition) may be necessary. Alternately, one may wish to purchase pre-sterilized hits which include media-impregnated pads and petri plates.

Grape Juice Agar

Grape juice agar is used as a general, nonselective medium for isolation and growth of yeasts. The medium is prepared by mixing 20 g agar in 1 L grape juice diluted 1 + 3. Dissolve the agar by holding in a boiling water bath, and sterilize by autoclaving.
Note: time and temperature regimes may vary with volume of media prepared. Check recommended protocol provided with individual autoclave used.

Wort Agar

Wort agar is also a good general growth medium for yeasts. Although it is commercially available, the following makeup procedure is provided (Difco 1977):

1. To approximately 800 mL distilled water, add the following:
   15.0 g malt extract
   0.78 g peptone
   12.75 g maltose
   2.75 g dextrin (Difco)
   2.35 g glycerol
   1.0 g dipotassium phosphate
   1.0 g ammonium choride
   15.0 g agar
2. Bring to a final volume of 1 L using distilled water.
3. Sterilize by autoclaving.

Malt Agar

Malt agar is another general growth medium for yeast and molds. It is also available commercially in dehydrated form. Analysts wishing to make their own media may use the following procedure:

1. In approximately 800 mL distilled water, dissolve the following (see comment below):
   30 g malt extract
   15 g agar
2. Bring to a final volume of 1 L using distilled water, and autoclave.

For maintenance of acid-producing microbes (e.g., *Brettanomyces*) on malt agar, it is recommended that calcium carbonate (2% wt/vol) be included in the above procedure. Even with the inclusion of carbonate, it is recommended that these cultures be transferred regularly to minimize toxicity due to production of acetic acid.

Modified Tanner-Vetsch Medium

The lactic acid bacteria are fastidious, microaerophilic–facultatively anaerobic organisms, which may be difficult to culture on synthetic media. Thus, most media include a combination of vegetable juice and other digests. (This is discussed in greater detail in Chapter 12.)

For best results, lactic bacteria should be cultured under conditions of low oxygen tension, such as in a candle jar or in specially designed incubators.

Several commercially prepared media are available for growth of lactic acid bacteria (e.g., Difco Tomato Juice Agar). The authors, however, have found the best growth to occur using the modified Tanner-Vetsch medium presented below:

1. Dilute 355 mL tomato juice to 1500 mL with distilled water, and filter it.
2. To 100 mL of the above solution, add:
   - 2 g agar
   - 2 g tryptone
   - 0.5 g peptone
   - 0.5 g yeast extract
   - 0.3 g glucose
   - 0.2 g lactose
   - 0.1 g liver extract
   - 0.1 mL 5% aqueous Tween 80
3. Heat on a boiling water bath until dissolution is complete. Adjust the acidity of the final *cooled* medium to pH 5.5 using concentrated HCl.
4. Autoclave according to the recommended format.

For cultivation of *Lactobacillus trichoides*, the above medium is recommended *without inclusion of agar*. After the broth has cooled, 10% (vol/vol) fortifying alcohol, as Neutral Spirits Fruit Grape (NSFG) is added.

Apple Rogosa Agar

1. Dilute 200 mL of apple juice (use either frozen juice or juice without preservatives) to 1 L using distilled water.
2. Transfer 700 to 800 mL of the diluted juice to a separate 1-L flask.

3. Dissolve the following into this volume:
   20 g tryptone
   5 g peptone
   5 g glucose
   5 g yeast extract
   20 g agar
4. Bring to volume with diluted apple juice.
5. Adjust the pH to 5.5.
6. Sterilize by autoclaving.

Acid Plates

It may be difficult to cultivate some acetic acid bacteria on routine laboratory media where pH levels exceed 5.0. To overcome this problem, acid plates often are employed. Media preparation is as follows:

1. In an appropriate volume of water, dissolve the following:
   5.0 g yeast extract
   20.0 g dextrose (glucose)
   5.0 g KH_2PO_4
   20.0 g agar
2. Hold in a boiling water bath until components are fully dissolved.
3. Bring to a final volume of 1 L with water.
4. Sterilize by autoclaving.
5. When the autoclave cycle is complete, and before pouring of plates, carefully adjust the acidity, using approximately 0.1 mL of concentrated HCl/100 mL of the medium, to a final pH of 4.0.

Calcium Carbonate–Ethanol Medium

Since acetic acid bacteria are capable of growth on glucose as a carbon-source, ethanol-supplemented media is of value only with respect to detecting formation of acetic acid in presumptive identifications. In alcoholic media, some acetic acid bacteria produce various organic acids that neutralize the $CaCO_3$ present, resulting in a zone of clearing around the colonies. Thus, these media may be used to differentiate acetic acid bacteria that may be present. Medium preparation is as follows:

1. In an appropriate volume of water, dissolve the following:
   20.0 g $CaCO_3$
   5.0 g yeast extract
   20.0 g agar
2. Dissolve components by holding in a boiling water bath.

3. Autoclave according to the defined procedure.
4. After the sterile medium has cooled (not solidified) sufficiently, add 3% (vol/vol) NSFG. *Note:* $CaCO_3$ should be uniformly suspended throughout prior to the pouring of plates.

Yeast Extract–Malt Extract (''YM'') Agar.

This medium is available from Difco and other supply houses.

Any of the above media may be made somewhat selective by inclusion of 10 to 100 mg/L cycloheximide (actidione) to inhibit most wine yeast and 10 mg/L biphenyl to inhibit mold growth. Cycloheximide, used at 100 mg/L, is described by Lodder (1970) as a test for resistance to the antibiotic. Both *Brettanomyces* and *Dekkera* are resistant to the compound at this level, so this may serve as a selection tool for these strains in mixed cultures.

SPORULATION MEDIA

Yeast identification relies on demonstration of the presence or absence of a sexual cycle—that is, the presence or absence of ascospores. Such demonstration is often difficult. The fact that ascospores are not observed does not necessarily mean that the yeast is an ''imperfect'' form; this may mean that it does not sporulate on the particular medium used, or that some other requisite conditions were not met. Further, the isolated strain may require conjugation with a compatible mating type prior to formation of ascospores.

A great variety of media have been used for sporulation, and several of the more important ones for the demonstration of ascospores in wine yeast are presented below. For details regarding other relavant sporulation media, the reader is referred to Lodder (1970).

Saccharomyces sp.

Demonstration of ascospore formation in *Saccharomyces* generally utilizes an acetic acid–supplemented substrate. Since sporulation in this genus does not occur under fermentative conditions, glucose concentrations are low (<1 g/L).

McClary's Acetate Agar:

1. In 900 mL distilled water, add:
   - 1 g glucose
   - 1.8 g KCl
   - 2.5 g yeast extract
   - 8.2 g sodium acetate ($3H_2O$)
   - 15 g agar

2. Dissolve the components, and bring to a final volume of 1 L using distilled water.
3. Sterilize by autoclaving.

Schizosaccharomyces sp.

For demonstration of sporulation in this group, malt extract agar is recommended. The product is available commercially in dehydrated form.

Pichia sp.

Sporulation has been achieved using acetate agar (see above), Gorodkowa agar, and yeast extract–malt extract (YM) agar. YM agar is commercially available, and preparation of the Gorodkowa medium is presented below.

Gorodkowa Agar (Modified)

1. In approximately 800 mL distilled water, add:
   10.0 g peptone
   1.0 g glucose
   5.0 g NaCl
   20.0 g agar
2. Dissolve components, and bring to 1 L final volume with distilled water.
3. Sterilize by autoclaving

Hanseniaspora sp.

Sporulation is demonstrated using Potato Dextrose Agar (PDA), which is commercially available.

Hansenula sp.

Malt extract or V8 juice has been used to sporulate *Hansenula.* Malt extract is commercially available.

V8 Vegetable Juice Agar

1. In a 1-L flask, collect approximately 350 mL V8 vegetable juice.
2. Add 5 g compressed yeast with thorough mixing.
3. Adjust the pH to 6.8.
4. Heat to near boiling. Cool, recheck pH, and adjust it to 6.8 if needed.
5. In a second flask, dissolve 20 g agar in distilled water.
6. Combine the contents of both flasks and bring to 1 L with distilled water.
7. Sterilize by autoclaving.

Dekkera sp.

Although the authors have observed sporulation of *Dekkera* in wine, laboratory demonstration of sporulation in this species usually is accomplished on YM agar that has been supplemented with vitamins. Sporulation is slow, taking two to three weeks to occur at room temperature. YM agar is commercially available, and the preparation of vitamin supplementation is presented below.

Stock Vitamin Solution

The preparation of stock vitamin solution is described by Van de Walt and Van Kerker 1961 as described by Lodder 1970.

1. The following list gives the final concentration per liter of vitamin solution made up in distilled water. Because of the relatively low concentrations of some components, multiple-strength solutions may be prepared and diluted as necessary.
   - 0.2 mg biotin
   - 0.2 mg folic acid
   - 20 mg *p*-aminobenzoic acid
   - 20 mg riboflavin
   - 40 mg niacin
   - 40 mg calcium pantothenate
   - 40 mg pyridoxine hydrochloride
   - 100 mg thiamine
   - 200 mg inositol
2. Sterilize by filtration (do not autoclave) into convenient-sized test tubes.
3. Refrigerate until use.

MEDIA FOR OBSERVATION OF FORMATION OF PSEUDOMYCELIA

Demonstration of mycelial or pseudomycelial formation in yeasts is often of diagnostic importance. The usual media recommended are either corn meal or potato dextrose agars. Both agars are available commercially, and purchase of prepared media is probably worth the additional expense.

Corn Meal Agar

1. To approximately 1 L distilled water, add 42 g yellow corn meal.
2. Heat for 1 hour in a 60°C water bath.
3. Filter, and bring to 1 L volume with water.
4. Autoclave for 15 minutes.

5. Cool and refilter.
6. Add 14 g agar, mix, and reautoclave.

LIQUID MEDIA

Apple Rogosa Broth

See preparation under "Apple Rogosa Agar" (above). In the case of the broth, omit the addition of agar.

Carbohydrate Utilization Broths

Final speciation in both yeast and bacteria generally requires profiling sugar utilization (either assimilation or fermentation). Fortunately for laboratory personnel in wineries, the numbers of species encountered is not great, so this degree of accuracy may not be necessary for routine work.

Wickerham's Basal Medium

1. In 1 L distilled water, add 7.5 g peptone and 4.5 g yeast extract.
2. Add enough bromothymol blue indicator to produce a green color.
3. Transfer 6 mL into test tubes of appropriate volume.
4. Sterilize by autoclaving.
5. When tubes are cool, add 3 mL of sterile-filtered sugar solution.

Sugars for fermentation tubes: Normally, sugars for fermentation/assimilation broths are prepared at 6% (wt/vol). In the case of raffinose, 12% (wt/vol) solutions are used. Prior to addition to sterile basal medium, sugar solutions should be sterile-filtered. Store under refrigeration until use.

Arginine Medium

1. Modify the procedure for preparation of "Apple Rogosa Broth" presented above to include 2% (wt/vol) glucose and 0.6% (wt/vol) L-arginine.
2. Dispense 2-mL aliquots into test tubes, and stopper the tubes.
3. Sterilize by autoclaving.

Fructose Medium

1. Modify the procedure for Apple Rogosa Broth" to include 2% (wt/vol) fructose instead of glucose.

2. Dispense 5-mL aliquots into appropriate-sized test tubes, and stopper the tubes.
3. Sterilize by autoclaving.

Vaspar Plugs

1. Vaspar is prepared by melting one part paraffin with six parts vaseline (wt/vol) and autoclaving.
2. Prior to use, tubes containing vaspar should be held in a water bath at a temperature sufficient to keep the product liquefied.

Nessler's Reagent

1. In a 1-L flask dissolve 50 g potassium iodide in 40 to 50 mL cold distilled water.
2. In a separate 1-L flask, prepare a saturated solution of mercuric chloride (approximately 11 g/100 mL) in distilled water.
3. Slowly add saturated HgCl to the solution of potassium iodide until a slight orange precipitate persists. This step should take 300 to 400 mL HgCl solution.
4. Add, with thorough mixing, 400 ml 50% (wt/vol) KOH.
5. Bring to volume with distilled water.
6. The above mixture should be allowed to sit overnight.
7. Decant and store the supernatant.

MICROBIOLOGICAL STAINS

Several dyes and stains commonly are used in microbiological examination of wine. In all cases, the goal of staining is to enhance differentiation of the object under investigation, or portions of the object, and the surrounding field. Several common dyes are discussed briefly in the following paragraphs.

Nigrosin

Commonly used as a background or counterstain, nigrosin is very effective in differentiating gross morphological differences between microorganisms (viz., separation of cocci from bacilli). In practice, any nonselective counterstain, such as India ink, may be substituted for nigrosin.

Nigrosin usually is prepared as a 10% (wt/vol) solution, by using water-soluble powder in distilled water. After dissolution is effected, the solution is placed in a boiling water bath for 20 minutes. After cooling to room temperature, 0.5 mL formalin is added (as a preservative), and the solution is membrane-filtered.

Methylene Blue

Methylene blue may exist in either of two forms, a colorless or leuco-form and the colored (blue) form. The presence of either in solution is dependent upon the redox potential of the medium, or the presence of active dehydrogenase enzyme systems in the living cell. Thus, methylene blue may serve as an E_h (oxidation-reduction) indicator or as a hydrogen acceptor in metabolic reactions.

Methylene blue can be used both as a direct counterstain and as a differential stain for distinguishing living from nonviable cells. In viable yeast cells, the indicator is absorbed by the cell and rapidly reduced to the leuco- or colorless form. Thus, viable yeast cells appear colorless against a blue background. Dead cells, by comparison, also absorb the stain but without subsequent reduction to the leuco-form; therefore, the presence of blue cells should be interpreted as nonviability. The winemaker, thus has a useful tool for monitoring the activity of starter cultures as well as the progress of fermentations, and so on. Note, however, that methylene blue is, itself, toxic to yeast cells. Therefore, slide preparations using this stain should be examined within 10 minutes to ensure proper interpretation (see Chapter 12 for more detail).

Methylene blue (methylthionine chloride) is prepared by dissolving 0.3 g of the powder in 30 mL 95% ethyl alcohol and 100 mL citrate buffer (pH 4.6). Prior to use, the dye should be membrane-filtered and examined for biological activity.

Gram Stain

One of the most commonly used staining procedures, Gram staining is used to differentiate bacteria based upon cell wall characteristics. The procedure calls for initial staining of cells with crystal violet and fixation with iodine. Subsequent removal of stain–iodine precipitate is effected from Gram negative cells with alcohol decolorizer and applied counterstain.

Cells retaining the primary stain are violet-purple in color and are described as Gram positive, whereas those retaining the secondary stain (appearing as pink-red) are termed Gram negative. The differentiation achieved by use of Gram stain is not absolute but is based on differences in the rate at which the primary stain is lost from the bacterial cell wall. Table 12.6 summarizes Gram reactions for common wine bacteria. The recommended procedure for Gram staining is presented below:

1. Smear a glass microscope slide with a specimen swab or suspension of the culture to be observed. Allow it to air-dry.
2. Fix the smear by *gently* applying heat from a bunsen burner to the opposite side of the slide.
3. Cool the slide.
4. Flood the slide with Gram crystal violet stain. Allow the stain to remain on the slide for 1 minute.

5. Rinse gently with water.
6. Flood the slide with Gram iodine solution. Allow it to remain on the slide for 1 minute.
7. Rinse gently with water.
8. Flood the slide with Gram decolorizer solution. Allow 15 to 10 sec for reaction to occur.
9. Rinse gently with water.
10. Flood the slide with Gram safranin stain. Allow the stain to remain on the slide for 1 minute.
11. Rinse gently with water. Allow the slide to air-dry and examine it under an oil immersion objective.

Ponceau-S Stain

As a rather specific stain for proteinaceous material, this procedure calls for initial staining of the field with Ponceau-S and subsequent rinses with 5% acetic acid. Acid treatment removes the stain from miscellaneous debris, leaving the yeast cells stained red (Kunkee and Neradt 1974). Since the technique does not distinguish living intact cells from dead intact cells (those that have not undergone autolysis), Kunkee recommends using Ponceau-S in conjunction with previously examined preparations of methylene blue. His differential technique is described below:

1. Ponceau-S is prepared by dissolving the following in approximately 70 mL distilled water:
   - 0.9 g Ponceau S
   - 13.4 g trichloroacetic acid
   - 13.4 g sulfosalicylic acid
2. When dissolution is effected, bring the solution to a final volume of 100 mL using distilled water.

Already prepared Ponceau S staining kits may be purchased from Millipore Corporation.

The procedure for use in conjuction with methylene blue is presented below:

1. Using an appropriate laboratory filter, collect a known volume of the suspect wine on a 1-μm membrane.
2. While it is still assembled, cover the membrane with a small volume of methylene blue (0.01%) prepared in citrate buffer (pH 4.6).
3. Hold 20 to 30 sec, and suck the stain through. Transfer the filter to a microscope slide and examine
4. Interpretation: Clear cells are viable, and blue-colored cells are dead.

5. Collect a fresh sample of the suspect wine, or transfer the previously stained membrane to an absorbent pad (or several thicknesses of laboratory filter paper) soaked with Ponceau-S stain.
6. Once stained, transfer the membrane to similar pads soaked with 5% acetic acid (decolorizer). Transfer it between pads at 3- to 4-minute intervals (total five to six transfers) until the membrane appears white to light pink.
7. Transfer the membrane to a microscope slide, quickly pass it over the flame of a Bunsen burner, and examine it under the microscope.
8. Subtract the results from methylene blue staining from the results of the Ponceau-S stain to obtain viable cells.

Many laboratories simply use Ponceau-S as a general staining tool without consideration of the above differential technique.

Walford's Stain

The Walford stain also may be used for viable yeast cell counting. In this case, dead cells absorb the stain and are black-purple in appearance, whereas living cells reject the stain and appear clear. The stain is prepared as follows:

1. Prepare three components:
   Component A: To 10 mL 95% ethanol, add 3 mL aniline. To the mixture, add 1.5 mL of methylene blue.
   Component B: To 65 mL 95% ethanol, add 1 g methylene blue indicator.
   Component C: In 100 mL 95% ethanol, dissolve 1 g basic fuchsin indicator.
2. Add 30 mL saturated methylene blue–ethanol solution (Component B) to Component A. To this solution, add 55.5 mL distilled water. Identify as Solution 1.
3. Prepare the final stain by mixing 10 mL Solution 1 with 2 mL Component C. Bring to 100 mL final volume with distilled water.
4. Filter the stain, and store in refrigerator when not in use. The stain should be prepared at three- or four-week intervals.

Procedure for Use:

1. Transfer equal volumes of stain and sample containing yeast to a test tube. Mix well. Prepare a wet mount, and evaluate it microscopically.
2. Interpretation: Living cells appear colorless, whereas dead cells are blue-purple.
3. Rapid quantification may be achieved by examining the well-mixed preparation on a cell-counting slide.

Chromatography Solvent

1. To an appropriate-sized separatory funnel, add the following reagents:
   100 mL *n*-butanol
   100 mL distilled water
   10.7 mL stock formic acid
   15 mL indicator solution prepared by dissolving 1 g water-soluble bromocresol green in 100 mL distilled water
2. Shake solvent mixture thoroughly by repeated inversion of separatory funnel. Allow for phase separation, and discard lower phase. *Note:* Solvent should be prepared on a weekly basis.
3. Standard acids: In a 100-mL volumetric flask, dissolve 0.3 g ACS grade acids in approximately 95 mL distilled water. Bring to volume at a defined temperature.

Bibliography

American Colloid Co. "Microfine Bentonite" Skokie, IL, 60077.

Amerine, M. A. 1954. Composition of wines. I. Organic constituents. In *Advances in Food Research* 5:354–466. New York: Academic Press.

Amerine, M. A. 1958. Composition of wines. II. Inorganic constituents. In *Advances in Food Research* 8:133–224. New York: Academic Press.

Amerine, M. A. 1965. *Laboratory Procedures for Enologists.* Davis, CA: University of California, Department of Viticulture and Enology.

Amerine, M. A. and Cruess, W. V. 1960. *The Technology of Winemaking,* 2nd ed. Westport, CT: AVI Publishing Co.

Amerine, M. A. and Joslyn M. A. 1970. *Table Wines, the Technology of Their Production,* 2nd ed. Berkeley, CA: The University of California Press.

Amerine, M. A. and Kunkee, R. E. 1968. The microbiology of winemaking. *Ann. Rev. Microbiol.* 22:323–358.

Amerine, M. A. and Ough, C. S. 1957. Studies on controlled fermentation. III. *Am. J. Enol.* 8:18–30.

Amerine, M. A. and Ough, C. S. 1974. *Wine and Must Analysis.* New York: John Wiley and Sons, Inc.

Amerine, M. A. and Ough, C. S. 1980. *Methods for Analysis of Musts and Wines.* New York: John Wiley and Sons, Inc.

Amerine, M. A. and Roessler, E. B. 1958a. Methods for determining field maturity of grapes. *Am. J. Enol. and Vitic.* 9:37–40.

Amerine, M. A. and Roessler, E. B. 1958b. Field testing of grape maturity. *Hilgardia* 28:93–114.

Amerine, M. A. and Roessler, E. B. 1976. *Wines: Their Sensory Evaluation.* San Francisco: W. H. Freeman and Co.

Amerine, M. A. and Thoukis, G. 1958. The glucose-fructose ratio in California grapes. *Vitis* 1:224–229.

Amerine, M. A. and Winkler, A. J. 1940. Maturity studies with California grapes. I. Balling-acid ratio of wine grapes. *Proc. Am. Soc. Hort. Sci.* 38:379–397.

Amerine, M. A. and Winkler, A. J. 1942. Maturity studies with California grapes: II. The titratable acidity, pH and organic acid content. *Proc. Am. Soc. Hort. Sci.* 40:313–324.

Amerine, M. A., Ough, C. S., and Bailey, C. B. 1959a. Suggested color standards for wines. *Food Tech.* 13:170–175.

Amerine, M. A., Roessler, E. B., and Filipello, F. 1959b. Modern sensory methods of evaluating wines. *Hilgardia* 28:447–567.

Amerine, M. A., Berg, H. W., and Cruess W. V. 1972. *The Technology of Winemaking,* 3rd ed. Westport, CT: AVI Publishing Co.

Amerine, M. A., Roessler, E. B., and Ough, C. S. 1965. Acids and the acid taste. I. The effect of pH and titratable acidity. *Am. J. Enol. and Vitic.* 16:29–37.

Amerine, M. A., Berg, H. W., Kunkee, R. E., Ough, C. S., Singleton, V. L., and Webb, A. D. 1980. *The Technology of Winemaking,* 4th ed. Westport, CT: AVI Publishing Co.

Ames Company. 1978. *Dextrocheck Test for Reducing Sugar: Information Summary.* Elkhart, IN: The Ames Company, A Division of Miles Laboratory, Inc.

Anelli, G. 1977. The proteins of must. *Am. J. Enol. and Vitic.* 28:200–203.

Arena, A. 1936. Alteraciones bacterianas de vinos Argentinos. *Rev. Facultad de Agric. y Vet. (Buenos Aires)* 8:155–315.

Arnold, R. A. and Noble, A. C. 1978. Bitterness and astringency of grape seed phenolics in model wine solutions. *Am. J. Enol. and Vitic.* 29:150–152.

Ashida, J., Higashi, N., and Kikuchi, T. 1963. An electronmicroscopic study on copper precipitation by copper-resistant yeast cells. *Protoplasma* 57:27–32.

Asai, T. 1968. *Acetic Acid Bacteria: Classification and Biochemical Activities.* Tokyo: University of Tokyo Press.

Atlas, R. M., and Bartha, R. 1981. *Microbial Ecology: Fundamentals and Applications,* Addison-Wesley Publ. Co., Reading, Mass.

Auerbach, R. C. 1959. Sorbic acid as a preservative in wine. *Wine and Vines* 40(8):26–28.

Balakian, S. and Berg, H. W. 1968. The role of polyphenols in the behavior of potassium bitartrate in red wines. *Am. J. Enol. and Vitic.* 19:91–100.

Bate-Smith, E. C. 1954. Flavonoid compounds in foods. In *Advances in Food Research* 5:262–295. New York: Academic Press.

Bate-Smith, E. C. and Morris, T. N. 1952. The nature of enzymatic browning. In *Food Science.* Cambridge, MA: Harvard University Press.

Bate-Smith, E. C. and Swan, T. 1953. Identification of leuco-anthocyanins as "tannins" in foods. *Chem. and Ind.* 377.

Bausch & Lomb Optical Company. *Abbe-56 Refractometer Reference Manual.* New York: Bausch & Lomb Optical Company.

Bayly, F. C. and Berg, H. W., 1967. Grape and wine proteins of white wine varietals. *Am. J. Enol. and Vitic.* 18:18–32.

Beckwith, J. 1936. Pure cultured yeast. *J. Dept. Agriculture S. Australia* 38:858–867.

Beech, F. W., Burroughs, L. F., Timberlake, C. F., Whiting, G. C. 1979. Progres recents sur l'aspect chimique et l'action anti-microbienne de l'anhydride sulfureux. *Bulletin of the O.I.V.* 586:1001–1022.

Beelman, R. B. 1982. Development and utilization of starter cultures to induce malolactic fermentations in red table wines. *Proc. University of California, Davis, Grape Wine Centennial Symp.* A. D. Webb (ed.), pp. 109–117. Dept. of Vitic. & Enology, University of Calif., Davis.

Beelman, R. B., Keen, R. M., Banner, M. J., and King, S. W. 1982. Interactions between wine yeast and malolactic bacteria under wine conditions. *Dev. Ind. Microbiol.* 23:107–121.

Bell, T. A., Etchellis, J. L., and Berg, A. F. 1958. Influence of sorbic acid on growth of certain species of bacteria, yeast, and filamentous fungi. Raleigh, NC: U.S. Food Laboratory, North Carolina Agr. Exp. Sta.

Berg, H. W. 1953. Special wine filtration procedures recommended for use following Cufex fining of wines. Wine Institute Tech. Advisory Committee.

Berg, H. W. 1959a. Investigations of defects in grapes delivered to California Wineries: 1958. *Am. J. Enol.* 10:61–69.

Berg, H. W. 1959b. The effects of several fungal pectic enzyme preparations on grape musts and wines. *Am. J. Enol. and Vitic.* 10:130–134.

Berg, H. W. 1960. Stabilization studies on Spanish sherry and on factors influencing KHT precipitation. *Am. J. Enol. Vitic.* 11:123–128.

Berg, H. W. and Akiyoshi, M. 1956. Some factors involved in browning of white wines. *Am. J. Enol.* 7:1–7.

Berg, H. W. and Akiyoshi, M. 1957. The effect of various must treatments on color and tannin content of red grape juices. *Food Res.* 22:373–383.

Berg, H. W. and Akiyoshi, M. 1961. Determination of protein stability in wine. *Am. J. Enol. and Vitic.* 12:107–110.

Berg, H. W. and Akiyoshi, M. 1962. Color behavior during fermentation and aging of wine. *Am. J. Enol. and Vitic.* 13:126–132.

Berg, H. W. and Akiyoshi, M. 1971. The utility of potassium bitartrate concentration product values in wine processing. *Am. J. Enol. and Vitic.* 22(3):127–134.

Berg, H. W. and Akiyoshi, M. A. 1975. On the nature of reactions responsible for color behavior in red wines: a hypothesis. *Am. J. Enol. and Vitic.* 26(3):134–143.

Berg, H. W. and Keefer, R. M. 1958. Analytical determination of tartrate stability in wine. I. Potassium bitartrate. *Am. J. Enol.* 9:180–193.

Berg, H. W. and Keefer, R. M. 1959. Analytical determination of tartrate stability in wine. II. Calcium tartrate. *Am. J. Enol.* 10:105–109.

Berg, H. W. and Marsh, G. L. 1954. Sampling deliveries of grapes on a representative basis. *Food Tech.* 9:104–108.

Berg, H. W., Filipello, F., Hinreiner, E., and Webb, A. D. 1955. Evaluation of thresholds and minimum difference concentrations for various constituents of wines: I. Water solutions of pure substances. *Food Tech.* 9:23–26.

Berg, H. W., DeSoto R. T., and Akiyoshi, M. 1968. The effect of refrigeration, bentonite clarification and ion exchange on potassium bitartrate behavior in wines. *Am. J. Enol. and Vitic.* 19:208–212.

Berg, H. W., Akiyoski, M., and Amerine, M. A. 1979. Potassium and sodium content of California wines. *Am. J. Enol. and Vitic.* 30(1):55–57.

Bergeret, J. 1963. Action de la gelatine et de la bentonite sur la couleur et l'astringence de quelques vins. *Ann. Technol. Agri. Paris* 12:15–25.

Bertrand, G. L., Carroll, W. R., and Foltyn, E. M. 1978. Tartrate stability of wines. I. Potassium complexes with pigments. *Am. J. Enol. and Vitic.* 29:25–29.

Bidan, P. and Andre, L. 1958. Sur la composition en acides amines de quelques vins. *Ann. Technol. Agr.* 7:403–432.

Bioletti, F. T. and Cruess, W. V. 1912. Enological investigations. *California Agr. Expt. Station Bull.* 230:1–118.

Blackburn, D. 1984. Malolactic bacteria: Research and application. *The Practical Winery. 5(3)*:44–48.

Blackburn, D. 1988. Use of sulfur dioxide. *The Practical Winery and Vineyard.* May/June:30–31.

Blouin, G., Guimberteau, G., and Audouit, P. 1982. Prevention des precipitations tartriques dans le vin par le procede contact. *Connaissance Vigne Vin* 16(1):63–67.

Boehringer Mannheim GMBH. 1980. Acetic Acid. UV-method for the determination of acetic acid in foodstuffs. Indianapolis, IN.

Bohringer, P. 1960. Technologie der hefen. In *Die Hefen,* pp. 157–270. Nurenberg: Verlag.

Bonastre, J. 1959. Contribution a l'etude des matieres minerales dans les produits vegetaux. Application au vin. *Ann. Technol. Agr.* 8:377–446.

Boulton, R. 1980a. The relationship between total acidity, titratable acidity and pH in wines. *Am. J. Enol. and Vitic.* 31:76–80.

Boulton, R. 1980b. The general relationship between potassium, sodium, and pH in grape juices and wines. *Am. J. Enol. and Vitic.* 31:182–186.

Boulton, R. 1980c. A hypothesis for the presence, activity and role of potassium/hydrogen adenosine triphosphatases in grapevines. *Am. J. Enol. and Vitic.* 31:283–287.

Boulton, R. 1980d. The relationship between total acidity, titratable acidity and pH in grape tissue. *Vitis* 19:113–120.

Boulton, R. 1980e. The nature of wine proteins. In *Proc. of the Sixth Wine Industry Tech. Symposium,* Santa Rosa, CA, WITS, J. L. Jacobs, editor pp 67–70.

Boulton, R. 1981. Advances in viticulture and enology. University of California Extension, 812E28.

Boulton, R. 1982. Total acidity, titratable acidity and pH for winemakers and grape growers. In *Proc. of the 1982 UCD Wine-Grape Day.*

Boulton, R. 1983. The conductivity method for evaluating the potassium bitartrate stability of wine, Part I. *Enology Briefs, 2(1)*:1–3. University of California Extension, G. Cooke, Ed., Davis, CA.

Boulton, R. 1985. Potassium Balance in Grapevines. *Practical Vineyard and Winery 5(5)*:40–41.

Bousbouras, G. E. and Kunkee, R. E. 1971. The effect of pH on malolactic fermentation in wine. *Am. J. Enol. and Vitic.* 22(3):121–126.

Braverman, J. B. S. 1953. The mechanism of interaction of sulfur dioxide with certain sugars. *J. Sci. Food Agr.* 4:540–547.

Braverman, J. B. S. 1963. *Introduction to the Biochemistry of Foods.* Amsterdam: Elsevier Publishing Co.

Brechot, P., Chauvet, J., Croson, M., and Irrman, R. 1966. Configuration optique de l'acide lactique apparu au cours de la fermentation malolactique pendant la vinification. *Comp. Rend. C* 262:1605–1607.

Brechot, P., Chauvet, J., Dupuy, P., Croson, M., and Rabatu, A. 1971. Acide oleanoique facteur de aroissance anaerobic de la levure du vin. *Cr. Acad. Sci.* 272:890–893.

Brenner, M. W., Owades, J. L., Gutcho, M., and Golyzniak, R. 1954. Determination of volatile sulfur compounds III. Determination of mercaptans. *Proc. Am. Soc. Brew. Chem.* 88–97.

Broome, G., Payne, M., and Scipione, J. 1985. Ultrafiltration physical stability of wine. *Proc. of the Australian Wine Research Institute,* Urrbrae, South Australia, Australian Industrial Publishers.

Buchanan, R. E. and Gibbons, N. E. 1974. *Bergey's Manual of Determinative Bacteriology,* 8th ed. Baltimore, MD: The Williams and Wilkins Co.

Buchanan, R. E. and Gibbons, N. E. 1984. *Bergey's Manual of Systematic Bacteriology,* 9th ed., ed. J. G. Holt. Baltimore, MD: The Williams and Wilkins Co.

Buechsenstein, J. W. and Ough, C. S. 1978. Sulfur dioxide determination by aeration-oxidation: a comparison with Ripper. *Am. J. Enol. and Vitic.* 29(3):161–164.

Burroughs, L. F. 1975. Determining free sulfur dioxide in red wines. *Am. J. Enol. and Vitic.* 26:25–29.

Burroughs, L. F. and Whiting, G. C. 1960. The sulfur dioxide combining power of cider. Ann. Report Agr. Hort. Res. Station, Long Ashton, Bristol, England, pp. 144–147.

California Department of Food and Agriculture. 1986. Wine grape inspection report.

Cano-Marota, C. R. and Ares Pons, J. 1961. Study of the malolactic fermentation in wines of Uruguay. *Rev. Soc. Quim. Mex.* 5:168.

Cantarelli, C. 1963. Prevention des precipitations tartriques. *Ann. Technol. Agr.* 12:343–357.

Capt, E. and Hammel, G. 1953. Le traitement des vins par l'acide carbonique. *Rev. Romande Agr., Vitricul., Arboricult.* 9:41–43.

Capucci, C. 1948. Observazone sulle prababite cause della unconsueta acidita volatile di alcuni vini delta regione Emiliano-Romagnola. *Riv. Viticolt. e Enol. (Conegliano)* 1:386–388.

Caputi, A., Jr. and Peterson, R. G. 1966. The browning problem of wines. *Am. J. Enol. and Vitic.* 16:9–13.

Caputi, A., Jr. and Ueda, M. 1967. The determination of copper and iron in wine by atomic absorption spectrophotometry. *Am. J. Enol. and Vitic.* 18:66–70.

Caputi, A., Jr. and Walker, D. R. 1987. Titrimetric determination of carbon dioxide in wine: collaborative study. *J.A.O.A.C.* 70(6):1060–1062.

Caputi, A., Jr., Ueda, M., and Brown, T. 1968. Spectrophotometric determination of ethanol in wine. *Am. J. Enol. and Vitic.* 19:160–165.

Caputi, A., Jr., Ueda, M., and Trombella, B. 1974. Determination of sorbic acid in wine. *J.A.O.A.C.* 57:951–953.

Carter, H. E. 1950. *Nitrogen Compounds of Yeasts. Proc. of the Symposium.* Champaign, IL: Garrard Press, pp. 5–8.

Carter, G. H., Nagel, C. W., and Clore, W. J. 1972. Grape sample preparation methods representative of must and wine analysis. *Am. J. Enol. Vitic.* 23(1):10–18.

Cartesio, M. S. and Campos, T. R. 1988. Malolactic fermentation in wine: improvement in paper chromatographic techniques. *Am. J. Enol. and Vitic.* 39(2):188–189.

Castelli, T. 1965. Ruolo della microbiologis nell'enolgia di oggi ed in quella di domani. *Atti. Accad. Ital. Vite Vino* 17:3–13.

Castino, M. 1965. L'azione riducente dell'acido ascorbico nella demetallizzazione dei vini con ferrocianuro potassico. *Ann. Accad. Ital. Viti. Vino* 17:143–151.

Castino, M., Usseglio-Tomasset, and Gandini, A. 1975. Factors Which Affect the Spontaneous Initiation of the Malo-lactic Fermentation in Wines, The Possibility of Transmission by Inoculation and its effect on Organopletic Properties. *In: Lactic Acid Bacteria in Beverages and Foods.* J. G. Carr, C. V. Cutting, and G. C. Whiting (Eds.) pp. 139–148, Academic Press, London.

Cheidelin, V. H. and King, T. E. 1953. Nutrition of microorganisms. *Ann. Rev. Microbiol.* 7:113–142.

Chlebek, R. W. and Lister, M. W. 1966. Ion pair effects in the reaction between potassium ferrocyanide and potassium persulfate. *Can. J. Chem.* 44:437–442.

Chow, H., and Gump, B. H. 1987. Phosphorus in wine: Comparison of atomic absorption spectrometry methods. *J. Assoc. Off. Anal. Chem.* 70(1):61–63.

Clark, W. G. and Geissman, T. A. 1949. Potentiation of effect of adrenaline by flavonoid (vitamin p-like) compounds. *J. Pharmacol. Expt. Therapy* 95:363.

Clark, J. P., Gump, B. H. and Fugelsang, K. C. 1985. Tartrate precipitation and stability. *Proc. of the Eleventh Wine Industry Tech. Symposium.* Wine Ind. Tech. Seminar. J. L. Jacobs (ed) Santa Rosa, CA.

Clark, J. P., Fugelsang, K. C., and Gump, B. H. 1988. Factors affecting induced calcium tartrate precipitation from wine. *Am. J. Enol. and Vitic.* 39(2):155–161.

Clementi, F. and Rossi, J. 1984. Research Note: Effect of drying and storage conditions on survival of *Leuconostoc oenos. Am. Soc. Enol. Vitic.* 35(3):183–186.

Cofran, D. R., and Meyer, J. 1970. The effect of fumaric acid on malolactic fermentation. *Am. J. Enol. Vitic.* 21:189–192.

Cone, C. 1987. Personal communication.

Cooke, G. M. and Berg, H. W. 1984. A re-examination of varietal table wine processing practices in California. II. Clarification, stabilization, aging and bottling. *Am. J. Enol. and Vitic.* 35(3):137–142.

Cootes, R. L. 1983. Grape juice aroma and grape quality assessment used in vineyard classification. *Proc. of the Fifth Australian Wine Industry Tech. Conf.*, pp. 275–292. Adelaid, Australia Australian Industrial Publisher

Cootes, R. L., Wall, P. J. and Nettelback, R. 1981. Grape quality assessment grape quality: Assessment from vineyard to juice preparation. *Australian Society of Viticulture and Oenology. Inc.* pp. 39–56. Glen Osmond; South Australia.

Cordonnier, R. 1966. Etude des proteines et des substances azotees. Leur evolution au cours des traitements oenologiques. Conditions de la stabilite proteique des vins. *Bull. O.I.V. (Office Intern. Vigne Vin)* 39:1475–1489.

Corison, C. A., Ough, C. S., Berg, H., and Nelson, K. E. 1979. Must, acetic acid and ethyl acetate as mold and rot indicators in grapes. *Am. J. Enol. and Vitic.* 30(2):130–134.

Costello, P. J., Morrison, G. L., Lee, T. H., and Fleet, G. H. 1983. Numbers and species of lactic acid bacteria in wines during vinification. *Food Tech. Australia.* 35:14–18.

Cowper, E. P. 1987. Volatile acidity in high Brix fermentations. Masters thesis, California State University, Fresno.

Crawford, C. 1951. Calcium in dessert wines. *Proc. Am. Society Enol.* 76–69.

Crowell, E. A. and Guymon, J. F. 1975. Wine constituents arising from sorbic acid addition and identification of 2-ethoxyhexa-3,5-diene as a source of geranium-like odor. *Am. J. Enol. and Vitic.* 26:96–102.

Crowell, E. A., Ough, C. S., and Bakalinsky, A. 1985. Determination of alpha amino nitrogen in musts and wines by TNBS method. *Am. J. Enol. and Vitic.* 36(2):175–177.

Cruess, W. V. 1918. The fermentation organisms of California grapes. *Univ. California Publ. Agric. Sci.* 4(1):1-66.

Dadic, M., Belleau, B. 1973. Polyphenolics and beer flavor. *Proc. Amer. Soc. Brew. Chem.*, p. 107-114.

Daudt, C. E., and Ough, C. S. 1973. A method for quantitative measurement of volatile acetate esters for wine. *Am. J. Enol. and Vitic.* 24(3):125-129.

Davis, C. R., Wibow, D., Eschenbruch, R., Lee, T. H., and Fleet, G. H. 1985. Practical implications of the malolactic fermentation: a review. *Am. J. Enol. and Vitic.* 36:290-301.

Davis, C. R., Wibowo, D., Fleet, G. H., and Lee, T. H. 1988. Properties of wine lactic bacteria: Their potential enological significance. *Am. J. Enol. Vit.* 39(2):137-142.

De Ley, J. 1958. Studies on the metabolism of *Acetobacter peroxydans*. Part I. General properties and taxonomic position of the species. *Antonie van Leeuwenhoek* 24:281-297.

De Ley, J. 1961. Comparative carbohydrate metabolism and a proposal for a phylogenetic relationship of the acetic acid bacteria. *J. Gen. Microbiol.* 24:31-50.

De Ley, J. and Dochy, R. 1960. On the localization of oxidase systems in *Acetobacter* cells. *Biochim. Biophys. Acta* 40:277-289.

De Ley, J. and Schell, J. 1959. Oxiation of several substrates by *Acetobacter aceti. J. Bacteriol.* 77:445-451.

DeMan, J. M. 1976. *Principles of Food Chemistry.* Westport, CT: AVI Publishing Co.

De Mora, S. J., Knowles, S. J., Eschenbruch R. and W. J. Torrey. 1987. Dimethyl sulfide in some Australian red wines. *Vitis* 26:79-84.

De Rosa, T., Margheri, G., Moret, I., Scarponi, G., and Versini, G. 1983. Sorbic acid as a preservative in sparkling wine. Its efficacy and adverse flavor effect associated with ethyl sorbate formation. *Am. J. Enol. and Vitic.* 34(2):98-102.

DeSoto, R. and Warkentine, H. 1955. Influence of pH and total acidity on calcium tolerances of sherry wine. *Food Res.* 20:301-309.

DeSoto, R. and Yamada, H. 1963. Relationship of solubility products to long range tartrate stability. *Am. J. Enol. and Vitic.* 14:43-51.

Desrosier, N. W. 1963. *The Technology of Food Preservation,* rev. ed. Westport CT: AVI Publishing Co.

Desrosier, N. W., and Desrosier, J. N. 1977. *The Technology of Food Preservation.* AVI Publishing Co., Inc. Westport, Conn.

DeVilliers, J. P. 1961. The control of browning in white table wines. *Am. J. Enol. and Vitic.* 12:25-30.

Difco Laboratories, 1977. *Manual of Dehydrated Culture Media and Reagents for Microbiological and Clinical Laboratory Procedure,* 9th ed. Detroit, MI.

Dimitriadis, E. and Bruer, D. R. G. 1984. Using the Markham still for the assay of terpene flavorants in grapes. *Australian Grape Grower and Winemaker* (Apr.):61-64.

Dimitriadis, E. and Williams, P. J. 1984. The development and use of a rapid analytical technique for estimation of free and potentially volatile monoterpene flavorants of grapes. *Am. J. Enol. and Vitic.* 35(2):66-71.

Dittrich, H. H. 1979. Anwendung von Trockenhefen bei der Weinbereitung. *Dtsch. Weinbau* 14:792-796.

Dittrich, H. H. and Kerner, E. 1964. Diacetyl als Weinfehler, Ursache und Beseitigung des Milchsauretones. *Wein-Wissen* 19:525-538.

Doelle, H. W. 1975. *Bacterial Metabolism,* 2nd ed. New York: Academic Press.

Drawert, F., Heimann, W., Emberger, R., and Tressl, R. 1966. Uber die Biogenese von Aromastoffen bei Pflanzen und Fruchten. II. Enzymatische Bildung von Hexen-2-al-1, Hexanal und deren Vorstufen. *Liebigs Ann. Chem.*, *694*:200.

Drawert, F., Tressl, R., Heimann, W., Emberger, R., and Speck, M. 1973. Uber die Biogenese von Aromastoffen bei Pflanzen und Fruchten. XV. Enzymatische-oxydative Bildung von C-6-Aldehyden und Alkoholen und deren Vorstufen bei Apfeln und Trauben. *Chem. Mikrobiol. Technol. Lebensm.* *2*:10-14.

Drysdale, G. S. and Fleet, G. H. 1985. Acetic acid bacteria in some Australian wines. *Food Tech. Aust.* 37:17–20.

Drysdale, G. S., and Fleet, G. H. 1988. Acetic acid bacteria in winemaking: a review. *Am. J. Enol. Vitic.* 39(2):143–54.

Drysdale, G. S., and Fleet, G. H. 1989. The growth and survival of acetic acid bacteria in wines at different concentrations of oxygen. *Am. J. Enol. Vitic.* 40(2):99–105.

Dubourdieu, D., Grassin, C., Deruche, C., and Ribereau-Gayon P. 1984. Mise au point d'une mesure rapide de l'activite laccase, dans les mouts et dans les vins par methode a la syringaldazine. Application a l'appreciation de l'etat sanitaire des vendanges. *Connaissance Vigne Vin* 18, N.4, 237–252.

Dunsford, P. 1979. The kinetics of potassium bitartrate crystallization from wine. Master's thesis. University of California, Davis.

Dunsford, P. and Boulton, R. 1981a. The kinetics of potassium bitartrate crystallization from table wines. I. Effect of particle size, particle surface area, and agitation. *Am. J. Enol. and Vitic.* 32:100–105.

Dunsford, P. and Boulton, R. 1981b. The kinetics of potassium bitartrate crystallization from table wines. II. Effect of temperatures and cultivar. *Am. J. Enol. and Vitic.* 32:106–110.

DuPlessis, C. S., and deWet, P. 1968. Browning in White Wines. I. Time and temperature effects upon tannin and leucoanthocyanidion uptake by musts from seeds and bushes. *S. African J. Agric. Sci.* 11:459–68.

DuPlessis, C. S. and A. L. Uys 1968. Browning in white wines. II. The effect of cultivar, fermentation, husk, seed and stem contact upon browning. *S. Afr. J. Agr. Sci.* 11:637–640.

Dupuy, P. 1957a. Les facteurs du development de l'acescence dans le vin. *Ann. Technol. Agr.* 6:391–407.

Dupuy, P. 1957b. Les *Acetobacter* du vin identification de quelques souches. *Ann. Technol. Agr.* 6:217–233.

Dupuy, P. and Maugenet, J. 1962. Oxidation de l'ethanol par *Acetobacter rancens. Ann. Technol. Agric. 11*:219-25.

Dupuy, P. and Maugenet, J. 1963. Metabolism de l'acid lactique par *Acetobacter rancens. Ann. Technol. Agr.* 12:5–14.

Dupuy, P., Nortz, M., and Puisais, J. 1955. Le vin et quelques causes de son enrichissement en fer. *Ann. Technol. Agr.* 4:101–112.

Edinger, W. D., and Splittstoesser, D. F. 1986. Production by lactic acid bacteria of sorbic alcohol, the precursor of the geranium odor compound. *Am. J. Enol. Vitic.* 37:34–38.

English, J. and Cassidy, H. G. 1956. *Principles of Organic Chemistry. An Introductory text in Oreganic Chemistry.* McGraw Hill Publish, New York.

Esau, P. 1967. Pentoses in wine. I. Survey of possible sources. *Am. J. Enol. and Vitic.* 18(4):210–216.

Esau, P. and Amerine, M. A. 1964. Residual sugar in wine. *Am. J. Enol. and Vitic.* 15:187–189.

Esau, P. and Amerine, M. A. 1966. Quantitative estimation of residual sugar in wine. *Am. J. Enol. and Vitic.* 17:265–267.

Eschenbruch, R. 1974. Sulfite and sulfide formation during winemaking—a review. *Am. J. Enol. and Vitic.* 25(3):157–161.

Eschenbruch, R., 1983. H_2S Formation—The Continuing Problem During Winemaking Fermentation Technology. *Austrailian Society of Viticulture and Oenology Proceedings.* T.H. Lee Editor, pp. 79–87. Glen Osmond. SA.

Eschenbruch, R., Bonish, P., and Fisher, B. M. 1978. Production of hydrogen sulfide by pure culture wine yeasts. *Vitis* 17:67–74.

Eschenbruch, R. and Dittrich, H. H. 1986. Metabolism of acetic acid bacteria in relation to their importance in wine quality. *Instit. Mikrobiol. Biochem., Forschungsanstalt Geisenheim, Postfach. 1154, D-2222. Geisenheim.*

Eschenbruch, R. and. P. H. Kleynhans 1974. The influence of copper containing fungicides on the copper content of grapes juice and hydrogen sulfide formation. *Vitis* 12:320–324.

Ethiraj, S. and E. R. Suresh 1988. *Pichia Membranafaciens:* a Benzoate Resistant Yeast from Spoiled Mango Pulp. *J. Food Sci. and Tech.* 25(2):63–66.

Ewart, A. J. 1984. A study of cold stability in Australian white table wines. *Australian Grape Grower and Winemaker* (Apr:)104–107.

Fairley, J. L. and Kilgour, G. L. 1966. *Essentials of Biological Chemistry,* 2nd ed. New York: Reinhold Publishing Co.

FEMS Microbiological Letter, Feb. 1988. Abstracted in: *The Australian Wine Res Inst Tech Review #52:* T. Lee Director

Ferenczi, S. 1966. Etude des proteines et des substances azotees. Leur evolution au cours des traitements oenologiques. Conditions de la stabilite proteique des vins. *Bull. O.I.V.* 39:1311–1336.

Fessler, J. H. 1961. Erythorbic acid and ascorbic acid as antioxidants in bottled wines. *Am. J. Enol. and Vitic.* 12(1):20–24.

Flanzy, M. and Dubois, P. 1964. Etude du dosage des tocopherols totaux. Application a l'huile de pepins de raisin. *Ann. Technol. Agr.* 13:67–75.

Flanzy, C., and Poux, C. 1960. Application spectrophotometrique a l'etude de la casse oxydasique des vins. *Compt. Rend.* 251(18):1910–11.

Flanzy, C., Poux, C., and Flanzy, M. 1964. Les levures alcooliques dans les vins. Proteolyse et proteogenese. *Ann. Technol. Agr.* 13:283–300.

Fleet, G. H. 1984. The physiology and metabolism of wine lactic acid bacteria. *Malolactic Fermentation.* Australian Society of Viticulture and Oenology, Inc.

Florenzano, G. 1949. La microflora blastomiceta de mostre e dei vini di alcune zona Toscane. *Ann. Sper. Agrar.* 3:887–918.

Fornachon, J. C. M. 1943. *Bacterial Spoilage of Fortified Wines.* Adelaide: Australian Wine Board.

Fornachon, J. C. M. 1957. The occurrence of malo-lactic fermentation in Australian wines. *Aust. J. Applied Sci.* 8:120–129.

Fornachon, J. C. M. 1964. A *Leuconostoc* causing malo-lactic fermentation in Australian wines. *Am. J. Enol. and Vitic.* 15:184–186.

Fornachon, J. C. M. 1965. Sulfur dioxide in winemaking. *Australian Wine Brew. Spirits Rev.* (May):20–26.

Fornachon, J. C. M. 1968. Influence of different yeasts on the growth of lactic bacteria in wine. *J. Sci. Food Agric.* 19:374–378.

Freeman, B. M. 1982. Regulation of potassium in grape vines (*Vitis vinifera*). Ph.D. thesis, University of California, Davis.

Freeman, B. M. 1983. Effects of irrigation and pruning of Shiraz grapevines on subsequent wine pigments. *Am. J. Enol. and Vitic.* 34:23–26.

Freeman, B. M. and Kliewer, W. M. 1983. Effect of irrigation, crop level and potassium fertilization on Carnignane vines. II. Grape and wine quality. *Am. J. Enol. and Vitic.* 34:197–207.

Fritz, J. S. and Schenk, G. H., Jr. 1974. *Quantitative Analytical Chemistry,* 3rd ed. Boston: Allyn and Bacon, Inc.

Fugelsang, K. C. 1974. Unpublished.

Fugelsang, K. C. 1987. Utilization of yeast starters in winemaking. *Practical Winery and Vineyard* (July-Aug.):28–32.

Fumi, M. D., Trioli, G., Colombi, M. G., and Colagrande, O. 1988. Immobilization of *Saccharomyces cerevisiae* in calcium alginate gel and its application to bottle-fermented sparkling wine production. *Am. J. Enol. Vitic.* 39(4):267–72.

Gahagan, R. 1988. Personal communication.

Gallander, J. and Stetson, J. 1976. Wine deacidification with mixed cultures of *Schizosaccharomyces* and *Saccharomyces cerevisiae.*

Gancedo, C., Gancedo, J. M., and Sols, A. 1968. Glycerol metabolism in yeasts. Pathways of utilization and production. *Eur. J. Biochem.* 5:165–172.

Gehman, H. and Osman, E. M. 1954. The chemistry of the sugar–sulfite reaction and its relationship to food problems. In *Advances in Food Research* 5:53–91. New York: Academic Press.

General Anilin and Film Corp. (GAF) 1975. Polyclar AT stabilizer in winemaking. Bulletin 9653-003 New York.

Gestrelium, S. 1982. Potential application of immobilized viable cells in the food fndustry: Malolactic fermentation of wine. *Enzyme Eng.* 6:245–250.

Gettler, A. O. and Goldbaum, L. 1947. Detection and estimation of microquantities of cyanide. *Anal. Chem.* 19:270–271.

Gilbert, E. 1976. Uberlegungen Zur Berechnung und Beurteilung des Restexraktgehaltes bei wein. *Die Wein Wirtschoft* 112(6):118–127.

Gillis, M., 1978. Intra- and intergeneric similarities of the rRNA cistrons of *Acetobacter* and *Gluconobacter. Antonie van Leeuwenhoek* 44:117–118.

Gillis, M. and De Ley, J. 1980. Intra- and intergeneric similarities of the ribosomal RNA cistrons of *Acetobacter* and *Gluconobacter. Int. J. Syst. Bacteriol.* 30:7–27.

Gini, B. and Vaughn, R. H. 1962. Characteristics of some bacteria associated with spoilage of California dessert wines. *Am. J. Enol. and Vitic.* 13:20–31.

Giudici, P., and Guerzoni, M. 1982. Sterol content as a character for selecting wine strains in Enology. *Vitis* 21:5–14.

Gnaegi, F. and Sozzi, T. 1983. Les bacteriophages de *Leuconostoc oenos* et leur importance oenologique. *Bulletin O.I.V. 56*:352–357.

Gnekow, B. and Ough, C. S. 1976. Methanol in wines and must: source and amounts. *Am. J. Enol. and Vitic.* 27(1):1–6.

Gomes, Marques J. V. 1973. Limitation de l'emploi de l'anhydride sulfureux compte tenu des beseins en matieres nutritives des levures pendant la fermentation. *Bull. O.I.V.* 46:316–329.

Goniak, O. J. and Noble, A. C. 1987. Sensory study of selected volatile sulfur compounds in white wine. *Am. J. Enol. and Vitic.* 38:223–227.

Gordon and Stewart, 1972. Effect of lipid status on cytoplasmic and mitochondrial protein synthesis in anaerobic cultures of *Saccharomyces cerevisiae. J. Gen. Microbiol.* 72:231–42.

Gorinstein, S., A. Goldblum, S. Kitov, J. Deutsch, C. Loinger, S. Cohen, H. Tabakman, A. Stiller and A. Zykerman. 1984. The relationships between metals, polyphenols nitrogenous substances and the treatment of red and white wines. *Am. J. Enol. Vitic.*, 35(1):9–15.

Gorinstein, S., Goldblum, A., Kitov, S., and Deutsch, J. 1984 Fermentation and post-fermentation changes in Israeli wines. *J. Food Sci.* 49(1):251–56.

Gottschalk, A. 1946. The mechanism of selective fermentation of d-fructose from invert sugar by sauternes yeast. *Biochem. J.* 40:621–626.

Gottschalk, A. 1947. The effect of temperature on fermentation of d-mannose by yeast. *Biochem. J.* 41:276–280.

Greenfield, S. and Claus, G. W. 1972. Nonfunctional tricarboxylic acid cycle and the mechanism of glutamate biosynthesis in *Acetobacter suboxydans. J. Bacteriol.* 112:1295–1301.

Greenshields, R. N. 1978. Acetic acid: vinegar. In *Primary Products of Metabolism, Economic Microbiology,* Vol. 2, ed. A. H. Rose, pp. 121–186. Academic Press London and New York.

Groat, M. and Ough, C. S. 1978. Effect of insoluble solids added to clarified musts on fermentation rate, wine composition, and wine quality. *Am. J. Enol. Vitic.* 29:112–119.

Guerzoni, M. E., Mattioli, R., and Giudici, P. 1981. Acetic acid degradation in wines by film-forming yeasts. Criteria for selection *Vigne et Vini, 8(11)*:43–7.

Guimberteau, G., Duboudreiu, D., Serrano, M., and Lefebure, A. 1981. Clarification et mise en bouteilles. In *Actualities oenological et viticoles,* eds. P. Ribereau-Gayon and P. Sudraud. Paris: Dunod.

Gump, B. H., and Kupina, S. A. 1979. "Analysis of Gluconic Acid in Botrytized Wine." In Liquid Chromatographic Analysis of Foods and Beverages, Vol 2. Academic Press, Inc. New York. pp. 331–52.

Gump, B. H., Saguandeekul, S., Murray, G., and Villar, J. T. 1985. Determination of malic acid in wines by gas chromatography. *Am. J. Enol. Vitic.* 36:248–51.

Handbook of Chemistry and Physics, 56th ed. 1975. Cleveland, OH: The Chemical Rubber Company Press.

Hahn, G. D. and Possman, P. 1977. Colloidal silicon dioxide as a fining agent for wine. *Am. J. Enol. and Vitic.* 28:108–112.

Hale, C. R. 1977. Relationship between potassium and the malate and tartrate content of grape berries. *Vitis* 16:9–19.

Harrison, J. J. and Graham, J. C. 1970. Yeasts in Distillary Practice. *In The Yeasts. Vol. III.* pp. 283–348. A. H. Rose and J. S. Harrison (ed).

Hathaway, D. E. and Seekins, J. W. 1957. Enzymic oxidation of catechin to a polymer structurally related to some phlobatannins. *Biochem. J.* 67:239–245.

Hauge, J. G., King, T. E., and Cheldelin, V. E. 1955. Oxidation of dihydroxyacetone via the pentose cycle in *Acetobacter suboxydans. J. Biol. Chem.* 214:11–26.

Hawker, J. S., Ruffner, H. P., and Walker, R. R. 1976. The sucrose content of some Australian grapes. *Am. J. Enol. and Vitic.* 27:125–129.

Heatherbell, D. A., Nagaba, P., Fombin, J., Watson, B., Garcia, Z., Flores, J., and Hsu, J. 1984. Recent developments in the application of ultrafiltration and protease enzymes to grape juice and wine processing. In *Proceedings of the International Symposium on Cool Climate Viticulture and Enology,* eds. D. A. Heatherbell, P. B. Lombard, F. W. Bodyfelt, and S. F. Price, pp. 418–445. Corvallis, OR: Oregon State Agricultural Experimental Station, Technical Publication No. 7628.

Heintze, K., 1976. *Ind. Obst. Gemueseverwert 61*:555–56 (referenced from Amerine and Ough, 1974). ''Wine and Must Analysis.'' NY: J. Wiley & Sons, Inc.

Hendrickson, J. B., Gram, D. J., and Hammond, G. S. 1970. *Organic Chemistry,* 3rd ed. New York: McGraw-Hill.

Hennig, K. 1958. Das chemische bild des Mostes und Weines. *Weinfach Kalender* 194–209.

Hennig, K. and Burkhardt, R. 1960. Detection of phenolic compounds and hydroxyacids in grapes, wines, and similar beverages. *Am. J. Enol. and Vitic.* 11:64–79.

Hill, G. and Caputi, A., Jr. 1970. Colorimetric determination of tartaric acid in wine. *Am. J. Enol. and Vitic.* 21:153–161.

Hodges, T. K. 1976. ATPases associated with membranes of plant cells. In *Encyclopedia of Plant Physiology,* Vol. 2A, eds. U. Luttge and M. G. Pitman, pp. 260–283. Berlin: Springer Verlag

Holzer, H., Bernhard, W., and Schneides, S. 1963. Zur Glycerinhldung in Backerhofe. *Biochem,* Z. 336(6):495–499.

Hood, A. V. 1984. Possible factors affecting SO_2 inhibition of lactic acid bacteria in newly-fermented wines. *Malolactic Fermentation.* Australian Society of Viticulture and Oenology Inc.

Horowitz, W. 1975. *Official Methods of Analysis of the Association of Official Analytical Chemists,* 12th ed. Washington, DC.

Hossack, J. A. and Rose, A. H. 1976. Fragility of the plasma membranes in *Saccharomyces cerevisiae* enriched with different sterols. *J. Bacteriol.* 127:67–75.

Hoynak, P. X. and Bollenback, G. N. 1966. *This is Liquid Sugar,* 2nd ed. New York: Refined Syrups and Sugars, Inc.

Hsia, C. L., Planck, R. W., and Nagel, C. W. 1975. Influence of must processing on iron and copper content of experimental wines. *Am. J. Enol. and Vitic.* 26:57–61.

Hsu, J. C. and Heatherbell, D. A. 1987. Heat-unstable proteins in wine. I. Characterization and removal by bentonite fining and heat treatment. *Am. J. Enol. and Vitic.* 38(1):6–10.

Hsu, J. C., Heatherbell, D. A., Flores, J. H., and Watson, B. T. 1987. Heat-unstable proteins in grape juice and wine. II. Characterization and removal by ultrafiltration. *Am. J. Enol. and Vitic.* 38:17–22.

Hubach, C. E. 1948. Detection of cyanides and ferrocyanides in wines. *Anal. Chem.* 20:115–16.

Hunter, K., and Rose, A. H. 1971. Yeast lipids and membranes. *In: The Yeasts, Vol II.* A. H. Rose and J. H. Harrison (eds). Academic Press, London.

Ingledew, W. M. and Kunkee, R. E. 1985. Factors influencing sluggish fermentations of grape juice. *Am. J. Enol. and Vitic.* 36:65–76.

Ingledew, W. M., Magnus, C. A., and Patterson, J. R. 1987. Yeast foods and ethyl carbamate formation in wine. *Am. J. Enol. and Vitic.* 38(4):332–335.

Ingraham, J. L. and Cooke, G. M. 1960a. A survey of the incidence of malo-lactic fermentation in California table wines. *Am. J. Enol. and Vitic.* 11:160–163.

Ingraham, J. L., Vaughn, R. H., and Cooke, G. M. 1960b. Studies on the malo-lactic organisms isolated from California wines. *Am. J. Enol. and Vitic.* 11:1–4.

Jakob, L. 1962. Bentotest. *Das Weinblatt* (No. 34/35, Sept.).

James, T. H. and Weissberger, A. 1939. Oxidation processes: XIII. The inhibitory action of sulfite and other compounds in auto-oxidation of hydroquinone and its homologs. *J. Am. Chem. Soc.* 61:442–450.

Jay, J. M. 1970. *Modern Food Microbiology.* New York: Van Nostrand Reinhold Co.

Johnson, T. and Nagel, C. W. 1976. Composition of central Washington grapes during maturation. *Am. J. Enol. and Vitic.* 27:15–20.

Jones, A. R., Dowling, E. J., and Skraba, W. J. 1953. Identification of some organic acids by paper chromatography. *Anal. Chem.* 3:35.

Jones, D. D. and Greenshields, R. N. 1971. *J. Inst. Brew.* 77:160.

Jones, R. S. and Ough, C. S. 1985. Variation in the percent ethanol (v/v) per Brix conversions of wines from different climatic regions. *Am. J. Enol. and Vitic.* 36(4):268–270.

Jordan, A. D. and Croser, B. J. 1983. Determination of grape maturity by aroma/flavor assessment. *Proc. of the Fifth Austrialian Wine Industry Technical Conference,* Adelaide, South Australia, Australian Industrial Publishers pp. 261–274.

Joslyn, M. A. 1950a. *Methods in Food Analysis.* New York: Academic Press, 528 pp.

Joslyn, M. A. 1950b. Hard chrome plating and avoidance of metal pick-up. *Wine and Vines* 31(4):67–69.

Joslyn, M. A. 1953. The theorectical aspects of clarification of wine by gelatin fining. *Proc. Am. Soc. Enol.* 4:39–68.

Joslyn, M. A. and Amerine, M. A. 1964. *Dessert, Appetizer and Related Flavored Wines; the Technology of Their Production.* Berkeley: Division of Agricultural Sciences, University of California, 482 pp.

Joslyn, M. A. and Braverman, J. B. S. 1954. The chemistry and technology of the pretreatment and preservation of fruit and vegetable products with sulfur dioxide and sulfites. In *Advances in Food Research* 5:97–160. New York: Academic Press.

Joslyn, M. A. and Dunn, R. 1938. Acid metabolism of wine yeasts. I: The relation of volatile acid formation to alcoholic fermentation. *J. Am. Chem. Soc.* 60:1137–1141.

Joslyn, M. A. and Goldstein, J. L. 1964. Astringency of fruit and fruit products in relation to phenolic content. In *Advances in Food Research* 13:179–217. New York: Academic Press.

Joyeux, A., Lafon-Lafourcade, S., and Ribereau-Gayon, P. 1984a. Evolution of acetic acid bacteria during fermentation and storage of wine. *Appl. Environ. Microbiol.* 48:153–156.

Joyeux, A., Lafon-Lafourcade, S., and Ribereau-Gayon, P. 1984b. Metabolism of acetic acid bacteria in grape must: consequences on alcoholic and malolactic fermentation. *Sciences Alim.* 4:247–55.

Jurd, L. 1964. Reactions involved in sulfite bleaching of anthocyanins. *J. Food Sci.* 29:16–19.

Jurd, L. 1969. Review of polyphenol condensation reactions and their possible occurrence in the aging of wines. *Am. J. Enol. Vitic.* 20:191–195.

Jurd, L. and Ansen, S. 1966. The formation of metal and "copigment" complexes of cyanidin-3-glucoside. *Phytochemistry* 5(6):1263–1271.

Kasimatis, A. 1984. Viticultural practices for varietal winemaking. University of California–Davis Extension Short Course Series.

Kean, C. E. 1954. Chemical composition of copper complexes causing cloudiness in various wines. Ph.D. thesis, University of California, Davis.

Kean, C. E. and Marsh, J. L. 1956a. Investigation of copper complexes causing cloudiness in wines. I: Chemical composition. *Food Res.* 21:441–447.

Kean, C. E. and Marsh, J. L. 1956b. Investigation of copper complexes causing cloudiness in wines. II: Bentonite treatment of wines. *Food Tech.* 10:355–359.

Kielhofer, E. 1941. Troubles albuminoides du vin. *Bull. Office Inter. Vin.* 16:7–10.

Kielhofer, E. 1960. Neue Erkenntnisse uber die Schweflige Saure im Wein und ihren Ersatz durch Ascorbinsaure. *Deut. Wein-Zkg.* 96:14–24.

Kielhofer, E. 1963. Etat et action de l'acide sulfureux dans les vins; regles de son emploi. *Ann. Technol. Agr.* 12:77–89.

Kielhofer, E. and Wurdig, G. 1960a. Die an Aldehyd gebundene Schweflige Saure im Wein. I. Acetaldehydbildung durch enzymatishce und nicht enzymatische Alkohol-Oxydation. *Weinberg Keller* 7:16–22.

Kielhofer, E. and Wurdig, G. 1960b. Die an aldehyd gebundene Schweflige Saure im Wein. II. Acetaldehydbildung bei der garung. *Weinberg Keller* 7:50–61.

King, S. W. and Beelman, R. B. 1986. Metabolic interactions between *Saccharomyces cerevisiae* and *Leuconostoc oenos* in a model grape juice/wine system. *Am. J. Enol. and Vitic.* 37(1):53–60.

King, T. E. and Cheldelin, V. H. 1952. Oxidations in *Acetobacter suboxydans. Biochim. Biophys. Acta* 14:108–116.

King, T. E., Kawasaki, E. H., and Cheldelin, V. H. 1956. Tricarboxylic acid cycle activity in *Acetobacter pasteurianum. J. Bacteriol.* 72:418–421.

Kliewer, W. M. 1967a. Glucose-fructose ratio of *Vitis vinifera* grapes. *Am. J. Enol. and Vitic.* 17:33–41.

Kliewer, W. M. 1967b. Concentration of tartrates, malates, glucose and fructose in fruits of the genus *Vitis. Am. J. Enol. and Vitic.* 18:87–96.

Kliewer W. M. and Benz, J. 1985. Personal Communication

Klingshirn, L. M., Liu, J. R., and Gallander, J. F. 1987. Higher alcohol formation in wines as it relates to particle size profiles of juice insoluble solids. *Am. J. Enol. and Vitic.* 38(3):207–210.

Koch, J. 1963. Proteines des vins blancs. Traitements des precipitations proteiques par chauffage et a l'aide de la bentonite. *Ann. Technol. Agr.* 12:297–313.

Kramling, T. E., and V. L. Singleton 1969. An estimate of the non-flavonoid phenols in wines. *Am. J. Enol. & Vitic.* 20:86–92.

Kraus, J. K., Scopp, R., and Chen, S. L. 1981. Effect of rehydration of dry wine yeast activity. *Am. J. Enol. and Vitic.* 32(2):132–134.

Kroll, T. 1963 *Weinburg U. Keller* 10:312.

Krumperman, P. H. and Vaughn, R. H. 1966. Some lactobacilli associated with decomposition of tartaric acid in wine. *Am. J. Enol. and Vitic.* 17(3):185–190.

Kunkee, R. E. 1967. Control of malo-lactic fermentation induced by *Leuconostoc citrovorum. Am. J. Enol. and Vitic.* 18:71–77.

Kunkee, R. E. 1968. Simplified chromatographic procedure for detection of malo-lactic fermentation. *Wine and Vines* 49(3):23–24.

Kunkee, R. E. and Amerine, M. A. 1970. Yeasts in winemaking. In *The Yeasts III,* eds. A. H. Rose and J. S. Harrison. New York: Academic Press.

Kunkee, R. E., and Neradt, F. 1974. A rapid method for detection of viable yeasts in bottled wine. *Wines & Vines* 55(12):36–39.

Kunkee, R. E., Ough, C. S., and Amerine, M. A. 1964. Induction of malo-lactic fermentation by inoculation of must and wine with bacteria. *Am. J. Enol. and Vitic.* 15:178–183.

Kupina, S. A. 1984. Simultaneous quantitation of glycerol, acetic acid and ethanol in grape juice by high performance liquid chromatography. *Am. J. Enol. and Vitic.* 35(2):59–62.

Labatut, E., Caruana, C., Bertrand, A., and Lafon-Lafourcade, S. 1984. Action des acides octanoique et decanoique a l'engard de quelquos levures en croissance dans le mout de raisin. *Institute d'Oenologie, Univ. de Bourdeaux II.*

Lafon-Lafourcade, S., Carre, E., Lanvaud-Funel, A. and Ribereau-Gayon, P. 1983. Induction de la fermentations malolactique des vins par inoculation d'une biomasse industrielle congelee de *L. oenos* apres reactuation. *Connaiss. Vigne Vin.* 17:55–71.

Lafon-Lafourcade, S. 1985. Role des microorganismes dans la formation de substances combinant le SO_2. Bull. OIV 652-53:590–604.

Lafon-Lafourcade, S., Carre, E., and Ribereau-Gayon, P. 1983. Occurrence of lactic acid bacteria at different stages of vinification and conservation of wine. *Appl. Environ. Microbiol.* 46:874-80.

Lafon-Lafourcade, S., Geneix, C., & Ribereau-Gayon, P. 1984. Inhibition of alcoholic fermentation of grape must by fatty acids produced by yeasts and their elimination by yeast ghosts. *Appl. Environ. Microbiol* 47:1246-1249.

Lafon-Lafourcade, S., and Ribereau-Gayon, P. 1976. Premiers observations sur l'utilisation des levures seches en vinification en blanc. *Conn. Vigne Vin* 10:277-92.

Larue, R. Lafon-Lafourcade, S., and Ribereau-Gayon, P. 1980. Relationship between the sterol content of yeast cells and their fermentation activity in grape must. *Appl. and Environ. Micro.* April:808-811.

Law, B. A. and Kolstad, J. 1983. Proteolytic systems in lactic acid bacteria. Antonie Van Leeuwenhoek 49:225-245.

Lea, A. G. H., and Arnold, G. M. 1978. The phenolics of ciders: Bitterness and astringency. *J. Sci. Food. Agric.* 29:478-483.

Lee, A. 1978. Bacteriophages Associates with Lactobacilli isolated from Wine. In *Proceedings of the 5th Oenol. Symposium Int. Assoc. Modern Winery Technol Mgt.* E. Lemperle & J. Frank (eds.) pp 287-295.

Lee, T. H. 1985. Protein instability: nature, characterization and removal by bentonite. Proceeding of the Austrialian Society of Viticulture and Oenology.

Lee, C. Y., Smith, N. L., and Nelson, R. R. 1979. Relationship between pectin methyl esterase activity and formation of methanol in concord grape juice. *Food Chem.* 4(2):143-148.

Lehninger, A. L., Reynafarje, B., and Alexandre, A. 1979. The stoichiometry of H^+ movements coupled to electron transport and ATP synthesis in mitochondria. In *Cation Flux Across Biomembranes.* eds. Y. Mukohata and L. Packer. pp. 243-54. New York, Academic Press.

Lewis, V. M., Esselen, W. B., and Fellers, C. R. 1949a. Nonenzymatic browning of foods. Production of carbon dioxide. *Ind. Eng. Chem.* 41:2587-2591.

Lewis, V. M., Esselen, W. B., and Fellers, C. R. 1949b. Nitrogen-free carboxylic acids in browning reactions. *Ind. Eng. Chem.* 41:2591-2594.

Lodder, J. 1970. *"The Yeasts: A Taxonomic Study,"* Second Edition. North Holland Publishing Co., Amsterdam

Long, Z. 1984a. Monitoring sugar per berry. *The Practical Winery and Vineyard* 5(2):52-54.

Long, Z. 1984b. Viticultural practices for varietal winemaking. *University of California Davis Extension Short Course.*

Long, Z. and Lindbloom, B. 1986. A report on Zelma Long's work at Semi. Juice oxidation experiment. *Wine and Vines* (Nov.):44-49.

Lucramet, V. 1981. Quelques Proprietes des Bacteries Lactiques. "Actualites Oenologiques et Viticoles." Ribereau-Gayon, P., Sudraud, P., eds. Universite de Bordeaux. Talence, France pp. 239-243.

Luthi, H. 1957. Symbiotic problems relating to the bacterial deterioration of wines. *Am. J. Enol. and Vitic.* 8:176-181.

Luthi, H. 1959. Microorganisms in noncitric juices. In *Advances in Food Research* 9:221-273. New York: Academic Press.

Machlis, L. and Torrey, J. 1956. *Plants in Action. A Laboratory Manual of Plant Physiology.* San Francisco: W. H. Freeman and Co.

Mangalith, P. Z. 1981. *Flavor Microbiology.* Springfield, IL: Charles C. Thomas.

Marsh, G. L. 1951. Calculation of proof gallon equivalents per ton of grapes. Wine Institute Technical Advisory Committee.

Marsh, G. L. 1952. New compound ends metal clouding. A report on the Fessler compound. *Wine and Vines* 33:19-21.

Marsh, G. L. and Guymon, J. F. 1959. Refrigeration in winemaking. *Am. Soc. Refrig. Eng. Data Book* Vol. 1, Chapter 10.

Mattick, L. R. 1983. A method for the extraction of grape berries used in total acid, potassium and individual acid analyses. Technical note. *Am. J. Enol. and Vitic.* 34(1):49.

Mayer, K. and Busch, I. 1963. Uber eine enzymatische Apfelsaurebestimmung in Wein und Traubensaft. *Mitt. Gebiete. Lebensm. Hyg.* 54:60–65.

McCloskey, L. P. 1974. Gluconic acid in California wines. *Am. J. Enol. and Vitic.* 25(4):198–201.

McCloskey, L. P. 1976. An acetic acid assay for wine using enzymes. *Am. J. Enol. and Vitic.* 27(4):176–180.

McCloskey, L. P. 1978. An enzymatic assay for glucose and fructose. *Am. J. Enol. and Vitic.* 29(3):226–227.

McCloskey, L. P. 1980. An improved enzymic assay for acetic acid in juice and wine. *Am. J. Enol. and Vitic.* 31(2):170–173.

McCord, J. D., Trousdale, E., and Ryu, D. D. Y. 1984. An improved sample preparation for the analysis of major organic components in grape must and wine by high performance liquid chromatography. *Am. J. Enol. and Vitic.* 35:28–29.

McLaren, A. D., Peterson, G. H., Barshad, I. 1958. The adsorption and reactions of enzymes and proteins on clay minerals. IV Kaolinite and Montmorillonite. *Soil Science Soc. of Am. Proceedings* 22:239–244.

McWilliams, D. J. and Ough, C. S. 1974. Measurement of ammonia in must and wine using a selective electrode. *Am. J. Enol. and Vitic.* 25(2):67–72.

Meidell, J. 1987. Unsuitability of fluorescence microscopy for rapid detection of small numbers of yeast cells on a membrane filter. *Am. J. Enol. Vitic.* 38(2):159–160.

Melamed, N. 1962. Determination des sucres residuels des vins. Leur relation avec la fermentation malolactique. *Ann. Technol. Agr.* 11:5–31.

Melnick, D. and Luckmann, F. H. 1954. Sorbic acid as a fungistatic agent in foods III. Spectrophotometric determination of sorbic acid in cheese and cheese wrappers. *Food Res.* 19:20–21.

Mesrob, B., Gorinova, N., and Tzakov, D. 1983. Characterization of the electrical properties and of the molecular weights of the proteins of white wines. *Nahrung* 27:727–733.

Miller, C. V. and Heilmann, A. S. 1952. Ascorbic acid and physiological breakdown in the fruits of the pineapple. *Science* 116:505.

Millies, K. 1975. Protein stabilization of wines using silica sol/gelatin fining. Mitteilungsblat der GDCH-Fachgruppe. *Lebensm. Gerich. Chemie* 29:50–53.

Minarik, E. and Nagyova, M. 1966. Mikroflora mustov a vin nitrianskej a podunajskej vinohradnickej oblasti. In *Prokrosky vo Vinohradnickom a Vinarskom Vyskume*, pp. 277–278. Bratislava: Vydavatel'stvo Slovenskej Akademie Vied.

Mobay. 1976. Baykisol 30. Wine and fruit juice fining with Baykisol 30. AC10022E.

Monk, P. R. 1986. Rehydration and propagation of active dry wine yeasts. *Australian Wine Ind. Journal* 1(1):3–5.

Monk, R. 1986. Rehydration and propagation of active dry wine yeast. *Australian Wine Industry J.* 3–5.

Moretti, R. H. and Berg, H. W. 1965. Variability among wines to protein clouding. *Am. J. Enol. and Vitic.* 16:69–78.

Morris, J. R., Sims, C. A., and Cawthon, D. L. 1983. Effects of excessive potassium levels on pH acidity and color of fresh and stored grape juice. *Am. J. Enol. and Vitic.* 34(1):35–39.

Muller-Spath, H., Moschtert, N., and Schafer, G. 1978. Observations in winemaking: the present state of the art. Seitz technical communication. Reprinted from *Die Weinwertschaft*, Vol. 36.

Munz, T. 1960. Die bildung des Ca-Doppelsalzes der Wein-und Apfelsaure die moglichkeiten seiner fallung durch $CaCO_3$ im Most. *Weinberg Keller* 7:239–247.

Munz, T. 1961. Methoden zur praktische Fallung der Wein-und Apfelsaure als Ca-Doppelsalz. *Weinberg Keller* 8:155–158.

Muraoka, H., Watabe, Y., Ogasawara, N., and Takahashi, H. 1983. Trigger damage by oxygen deficiency to the acid production system during submerged acetic acid fermentation with *Acetobacter aceti*. *J. Ferm. Tech.* 61:89–93.

Nagel, C. W. and Graber W. R. 1988. Effect of must oxidation on quality of white wines. *Am. J. Enol. Vitic.* 39(1):1–4.

Nelson, K. E. 1951. Effect of humidity on infection of table grapes by *Botrytis cinerea*. *Phytopathology* 41:859–864.

Neradt, F. 1977. A reliable new tartrate stabilization process. Presented at the Annual Meeting of the Am. Soc. for Enol. and Vitic.

Nel, L., Wingfield, B. D., Van der meer, L. S., and Van Vuuren, H. J. J. 1987. Isolation and Characterization of *Leuconostoc Oenos* Bacterophages from Wine and Sugar Cane.

Nishimura & Masuda, 1983. *In Proceedings of the American Chemical Society 22:* Seattle, Washington.

Nishino, H., Miyazaki, S., and Tohjo, K. 1985. Effect of osmotic pressure on the growth rate and fermentation activity of wine yeasts. *Am. J. Enol. and Vitic.* 36:170–174.

Noble, A. C., Arnold, R. A., Masuda, B. M., Pecore, S. D., Schmidt, J. O., and Stern, P. M. 1984. Progress towards a standardized system of wine aroma terminology. *Am. J. Enol., Vitic 35*:107-9.

Noble, A. C., Arnold, R.A., Buechenstein, J., Leach, E. J., Schmidt, J. O., and Stern, P. M. 1987. Research note: Modification of a standardized system of aroma terminology. *Am. J. Enol. and Vitic.* 38(2):143–146.

Nordstrom, K. 1963. Formation of esters from acids by brewers yeast. I: Kinetic theory and basic experiments. *J. Inst. Brew.* 69:310–322.

Nordstrom, K. 1965a. Formation of esters from lower fatty acids by various yeasts species. *J. Inst. Brew.* 72:38–40.

Nordstrom, K. 1965b. Formation of volatile esters by brewers yeast. *Brewers Dig.* 40(11):60–67.

Norton, K. M. and Heatherbell, D. A. 1988. A rapid, simple method for the estimation of malate and tartrate in grape juice. *Proceedings of the Second International Symposium for Cool Climate Viticulture and Oenology.* Edited by Smart R., Thornton S., Rodriguez, S., and Young, J. pp. 193–196.

Official Journal of the European Communities (English ed.). 1978. Vol. 21, p. 28.

Official Methods of Analysis of the Association of Official Analytical Chemists. 1984. 12th ed., ed. W. Horwitz. Washington, DC.

Olijve, W. and Kok, J. J. 1979. Analysis of growth of *Gluconobacter oxydans* in glucose containing media. *Arch. Microbiol.* 121:283–290.

Olphen, H. 1963. *An Introduction to Clay Colloid Chemistry.* New York: Interscience Publishers.

Olsen, E. 1948. Studies of bacteria in Danish fruit wines. *Antonie van Leeuwenhoek J. Microbiol. Serol.* 14:1–28.

Ong, B. Y. and Nagel, C. W. 1978. Hydroxycinnamic acid–tartaric acid ester content in mature grapes and during maturation of White Riesling grapes. *Am. J. Enol. Vitic.* 29(4):277–281.

Organization of American States. 1975. Paper and thin-layer chromatography (Monograph No. 16). Washington, DC: Regional Programs of Scientific and Technological Development, Department of Scientific Affairs.

Oszminanski, J., Romeyer, F. M., Sapis, J. C. and Macheix, J. J. 1986. Grape seed phenolics: Extraction as effected by some conditions occurring during wine processing. *Am. J. Enol. Vitic.* 37(1):7–12.

Ough, C. S. 1960. Gelatin and polyvinylpyrrolidone compared for fining red wines. *Am. J. Enol. and Vitic.* 11:170–173.

Ough, C. S. 1968. Proline content of grapes and wine. *Vitis* 7:321–331.

Ough, C. S. 1969. Substances extracted during skin contact with white must. I. General wine composition and quality changes with contact time. *Am. J. Enol. and Vitic.* 20:93–100.

Ough, C. S. and Amerine, M. A. 1962. Studies with controlled fermentation. VII. Effect of antefermentation blending of red must and white juice on color, tannins, and quality of Cabernet Sauvignon wine. *Am. J. Enol. and Vitic.* 13:181–188.

Ough, C. S. and Anelli, G. 1979. Zinfandel grape juice and amino acid makeup as affected by crop level. *Vitis* 30:8–10.

Ough, C. S. and H. W. Berg 1971. Simulated mechanical harvest and gondola transport. II. Effect of temperature, atmosphere, and skin contact on chemical and sensory qualities of white wines. *Am. J. Enol. Vitic.* 22:194–198.

Ough, C. S., Caputi, A., Jr., and Groat, M. 1979. A rapid colorimetric calcium method. *Am. J. Enol. and Vitic.* 30:8–10.

Ough, C. S., and Crowell, E. A. 1987. Use of sulfur dioxide in winemaking. *J. Food Science* 52(2):386–89.

Ough, C. S. and Ingraham, J. L. 1960. Use of sorbic acid and sulfur dioxide in sweet table wines. *Am. J. Enol. and Vitic.* 11:117–122.

Ough, C. S. and Kunkee, R. E. 1974. The effect of fumaric acid on malo-lactic fermentation in wine from warm areas. *Am. J. Enol. Vitic.* 25:188–190.

Ough, C. S. and R. M. Stashak 1974. Further studies on proline concentration in grapes and wines. *Am. J. Enol. Vitic.* 25:7–12.

Ough, C. S., Fong, D. and Amerine, M. A. 1972. Glycerol in wine: determination and some factors affecting formation. *Am. J. Enol. and Vitic.* 23:1–5.

Oura, E. 1977. Reaction products of yeast fermentation. *Process Biochem.* 12(3):19–23.

Pallota, U., and Cantarelli, C. 1979. Le Catechine: Loro importanza sulla qualita dei vini bianchi. *Estratto da Vignevini, Gruppo Giornalistico Edagricole* 4:19–46.

Perin, J. 1977. Compte rendu de quelques essais de refrigeration des vins. *Le Vigneron Chapenois* 98(3):97–101.

Peterson, R. G., Joslyn, M. A., and Durbin, P. W. 1958. Mechanism of copper formation in white table wines. III. Source of the sulfur sediment. *Food Res.* 23:518–524.

Peynaud, E. 1937. Etudes sur les phenomenes d'esterfication. *Rev. Viticult.* 86:209–475.

Peynaud, E. 1939–1940. Sur la formation et la dimunition des acides volatile pendant la fermentation alcoolique en anaerobiose. *Ann. Ferment.* 5:321–327, 385–402.

Peynaud, E. 1947. Contribution a l'etude biochimique de la maturation du raisin et de la composition des vins. *Inds. Agr. et Aliment.* 64:87–414.

Peynaud, E. 1956. New information concerning biological degradation of acids. *Am. J. Enol.* 7:150–156.

Peynaud, E. 1984. *Knowing and Making Wine.* New York: John Wiley and Sons, 391 pp.

Peynaud, E. and Domerco, S. 1953. Etude des levures de la gironde. *Ann. Technol. Agr.* 2:265–300.

Peynaud, E., and G. Guimberteau 1961. Recherches sur la constitution et l' efficacite' anticristallisante de l' acide metatartarique. *Ind. Alimant. Agr. (Paris)* 78:131–135, 413–418.

Peynaud, E. and Maurie, A. 1953. Sur l'evolution de l'azote dans les differentes parties du raisin au cours de la maturation. *Ann. Technol. Agr.* 2:15–25.

Peynaud, E. and Subraud, P. 1964. Utilisation de l'effet desacidifiant des *Schizosaccharomyces* en vinification de raisins acides. *Ann. Technol. Agr.* 13:309–328.

Peynaud, E., Giumberteau, G., and Blouin, J. 1964. Die Loslichkeitsgleichgewichte von Kalzium and Kalium in wein. *Mitt. Rebe u. Wein Serie A (Klosterneuburg)* 14:176–186.

Pfeffer, T. E., Clary, C. D., and Petrucci, V. E. 1985. Adaptation of HPLC to wine grape inspection. Report to the Wine Grape Inspection Advisory Committee, CDFA, Viticulture Research Center, California State University, Fresno.

Phaff, H. J., Miller, M. W., and Mrak, E. M. 1978. *The Life of Yeasts,* 2nd ed. Cambridge, MA: Harvard University Press, 154 pp.

Pilone, G. J. 1967. Effect of lactic acid on volatile acid determination of wine. *Am. J. Enol. and Vitic.* 18:149–156.

Pilone, G. J. 1979. Technical note: Preservation of wine yeast and lactic acid bacteria. *Am. J. Enol. and Vitic.* 30(4):326.

Pilone, G. J. and Berg, H. W. 1965. Some factors affecting tartrate stability in wine. *Am. J. Enol. and Vitic.* 16:195–211.

Pilone, G. J. and Kunkee, R. E. 1972. Characterization and energetics of *Leuconcstoc oenos* ML-34. *Am. J. Enol. and Vitic.* 23(2):61–69.

Pilone, G. J., Kunkee, R. E., and Webb, A. D. 1966. Chemical characterization of wines fermented with various malolactic bacteria. *Appl. Microbiol.* 14:608–615.

Pilone, G. J., Rankine, B. C., and Pilone, A. D. 1974. Inhibiting malolactic fermentations in Australian dry red wines by adding fumaric acid. *Am. J. Enol. and Vitic.* 25(2):99–107.

Plane, R. A., Mattick, L. R., and Weirs, L. D. 1980. An acidity index for the taste of wines. *Am. J. Enol. and Vitic.* 31:265–268.

Pocock, K. F. and Rankine, B. C. 1973. Heat test for detecting protein instability in wine. *Australian Wine Brew. Spirits Rev.* 91(5):42–43.

Pointing, J. D. and Johnson, G. 1945. Determination of sulfur dioxide in fruits. *Ind. Eng. Chem. (Anal. Ed.)* 17:682–686.

Porter, L. J. and Ough, C. S. 1982. The effects of ethanol, temperature, and dimethyl dicarbonate on viability of *Saccharomyces cerevisiae* Montrachet No. 522 in wine. *Am. J. Enol. and Vitic.* 33(4):222–225.

Postel, W. 1983. La solubilite et la cinetique de crystalisation du tartrate de calcium dans de vin. *Bull. O.I.V.* 629–630.

Postel, W. and Prasch, E. 1977. Untersuchungen zur Weinsteinstabilizierung von Wein durch Elektrodialyse. I. Mitteilung. Absenkung der Kalium un Weinsaurekonzentration. *Weinwirtschaft* 113(45):1277–1283.

Quinsland, D. 1978. Technical note. Identification of common sediments in wine. *Am. Soc. of Enol. and Vitic.* 29(1):70–71.

Quinsland. 1978. Remedial

Radler F. 1965. The main constituents of the surface waxes of varieties and species of the genus *Vitis. Am. J. Enol. Vitic.* 16:159–167.

Radler, F. 1968. Bakterieller Apfelsaureabbau in Deutschen Spitzenweinen. *Z. Lebensm. Untersuch. Forsch.* 138:35–39.

Radler, F. and Schutz, H. 1982. Glycerol production from various strains of *Saccharomyces. Am. J. Enol. and Vitic.* 33(1):36–40.

Ramey, D., Bertrand, A., Ough, C. S., Singleton, V. L., and Sanders, E. 1986. Effects of skin contact temperature on Chardonnay must and wine composition. *Am. J. Enol. and Vitic.* 37(2):99–106.

Rankine, B. C. 1955. Quantitative differences in products of fermentation by different strains of wine yeasts. *Am. J. Enol. and Vitic.* 6(1):1–10. See also *Aust. J. Applied Sci.* 4:590–602 (1953).

Rankine, B. C. 1963. Nature, origin, and prevention of H_2S aroma in wines. *J. Sci. Food Agr.* 14:75–91.

Rankine, B. C. 1966a. Decomposition of L-malic acid by wine yeasts. *J. Sci. Food Agr.* 17:312–316.

Rankine, B. C. 1966b. *Pichia membrafaciens,* a yeast causing film formation and off flavor in table wine. *Am. J. Enol. Vitic.* 17:302–307.

Rankine, B. C. 1972. Influence of yeast strain and malolactic fermentation on the composition and quality of table wines. *Am. J. Enol. and Vitic.* 23(4):152–158.

Rankine, B. C. 1977. Developments in malo-lactic fermentations in Australian table wines. *Am. J. Enol. Vitic.* 28:27–33.

Rankine, B. C. 1984. Use of Isinglass to fine wines. *Australia Grape Grower and Winemaker.* 249:16.

Rankine, B. C. and Bridson, D. A. 1971. Glycerol in Australian wines and factors influencing its formation. *Am. J. Enol. and Vitic.* 22:6–12.

Rankine, B. C. and Pocock, K. F. 1971. A new method for detecting protein instability in white wines. *Australian Wine Brew. Spirits Rev.* 89:61.

Rao, M. R. R. 1957. The acetic acid bacteria. *Ann. Rev. Microbiol.* 11:317–338.

Reazin, G., Scales, H., and Andrease, A. 1970. Mechanism of major cogener formation in alcoholic grain fermentations. *J. Agr. Food* 18(4):585.

Reed, G. 1966. *Enzymes in Food Processing.* New York: Academic Press.

Reed, G. and Chen, S. L. 1978. Evaluating commercial active dry wine yeasts by fermentation activity. *Am. J. Enol. and Vitic.* 29:165–168.

Reed, G. and Peppler, H. J. 1973. *Yeast Technology.* Westport, CT: AVI Publishing Co.

Rentschler, H. and Tanner, H. 1951a. Das Bitterwerden der Rotweine. Beitsag zur kennfinis des vorkommens von Acroleins in Getranken und seine Bezihung zum Bitterwerden der Weine. *Mitt. Gebiete Lebenson Hyg.* 42:463–475.

Rentschler, H. and Tanner, H. 1951b. Uber die Kupfersulfittrubung von Weiss Weinen, und Sussmosten. *Schweiz. Z. Obst-Weinbau* 60:298–301.

Rhein, O. and Neradt, F. 1979. Tartrate stabilization by the contact process. *Am. J. Enol. and Vitic.* 30(4):265–271.

Ribereau-Gayon, J. 1933. Contribution a l'etude des oxidations et reductions dans les vin. Application a l'etude du vieillissement et des casses, 2nd ed. Bordeaux, France.

Ribereau-Gayon, J. 1961. La composition chimique des vins. In *Traite d'Oenologie.* Paris: Librairie Polytechnnique Ch. Beranger.

Ribereau-Gayon, J. 1963. Phenomenon of oxidation and reduction in wines and applications. *Am. J. Enol. and Vitic.* 14:139–143.

Ribereau-Gayon, J. 1965. Identification d'esters des acides cinnamiques et l'acide tartique dans les limbes et les baies de *V. vinifera, C.R. Acad. Sci., Paris* 260:341–343.

Ribereau-Gayon, P. 1972. Evolution des composes phenoliques au cours de la maturation du raisin. II. Discussion des resultats obtenus en 1969, 1970 et 1971. *Conn. Vigne Vin* 2:161–175.

Ribereau-Gayon, P. 1973, Interpretation chemique de la couleur des vins rouges. *Vitis* 12:119–142.

Ribereau-Gayon, P. 1974. The chemistry of red wine color. In *The Chemistry of Winemaking,* ed. A. D. Webb. Advances in Chemistry 137. Washington, DC: American Chemical Society.

Ribereau-Gayon, P., 1985. New development in wine microbiology. *Am. J. Enol. Vitic.* 36:1–10.

Ribereau-Gayon, P. 1988 Botrytis: Advantages and disadvantages for producing quality wines. *Proceedings of the Second International symposium for Cool Climate Viticulture and Enology.* New Zealand Society for Oenology and Viticulture. Editors: Smart, R., Thornton, R, Rodriquez, S. and J. Youg, P 319–323.

Ribereau-Gayon, P., Lafon-Lafourcade, S., Dubourdieu, D., Lucmaret, V., and Larue, F. 1979. Metabolisme de *Saccharomyces cerevisiae* dans les monts de raisins parasites par *Botrytis cinerea.* Inhibition de la fermentation; formation d'acide acetique et du glycerol. C.R. Acad. Sci. *289*:441–44.

Ribereau-Gayon, J. and Peynaud, E. 1958. *Analyses et Controle des Vins,* 2nd ed. Paris and Liege: Libraire Polytechnique.

Ribereau-Gayon, J. and Peynaud, E. 1961. *Traite d'Oenologie.* Paris: Libraire Polytechnique.

Ribereau-Gayon, J., Peynaud, E., and Lafon, M. 1956. Investigations of the origin of secondary products of alcoholic fermentation Part I and II. *Am. J. Enol. and Vitic.* 7:53–61; 112–118.

Ribereau-Gayon, J., Peynaud, E., Sudraud, P., and Ribereau-Gayon, P. 1972. *Sciences et Techniques du Vin,* Vol. I, 471–514. Paris: Dunod.

Ribereau-Gayon, J., Peynaud, E., Ribereau-Gayon, P., and Sudraud, P. 1976. *Traite d'Oenologie. Science et Techniques du Vin.* Vol. III: *Vinifications, Transformations du Vin.* Paris: Dunod.

Ribereau-Gayon, J., Ribereau-Gayon, P., and Seguin, G. 1980. *Botrytis cinerea* in enology. In *The Biology of Botrytis,* eds. J. R. Cooley-Smith, K. Verhoeff, and W. R. Jarvis. London: Academic Press, 262 pp.

Ribereau-Gayon, P. and Glories, Y. 1987. Phenolics in grapes and wines. *Proc. of the Sixth Australian Wine Industry Technical Conference,* ed. T. Lee, pp. 247–256. Adelaide, South Australia: Australian Industrial Publishers.

Ribereau-Gayon, P. and Sudraud, P. 1981. *Actualites Oenologiques et Viticoles.* Paris: Dunod, 395 pp.

Rice, A. C. 1965. The malolactic fermentation in New York State wines. *Am. J. Enol. and Vitic.* 16:62–68.

Riddle, C., and Turek, A 1977. *Anal. Chim. Acta* 92.49–53. An indirect method for the sequential determination of silicon and phosphorus in rock analysis by atomic absorption spectrometry.

Riese, H. 1980. Flow, a new process for repid and continuous tartrate stabilization of wine. Paper presented 31st Annual meeting of American Society for Enology and Viticulture.

Riwaw, H. 1980. Crystal-flow, a new process for rapid and continuous tartrate stabilization of wine. Presented at the 31st Annual Meeting of the Am. Soc. for Enol. and Vitic.

Roberts, S. 1988. Personal communication.

Rodriquez, S. B. 1987. A system for identifying spoilage yeast in packaged wine. *Am. J. Enol. and Vitic.* 38(4):237–276.

Rose, A. H. and Harrison, J. S., eds. 1970, 1971. *The Yeasts,* Vols. II and III. London and New York: Academic Press.

Rosell P. F., Ofria, H. V., and Palleroni, N. J. 1968. Production of acetic acid from ethanol by wine yeasts. *Am. J. Enol. and Vitic.* 19(1):13–16.

Rossi, J. and Clementi, F. 1984. L-Malic acid catabolism by polyacrylamide gel entrapped *Leuconostoc oenos. Am. J. Enol. and Vitic.* 35(2):100–102.

Rossi, J. A., Jr. and Singleton, V. L. 1966. Flavor effects and adsorptive properties of purified fractions of grape seed phenols. *Am. J. Enol. and Vitic.* 17:240–246.

Saavedra, I. J. and Garrido, J. M. 1963. La levadura de "flor" en la crianza del vino. *Rev. Cienc. Apl.* 17:497–501.

Sall, M. A., Teviotdale, B. L., and Savage, S. D. 1982. Bunch rot grape pest management. Div. Agric. Sciences, University of California, Pub. no. 4105.

Saller, W. 1957. Die spontane-Sprosspilzflora frische gepresster Traubensafe und die Reinhefegarung. *Mitt. Rebe u. Wein Serie A (Klosterneuburg)* 7:130–138.

Schanderl, H. 1955. Über Storungsfaktoren bei umgarungen einschlie blesh Schaumweingarungen. *Weinberg und Keller* 2:313–30.

Schanderl, H. 1959. *Die Mikrobiologie des Mostes und Weines,* 2nd ed. Stuttgart: Eugen Ullmer.

Schanderl, H. and Staudenmeyer, T. 1964. Uber den einfluss den schwefligen Saure auf die Acetaldehydbildung verschiedener hefen bei Most- und Schaumweiningarungen. *Mitt. Rebe u. Wein Serie A (Klosterneuburg)* 14:267–281.

Schmidt, T. R. 1987. Potassium sorbate or sodium benzoate. *Wine and Vines* (Nov.):42–44.

Schmitthenner, J. 1950. Die Werkung der Kohlensaure aus Hefen und Bakerterien. *Seitz-Werke, Bad Kreuznach.*

Schneyder, J. and Vlcek, G. 1977. Mitt hocheren Bundeslehr-Versuchsanst. *Wein-Obstbau (Klosterneuburg)* 27:87–88.

Schroeter, L. C. 1966. *Sulfur Dioxide: Applications in Foods, Beverages, and Pharmaceuticals.* New York: Pergamon Press.

Schug, W. 1982. Vinification of fine wine in California. Institute of Masters of Wine's International Symposium on Viticulture, Vinification, and the Treatment and Handling of Wine. Oxford, England.

Schutz, M. and Kunkee, R. E. 1977. Formation of hydrogen sulfide during fermentation by wine yeasts. *Am. J. Enol. and Vitic.* 28:137–140.

Scott Laboratories, 1967. Production notes. Preparation and use of casein. San Rafael, CA.

Scott, R. S. 1967. Clarification—the better half of filtration. *Wine and Vines* 48(10):29–30.

Scott, R. S., Anders, T. G., and Hums, N. 1981. Rapid cold stabilization of wine by filtration. *Am. J. Enol. and Vitic.* 32(2):138–143.

Seitz Werke GMBH (SWK Machines Inc.) Technical File. 1978. Methods to determine and evaluate KHT stability by means of the contact method on a laboratory scale. Bath, New York.

Shimazu, Y., and Watanabe, M. 1981. Determination of organic acids in grape must and wine by HPLC. Nippon Jozo Kyokai Zasshi 76:418–423.

Shimizu, K., Adachi, T., Kitano, K., Shimazaki, T., Totsuka, A., Hara, S., and Dittrich, H. H. 1985. Killer properties of wine yeasts and characterization of killer wine yeasts. *J. Ferment. Technol.* 63:421–429.

Silver, J. and Leighton, T. 1981. Control of malolactic fermentation in wine. II. Isolation and characterization of a new malolactic organism. *Am. J. Enol. and Vitic.* 32:64–72.

Simi Winery 1983. Personal communication from Zelma Long.

Simpson, E. and Tracey, R. P. 1986. Microbiological quality, shelf life and fermentation activity of active dried yeast. *S. African J. Enol. and Vitic.* 7(2):61–65.

Singleton, V. L. 1967. Adsorption of Natural Phenols from beer and wine. *Tech. Quart. Master Brewers Assoc. America* 4(4):245–253.

Singleton, V. L. 1980. Grape and wine phenolics; background and prospects. *Proc. of Grape and Wine Centennial Symposium,* ed. A. D. Webb, pp. 215-227. University of California, Davis.

Singleton, V. L. (1982a) Grape and wine phenolics: background and prospects. In: *Proc. of The Univ. of Calif., Davis, Grape and Wine Centennial Symposium.* A. D. Webb (ed) pp. 215–27. Dept. of Vitic. and Enology, Univ. of California, Davis.

Singleton, V. L., 1982b. Oxidation of Wine. *Proc. of the International Symposium on Viticulture, Vinification, and Treatment and Handling of Wine.* Oxford, England. p 1

Singleton, V. L., 1985. Recent Conclusions in Wine Oxidation. *In Proc. of the eleventh Wine Industry Tech Symp.* pp. 17–24, 43–48. Wine Industry Symposium, San Francisco.

Singleton, V. L. 1987. Oxygen with phenols and related reactions in musts, wines, and model system: observations and practical implications. *Am. J. Enol. and Vitic.* 38(1):69–77.

Singleton, V. L. and Draper, D. E. 1962. Adsorbents and wines. I. Selection of activated charcoals for treatment of wines. *Am. J. Enol. and Vitic.* 13:114–125.

Singleton, V. L. and Guymon, J. F. 1963. A test of fractional addition of wine spirits to red and white port wines. *Am. J. Enol. and Vitic.* 14:129–146.

Singleton, V. L. and Esau, P. 1969. Phenolic Substances in Grapes and Wines and Their Significance. New York: Academic Press.

Singleton, V. L. and Kramling, T. E. 1976. Browning in white wines and an accelerated test for browning capacity. *Am. J. Enol. and Vitic.* 27:157–160.

Singleton, V. L. and Noble, A. C. 1976. Wine flavor and phenolic substances. In *Phenolic, Sulfur, and Nitrogen Compounds in Food Flavors,* eds. G. Charalambous and I. Katz. A.C.S. Symposium Series 26:47–70.

Singleton, V. L. and Rossi, J. A., Jr. 1965. Colorimetry of total phenolics with phosphomolybdic-phosphotungstic acid reagents. *Am. J. Enol. and Vitic.* 16:144–158.

Singleton, V. L., Berg, H. W., and Guymon, J. F. 1964. Anthocyanin color level in port-type wines as affected by the use of wine spirits containing aldehydes. *Am. J. Enol. and Vitic.* 15:75–81.

Singleton, V. L., Salgues, M., Zaya, J. and Trousdale, E. 1985. Caftaric acid disappearance and conversion to products of enzymic oxidation in grape must and wine. *Am. J. Enol. Vitic.* 36(1):50–56.

Singleton, V. L., Sullivan, A. R., Kramer, C. 1971. An analysis of wine to indicate aging in wood or treatment with wood chips or tannic acid. *Am. J. Enol. Vitic.* 22(3):161–66.

Singleton, V. L., Zaya, J., and Trousdale, E. 1980. White table wine quality and polyphenol composition as affected by must sulfur dioxide content and pomace contact time. *Am. J. Enol. and Vitic.* 31(1):14–20.

Skoog, D. A. and West, D. M. 1971. *Principles of Instrumental Analysis.* New York: Holt, Reinhart and Winston, Inc.

Smart, R. E., Dry P. R., and Brien, D. R. G. 1977. Field temperature of grape berries and implications for fruit composition. *Int. Symp. on Quality of the Vintage,* pp. 227-231, Capetown, South Africa.

Smith, C. 1982. Review of basics on sulfur dioxide—Part II. *Enology Briefs* 1(2):1–3. Cooperative Extension, University of California.

Snow, R. 1979. Toward genetic improvement of wine yeasts. *Am. J. Enol. and Vitic.* 30(1):33-36.

Snow, P. G. and Gallander, J. F. 1979. Deacidification of white table wines through partial fermentation with *Schizosaccharomyces pombe. Am. J. Enol. and Vitic.* 30(1):45-48.

Sols, A., Gancedo, C., and Delafuente, G. 1971. Energy yielding metabolism in yeasts. In *The Yeasts II,* eds. A. H. Rose and J. S. Harrison, pp. 271-303. New York: Academic Press.

Somattmadja, D., Powers, J. J., and Handy, M. K. 1964. Anthocyanins VI. Chelation studies on anthocyanins and other related compounds. *J. Food Sci.* 29:655-660.

Somers, T. C. 1978. Interpretation of color composition in young red wines. *Vitis* 17:161-167.

Somers, T. C. 1987. Assessment of phenolic components in viticulture and oenology. *Proc. of the Sixth Australian Wine Industry Technical Conference,* ed. T. Lee. Adelaide, Australia: Australian Industrial Publishers. pp. 257-61.

Somers, T. C., and Evans, M. E. 1977. Spectral evaluation of young red wines: Anthocyanin equilibria, total phenolics, free and molecular SO_2. *J. Sci. Food Agric.* 28:279-287.

Somers, T. C. and Ziemelis, G. 1972. Interpretations of ultraviolet absorption in white wines. *J. Sci. Food Agr.* 23:441-453.

Somers, T. C. and Ziemelis, G. 1973a. The use of gel column analysis in evaluation of bentonite fining procedures. *Am. J. Enol. and Vitic.* 24(2):51-54.

Somers, T. C. and Ziemelis, G. 1973b. Direct determination of wine proteins. *Am. J. Enol. and Vitic.* 24(2):47-50.

Somers, T. C., and Ziemelis, G. 1985. Spectral evaluation of total phenolic components in *Vitis viniferia:* grapes and wine'' *J. Sci. Food Agric.* 36:1275-1284.

Spedding, D. J. and Raut, P. 1982. The influence of dimethyl sulfide and carbon disulfide on the bouquet of wines. *Vitis.* 21:240-46.

Spettoli, P., Bottacin, A., Nuti, M. P., and Zamorani, A. 1982. Immobilization of *Leuconostoc oenos* ML-34 in calcium alginate gels and its application to wine technology. *Am. J. Enol. and Vitic.* 33(1):1-5.

Splittstoesser, 1981. Preservation of fresh grape juice. *Proceedings of Ohio grape and wine short course.*

Splittstoesser, D. F. and Wilkinson, M. 1973. Some factors affecting the activity of DEPC as a sterilant. *Appl. Microbiol.* 25:853-857.

Sponholz, W. R. and Dittrich, H. H. 1985. Uber die herkunft von gluconsaure, 2- und 5-oxo gluconsaure sowie glucuron- und galacturonsaure in mosten un weinen. *Vitis* 24:51-58.

Stafford, P. A. and Ough, C. S. 1976. Formation of methanol and ethyl methyl carbonate by dimethyl dicarbonate in wine and model solutions. *Am. J. Enol. Vitic.* 27:7-11.

Stahl, E. 1967. *Thin-Layer Chromatography. A Lab Handbook.* Singapore: Toppar Printing Co.

Stanier, R. Y., Adelberg, E. A., and Ingraham, J. L. 1976. *The Microbial World 4th Ed.* Prentice-Hall, Englewood Cliffs, N.J.

State of California. 1973. California Administrative Code, Title 17: Public Health, pp. 431-443. Bureau of Printing (Documents Division).

Steele, J. T. and Kunkee, R. E. 1978. Deacidification of musts from western United States by calcium double salt precipitation process. *Am. J. Enol. and Vitic.* 29(3):153-160.

Steele, J. T. and Kunkee, R. E. 1979. Deacidification of high acid California wines by calcium double salt precipitation. *Am. J. Enol. and Vitic.* 30(3):227-231.

Steinschreiber, P. 1984. ''Diacetyl in Wines. A Varietal, pH, and Inhibition Study.'' Masters Thesis, C.S.U. Fresno.

Stern, P. 1983. Technical projects workshop. American Society of Enologists Annual Meeting, Reno, NV.

Su, C. T., and Singleton, V. L. 1969. Identification of three flavan-3-ols from grapes. *Phytochem.* 8:1553-58.

Sudraud, P. 1958. Interpretation des courbes de'absorption des vins rouges. *Ann. Techol. Agric.* 7:203-208.

Suomalainen, H. and Oura, E. 1971. *Yeast Nutrition and Solute uptake in The yeasts.* Vol. II, A. H. Rose and J. S. Harrison (Ed). Academic Press, London.

Suzzi, G., Romano, P., and Zambonelli, C. *Saccharomyces* strain selection in minimizing sulfur dioxide requirements during vinification. *Am. J. Enol. and Vitic.* 36(3):199–202.

Tagunkov, Y. D. 1966. Snizhenie Aktevnosti katekholoksidazy sulsa adsorbentami, Vindelie i Vinogradarstro *SSSR* 26(6):20–23.

Tanner, H. and Vetsch, U. 1956. How to characterize cloudiness in beverages. *Am. J. Enol.* 7:142–149.

Thoukis, G. and Amerine, M. A. 1956. Fate of copper and iron during fermentation of grape musts. *Am. J. Enol.* 7:45–52.

Thoukis, G. and Stern, L. A. 1962. A review of some studies of the effect of sulfur on formation of off odors in wine. *Am. J. Enol. and Vitic.* 13:133–140.

Thoukis, G., Ueda, M., and Wright, D. 1965. Formation of succinic acid during alcoholic fermentation. *Am. J. Enol. and Vitic.* 16:1–8.

Timberlake, C. F. and Bridle, P. 1967. Flavylium salts, anthocyanidins and anthocyanins. II. Reactions with sulfur dioxide. *J. Sci. Food Agr.* 18:479–485.

Timberlake, C. F. and Bridle, P. 1976. Interactions between anthocyanins, phenolic compounds, and acetaldehyde and their significance in red wines. *Am. J. Enol. and Vitic.* 27:97–105.

Tipper, D. J., and Bostian, K. A. 1984. Double-stranded ribonucleic acid killer systems in yeasts. *Microbiol. Revs.* 48:125–56.

Tomada, Y. 1927. On the production of glycerine by fermentation. IV: Dissociation of acetaldehyde-bisulfite complex in alkaline solution. *J. Soc. Chem. Ind. (Japan)* 30:747–759.

Tromp, A. and Agenbach, W. A. 1981. Sorbic acid as a preservative. Its effiacacy and organoleptic thresholds. *S. Afr. J. Enol. and Vit.* 2:1–5.

Troost, G., Die Technologie des Weines, Ulmer, Stuttgart 1961.

Troost, G. 1972. *Die Technologie des Weines,* 4th ed. Stuttgart: E. Ullmer.

Tyson, P. J., Luis, E. S., Day, W. R., and Lee, T. H. 1981. Estimation of soluble proteins in must and wine by high performance liquid chromatography. *Am. J. Enol. and Vitic.* 32(3):241–243.

Tryon, C. R., Edwards, P. A., and Chisholm, M. G. 1988. Determination of the phenolic content of some French-American hybrid white wines using ultraviolet spectroscopy. *Am. J. Enol. and Vitic. 39(1)*:5–10.

United States Bureau of Alcohol, Tobacco and Firearms. Wine: Part 240 of Title 27, Code of Federal Regulations.

U.S. Bureau of Standards. "Standard Density and Volumetric Tables (Circular 19)." U.S. Govt. Printing Office, Washington D.C. 1924.

U.S. Internal Revenue Service. "Regulations No. 7, Wine." U.S. Govt. Printing Office, Washington D.C. 1945.

Urlaub, R. 1985. Benefits of combined use of a protease and a pectic enzyme in white wine processing. *Proc. of the Australian Society of Viticulture and Oenology.* Glen Osmond, Australia. Australian Industrial Publishers.

Vahl, J. M. 1979. Relative density-extract-ethanol nomograph for table wines. *Am. J. Enol. and Vitic.* 30(3):262–263.

Van Buren, J. P., Hrazdina, G., and Robinson, W. B. 1974. Color of anthocyanin solutions expressed in lightness and chromaticity terms. Effect of pH and the type of anthocyanin. *J. Food Sci.* 39:325–328.

VanVuuren, H. J. J., and Wingfield, B. D. 1986. Killer yeasts—cause of stuck fermentations in a wine cellar. *S. African J. Enol. & Vitic.* 7(2):113–118.

Vaughn, R. H. 1938. Some effects of association and competition on *Acetobacter. J. Bact.* 36:357–367.

Vaughn, R. H. 1955. Bacterial spoilage of wine with special reference to California conditions. In *Advances in Food Research* 6:67–108. New York: Academic Press.

Vaughn, R. H. and Douglas, H. C. 1938. Some lactobacilli encountered in abnormal musts. *J. Bact.* 36:318-319.

Vaughn, R. H. and Tchelistcheff, A. 1957. Studies on the malic acid fermentation of California table wines. I. An introduction to the problem. *Am. J. Enol. and Vitic.* 8:74-79.

Vas, K. and Ingraham, M. 1949. Preservation of fruit juices with less sulfur dioxide. *Food Manuf.* 24:414-416.

Vetch, U. and Luthi, H. 1964. Farbstoffverlustewahrend des biologischen Saureabbaues. *Schweiz. Z. Obst-Weinbau* 73:124-126.

Vetsch, U. and Mayer K. 1978. Entstehung bakterieller Nachtrubungen in Flaschenweinen. *Schweiz. Z. Obst. Weinbau.* 114:285-294.

Vine, R. P. 1981. *Commercial Winemaking: Processing and Controls.* Avi. Pub. Co., Inc., Westport, Conn.

Vos, P. J. A. and Gray, R. S. 1979. The origin and control of H_2S during fermentation of grape must. *Am. J. Enol. and Vitic.* 30(3):187-196.

Wahab, A., Witzke, W., and Cruess, W. V. 1949. Experiments with ester-forming yeast. *Fruit Prod. J.* 28:198-219.

Wahlstrom, V. L. and Fugelsang, K. C. 1988. Utilization of yeast hulls in winemaking. California Agricultural Tech. Inst. Bull. 880103.

Wang, Ling-Feng. 1985. Off-flavor development in white wine by *Brettanomyces* and *Dekkera.* Masters Thesis, C.S.U. Fresno.

Warkentine, H. and Nury, M. S. 1963. Alcohol losses during fermentation of grape juice in closed containers. *Am. J. Enol. and Vitic.* 14:68-74.

Weeks, C. 1969. Production of sulfur dioxide-binding compounds and sulfur dioxide by two *Saccharomyces* yeast. *Am. J. Enol. Vitic.* 20:32-9.

Weetall, H. H., Zelko, J. T., and Bailey, L. F. 1984. A new method for stabilization of white wine. *Am. J. Enol. and Vitic.* 35(4):212-215.

Weenk, G., Olijve, W., and Harder, W. 1984. Ketogluconate formation by *Gluconobacter* species. *Appl. Microbiol. Biotech.* 20:400-405.

Weiller, H. G. and Radler, F. 1972. Vitamin-und Aminosaure-Bedarf Von Milchsaurebakterien aus wein und von Rebenblatterm. Mitt. Hoeheren Bundeslehr Versuchsanst. *Wein Obst. Kloesterneuburg* 22:4-18.

White, B. B., and Ough, C. S. 1973. Oxygen uptake studies in grape juice. *Am. Soc. Enol. Vitic.* 24(4):148-152.

Whiting, G. C., and Coggins, R. A. 1971. The role of quinate and shikmate in metabolism of lactobacilli. *Antonie van Leeuwenhoek J. Microbiol. Serol.* 37:33-49.

Wildenradt, H. L. and Singleton, V. L. 1974. The production of aldehydes as a result of oxidation of polyphenolic compounds and its relation to wine aging. *Am. J. Enol. and Vitic* 25(2):119-126.

Williams, J. T., Ough, C. S., Berg, H. W. 1978. White wine composition and quality as influenced by methods of must clarification. *Am. J. Enol. Vitic.* 29(1):92-96.

Winkler, A. J., Cook, J. A., Kliewer, W. M., and A. L. Linder 1974. "General Viticulture." University of California Press. Berkeley

Woidick, H., and Pfannhauser, W. 1974. Zur Gaschromatogrpahischen Analyse von Branntweinen: Quantitative bestimming von Acetaldehyd, Essigsauremethylester, Essigsaureathylester, Methanol, Butanol-(1), Butanol-(2), Propanol-(1), 2-Methylpropanol-(1), "Amylalkoholen' Und hexanol-(1). Mitt. Hoehesen Bundeslehr-Versuchsanst. *Wein Obstbau. Klosterneuburg* 24:155-166.

Wong, G. and Caputi, A. Jr. 1966. A new indicator for total acid determination in wines. *Am. J. Enol. and Vitic.* 17(3):174-177.

Wucherpfennig, K. 1978. Possible applications of procedures using membranes for the stabilization of wines. Ann. Technol. Agric. 27(1):319-31.

Wucherpfennig, K. and Ratzka, P. 1967. Über die verzogerung der Weinsteinausscheidung durch polymere Substanzen des Weines. *Weinberg Keller* 14:499–509.

Wucherpfenning, K., and Semmler, G. 1973. Uber den SO_2-bedarf der Wein aus verschiedenen Weinbau gebieten der Welt und dessen Abhangigkeit von der bildung von Acetaldehyd im verlauf der garung. *Deut. Weinbau* 28:851–55.

Wurdig, G., Schlotter, H. A. and Klein, E. 1974. Uber die Ursachen des sogenannten Geranientones. *Allg. Deut. Weinfachztg.* 110:578–83.

Yamada, H. and Desoto, R. T. 1963. Relationship of solubility products to long-range tartrate stability. *Am. J. Enol. and Vitic.* 14:43–51.

Yamada, S., Nabe, K., Izuo, M., and Chibata, I. 1979. Enzymic production of dihydroxyacetone by *Acetobacter suboxydans* ATCC 621. *J. Ferment. Tech.* 57:321–326.

Ziemelis, T., and Somers, T. C. 1978. Rapid determination of sorbic acid in wine. *Am. J. Enol. Vitic.* 29:217–219.

Zimmerman, H. W. 1963. Studies on the dichromate method of alcohol determination. *Am. J. Enol. and Vitic.* 14:205–213.

Zoecklein, B. W. 1984. Bentonite fining. *The Practical Winery* 3(5) (May/June):84–91.

Zoecklein, B. W., 1986. Vineyard sampling. Virginia Polytechnic Inst. Cooperative Extension Bull.

Zoecklein, B. W. 1987. pH imbalance in Cabernet sauvignon. *Proceedings of the Am. Soc. of Enol. and Vitic. Eastern Section Regional Meeting.* 26–37.

Zoeklein, B. W. 1988. A review of Potassium bitartrate stabilization of wines. *Virginia Cooperative Extension. VPI-SU No. 463-013.*

Index

Note: This general index does not include *all* references to specific chromatographic applications, which are included in the *Procedure* sections at the end of each chapter.